Biofouling

Edited by

Simone Dürr
School of Natural Sciences and Psychology, Liverpool John Moores University
Liverpool, UK

Jeremy C. Thomason
School of Biology, Newcastle University
Newcastle-upon-Tyne, UK

WILEY-BLACKWELL

A John Wiley & Sons, Ltd., Publication

This edition first published 2010
© Blackwell Publishing Ltd

Blackwell Publishing was acquired by John Wiley & Sons in February 2007. Blackwell's publishing programme has been merged with Wiley's global Scientific, Technical, and Medical business to form Wiley-Blackwell.

Registered office
John Wiley & Sons Ltd, The Atrium, Southern Gate, Chichester, West Sussex, PO19 8SQ, United Kingdom

Editorial office
9600 Garsington Road, Oxford, OX4 2DQ, United Kingdom
2121 State Avenue, Ames, Iowa 50014-8300, USA

For details of our global editorial offices, for customer services and for information about how to apply for permission to reuse the copyright material in this book please see our website at www.wiley.com/wiley-blackwell.

The right of the author to be identified as the author of this work has been asserted in accordance with the Copyright, Designs and Patents Act 1988.

Library of Congress Cataloging-in-Publication Data

Biofouling / edited by Simone Dürr, Jeremy C. Thomason. – 1st ed.
    P. cm.
    Includes bibliographical references and index.
    ISBN 978-1-4051-6926-4 (hardback : alk. paper)   1. Fouling organisms.   2. Fouling.   I. Dürr, Simone.
II. Thomason, Jeremy.
    TD427.F68B56 2010
    628.9′6–dc22   2009023941

A catalogue record for this book is available from the British Library.

Set in 10/12.5 pt Times by Aptara® Inc., New Delhi, India

1   2010

Contents

Colour plates appear between pages 332 and 333

Contributors

Brigitte Behrends, Elsfether Zentrum für maritime Forschung GmbH, An der Weinkaje 4, 26931 Elsfleth, Germany

Alan J. Butler, Wealth from Oceans National Research Flagship, CSIRO Marine and Atmospheric Research, Hobart, Tasmania 7000, Australia

João Canning-Clode, IFM-GEOMAR – Leibniz Institute for Marine Sciences, Düsternbrooker Weg 20, 24105 Kiel, Germany

Ilona Cheyne, Newcastle Law School, Newcastle University, Newcastle upon Tyne, NE1 7RU, UK

Bret J. Chisholm, Center for Nanoscale Science and Engineering, North Dakota State University, PO Box 6050, Department 4310, Fargo, ND 58108, USA

Ashley D.M. Coutts, Aquenal Pty Ltd, 244 Summerleas Road, Kingston, Tasmania 7050, Australia

Phillip R. Cowie, University of London Marine Biological Station, Millport, Isle of Cumbrae, KA28 OEG, UK

Rocky de Nys, Head of Discipline – Aquaculture, School of Marine and Tropical Biology, James Cook University, Townsville, Queensland 4811, Australia

Sergey Dobretsov, Department Marine Science and Fisheries, College of Agricultural and Marine Sciences, Sultan Qaboos University, Al Khoud 123, PO Box 34, Muscat, Sultanate of Oman

Jenifer E. Dugan, Marine Science Institute, University of California, Santa Barbara, California 93106, USA

Simone Dürr, School of Natural Sciences and Psychology, Liverpool John Moores University, Byrom Street, Liverpool, L3 3AF, UK

Robert Edyvean, Department of Chemical and Process Engineering, The University of Sheffield, Mappin Street, Sheffield, S1 3JD, UK

Marco Faimali, CNR – Istituto di Scienze Marine (ISMAR) – Sezione Tecnologie Marine Genova, Via De Marini, 6 - IV P., 16149 Genova, Italy

Alistair A. Finnie, Technology Centre – M&PC, International Paint Ltd, Stoneygate Lane, Felling, Gateshead, Tyne and Wear, NE10 0JY, UK

Jana Guenther, Centre for Research-based Innovation in Aquaculture Technology, SINTEF Fisheries and Aquaculture, Brattørkaia 17B, 7465 Trondheim, Norway

Jon N. Havenhand, Department of Marine Ecology – Tjärnö, University of Gothenburg, Tjärnö, 452 96 Strömstad, Sweden

Peter Henderson, PISCES Conservation Ltd, IRC House, The Square, Pennington, Lymington, Hants, SO41 8GN, UK

Dickon Howell, Newcastle University, c/o 11a Northumberland Terrace, Tynemouth, Tyne and Wear, NE30 4BA, UK

Stuart R. Jenkins, School of Ocean Sciences, Bangor University, Menai Bridge, Anglesey, LL59 5AB, UK

John A. Lewis, ES Link Services Pty Ltd, 1 Queensberry Place, North Melbourne, Victoria 3051, Australia

Gustavo M. Martins, Departamento de Biologia, Secção de Biologia Marinha, Universidade dos Açores, Rua da Mãe de Deus, 52, 9500 Ponta Delgada, São Miguel – Azores, Portugal

Christopher D. McQuaid, Department of Zoology and Entomology, Rhodes University, Grahamstown 6140, South Africa

Karen Miller, Institute of Antarctic and Southern Ocean Studies, University of Tasmania, Private Bag 77, Hobart, Tasmania 7000, Australia

Henry M. Page, Marine Science Institute, University of California, Santa Barbara, California 93106, USA

Fred Piltz, Minerals Management Service, Pacific OCS Region 770, Paseo Camarillo, Camarillo, California 99010, USA

Elvira S. Poloczanska, Climate Adaptation Flagship, CSIRO Marine and Atmospheric Research, PO Box 120, Cleveland, QLD 4163, Australia

Jonathan Pratten, Microbial Diseases, UCL Eastman Dental Institute, 256 Gray's Inn Road, London, WC1X 8LD, UK

Gabrielle S. Prendergast, School of Biology, Division of Biology and Psychology, Newcastle University, Newcastle upon Tyne, NE1 7RU, UK

Derren Ready, Microbial Diseases, UCL Eastman Dental Institute, 256 Gray's Inn Road, London, WC1X 8LD, UK

Dan Rittschof, Duke University Marine Laboratory, Nicholas School of the Environment, 135 Duke Marine Lab Road, Beaufort, North Carolina 28516-9721, USA

Anna M. Romaní, Institute of Aquatic Ecology and Department of Environmental Sciences, University of Girona, Campus de Montilivi, 17071 Girona, Spain

David A. Spratt, Microbial Diseases, UCL Eastman Dental Institute, 256 Gray's Inn Road, London, WC1X 8LD, UK

Craig Styan, Environment – Energy & Resources, RPS Australia/SE Asia, PO Box 465, Subiaco, Western Australia 6904, Australia

Cato C. ten Hallers-Tjabbes, Oosterweg 1, 9995 VJ Kantens, The Netherlands

Antonio Terlizzi, Laboratory of Zoology and Marine Biology, Department of Biological and Environmental Science and Technologies (DiSTeBA), University of Salento, I-73100 Lecce, Italy

Jeremy C. Thomason, School of Biology, Newcastle University, Newcastle upon Tyne, NE1 7RU, UK

Iosune (Maria J.) Uriz, Consejo Superior de Investigaciones Cientificas CSIC, Centre d'Estudis Avançats de Blanes CEAB, Accés a la Cala St. Francesc, 14, 17300 Blanes, Girona, Spain

Martin Wahl, IFM-GEOMAR – Leibniz Institute for Marine Sciences, Düsternbrooker Weg 20, 24105 Kiel, Germany

Simon Walmsley, 4 Farnham Park, Drive, Upper Hale Farnham, Surrey, GU9 0HS, UK

Douglas I. Watson, Aquaculture and Fisheries Development Centre, Department Zoology, Ecology & Plant Science, University College Cork, Cooperage Building, Distillery Fields, North Mall, Cork, Ireland

Dean C. Webster, Department of Coatings and Polymeric Materials, North Dakota State University, PO Box 6050, Department 2760, Fargo, ND 58108, USA

David N. Williams, Technology Centre – M&PC, International Paint Ltd, Stoneygate Lane, Felling, Gateshead, Tyne and Wear, NE10 0JY, UK

Preface

Fouling is the process of accumulation of unwanted material at an interface[1]. The material that accumulates may be mostly non-living, comprising detritus and organic or inorganic compounds, but it may also include organisms which may range in size from tiny viruses up to giant kelps, which can go on to form complex multi-species, multi-dimensional communities. By using the prefix *bio*, we are knowingly focussing the scope of this book on the processes that lead to the accrual of a biological community at an interface. This deliberately excludes a considerable body of work on fouling in its broadest sense, for example on the chemical fouling of reverse osmosis membranes in desalination plants, but the reach of the book is already large enough. Much of what is encompassed by this book is to do with biofouling in a marine context where the interface is between a hard surface and sea water where it is often associated with huge monetary losses in industries such as shipping, offshore oil or aquaculture. This is, however, not the full scope of biofouling and neither does it cover the variety of interfaces (solid/liquid, solid/gas and liquid/liquid) that may become biofouled. Thus this book with its simple title is still very broad in extent.

A cursory list of words relevant to biofouling includes *foul, fouled, fouling*[2], *foulant, foulage, antifouling, antifouling coating* and *paint*, as well as more strictly defined words such as *epibiont, basibiont, epibiosis* and *biofilm. Epibiosis* refers to the biofouling of living organisms, where the organism being fouled is the *basibiont* and the organism doing the fouling is the *epibiont*. Its use seems to be limited to marine and freshwater organisms, though it could be used for all medical biofouling where by definition the basibiont is a human. *Biofilm* refers solely to a microbial biofouling community and is a fairly recent neologism[3]. Thus this book with its title *Biofouling* covers both the processes of biofouling and anti(bio)fouling and the devices to stop the biofouling process, i.e. anti(bio)fouling coatings and paints, as well as the consequences of the biofouling process, biofouling communities.

This book on Biofouling started life as an off-shoot of the European Union-funded CRAB project when the field trials were being coordinated by us when we both worked in the School of Biology at Newcastle University. It is apt that this book was conceived next to the River Tyne as the Tyneside area has been a focus for biofouling and antifouling research and the manufacture of antifouling paints for over a century, with academics researching biofouling

---

[1] The *Oxford English Dictionary (OED)* lists 14 different meanings dating back to the ninth century for the verb *foul*, only one of which has a nautical derivation.

[2] The use of the word *fouling* both as a noun and a verb is now common, though strictly speaking it is used incorrectly as a noun.

[3] The earliest use of *biofilm* listed by ISI Web of Knowledge is 1976 but it does not appear in the *OED* until 1981. The *OED* does not have a listing for *biofouling* which appears on Web of Knowledge in 1975.

at Newcastle University and a company, now known as International Paint Ltd, producing antifouling paints and coatings on the south bank of the Tyne at Felling. This focus developed as a consequence of regional development when Newcastle on the River Tyne, and Sunderland on the River Wear just a few miles away, were at the heart of the global ship building and armaments industries throughout the first half of the twentieth century. Whilst the shipping industry has relocated largely to Asia, a regional maritime emphasis on Tyneside remains. The focal dynamism this regionalism engenders ensures that robust and evolving linkages span not only the UK, but also Europe and the rest of the world, and this we hope is reflected in this book with its regional and global spread of contributors. We also hope that we have covered the field of biofouling in its widest sense by bringing together a suite of reviews that range from basic and applied research, relevant industries, environmental law and governance, and all the way through to horizon scanning. We also wanted to give space to some of the academics who are not so widely known at the biennial Biofouling Congresses, because perhaps they see their work as more mainstream ecology, as well as to those with a non-academic voice, who work at the forefront of dealing with the problems caused by biofouling in shipping and other industries. This we hope has enabled this volume to present some different views and fresh ideas to the field as well as giving space for people to be controversial: it is clear that quite a few of the contributors would not agree with each other on a variety of issues, but no heavy-handed editing has been done to obtain a bland consensus within the book. Science needs its mavericks and their non-conformist views and we have intentionally encouraged this. Similarly, we have let the authors use the terminology they are most comfortable with, and thus where they have used *biofouling* or *fouling* as a noun or the oddly constructed compound term *foul release coating(s)*[4], for example, we have used a light editorial hand and left it as such. The reader thus needs to be aware that the terminology used changes from chapter to chapter according to the authors' preferences.

Biofouling is governed by the same drivers that affect natural ecological communities and thus at the beginning of this book there are chapters by Havenhand and Styan (Sweden and Australia) on invertebrate reproductive processes, McQuaid and Miller (South Africa and Australia) on larval supply and dispersal, Prendergast (Newcastle) on larval settlement and behaviour, and Jenkins and Martin (UK and Azores) on community succession. These are rounded off by chapters from Canning-Clode and Wahl (Germany), with a review of spatial patterns of biofouling with a global perspective, and Cowie (UK) who shows us the importance of depth in marine biofouling. The differences between a natural ecological community on a hard substratum, such as rocky shore, and one developing on an artificial substratum are described in the linking chapter provided by Terlizzi and Faimali (Italy).

As Finnie and Williams (Felling) point out in their chapter on antifouling coatings and paints, one of the main drivers for research and development of novel coatings is the economics of the shipping industry unremittingly trying to reduce costs: this has always been the case. Working at Felling during the latter half of the twentieth century was Alex Milne, who has been a key contributor of innovation in the antifouling industry, having had a role in developing

---

[4] This terminology is difficult as *foul* is not normally used as a noun in this context, and the alternative *fouling release coatings* uses the active participle of the verb *foul* as a noun. It might be better to call them *foulage release coatings*, though the term *antifoulant* describing a compound with a specific effect suggests that *foulant release coatings* might also do.

both tin-based self-polishing coatings (TBT-SPCs) and biocide-free fouling release (FRC) coatings[5].

"Fouling prevention is a problem in economics and the environment where 1974 represented an impasse. The top grade antifouling of those years was USN 121 antifouling, a vinyl rosin high copper composition which lasted for only 7 to 14 months. But the rules of shipbuilding changed; instead of annual docking vessels were allowed to stay at sea for thirty months (eventually sixty months). This was also the year of the first oil shock when the price of Heavy Fuel Oil went from US$100 per tonne to US$350 per tonne and it doubled again in 1979. The antifouling industry came under extreme pressure to improve its act by 300 % or more and this led to much innovation exemplified by the self-polishing organotin antifoulings."[6]

Environmental issues also came to the fore from the 1970s onwards reaching a crescendo in the 1980s with the revelation of the substantial environmental impact of leached TBT. However, as Alex notes:

"Tributyltin was not the 'best'. It was significantly inferior to the organo-mercurials or the organo-arsenicals. The norm against which all biocides must be measured is phenoxarsine oxide. It would be deceitful to pretend that we [the antifouling industry] gave up organo-mercurials for environmental reasons; they became too expensive. We gave up the arseni-cals also. When airless spray became the norm it was impossible to continue with organo-arsenicals; they were the sternutators and lachrymators of the First World War. Not lethal, just intolerable at 10^{-6} of the damaging dose."

The final nail in the coffin of TBT-SPCs has now been hammered in and the legislation that led to its demise and the future control of product development is reviewed here by Cheyne (Newcastle) with ten Hallers-Tjabbes and Walmsley (Netherlands and UK) giving us what must surely be the final word on the environmental impact of TBT. Taking us forwards are Howell and Behrends (Newcastle) who review the potential unwelcome legacy of the other biocides currently in use. The collateral impact of biocides also provides a large stick for industry to develop non-toxic coatings, and as Finnie and Williams remark, although this can be traced back at least 50 years, the research was not really successful until Alex Milne developed a silicone fouling release coating (FRC) in the 1970s:

"The universal recommendation at the time was to try the 'non-stick frying pan'. Curiously, it did not work. This was, in any case, of little use to a marine compositions manufacturer in need of a liquid sprayable polymer. Curable silicones could be sprayed. The absence of the self-polishing effect was compensated by the outstanding flow and levelling of the liquid silicones which meant that the initial roughness of vessels should be less."

There has been an almost mythical belief in the superiority of TBT-SPCs with a concomitant wariness of their replacements, but as Thomason (Newcastle) shows in his analysis of the world's largest database of fouling on ships, this was unfounded and that the response of the industry was to provide other efficacious solutions.

---

[5] Milne, A. (1975) *Coated marine Surfaces*. Patent GB1470465A, International Paint Co. Ltd.; Milne, A. (1977) *Anti-fouling marine compositions*. Patent US4025693, International Paint Co. Ltd.; Milne, A. & Hails, G. (1976) *Marine Paint*. Patent GB1457590, International Paint Co. Ltd.

[6] Written in Newcastle, November 2008. This is Alex Milne's personal opinion and does not reflect that of any company he used to work for.

Alex Milne found inspiration in the natural world around him and was one of the first to encourage academics to explore the world of natural antifouling mechanisms:

"I have been to the sea shore and admired its variety. But natural fouling prevention is a vast and only partly explored field. From 'tide-boat research', children or grandchildren a suitable disguise, one notes that echinoderms are conspicuously clean, as are beadlet anemones."

This arms race between epibiont and basibiont is as old as life itself with biofilms being recorded from the 3400 to 3500 Ma Warrawoona chert[7] and the earliest record of an organism living on another is from the 1400 to 1500 Ma Gaoyuzhuang Formation[8]. In this book our understanding of epibiosis and how the organisms that are affected deal with the threat of biofouling are reviewed in detail by Wahl (Germany) and de Nys, Guenther and Uriz (Australia, Norway and Spain).

Although the impact of biofouling is often emphasised by the consequences for the marine shipping industry, herein reviewed by Edyvean (UK), there are also less obvious but very important areas in our lives where we are affected by biofouling. The dramatic growth of the aquaculture industry has reached over 60 million tonnes with a value of over US\$70 billion and this has resulted in a demand for specific solutions to biofouling. Dürr and Watson (Newcastle and Ireland) review the impacts caused by biofouling in this industry, which affect not only the infrastructure but also the organisms being cultured, and compare the specific antifouling needs of this sector with respect to the mainstream maritime industry. Likewise, Page, Dugan and Piltz (USA) and Henderson (UK) thoroughly review biofouling in the oil, gas and other industries, which also require industry-specific antifouling solutions.

Historically, most of the issues with fouling in maritime and other industries have been at the macro-scale, i.e. dealing with barnacles, mussels and algae, but the recent success of the latest generation of antifouling coatings has focused research efforts on reducing the impact of microbial fouling, the biofilms. Thus, we provide in this book three chapters to comprehensively cover biofilms, with Dobretsov (Oman) and Romaní (Spain) giving us comparative works on marine and freshwater biofilms, respectively. Outside of medicine, the impact of biofilms on implanted devices and prosthetics is largely overlooked and thus to remedy this we have included the gruesomely detailed chapter by Spratt, Ready and Pratten (UK).

The final section of the book is broadly defined as horizon scanning where we aim to give the reader information not only about where legislation (Cheyne, Newcastle), technology (Webster and Chisholm, USA) and research (Rittschof, USA) should lead us in the coming decade, but also where global climate change (Poloczanska and Butler, Australia) and the impact of not using adequate antifouling strategies will take us (Lewis and Coutts, Australia). Also included in the ultimate section is a summary of biofouling measurement techniques that are, or may be, useful across the spectrum of biofouling research.

We hope that by providing our reader with such a broad range of information from diverse fields this book bridges gaps, and thus is not simply a collection of scientific articles from any single individual's view point, but is an authoritative international multi-author compendium

---

[7] Schopf, W., Bonnie, M. & Packer, B.M. (1986) Newly discovered early Archean (3.4-3.5 Ga Old) microorganisms from the Warrawoona Group of Western Australia. *Origins of Life and Evolution of Biospheres*, **16**, 339–340.

[8] Seong Joo, L., Golubic, S. & Verrecchia, E. (1999) Epibiotic relationships in Mesoproterozoic fossil record: Gaoyuzhuang Formation, China. *Geology*, **27**, 1059–1062.

on biofouling and antifouling strategies suitable for the specialist in this field, as well as for non-specialists in biology, chemistry, medicine, law, industry and government.

Simone Dürr (Newcastle & Liverpool) and Jeremy C. Thomason
(Newcastle & Paris)

Chapter 1
Reproduction and Larvae/Spore Types

Jon N. Havenhand and Craig A. Styan

We describe the diversity of traits and adaptations that biofouling organisms have evolved to ensure reproductive success. Our chapter focuses on the processes of propagule production through sexual reproduction (fertilisation of gametes leading to a larval stage that may or may not disperse) or asexual reproduction (fragmentation of the embryo, larva or adult, or by production of diploid spores). We focus on key issues in broadcast spawning of gametes and fertilisation, adaptations to overcome Allee effects, and so-called 'carry-over' effects. In so doing we have not provided an exhaustive review of spore – or larval – types, but rather attempt to focus on areas that are currently at the forefront of research while highlighting avenues for further research.

1.1   Introduction

Concerns about biofouling organisms mostly focus on the rate and form of growth that allows organisms to spread and take up space on a surface. Yet isolated surfaces cannot be overgrown, and most unitary biofouling organisms like barnacles or polychaetes do not reproduce vegetatively. In these cases new biofouling must occur by way of some form of dispersive propagule – which for most animals and algae usually involves sexual reproduction (see Plate I A). This involves a cascade of processes, all of which need to happen successfully:

- Gametes (or spores) have to be produced and released at the right time.
- The sperm has to meet, fuse with and fertilise an egg.
- The resultant zygote or spore must be able to develop, perhaps feed, and disperse, whilst not getting eaten or diseased.
- The resulting competent larva/sporeling/germling must settle and recruit onto a surface.

In a biofouling context, understanding all the processes involved in this cascade is important because disruption to just one of them can provide a means to control the supply of potential recruits (see also Chapters 2 and 3). Clearly, this is a substantial research area that includes a wide range of organisms with diverse life histories. It is not our intent to review this topic comprehensively, but rather to provide an introduction to the main issues, emerging topics and key literature.

1.2 Some terminology

Most organisms which are important as biofoulers have a complex life cycle, involving at least one dispersive, usually microscopic, stage that does not look much like the adult sessile life form. A range of different, sometimes confusing, terminologies are used by zoologists and phycologists to refer to these propagules and the different processes that are employed to produce them. We define some of them here that we use below:

Broadcast spawning: Both male and female gametes are released and fertilisation occurs externally, in the sea (some authors also designate this as free spawning, but see below).

Spermcast mating: Male gametes are released and fertilise eggs retained by maternal individuals (also designated as free spawning egg brooders).

Free spawning: Considerable ambiguity has surrounded the definition of free spawning (e.g. [1]): Some authors have used this term to describe the release of sperm, independently of whether the eggs are released or retained (e.g. [2–4]), and others to describe the release of both sperm and eggs, i.e. identical to broadcasting (e.g. [5, 6]). We therefore restrict ourselves to using the terms 'broadcasting' and 'spermcasting', and avoid the ambiguous meaning of this term.

Copulatory fertilisation: Direct delivery of sperm to maternal individuals (by copulation or pseudo-copulation).

Egg brooding: Retention of eggs in the maternal individual until fertilisation.

Embryo or larval brooding: Retention of embryos following fertilisation up to a more advanced stage at which they are released, and which may still include a short pelagic phase or just direct settlement.

Encapsulation: Retention of embryos and/or larvae within protective benthic structures until they hatch, either as swimming larvae or as crawl-away juveniles.

Larvae: Potentially dispersive (multicellular) propagules produced by animals, most often by sexual reproduction. There is a huge range of different larval types and forms across different taxonomic groupings. One of the best overviews of the incredible diversity of larvae is the recent compendium edited by Young *et al.* [7]. More broadly, larvae are often classified according to whether they feed (planktotrophic) or do not feed (lecithotrophic) before they settle and metamorphose into a juvenile/adult form.

Zygote: The product of two gametes uniting to form a diploid cell. In animals, this develops into a larva. In algae, the zygote will settle and germinate into a diploid vegetative plant. A settling algal zygote will sometimes be called a *germling*.

Spores: Formally, a spore is defined as a unicellular reproductive unit produced by an (algal) sporophyte [8]. Algal biologists base their definitions of spores and gametes on whether a unicellular propagule settles and develops into a vegetative unit in its own right (a *spore*) or if it must first unite with another (*gamete*) to form a zygote which then settles and grows into a vegetative unit; unfortunately, the terms sometimes seem to be misapplied interchangeably. A spore can be produced mitotically (*mitospore*) or meiotically (*meiospore*), and be haploid or diploid, depending on the particular life history pattern of the alga in question – which is often exceedingly complex (see below). Perhaps further confusing zoologists, the prefix 'zoo' (*zoospore*) or the term 'swarmer' is sometimes used to indicate that a spore is motile.

1.3   Asexual production of propagules

Fragments of vegetatively produced (clonal) growth that break off from parental colonies can often function as dispersive propagules. Colonisation of new surfaces via these fragments can result from two processes: by successful settlement of the fragment, or by the release of larvae or gametes by the fragment. Settlement and re-establishment of fragments appears common among a range of groups including sponges, corals, ascidians, bryozoans and algae and can be a very important demographic process [9]. For example, an invading didemnid ascidian is presently causing major biofouling problems for mussel aquaculture in North America and, indeed, seems to be overgrowing and smothering virtually everything in its path [10]. The rope-like growth form of this and related species can result in lobe-shaped fragments being broken off naturally or as a result of human activities such as cleaning of mussel lines or scallop dredging. Thus, cleaning mussel lines of such ascidians may actually exacerbate biofouling because the new fragments which fall back into the water are capable of reattaching and growing if they lodge in an appropriate place [10].

The role of dispersive fragments as vehicles carrying gametes that are then shed in new locations is less well established, but in some cases gametes or larvae can develop and be released from surprisingly small fragments. For example, some sponges can release embryos from fragments only a few millimetres in diameter [11]. In general, the potential importance of rafting as a source of colonists and/or gametes has been largely overlooked until recently [12] and yet this could be a very important dispersal mechanism, particularly for large macroalgae [13, 14].

Several taxa produce asexual dispersive propagules. Amongst the more intriguing of these is the parthenogenetic production of larvae found in the *Lasaea* complex of clams in northeast Pacific. Most species of *Lasaea* produce outcrossed sexually derived larvae (see below), but a number of strictly asexual lineages produce larvae parthenogenetically and have apparently done so for a long period of evolutionary time [15]. These lineages do not self-fertilise in the usual sense; rather, the development of (diploid) eggs into parthenogenetic larvae is triggered by fertilisation with autosperm; the sperm then disintegrate within the egg cortex, rather than fusing with the egg 'pronucleus' [16]. Exclusively parthenogenetic reproductive strategies like this are rare amongst typical biofouling taxa, but facultative parthenogenesis, in which male and/or female gametes can develop into mature reproductive thalli if fertilisation does not take place, is recorded in a number of algae [17], including common biofoulers such as *Ulva* and *Codium* [18]. By providing reproductive assurance when population sizes are small and mate availability is low, facultative parthenogenesis may be an important form of reproduction in some taxa.

Much more commonly, organisms combine alternating periods of sexual and asexual reproduction. The most obvious and widespread examples of these are the various haploid–diploid life cycles of many macroalgae. For example, 'diplohaplont' life histories are common within the green and brown algae and involve the alternation of both haploid and diploid vegetative stages. The basic pattern is that meiotically derived (haploid) motile spores are produced by sporophytes and germinate into gametophytes. These in turn produce gametes mitotically, which then fuse (syngamy) to form a diploid zygote that germinates into the mature sporophyte, completing the cycle. There are, however, many variations on this theme and probably many more yet to be fully described [18–20]. For example, diploid sporophytes of the brown algae *Ectocarpus* can produce haploid zoospores via meiosis that will settle and grow into haploid gametophytes or they can produce diploid zoospores by mitosis that will settle and grow into new diploid sporophytes [21]. Even more complex, the red algae (Rhodophyta) can incorporate a third life history phase, the carposporophyte, which is unique in that it is formed on the female gametophyte. In this context, sexual reproduction in red algae is more similar to brooding (spermcasting) animals than to the broadcast spawning strategies in other algal groups (see below). Algal life histories are exceedingly complex and we refer readers to other sources for much better overviews of these [3, 18–20]. Whether produced asexually or sexually, however, spores and gametes of algae are almost always unicellular (but see [22]).

Animals too can alternate between sexual and clonal production, but the asexual propagules produced by animals are almost always multicellular. Colonial taxa such as corals, freshwater bryozoan and sponges are all well known for their ability to produce asexually derived larvae/propagules at various times, often depending on demographic or environmental conditions (e.g. [23–26]). For example, freshwater bryozoans produce desiccation-resistant asexual statoblasts during times of drying, which may be an important source for colonisation when conditions again become suitable [27]. Other animal life histories include an obligatory asexually produced stage (e.g. some hydroid medusae [28] – which in most species disperses freely and in all cases produces freely spawned gametes which unite to form a larval stage that then settles and metamorphoses to a juvenile). Similarly, some unitary animals produce asexual clones of themselves which can disperse short distances via crawling or drifting but at other times release gametes to produce sexually derived larvae capable of much wider dispersal. Anemones (anthozoans) are well known for this and can reproduce asexually using a variety of mechanisms including longitudinal (or even transverse) fission or through pedal laceration [29]. Equally, some tube-building serpulid and sabellid polychaetes bud off miniature versions of themselves from their posterior segments – a strategy which could lead to an obvious problem for the newly cloned individual trapped at the end of a tube behind its parent: how to get out? In at least some species of *Filograna* and *Salmacina*, adults build 'escape hatches' into the solid calcareous tubes as a way for budded-off children to escape and then start building their own tubes [30].

Finally, there are some truly strange life history strategies in which larvae produced are a combination of both sexual and asexual reproduction. Polyembryony is apparently used by all cyclostome bryozoans and involves the asexual clonal production (or twinning) of a sexually generated larva [5]. This strategy has been seen as something of a paradox in an evolutionary sense as it apparently combines the potential negative consequences of both fully outcrossed sexual reproduction and asexual reproduction [31]. One suggestion [32] is that polyembryony may provide a form of reproduction assurance when sperm are rare and/or quickly diluted in the sea [33] and there would not otherwise be enough sperm available to fertilise multiple eggs. Similar clonal splitting of sexually produced larvae is known in other free-living groups (e.g. echinoderms [34]), but this is an apparently rare strategy in sessile biofouling organisms. Even more bizarre are chimeric sponge larvae, which can be formed from the fusion of independently and sexually produced larvae [35]. In contrast to quite high relatedness often needed for fusion in adult sponges, allorecognition systems do not appear to be as well developed in the larvae of sponges such as *Haliclona* sp. Consequently, non-related larvae fuse quite readily to form larger larvae that swim and can still metamorphose and develop into adult sponges with a mosaic genetic identity. Perhaps because larger larvae may be more successful competitors, fusion might give chimeric larvae size advantages during settling and recruitment (though see [36]). Larvae also behave in an aggregative way in the laboratory that would apparently increase the opportunities for fusion to happen, though if or how often this happens in natural field populations is unclear [35]. As an aside, genetically mosaic individuals (chimaeras) appear to be much more common than often appreciated, and examples are known across a wide range of colonial biofouling taxa such as sponges [37], ascidians [38, 39], hydroids [40] and red algae [41,42]. Just how multiple and presumably interacting genotypes within a single demographic 'individual' influences allocation of resources to asexual or gametic reproduction and then affects sexual reproduction is a fascinating area for future research [43].

1.4   Sexual reproduction – mechanisms facilitating fertilisation

Sexual reproduction involves the union of haploid gametes, nominally a sperm from a male and an egg from a female, to form a diploid zygote (i.e. syngamy). Of course many organisms are hermaphroditic and so can act as a male (producing sperm) and female (producing eggs) either simultaneously or at different times, sequentially – first as a male (protandrous) or as a female first (protogynous). Thus, in many cases the terms male and female below refer to whether individuals are releasing sperm or eggs. In algae the situation can be even less distinct. Where gametes are isogamous (i.e. similar sized and functionally equivalent) but gametes from two different genetic individuals are required for fertilisation, gametes are referred to as +ve and −ve, indicating that complementary gametes are involved; the green alga *Ulva* is a well-known example where this occurs [20].

   Where sexual reproduction takes place, there are three basic mechanisms which aquatic organisms use to get sperm and egg together:

1. There is some form of direct coupling during which males directly transfer sperm to the female and fertilisation takes place within the female (internal fertilisation).
2. Males release sperm freely into the water, allowing water movements to disperse the sperm to eggs which are retained/brooded within the female and fertilisation takes place internally (spermcasting).
3. Both sperm and eggs released freely and fertilisation takes place externally, in the open water column (broadcasting).

   Perhaps for biophysical reasons [44], there appears to be no biological intermediary that facilitates the transfer sperm (or eggs) for fertilisation in aquatic systems; that is, there is no aquatic equivalent to the pollinator-mediated systems of terrestrial flowering plants [45].

   Internal fertilisation is largely restricted to mobile animals, which are of little concern in biofouling (except perhaps as predators or grazers of biofoulers). The obvious exception to this pattern is barnacles, which fertilise internally by means of a famously long (relative to body size) penis. In the common European barnacles *Semibalanus balanoides* and *Chthamalus montagui*, penis length is up to 25 mm [46]; body size varies and is typically 8–12 mm. This constraint on fertilisation seems to drive larval settlement patterns: in a recent study, Kent *et al.* [46] found that larvae of *S. balanoides*, an obligate outcrosser, were far more likely to settle within one penis length of a conspecific than larvae of *C. montagui*, a facultatively selfing species. Although the issue has been discussed for a long time, the intriguing possibility that settlement distances in barnacles might be determined not only by penis length but also by capacity for selfing requires further investigation – not least because this will influence both mating system and reproductive success. A few studies have reported the capacity for selfing in barnacle species [47, 48], but we still know very little about the relative frequencies of selfing versus outcrossing in barnacles.

In contrast to internal fertilisation, spermcasting provides an obvious solution for transferring sperm in sessile taxa: where males cannot move to deliver sperm directly to females, they release it freely so that the sperm are advected to females. Spermcasting is used by many biofouling species, notably bryozoans, colonial ascidians and some species of red algae. Bishop and Pemberton [5] listed six generalisations concerning reproductive traits of spermcasters:

1. Sperm are relatively long lived once diluted in sea water.
2. Only very dilute sperm suspensions are needed to ensure fertilisation.
3. Sperm can be stored by the recipient.
4. There is extensive contact between allosperm and the somatic tissue of the recipient.
5. Gamete compatibility systems are often involved (although this is also true of many – perhaps all – other sexually reproducing taxa: see below).
6. Receipt of compatible sperm may trigger female investment.

The first three of these traits are adaptations that minimise sperm limitation effects (see below), whereas the last three allow for differential provisioning of larvae with maternal resources, depending on the genetic make-up of the fertilising sperm (see below). All of these generalisations are relevant to taxa such as bryozoans and colonial ascidians; however, not all spermcasting species show all of these adaptations. For example, in ostreid oysters, serpulid and spirorbid polychaetes, and some ascidians (e.g. *Corella*), fertilisation takes place within the mantle/atrial cavity, i.e. outside the body. In these cases the eggs are indeed retained by the mother, and it is probable that maternal control of water flow through the mantle/atrial cavity mediates fertilisation success, but it is not clear that sperm can be stored, or indeed that there is extensive contact between the sperm and the somatic tissue of the recipient. Moreover, eggs are fully provisioned before sperm are received and therefore there is no opportunity for sperm-induced maternal investment. This suggests that rather than spermcasting being a wholly distinct and separate category, it may rather lay at the end of a continuum of traits from broadcasting to 'fully' spermcasting. The processes that control sperm release, acquisition and fertilisation in spermcasters are very poorly understood, and this promises to be an exciting and productive research area.

For broadcast spawners, a key distinction is that there is little or no direct control over fertilisation: eggs and sperm are released into the water column where fertilisation occurs. Nonetheless, a series of adaptations have evolved to ensure fertilisation success in broadcast spawning species. These include several mechanisms to increase gamete concentration, such as spatial and/or temporal aggregation at spawning (simultaneous or 'mass' spawnings), gamete buoyancy and swimming behaviour, gamete longevity, chemical attraction of sperm to eggs and the retention of 'sticky' eggs, as well as adaptations to ensure that eggs are not fertilised by too many sperm ('polyspermy'). We need to be careful to take variation in these traits into account when attempting to estimate the success of broadcast spawning: these traits vary widely not only between taxa, but also between populations, individuals, seasons and times of day, so that even comparisons within species can be difficult [49].

1.5   Demographic effects on fertilisation success – Allee effects

A critical requirement for sexual reproduction is the availability of a partner, and consequently sexual reproduction can be risky when the chance of finding a mate is low. Interestingly, reproductive rate scales with population density, whereas death rate often does not. The result is that below critical population densities, reproductive success is lower than the death rate, leading to extinction of that population (if rates do not change). This effect – the 'Allee effect' – is non-linear, and does not only apply to birth/death rates: it can result in reduced choice and/or lower quality of mates and correspondingly reduced or zero fitness [50].

The Allee effect is particularly relevant for invading species – in nearly all natural circumstances it does not just take two individuals to establish a species (Noah was seriously misguided!). Invading individuals need to be close enough to each other, spawn at the same time and be compatible for the inoculation to lead to successful reproduction. For this to occur, sufficient 'propagule pressure' (enough individuals recruiting and making it through to sexual reproduction) is needed to establish a novel population [51]. The potentially valuable corollary of this is that if we reduce population size below a critical level, we may be able to prevent (or at least limit) the number of sexually produced propagules and thus limit the recruitment of larvae/germlings to surfaces.

In the last 20 years there has been a lot of interest in the potential impacts of Allee effects in broadcast spawners. Specific attention has focused on Allee effects and sperm limitation – the condition in which a fraction of released eggs remain unfertilised because sperm concentrations are too low. Sperm limitation is probably common: gametes are relatively short-lived, and they disperse and dilute quickly (especially in strong flows). Consequently, despite many adaptations to increase likelihood of gamete contact (reviewed by Serrão and Havenhand [49]), gametes may often expire before ever having the chance to meet. Clearly, this could be a major constraint to reproduction, but how often does this actually happen?

Field experiments to measure fertilisation success suggest that sperm limitation could be important in many situations; however, there are many potential artefacts that limit our ability to measure fertilisation success experimentally. Yund and Meidel [52] have shown that the 'broadcast' spawning of urchins may actually involve the release of eggs in viscous fluids that retain them on the urchin surface, where they effectively 'filter' sperm from the passing water mass, thereby integrating sperm concentrations across a long period and elevating fertilisation rates over those expected from simultaneous unencumbered release of gametes. Similar release of eggs in mucus strings, and successful fertilisation within those strings, has been reported from the laboratory and field for the tunicate *Ciona intestinalis* [53]. Whether such adaptations are common among marine invertebrates is not known: relatively few species have been observed spawning in the field; however, the release of gametes in mucilage occurs commonly in algae, both at low tides (e.g. fucoids) and subtidally (e.g. rhodophytes). Indeed, it has been hypothesised that in red algae, the release of spermatia in mucus strings will greatly facilitate capture by the trichogynes (spermatia are aflagellate and cannot swim), thereby enhancing fertilisation success (reviewed by Brawley and Johnson [18]). Even the briefest analysis shows that the potential benefits of this behaviour are considerable, and it is therefore surprising that this has not been reported more frequently (while noting the obvious caveat that absence of evidence is not evidence of absence). Indeed, there are very few observations of spawning in marine organisms in general, and data on the degree of spawning synchrony, timing of spawning and correlations with environmental variables, and rates and variance in fertilisation success are all badly needed.

Notwithstanding these constraints, one generalisation that can be made is that the mating system (internal fertilisation, spermcasting, broadcasting) does seem to make a difference to fertilisation success. The very few data available suggest that internal fertilisers (such as barnacles) are not sperm-limited (although to our knowledge this has not been investigated and because sperm limitation would probably manifest as reduced reproductive frequency, this could have been overlooked and/or misinterpreted in previous studies). Similarly, spermcasters seem to be reasonably buffered against sperm limitation: genetic studies of ascidians and bryozoans show multiple paternity (in addition to reproductive assurance by selfing) and that reproductive success is largely independent of population density [4,5]. In contrast, for a great many species broadcasting remains a risky, though probably unavoidable, alternative.

1.6   Environmental factors affecting fertilisation

A wide range of environmental factors can directly or indirectly affect the processes that lead to larval/spore production. Salinity, dissolved gases, UV radiation, temperature and pH can all affect various chemical processes involved in sperm viability and motility, egg/zygote viability or the actual process of fertilisation itself [54]. Similarly, a range of physical processes can affect gametes and whether they get or stay together. Turbulence is critically important in determining not only rates of mixing of gametes, but also rates of dilution and dispersion – which may happen before gametes have had a chance to meet [54,55]. Gametes of some algae are positively phototactic leading them to move towards the surface, thereby concentrating them and thus increasing the likelihood of contact and fertilisation with other gametes [56–58]. There is also a plethora of ecotoxicological studies on the effects of various specific toxicants on sperm, eggs and developing zygotes/larvae. Again, toxicants can all potentially affect sperm and/or eggs and/or the developing zygote and/or the process of fertilisation itself [59] and, much less often investigated, multiple toxicants/physical stressors probably often do so in synergistic ways [60].

The impacts of environmental variables on fertilisation are less well investigated but can be critically important. For example, the membrane-based 'fast' blocks to polyspermy in most taxa involve Na^+/K^+ ion channels that alter permeability of the membranes. In fucoid algae, these polyspermy blocks are efficient at preventing penetration by supernumerary sperm at typical marine salinities, but in the Baltic Sea and/or estuarine areas where salinities are lower, higher rates of polyspermy ensue, presumably because ion exchange and thus fast-electrical blocks work less well [61]. Equivalent data for marine invertebrates are lacking.

1.7   Links between fertilisation and subsequent larval attributes

For broadcast spawners, egg-size dependent fertilisation success can influence the average size and quality of larvae produced [62]. It has long been known that the cross-sectional area of an egg in part determines how easily an egg will be found by searching sperm [63, 64]. Small eggs represent small targets for randomly swimming sperm, so when sperm are limited smaller eggs are less likely to be found and are thus more likely to remain unfertilised. If egg size varies within a brood (which abundant evidence now indicates is the case, e.g. [65]), then at low sperm concentrations, only the larger eggs will be fertilised. For the same reasons, when sperm are very abundant, larger eggs will be more likely to be fertilised by more than one sperm (before the polyspermy blocks can prevent further sperm penetration). In that situation only smaller eggs within a brood will be fertilised. This theoretical effect [66] has been demonstrated in the laboratory [62] and has been shown to have further 'carry-over' effects later in the life cycle (see below). The extent to which these processes operate in biofouling communities in the field is not known. Nor is it known whether such processes play any role at all in spermcasting and/or internally fertilising species (although the possibility to physically direct and manipulate sperm in such mating systems makes it unlikely that such 'random' processes play a large role).

Fertilisation success and subsequent zygote/larval development depends also on the genetic constitution of the egg and sperm. Firstly, sperm/egg binding proteins show considerable variation so that some genotypes will bind and fertilise more easily than others [67, 68]. This imposes yet another restriction on fertilisation success: even if sperm and egg meet, the likelihood that the sperm can penetrate and fertilise the egg may vary markedly. Consequently, we would predict that spawnings involving a greater number of individuals (more gamete genotypes in a mixture) would be more likely to generate a compatible match. A corollary of this is that such mass spawnings are also more likely to lead to high sperm concentrations, and hence polyspermy. The obvious trade-off between the risks of polyspermy and unfertilised eggs has been proposed to be one mechanism driving the evolution of variability in gamete recognition systems [67] in a manner analogous to the large versus small egg size trade-off discussed earlier. The mathematically self-evident benefits accruing from polyandrous spawnings have been demonstrated by Evans and Marshall [69], although they showed this at constant sperm concentrations (thereby excluding the increased risk of polyspermy that polyandrous spawnings imply).

Post-fertilisation compatibility issues will also limit the development of the zygote/larva. Perhaps the most obvious example is that of selfing versus outcrossing – an issue of particular relevance to clonal biofouling species. Many simultaneous hermaphrodites, such as the ascidian *Ciona*, have partial blocks to self-fertilisation. Yet these blocks can change with gamete age, and vary markedly between species [70–72]. In contrast, some species show no disadvantage from inbreeding (e.g. [73]), and indeed outbreeding depression may exist (e.g. [74]), although this is rare.

As a general note, selfing/inbreeding is probably more common than we tend to think. Alteration of algal life stages and ploidy levels, combined with limited dispersal of algal spores, almost inevitably leads to a large degree of inbreeding in many circumstances. Similarly, the universality of 'blocks' to selfing has been questioned, and several notionally 'self-incompatible' species have been shown to self, while other species known to be able to self have been shown to preferentially outcross – even at the gamete-recognition level [75]. Equivalent results relating to the probability of egg/larval provisioning have been shown for spermcasters [76, 77].

1.8 Hatching and development

One of the most obvious differences between brooders and broadcasters is the location of embryonic development. Brooders (a category that includes all spermcasters, and some – though certainly not all – internal fertilisers) generally retain the embryos until they have hatched as larvae and are capable of independent swimming and feeding. In many mobile internal fertilisers, encapsulation of eggs/embryos in benthic masses is an important alternative 'brooding-like' strategy [78]; however, this is not relevant for most biofoulers.

The developing embryos of broadcasters, in contrast, are exposed to the environment throughout their embryonic development, and only when they reach the larval stage can they begin to swim, feed and benefit from defensive behaviours. Staver and Strathmann [79] have shown that early development rates of such 'unprotected' embryos are significantly faster than those of encapsulated or brooded embryos, suggesting strong selective pressures to minimise early development times. The defensive capacities of larvae (as distinct from embryos) have been known for a long time [80, 81]; see review by Morgan [82]. Nonetheless, larval mortality rates can be extremely high, and are often ten times those of benthic encapsulated embryos [82]. The large numbers of eggs typically released by broadcasters help to overcome this mortality.

1.9   Mobility and survival of larvae/spores

The mobility of larvae and spores is generally very limited, but nonetheless plays a major role in their dispersal. Most propagules are ciliated, flagellated or swim by muscular movement of appendages [83]. This topic has been reviewed extensively by Chia *et al.* [83] and Young [84], and is typically cited for highlighting the gross disparities between larval swimming speeds ($<$cm s^{-1}) and near-shore currents (often $>$m s^{-1}). Nonetheless, recent work has shown the capacity of slow-swimming bivalve larvae to significantly influence their distribution in ocean currents (e.g. [85]) and to settle in rapid flows (reviewed by Koehl [86]).

The fate of larvae and spores during the dispersive phase – especially the impacts upon them of nutritional/light regimes that influence their energy reserves, and hence their ability to endure prolonged dispersal and searching for suitable settlement sites – is almost wholly unknown. The difficulties of tracing microscopic larvae in open water masses are obvious and despite several significant advances in technology that aid such measurement (e.g. [87, 88]), this still remains a very poorly understood – yet vital – component of the life history.

1.10   Carry-over effects

One aspect of larval/spore experience that has been investigated closely is so-called 'carry-over' effects. These are effects of processes that happen during one life stage but which are expressed in later life stages [89]. Perhaps the most obvious aspect of this is the respective survival and settlement probabilities of larvae that have received plentiful, versus scant, provisioning over the larval period. However, carry-over effects can be far more subtle. There is increasing evidence that events during egg provisioning, fertilisation and embryonic and larval development can have profound influences on settlement and post-settlement success [89–92]. These events may have positive and/or negative impacts, raising again the importance of Fu-Shiang Chia's exhortation that larvae should not be studied in isolation, but in terms of the whole life cycle [93]. To our knowledge, equivalent investigations have not been undertaken with algal spores and gametes; however, it is to be expected that similar results will be found.

If this turns out to be important, then a possible focus for future biofouling control methods is to develop strategies that influence the quality of the very earliest life stages of biofouling organisms so that gamete viability, fertilisation success and/or successful spore/larval development are compromised.

1.11 Conclusions

- Biofouling organisms utilise a wide variety of mechanisms to produce dispersive propagules; both sexual and asexual reproduction is common.
- With the obvious exception of barnacles, sexual reproduction in biofouling organisms nearly always involves the release of sperm (only) or broadcast spawning both eggs and sperm.
- There are many apparent adaptations to overcome potential Allee effects in biofouling organisms. Selection is likely for spawning behaviours that increase gamete encounter rates during suitable environmental conditions and on gametic traits themselves that would increase the likelihood of successful fertilisation. Similarly, the ability to self or reproduce parthenogenetically in some organisms may allow for reproductive assurance when population sizes are low.
- Larval/germling traits can be affected by processes before, during and after fertilisation. Such effects may be carried over through metamorphosis and have impacts on later life history stages.

References

1. Giese, A.C. & Kanatani, H. (1987) Maturation and spawning. In: *Reproduction of Marine Invertebrates*, Vol. **9** (eds A.C. Giese, J.S. Pearce & V. B. Pearce), pp. 251–329. Blackwell Scientific, Palo Alto.
2. Levitan, D.R. (1998) Sperm limitation, gamete competition, and sexual selection in external fertilizers. In: *Sperm Competition and Sexual Selection* (eds T.R. Birkhead & A.P. Moller), pp. 175–217. Academic Press, London.
3. Santelices, B. (2002) Recent advances in fertilization ecology of macroalgae. *Journal of Phycology*, **38** (1), 4–10.
4. Yund, P.O. (2000) How severe is sperm limitation in natural populations of marine free-spawners? *Trends in Ecology & Evolution*, **15** (1), 10–13.
5. Bishop, J.D.D. & Pemberton, A.J. (2006) The third way: spermcast mating in sessile marine invertebrates. *Integrative and Comparative Biology*, **46** (4), 398–406.
6. Pemberton, A.J., Hughes, R.N., Manriquez, P.H. & Bishop, J.D.D. (2003) Efficient utilization of very dilute aquatic sperm: sperm competition may be more likely than sperm limitation when eggs are retained. *Proceedings of the Royal Society of London Series B – Biological Sciences*, **270**, S223–S226.
7. Young, C.M., Sewell, M.A. & Rice, M.E. (2002) *Atlas of Marine Invertebrate Larvae*. Academic Press, London.
8. Womersley, H.B.S. (1994) *The Marine Benthic Flora of Southern Australia. Part IIIa*. Australian Biological Resources Study, Canberra, Adelaide, South Australia.
9. Wright, J.T. & Davis, A.R. (2006) Demographic feedback between clonal growth and fragmentation in an invasive seaweed. *Ecology*, **87** (7), 1744–1754.

10. Bullard, S.G., Sedlack, B., Reinhardt, J.F., Litty, C., Gareau, K. & Whitlatch, R.B. (2007) Fragmentation of colonial ascidians: differences in reattachment capability among species. *Journal of Experimental Marine Biology and Ecology*, **342** (1), 166–168.

11. Maldonado, M. & Uriz, M.J. (1999) Sexual propagation by sponge fragments. *Nature*, **398** (6727), 476.

12. Thiel, M. & Haye, P.A. (2006) The ecology of rafting in the marine environment. III. Biogeographical and evolutionary consequences. *Oceanography and Marine Biology – An Annual Review*, **44**, 323–429.

13. Macaya, E.C., Boltana, S., Hinojosa, I.A., et al. (2005) Presence of sporophylls in floating kelp rafts of *Macrocystis* spp. (Phaeophyceae) along the Chilean Pacific coast. *Journal of Phycology*, **41** (5), 913–922.

14. Hernandez-Carmona, G., Hughes, B. & Graham, M.H. (2006) Reproductive longevity of drifting kelp *Macrocystis pyrifera* (Phaeophyceae) in Monterey Bay, USA. *Journal of Phycology*, **42** (6), 1199–1207.

15. O'Foighil, D. & Smith, M.J. (1995) Evolution of asexuality in the cosmopolitan marine clam *Lasaea*. *Evolution*, **49** (1), 140–150.

16. O'Foighil, D. & Thiriotquievreux, C. (1991) Ploidy and pronuclear interaction in northeastern Pacific *Lasaea* clones (Mollusca, Bivalvia). *Biological Bulletin*, **181** (2), 222–231.

17. Clayton, M.N., Kevekordes, K., Schoenwaelder, M.E.A., Schmid, C.E. & Ashburner, C.M. (1998) Parthenogenesis in *Hormosira banksii* (Fucales, Phaeophyceae). *Botanica Marina*, **41** (1), 23–30.

18. Brawley, S.H. & Johnson, L.E. (1992) Gametogenesis, gametes and zygotes – an ecological perspective on sexual reproduction in the algae. *British Phycological Journal*, **27** (3), 233–252.

19. Clayton, M.N. (1988) Evolution and life histories of brown algae. *Botanica Marina*, **31** (5), 379–387.

20. Santelices, B. (1990) Patterns of reproduction, dispersal and recruitment in seaweeds. *Oceanography and Marine Biology – An Annual Review*, **28**, 177–276.

21. Peters, A.F., Marie, D., Scornet, D., Kloareg, B. & Cock, J.M. (2004) Proposal of *Ectocarpus siliculosus* (Ectocarpales, Phaeophyceae) as a model organism for brown algal genetics and genomics. *Journal of Phycology*, **40** (6), 1079–1088.

22. Keum, Y.S., Oak, J.H., van Reine, W.F.P. & Lee, I.K. (2003) Comparative morphology and taxonomy of *Sphacelaria* species with tribuliform propagules (Sphacelariales, Phaeophyceae). *Botanica Marina*, **46** (2), 113–124.

23. Uotila, L. & Jokela, J. (1995) Variation in reproductive characteristics of colonies of the freshwater bryozoan *Cristatella mucedo*. *Freshwater Biology*, **34** (3), 513–522.

24. Ereskovsky, A.V. & Tokina, D.B. (2007) Asexual reproduction in homoscleromorph sponges (Porifera; Homoscleromorpha). *Marine Biology*, **151** (2), 425–434.

25. Corriero, G., Liaci, L.S., Marzano, C.N. & Gaino, E. (1998) Reproductive strategies of *Mycale contarenii* (Porifera: Demospongiae). *Marine Biology*, **131** (2), 319–327.

26. Ayre, D.J. & Resing, J.M. (1986) Sexual and asexual production of planulae in reef corals. *Marine Biology*, **90** (2), 187–190.

27. Callaghan, T.P. & Karlson, R.H. (2002) Summer dormancy as a refuge from mortality in the freshwater bryozoan *Plumatella emarginata*. *Oecologia*, **132** (1), 51–59.

28. Boero, F. & Bouillon, J. (1993) Zoogeography and life cycle patterns of Mediterranean hydromedusae (Cnidaria). *Biological Journal of the Linnean Society*, **48** (3), 239–266.

29. Fautin, D.G. (2002) Reproduction of cnidaria. *Canadian Journal of Zoology – Revue Canadienne De Zoologie*, **80** (10), 1735–1754.

30. Pernet, B. (2001) Escape hatches for the clonal offspring of serpulid polychaetes. *Biological Bulletin*, **200** (2), 107–117.

31. Craig, S.F., Slobodkin, L.B., Wray, G.A. & Biermann, C.H. (1997) The 'paradox' of polyembryony: a review of the cases and a hypothesis for its evolution. *Evolutionary Ecology*, **11** (2), 127–143.

32. Hughes, R.N., D'Amato, M.E., Bishop, J.D.D. *et al.* (2005) Paradoxical polyembryony? Embryonic cloning in an ancient order of marine bryozoans. *Biology Letters*, **1** (2), 178–180.

33. Levitan, D.R. & Petersen, C. (1995) Sperm limitation in the sea. *Trends in Ecology & Evolution*, **10** (6), 228–231.
34. McEdward, L.R. & Miner, B.G. (2001) Larval and life-cycle patterns in echinoderms. *Canadian Journal of Zoology – Revue Canadienne De Zoologie*, **79** (7), 1125–1170.
35. McGhee, K.E. (2006) The importance of life-history stage and individual variation in the allorecognition system of a marine sponge. *Journal of Experimental Marine Biology and Ecology*, **333** (2), 241–250.
36. Maldonado, M. (1998) Do chimeric sponges have improved chances of survival? *Marine Ecology – Progress Series*, **164**, 301–306.
37. Curtis, A.S.G., Kerr, J. & Knowlton, N. (1982) Graft rejection in sponges – genetic structure of accepting and rejecting populations. *Transplantation*, **33** (2), 127–133.
38. Sommerfeldt, A.D. & Bishop, J.D.D. (1999) Random amplified polymorphic DNA (RAPD) analysis reveals extensive natural chimerism in a marine protochordate. *Molecular Ecology*, **8** (5), 885–890.
39. Stoner, D.S., Rinkevich, B. & Weissman, I.L. (1999) Heritable germ and somatic cell lineage competitions in chimeric colonial protochordates. *Proceedings of the National Academy of Sciences of the United States of America*, **96** (16), 9148–9153.
40. Fuchs, M.A., Mokady, O. & Frank, U. (2002) The ontogeny of allorecognition in a colonial hydroid and the fate of early established chimeras. *International Journal of Developmental Biology*, **46** (5), 699–704.
41. Santelices, B., Aedo, D., Hormazabal, M. & Flores, V. (2003) Field testing of inter- and intraspecific coalescence among mid-intertidal red algae. *Marine Ecology – Progress Series*, **250**, 91–103.
42. Santelices, B. (2004) A comparison of ecological responses among aclonal (unitary), clonal and coalescing macroalgae. *Journal of Experimental Marine Biology and Ecology*, **300** (1–2), 31–64.
43. Santelices, B. (2004) Mosaicism and chimerism as components of intraorganismal genetic heterogeneity. *Journal of Evolutionary Biology*, **17** (6), 1187–1188.
44. Van Der Hage, J.H.C. (1996) Why are there no insects and so few higher plants in the sea? New thoughts on an old problem. *Functional Ecology*, **10**, 546–547.
45. Strathmann, R.R. (1990) Why life histories evolve differently in the sea. *American Zoologist*, **30** (1), 197–207.
46. Kent, A., Hawkins, S.J. & Doncaster, C.P. (2003) Population consequences of mutual attraction between settling and adult barnacles. *Journal of Animal Ecology*, **72** (6), 941–952.
47. Furman, E.R. & Yule, A.B. (1990) Self fertilization in *Balanus improvisus* Darwin. *Journal of Experimental Marine Biology and Ecology*, **144** (2–3), 235–239.
48. Barnes, H. & Crisp, D. (1956) Evidence of self fertilization in certain species of barnacles. *Journal of the Marine Biological Association of the United Kingdom*, **35**, 631–639.
49. Serrão, E.A. & Havenhand, J.N. (2009) Fertilization strategies. In: *Marine Hard Bottom Communities: Patterns, Dynamics, Patterns, and Change* (ed. M. Wahl). Springer, Heidelberg.
50. Stephens, P.A. & Sutherland, W.J. (1999) Consequences of the Allee effect for behaviour, ecology and conservation. *Trends in Ecology & Evolution*, **14** (10), 401–405.
51. Lockwood, J.L., Cassey, P. & Blackburn, T. (2005) The role of propagule pressure in explaining species invasions. *Trends in Ecology & Evolution*, **20** (5), 223–228.
52. Yund, P.O. & Meidel, S.K. (2003) Sea urchin spawning in benthic boundary layers: are eggs fertilized before advecting away from females? *Limnology and Oceanography*, **48** (2), 795–801.
53. Svane, I. & Havenhand, J.N. (1993) Spawning and dispersal in *Ciona intestinalis* (L). *Marine Ecology – Pubblicazioni Della Stazione Zoologica Di Napoli I*, **14** (1), 53–66.
54. Levitan, D.R. (1995) The ecology of fertilization in free-spawning marine invertebrates. In: *Ecology of Marine Invertebrates* (ed. L.R. McEdward), pp. 123–156. CRC Press, Boca Raton.
55. Denny, M.W. & Shibata, M.F. (1989) Consequences of surf zone turbulence for settlement and external fertilization. *American Naturalist*, **134** (6), 859–889.
56. Togashi, T., Motomura, T., Ichimura, T. & Cox, P.A. (1999) Gametic behavior in a marine green alga, *Monostroma angicava*: an effect of phototaxis on mating efficiency. *Sexual Plant Reproduction*, **12** (3), 158–163.

57. Reed, D.C., Amsler, C.D. & Ebeling, A.W. (1992) Dispersal in kelps – factors affecting spore swimming and competence. *Ecology*, **73** (5), 1577–1585.
58. Clifton, K.E. & Clifton, L.M. (1999) The phenology of sexual reproduction by green algae (Bryopsidales) on Caribbean coral reefs. *Journal of Phycology*, **35** (1), 24–34.
59. Marshall, D.J. (2006) Reliably estimating the effect of toxicants on fertilization success in marine broadcast spawners. *Marine Pollution Bulletin*, **52** (7), 734–738.
60. Harrison, P.L. & Ward, S. (2001) Elevated levels of nitrogen and phosphorus reduce fertilisation success of gametes from scleractinian reef corals. *Marine Biology*, **139** (6), 1057–1068.
61. Serrão, E.A., Brawley, S.H., Hedman, J., Kautsky, L. & Samuelson, G. (1999) Reproductive success of *Fucus vesiculosus* (Phaeophyceae) in the Baltic Sea. *Journal of Phycology*, **35** (2), 254–269.
62. Marshall, D.J., Styan, C.A. & Keough, M.J. (2002) Sperm environment affects offspring quality in broadcast spawning marine invertebrates. *Ecology Letters*, **5** (2), 173–176.
63. Vogel, H., Czihak, G., Chang, P. & Wolf, W. (1982) Fertilization kinetics of sea urchin eggs. *Mathematical Biosciences*, **58** (2), 189–216.
64. Rothschild, L. & Swann, M.M. (1951) The fertilization reaction in the sea urchin: the probability of a successful sperm-egg collision. *Journal of Experimental Biology*, **28**, 403–416.
65. Marshall, D.J., Styan, C.A. & Keough, M.J. (2000) Intraspecific co-variation between egg and body size affects fertilisation kinetics of free-spawning marine invertebrates. *Marine Ecology – Progress Series*, **195**, 305–309.
66. Styan, C.A. (1998) Polyspermy, egg size, and the fertilization kinetics of free-spawning marine invertebrates. *American Naturalist*, **152** (2), 290–297.
67. Levitan, D.R. & Ferrell, D.L. (2006) Selection on gamete recognition proteins depends on sex, density, and genotype frequency. *Science*, **312** (5771), 267–269.
68. Palumbi, S.R. (1999) All males are not created equal: fertility differences depend on gamete recognition polymorphisms in sea urchins. *Proceedings of the National Academy of Sciences of the United States of America*, **96** (22), 12632–12637.
69. Evans, J.P. & Marshall, D.J. (2005) Male-by-female interactions influence fertilization success and mediate the benefits of polyandry in the sea urchin *Heliocidaris erythrogramma*. *Evolution*, **59** (1), 106–112.
70. Marino, R., De Santis, R., Giuliano, P. & Pinto, M.R. (1999) Follicle cell proteasome activity and acid extract from the egg vitelline coat prompt the onset of self-sterility in *Ciona intestinalis* oocytes. *Proceedings of the National Academy of Sciences of the United States of America*, **96** (17), 9633–9636.
71. Murabe, N. & Hoshi, M. (2002) Re-examination of sibling cross-sterility in the ascidian, *Ciona intestinalis*: Genetic background of the self-sterility. *Zoological Science*, **19** (5), 527–538.
72. Byrd, J. & Lambert, C.C. (2000) Mechanism of the block to hybridization and selfing between the sympatric ascidians *Ciona intestinalis* and *Ciona savignyi*. *Molecular Reproduction and Development*, **55** (1), 109–116.
73. Cohen, C.S. (1996) The effects of contrasting modes of fertilization on levels of inbreeding in the marine invertebrate genus *Corella*. *Evolution*, **50** (5), 1896–1907.
74. Grosberg, R.K. & Quinn, J.F. (1986) The genetic control and consequences of kin recognition by the larvae of a colonial marine invertebrate. *Nature*, **322** (6078), 456–459.
75. Jiang, D. & Smith, W.C. (2005) Self- and cross-fertilization in the solitary ascidian *Ciona savignyi*. *Biological Bulletin*, **209** (2), 107–112.
76. Bishop, J.D.D., Manriquez, P.H. & Hughes, R.N. (2000) Water-borne sperm trigger vitellogenic egg growth in two sessile marine invertebrates. *Proceedings of the Royal Society of London Series B – Biological Sciences*, **267** (1449), 1165–1169.
77. Yund, P.O., Johnson, S.L. & Connolly, L.E. (2005) Multiple paternity and subsequent fusion/rejection interactions in a colonial ascidian. *Integrative and Comparative Biology*, **45** (6), 1101.
78. Pechenik, J.A. (1979) Role of encapsulation in invertebrate life histories. *American Naturalist*, **114** (6), 859–870.

79. Staver, J.M. & Strathmann, R.R. (2002) Evolution of fast development of planktonic embryos to early swimming. *Biological Bulletin*, **203** (1), 58–69.
80. Pennington, J.T., Rumrill, S.S. & Chia, F.S. (1986) Stage-specific predation upon embryos and larvae of the Pacific sand dollar, *Dendraster excentricus*, by 11 species of common zooplanktonic predators. *Bulletin of Marine Science*, **39** (2), 234–240.
81. Cowden, C., Young, C.M. & Chia, F.S. (1984) Differential predation on marine invertebrate larvae by 2 benthic predators. *Marine Ecology – Progress Series*, **14** (2–3), 145–149.
82. Morgan, S.G. (1995) Life and death in the plankton: larval mortality and adaptation. In: *Ecology of Marine Invertebrate Larvae* (ed. L.R. McEdward), pp. 279–321. CRC Press, Boca Raton.
83. Chia, F.S., Bucklandnicks, J. & Young, C.M. (1984) Locomotion of marine invertebrate larvae – a review. *Canadian Journal of Zoology – Revue Canadienne De Zoologie*, **62** (7), 1205–1222.
84. Young, C.M. (1995) Behavior and locomotion during the dispersal phase of life. In: *Ecology of Marine Invertebrate Larvae* (ed. L.R. McEdward), pp. 249–277. CRC Press, Boca Raton.
85. Shanks, A.L. & Brink, L. (2005) Upwelling, downwelling, and cross-shelf transport of bivalve larvae: test of a hypothesis. *Marine Ecology-Progress Series*, **302**, 1–12.
86. Koehl, M.A.R. (2007) Mini review: hydrodynamics of larval settlement into fouling communities. *Biofouling*, **23** (5), 357–368.
87. Webb, K.E., Barnes, D.K.A., Clark, M.S. & Bowden, D.A. (2006) DNA barcoding: a molecular tool to identify Antarctic marine larvae. *Deep-Sea Research Part II – Topical Studies in Oceanography*, **53** (8–10), 1053–1060.
88. Levin, L.A. (2006) Recent progress in understanding larval dispersal: new directions and digressions. *Integrative and Comparative Biology*, **46** (3), 282–297.
89. Pechenik, J.A. (2006) Larval experience and latent effects – metamorphosis is not a new beginning. *Integrative and Comparative Biology*, **46** (3), 323–333.
90. Marshall, D.J., Pechenik, J.A. & Keough, M.J. (2003) Larval activity levels and delayed metamorphosis affect post-larval performance in the colonial ascidian *Diplosoma listerianum*. *Marine Ecology – Progress Series*, **246**, 153–162.
91. Marshall, D.J., Bolton, T.F. & Keough, M.J. (2003) Offspring size affects the post-metamorphic performance of a colonial marine invertebrate. *Ecology*, **84** (12), 3131–3137.
92. Marshall, D.J. & Keough, M.J. (2006) Complex life cycles and offspring provisioning in marine invertebrates. *Integrative and Comparative Biology*, **46** (5), 643–651.
93. Chia, F.S. (1974) Classification and adaptive significance of developmental patterns in invertebrates. *Thalassia Jugoslavica*, **10**, 267–284.

Chapter 2
Larval Supply and Dispersal

Christopher D. McQuaid and Karen Miller

This chapter addresses the fundamental importance of the dispersal and supply of propagules both in the broad context of marine ecology and to issues of biofouling specifically. We deal firstly with the intrinsic and extrinsic factors that influence scales of propagule dispersal, including large and small scale oceanographic factors and the more recent recognition of the importance of larval behaviour. In most cases, directly estimating the scales of propagule dispersal is difficult or impossible and instead estimation generally relies on indirect methods such as genetic and other tags that can be used to identify the origins of propagules. Finally we discuss the practical implications of different dispersal scales, including their relevance to problems of biofouling.

2.1 Introduction

The adult stages of biofouling species are sessile or sedentary, so that adult dispersal is generally negligible or non-existent. Consequently larvae and juveniles are the key life-history stages for recruitment to and dispersal among populations. The concepts of supply and dispersal are two critical components of the ecology of benthic marine invertebrates including biofouling species. The idea of larval supply to a population is intuitively obvious, but a diverse terminology has been used to describe the movement of propagules among populations. The problem of loose terminology has recently been addressed by Pineda *et al.* [1], who define larval transport as the horizontal translocation between two points and dispersal as the spread of propagules from their source populations to the place where they settle to a benthic life style. Obviously the concepts of supply and dispersal are closely linked, and an understanding of both the scale/direction of larval dispersal and of the numbers of larvae arriving into populations remains among the major challenges in marine ecology.

2.2 The significance of scales of larval dispersal

The widespread occurrence of feeding larval forms suggests that they are both ancient and conserved [2]. Hadfield *et al.* [3] proposed that convergent evolution across diverse taxa of the ability for rapid larval metamorphosis (mainly involving the loss of larval characteristics with little de novo gene action) occurred because it allows larvae to feed planktonically, while being able to transpose extremely rapidly to a benthic life style. This is in contrast to the gradual, hormonally regulated metamorphosis of groups such as insects and means that marine larvae are able to maximise their dispersal, while being able to respond rapidly to settlement cues. This implies enormous selective pressure for rapid metamorphosis and for enhancing the potential of dispersal, though it can be argued that dispersal is a consequence of the separation of benthic adult and planktonic larval stages, rather than a driving force [4, 5]. Whatever the selective pressures in favour of planktonic larvae, their scales of dispersal dictate the connectivity of populations and so have fundamental implications for gene flow as well as community structure and population regulation [6], for example whether population sizes are regulated by the supply of propagules, or by intraspecific competition resulting from oversupply of propagules [7, 8]. Scales of dispersal also influence the relationship between how abundant a species is and the area over which it occurs [9] and which species are likely to be problematic as fouling organisms. For many sessile marine invertebrates, including biofouling species, the rate at which larvae settle in the adult habitat has a huge influence on adult abundances. The success of biofouling species will rely on their ability to colonise, which depends on successful larval dispersal, but, because low dispersal ranges are linked to better local retention [9], local population abundances can be high for species with low dispersal.

Larval dispersal is also critical in determining both ecological and genetic connectedness as well as the ability to colonise new habitats; effectively, scales of larval dispersal dictate how we define metapopulations [10], meaning assemblages of spatially delimited sub-populations connected by some level of migration.

Levin [11] has recently provided an excellent review of the literature on larval dispersal and identified several critical points that are relevant to dispersal scales. One concerns the fact that scales of larval dispersal determine the openness of populations or the degree to which larvae will be exchanged among populations. This has profound ecological implications. How easy is it to colonise new habitats or to re-establish populations extinguished by gradual attrition or some extreme event? To what degree can populations be sustained by immigration of larvae in the event of poor reproductive output or reproductive failure? For example, can a species maintain populations beyond the geographic range within which it is capable of reproducing? Do populations experience saturation recruitment, in which case intraspecific competition may become important or are they recruit-limited? How fast are population turnover rates? Are fluctuations in population size reduced by having planktonic larvae? There are also important genetic and evolutionary implications. How strong is gene flow among populations? How much genetic variability is there within populations and does this offer a buffer against changing conditions? How does speciation occur in marine organisms and is it influenced by larval dispersal? The answers to these and other questions are linked to larval supply, which in turn is intimately linked to scales of larval dispersal.

A second point identified by Levin is that our assumption that larvae essentially act as passive particles and that their behaviour is largely irrelevant has been radically revised. This is critical, especially in the context of dispersal scales, and is discussed below. Thirdly, Levin notes the results of studies by Shanks *et al.* [12] and Siegel *et al.* [13], both of which show what one would expect a strong relationship between dispersal distance and the period for which propagules are planktonic, though both also suggest that few propagules disperse between 1 and 20 km. Palaeological studies have shown dispersal to be related to species longevity [14] and distributional range [15]. That the relationship with distributional range often breaks down in studies of contemporary populations in large geographic areas may be largely artefactual.

Paulay and Meyer [16] analysed the relationship for Indo-West Pacific cowries and found that correcting for intraspecific variation, remote endemics and poor taxonomy substantially improved the correlation. Dispersal also profoundly affects speciation and the distribution of biodiversity. Gastropod groups with good potential for dispersal are better represented on oceanic islands of the Pacific than those with poor dispersal potential, while the latter show a more rapid drop in diversity as one moves away from the centre of marine diversity in the Indo-West Pacific [16].

There can be no question that understanding scales of larval dispersal and the control of larval supply is fundamental to understanding marine populations. However, it is important to recognise that in discussing larval dispersal it is not possible to be precise and there is considerable controversy over the broad scales over which larvae can disperse. Because this is influenced by so many factors, including large-scale [17] and small-scale [18] water movement, duration of the larval phase [13] and, as is increasingly recognised, larval behaviour [11], there is no doubt that these scales differ geographically, among taxa and even at different times within a taxon and location. Generally it appears that many algae disperse over small (1–100 m) scales [19], while the larvae of animals can disperse on scales ranging from tens to hundreds of metres, such as ascidians [20], to thousands of kilometres, such as oceanic molluscs [21]. However, we need to recognise that even within a species, dispersal scales can vary tremendously through effects other than hydrodynamics. Even larvae from the same species, but different populations, can show different vertical migration behaviour, resulting in different dispersal [22], and even within a population, it is possible that behaviour may differ among dates (cf. [23]).

2.3   Factors influencing dispersal scales

The distance a marine larva or propagule might successfully disperse will be tightly linked to the biology of the organism (i.e. larval developmental mode and behaviour, see Chapter 1) as well as environmental factors, including water movement and the availability of suitable settlement substrata. Pineda [24] portrayed the mechanisms influencing larval supply and dispersal as a series of funnels, with large-scale physical transport processes operating on the available larval pool to determine the supply of larvae to a settlement site, while microhydrodynamic effects and larval behaviour influence settlement rates at very small spatial scales. Because they operate on larger numbers of larvae, large-scale, offshore effects can have a greater influence on final settlement rates. Also, because larvae must pass successfully between the various funnels, small changes in the proportions that do so can have strong effects on population size. These concepts fit well with the idea that passive processes dominate at the larger scales (tens to hundreds of kilometres), while active mechanisms become important at smaller scales [25].

As discussed in Chapter 1, marine invertebrate larvae can include a range of forms and functions and this variety of developmental modes will likely influence their dispersal potential. For example, long-lived planktonic larvae may be transported many hundreds of kilometres by ocean currents, whereas short-lived benthic larvae may crawl only a very short distance from the parent before settling. Although there are unexpected findings, such as the ability of littorinid snails with direct development being able to colonise new habitats as quickly as those with planktonic larvae [26], there does seem to be a strong and predictable relationship between larval type, time spent in the water column and dispersal distance for some species [12, 13, 27]. Importantly, though, it is the interaction between the biology and behaviour of the larva and the physical environment that is probably the most critical for dispersal, and the major factor that prevents any simple correlation between larval type and dispersal ability [28, 29].

2.3.1 The role of behaviour in dispersal

Because most larvae are ineffectual swimmers, it has been assumed that they are dispersed largely as passive particles. However, behaviour has been shown to have the potential to affect dispersal at all stages in a species' life cycle, from the timing of propagule release through to larval settlement. Many species will spawn gametes or release propagules to coincide with specific tidal movements or seasonal influences including photoperiod, water temperature and salinity, and these can strongly influence dispersal [30]. For example, the coral *Goniastrea favulus* at One Tree Reef on the southern Great Barrier Reef, Australia, spawns predictably at low tide when tidal currents are minimal. The negatively buoyant eggs and sperm are subsequently retained on natal reef flats by the incoming tide, which promotes fertilisation success [31], and acts to retain larvae on natal reefs for the 2–3-day period prior to settlement competency [32]. Certainly genetic studies of this species reveal population structures consistent with limited larval dispersal and self-seeding of reefs [33]. Similar specialised spawning behaviours likely to affect dispersal have been observed in a range of marine taxa from ascidians [34] to macroalgae [35].

The growing literature on the role of larval behaviour in dispersal has led to a perception that larval retention may be much more common than previously believed, and can result in highly localised dispersal or self-seeding of populations, even in taxa with long-lived, supposedly highly dispersive larvae [36]. A particularly useful distinction is made by Sponaugle [37] between physical retention of larvae, which depends solely on physical effects due to departures from conditions of uniform flow, and biophysical retention, which depends on the exploitation of physical effects by larvae through their behaviour. Probably one of the best examples of biophysical retention is the vertical migration of decapod larvae in estuaries. By altering their position within the water column, larvae can migrate upstream with the incoming tide if they are near the water's surface, but can retain their position in an estuary by migrating to the bottom during the outflowing tide (tidal migration). Similar behaviour has been reported in oceanic species whereby vertical migration is used to assist onshore/offshore transport of larvae associated with wind and currents and to concentrate advanced larvae in inshore areas where suitable settlement substrata are abundant (reviewed by Queiroga *et al.* [38]). In this way, by simply altering their vertical position within the water column, larvae can exert considerable control over the direction and distance of dispersal.

Larvae of species as diverse as echinoderms, ascidians, bivalves, barnacles and polychaetes have also been shown to display behaviours that influence their position in the water column in relation to physical features such as fronts, haloclines and thermoclines, as well as in response to food availability [39, 40]. Certainly larval condition will play an important role both in the ability of a larva to survive planktonic dispersal as well as its ability to settle and metamorphose at the end of its journey [41]. Additionally behavioural responses including phototaxis, geotaxis and chemotaxis have been demonstrated to exist in the larvae of many marine species including echinoderms [42], corals [43], polychaetes [44], bivalves [45] and ascidians [46]. Although the role of these behaviours in dispersal is not well understood (reviewed in [47,48]), it is clear that they may have profound consequences. Even in species that have no apparent capacity for behaviour within the water column, settlement preferences may well influence their effective dispersal. For example, in the case of the green alga *Enteromorpha*, the settlement of zoospores is positively linked to the abundance of specific strains of bacteria in the biofilm [49]. Obviously in such instances, dispersal will be ineffectual if a propagule arrives into a habitat that lacks suitable settlement cues (see Chapter 3).

The evidence for self-recruitment is perhaps strongest in the case of the relatively large and sophisticated larvae of reef fish. The degree to which reef fish populations are open or closed has been unclear [50,51], but fish larvae have sufficient swimming power to affect their dispersal to a degree similar to that of ocean currents [52] and there is a growing consensus that their behaviour promotes larval retention [53,54] and that typical larval dispersal scales are of the order of only 10–100 km [55].

2.3.2   Extrinsic factors and dispersal

Although the ocean may appear to be a large and open system, smaller scale oceanographic features such as fronts, eddies and convergences will act both to concentrate larvae and to restrict the dispersal even of long-lived planktonic forms [37]. For example, larvae of crabs, barnacles and mussels have been shown to accumulate in the lee of headlands, although such accumulation varies spatially and temporally [56,57].

Even though larvae have the potential to influence their dispersal through behaviour, the swimming speeds and abilities of most invertebrate larvae are very poor, and many species may still disperse as passive particles. In such cases, dispersal direction and distance may be at the mercy of physical and environmental factors such as wind, tide and currents. The rafting of species on either natural or man-made substrata clearly demonstrates this effect. Drift trajectories of pumice from a volcanic eruption near Fiji in 2001 closely followed known current systems, including the East Australian Current. Although the pumice took over a year to reach the eastern shores of Australia from Fiji, it brought with it a variety of fouling organisms including algae, barnacles, serpulid worms and corals, emphasising the importance of rafting for long-distance dispersal of many species [58]. Similarly transport through anthropogenic sources such as on hulls and in ballast water of ships may result in long-distance dispersal even of species that have short-lived larvae [59].

The difference between the potential and the realised dispersal distance of marine larvae may be great. Kendrick and Walker [60] stained propagules of the alga *Sargassum spinuligerum* and found that even these passive particles dispersed only a few metres from the parent thalli. Mortality will also play an important role in dispersal. Olson and McPherson [61] followed ascidian larvae in situ, and found that although they have a potential dispersal distance of several hundred metres based on swimming speed and duration, >75% of larvae were eaten and of the remaining 25%, most settled within 10 m of the parent. In addition, a recent study by Allen and McAlister [62] indicated that larval mortality will be higher in benthic than in planktonic larvae, suggesting the likelihood of effective dispersal is low, irrespective of larval type.

Clearly the range of larval types, behaviours and physical factors that might influence the dispersal of benthic marine invertebrate larvae is extensive. Importantly the interaction between all these factors will result in highly variable and unpredictable patterns of dispersal in most species.

2.4   Estimating dispersal

Dispersal is difficult to estimate because it is (usually) impossible to track larvae so, while one can tell where an organism is, one cannot tell where it came from. Apart from modelling, the most effective ways of estimating dispersal scales are based on colonisation events and genetic effects.

A wide range of genetic approaches have been used to study large-scale (e.g. phylogeographic comparisons based on DNA sequence data) and small-scale connectivity (e.g. among local populations using co-dominant markers such as microsatellite DNA or allozymes; reviewed by Hellberg *et al.* [63]). The limitation with genetic estimates is that it is often difficult to differentiate between the genetic signature of past gene flow (on evolutionary time scales) and present day connectivity (ecological time scales) and measures of gene flow such as the number of migrants per generation (N_em) rarely reflect a true estimate of dispersal. Recent advances in molecular methods and specifically data analysis approaches such as coalescent and assignment tests are now enabling a better distinction between historic and recent gene flow in studies of connectivity [64, 65], although quantifying migration among populations based solely on genetic data remains elusive. Certainly the most reliable means of estimating dispersal will involve a combination both of genetic tests of connectivity and empirical studies of larval ecology [66].

Direct observation of larval dispersal is only feasible if the larvae are large enough to see and in the water column for a short period, such as ascidian larvae [67], though there are occasional reports of larval masses so dense as to be visible, such as slicks of coral eggs and embryos [68]. However, even when larvae can be found in the water column, identification even to family or genus is often challenging as many larvae lack specific taxonomic characters. Recent developments in the molecular identification of marine larvae (e.g. using oligonucleotide probes [69,70]) have made the goal of tracking larvae across oceanic expanses more attainable, although still largely out of reach. Consequently most approaches to estimating scales of larval dispersal are necessarily indirect. McQuaid and Phillips [71] monitored the density of mussel larvae over a grid of hundreds of metres and, by rapid, repeated sampling tracked the movement of dense clouds of larvae. This was correlated with wind-driven surface currents and allowed them to estimate dispersal scales that matched the observed spread of the invasive *Mytilus galloprovincialis* from a unique, known source population. However, such opportunities are rare and most studies are even less direct. Modelling studies typically use a Lagrangian approach [13], in which the trajectories of individual particles are calculated, and tend to focus on the concept of a dispersal kernel [29]. This is a probability density function that describes the likelihood of a single larva dispersing over a given distance before it settles [72]. However, most models assume larvae will act as passive particles and few modelling studies to date have incorporated aspects of larval behaviour, which currently limits the interpretation of dispersal from modelling exercises.

Alternative approaches to understanding dispersal try to identify the origins of settlers, primarily using some form of environmental marker. These include morphological and parasite markers [73, 74] and natural tags such as trace elements in larval shells [75]. For example, new recruits of mussels from bay and open coast populations in California differ in their chemical composition [76]. An early example of this approach, used for a different purpose, was the work of Killingley and Rex [77], who analysed the stable isotope signatures of adult and larval shells of deep-sea gastropods to determine where in the water column the larvae of lecithotrophic and planktotrophic species developed. Stable isotope signatures analysed using fine-scale laser ablation techniques on statolith composition have more recently been used to track ontogenetic changes in habitat use by squid [78], but these techniques currently appear to have limited application in benthic marine invertebrates such as bivalves [79].

In some studies, larvae have been marked artificially. This has been done, for example, by immersion of a variety of invertebrates in solutions of rare earths [80], by feeding selenium to crabs and barnacles [81], by incubating fish larvae in tetracycline [82] and bivalve veligers in the fluorochrome calcein (see Plate I B) [83]. Unfortunately such approaches appear to work only over relatively small scales of time (hours to days) and space (metres to kilometres), otherwise the dilution of marked larvae makes their recovery difficult or impossible.

2.5   Practical consequences of dispersal scales

How far larvae disperse and whether they settle and recruit successfully following dispersal will have important consequences at the population and species level, including the size of neighbourhoods and the geographic range of breeding units, as well as ecological effects ranging from maintenance of genetic diversity to range extension. There are also community-level consequences in terms of the spatial distribution of species and the proportion of the regional species pool that will be represented within a species assemblage [84].

2.5.1   Localised dispersal

Mating among close relatives (inbreeding) is a likely consequence of localised dispersal whereby larvae settle close to parents or within the natal habitat; and without migration into a population from outside, all individuals will become related to each other through time. Although limited dispersal coupled with rapid settlement is likely to increase the chance of survival of larvae and will ensure that larvae settle in a habitat suitable for growth and re-production, there will be trade-offs associated with increased competition among conspecifics and potential loss of genetic variation and fitness through inbreeding. Interestingly, although inbreeding is generally thought to be maladaptive leading to reduced reproductive success and a reduction in heterozygosity (inbreeding depression), inbreeding may persist in sessile marine invertebrates simply as a consequence of the limited mobility of adults and juve-niles/larvae [85]. In addition, many sessile marine species are hermaphroditic and capable of self-fertilisation (the most extreme form of inbreeding). Self-fertilisation may well be advantageous where population densities are low (e.g. colonisation of new substrata) or sperm limiting [86]. Certainly genetic studies of sessile marine invertebrates have revealed heterozygote deficiency to be a widespread phenomenon [87], which may be associated with localised dispersal and the prevalence of inbreeding.

Probably one of the most important consequences of localised dispersal is in the maintenance or establishment of new populations. Based on the corroboration of Lagrangian dispersal models by observed rates of fish recruitment, Cowen *et al.* [88] conclude that marine populations are dependent on effects that increase the likelihood of self-recruitment rather than depending on recruits from distant populations. This of course has implications for the ability of populations to recover from catastrophe if the adult populations are eliminated or drastically reduced. For example, the failure of populations of two species of abalone, *Haliotis*, to respond to a closure of the fishery until adults were experimentally re-introduced [89]. Equally, if populations are largely self-recruiting, then the likelihood of the colonisation of new areas via larval dispersal must be low, although once a few individuals establish, self-recruitment will result in the rapid establishment of a population [26]. This will be especially relevant for biofouling species.

2.5.2   *Long-distance dispersal*

In contrast to species with localised recruitment of larvae, it is generally assumed that, for species with high dispersal, larvae will be thoroughly mixed in the plankton, levels of gene flow should be high across wide geographic scales, and consequently the population genetic structure of these species will reflect panmixis. This has been demonstrated using genetics for a wide range of species [27, 28], but may well be an oversimplification. Acorn barnacles have a long larval phase but using microsatellite DNA markers, Veliz *et al.* [90] showed kin aggregation in cohorts of acorn barnacles in the Gulf of St Lawrence, Canada, and concluded dispersal was in fact non-random even for this planktonic disperser and despite high dispersal potential. This observation is most likely related to a combination of gregarious adult populations, internal fertilisation by adjacent males, synchronous release of siblings and stable ocean currents in the region. As a consequence, the genetic structure observed in these populations does not reflect panmixis and indeed displays heterozygote deficits as might be expect in species with localised dispersal. The effects of localised selection may also affect population genetic structure in species such as acorn barnacles, as well as other marine invertebrate species even when dispersal potential is high [91, 92]. In such instances, the net result of environmental factors and selection is strong population structuring despite apparently high dispersal ability.

2.5.3   *Dispersal polymorphisms*

While the advantages of long-distance dispersal may be great, including wide geographic range, high genetic diversity resulting in an ability to adapt to a wide range of environmental conditions, potential for allopatric speciation and reduced rates of localised extinctions, the costs may be great. As highlighted earlier, mortality of long-lived larvae is likely to be high and the probability of successful dispersal very low.

Interestingly, many sessile marine species have developed reproductive strategies that enable them to take advantage of long-distance dispersal while also maintaining local levels of recruitment. Such bet-hedging strategies have been described in corals [32], the mollusc *Alderia modesta* [93], ascidians and echinoderms [94] and tubeworms [95]. Additionally, many sessile marine species have complex life cycles combining both a sexual and an asexual mode of reproduction, e.g. corals, anemones, ascidians, sponges, bryozoans and algae. Empirical theories such as the strawberry-coral or aphid-rotifer models [96] predict that for such organisms, asexual reproduction will produce larvae or juveniles that will be used to build local populations of clones rapidly, while sexual reproduction will produce genetically diverse and widely dispersed offspring. Although there are few tests of these predictions in marine species, most genetic studies have shown these generalisations to be broadly true (e.g. cnidarians [97–100], but see [101–103]; sponges [104, 105]; bryozoans [106]).

2.5.4   Supply, dispersal and biofouling

The variation in the dispersal, supply and subsequent recruitment of larvae both in space and through time is well recognised [107], and clearly there are many biological and environmental factors that can affect dispersal in marine species. Such variation has important ecological ramifications in terms of the establishment and maintenance of populations. In the context of biofouling species, the likelihood that new populations will establish is linked in part to the dispersal potential of the organism and the distance from existing populations. Additionally once a few individuals settle, localised dispersal and self-recruitment are likely to facilitate rapid expansion of a population. Consequently the biological mechanisms of control of populations of biofouling organisms (e.g. through supply of recruits, intra- or interspecific competition) will be strongly influenced by scales of dispersal. There is unlikely to be a clear or simple relationship between a biofouling organism's larval biology and the likelihood that it will colonise or establish on available habitats, but knowledge of larval dispersal/retention and the control of successful settlement is one of the critical components in understanding the dynamics of biofouling communities.

2.6   Conclusions

- Supply and dispersal of propagules are intimately interlinked and critical to both the ecology and the evolution of biofouling organisms. In most cases dispersal can only be estimated indirectly (e.g. using genetic or other markers) or through modelling.
- Recent studies indicate that, although most propagules have limited swimming abilities, their behaviour can dramatically affect dispersal.
- Consequently interactions between biology/behaviour and the physical environment prevent simple correlations between larval type/duration and scales of dispersal.
- Scales of dispersal tend to be limited for algal sporelings (metres to hundreds of metres), but range over many orders of magnitude (tens of metres to thousands of kilometres) for the larvae of animals. This depends partly on taxon, but can differ dramatically among populations of conspecifics.
- Self-recruitment may be more common than previously believed and heterozygote deficiency is widespread among marine organisms.

References

1. Pineda, J., Hare, J.A. & Sponaugle, S. (2007) Larval transport and dispersal in the coastal ocean and consequence for population connectivity. *Oceanography,* **20** (3), 22–39.
2. Strathmann, R.R. (1985) Feeding and non-feeding larval development and life-history evolution in marine invertebrates. *Annual Review of Ecology & Systematics,* **16**, 339–361.

3. Hadfield, M., Carpizo-Ituarte, E.J., Del Carmen, K. & Nedved, B.T. (2001) Metamorphic competence, a major adaptive convergence in marine invertebrate larvae. *American Zoologist*, **41**, 1123–1131.

4. Palmer, A.R. & Strathmann, R.R. (1981) Scale of dispersal in varying environments and its implications for life histories of marine invertebrates. *Oecologia*, **48**, 308–318.

5. Todd, C.D., Lambert, W.J. & Thorpe, J.P. (1998) The genetic structure of intertidal populations of two species of nudibranch molluscs with planktotrophic and pelagic lecithotrophic larval stages: are pelagic larvae 'for' dispersal? *Journal of Experimental Marine Biology and Ecology*, **228**, 1–28.

6. Eckert, G.L. (2003) Effects of the planktonic period on marine population fluctuations. *Ecology*, **84** (2), 372–383.

7. Connell, J.H. (1985) The consequences of variation in initial settlement *vs* post-settlement mortality in rocky intertidal communities. *Journal of Experimental Marine Biology and Ecology*, **93**, 11–45.

8. Connolly, S.R., Menge, B.A. & Roughgarden, J. (2001) A latitudinal gradient in recruitment of intertidal invertebrates in the northeast Pacific Ocean. *Ecology*, **82** (7), 1799–1813.

9. Foggo, A., Bilton, D.T. & Rundle, S.D. (2007) Do developmental mode and dispersal shape abundance-occupancy relationships in marine macroinvertebrates? *Journal of Animal Ecology*, **76**, 695–702.

10. Ellien, C., Thiebaut, E., Barnay, A.S., Dauvin, J.C., Gentil, F. & Salomon, J.C. (2000) The influence of variability in larval dispersal on the dynamic of a marine metapopulation in the eastern Channel. *Oceanologica Acta*, **23** (4), 423.

11. Levin, L.A. (2006) Recent progress in understanding larval dispersal: new directions and digressions. *Integrative and Comparative Biology*, **46** (3), 282–297.

12. Shanks, A.L., Grantham, B.A. & Carr, M.H. (2003) Propagule dispersal distance and the size and spacing of marine reserves. *Ecological Applications*, **13** (Suppl. 1), 159–169.

13. Siegel, D.A., Kinlan, B.P., Gaylord, B. & Gaines, S.D. (2003) Lagrangian descriptions of marine larval dispersion. *Marine Ecology Progress Series*, **260**, 83–96.

14. Jeffery, C.H. & Emlet, R.B. (2003) Macroevolutionary consequences of developmental mode in temnopleurid echinoids from the tertiary of southern Australia. *Evolution*, **57**, 1031–1048.

15. Jablonski, D. & Lutz, R.A. (1983) Larval ecology of marine invertebrates: paleobiological implications. *Biological Reviews*, **53**, 21–89.

16. Paulay, G. & Meyer, C. (2006) Dispersal and divergence across the greatest ocean region: Do larvae matter? *Integrative and Comparative Biology*, **46** (3), 269–281.

17. Roughgarden, J., Pennington, J.T., Stoner, D., Alexander, S. & Miller, K. (1991) Collisions of upwelling fronts with the intertidal zone: the cause of recruitment pulses in barnacle populations of central California. *Acta Oecologica*, **12** (1), 35–51.

18. Porri, F., McQuaid, C.D. & Radloff, S. (2006) Spatio-temporal variability of larval abundance and settlement of *Perna perna*: differential delivery of mussels. *Marine Ecology Progress Series*, **315**, 141–150.

19. Kinlan, B.P., Gaines, S.D. & Lester, S.E. (2005) Propagule dispersal and the scales of marine community process. *Diversity and Distributions*, **11** (2), 139–148.

20. Stoner, D.S. (1990) Recruitment of tropical colonial ascidians: relative importance of pre-settlement *vs.* post-settlement processes. *Ecology*, **71**, 1682–1690.

21. Scheltema, R.S. & Williams, I.P. (1983) Long-distance dispersal of planktonic larvae and the biogeography and evolution of some Polynesian and western Pacific molluscs. *Bulletin of Marine Science*, **33**, 545–565.

22. Manuel, J.L., Gallager, S.M., Pearce, C.M., Manning, D.A. & O'Dor, R.K. (1996) Veligers from different populations of sea scallop *Placopecten magellanicus* have different vertical migration patterns. *Marine Ecology Progress Series*, **142**, 147–163.

23. Osgood, K.E. & Frost, B.W. (1984) Ontogenetic diel vertical migration behaviors of the marine planktonic copepods *Calanus pacificus* and *Metridia lucens*. *Marine Ecology Progress Series*, **104**, 13–25.

24. Pineda, J. (1994) Spatial and temporal patterns in barnacle settlement rate along a southern California rocky shore. *Marine Ecology Progress Series*, **107**, 125–138.
25. Bradbury, I.R. & Snelgrove, P.V.R. (2001) Contrasting larval transport in demersal fish and benthic invertebrates: the role of behaviour and advective processes in determining spatial pattern. *Canadian Journal of Fisheries and Aquatic Science*, **58**, 811–823.
26. Johannesson, K. & Warmoes, T. (1990) Rapid colonization of Belgian breakwaters by the direct developer *Littorina saxatilis* (Olivi) (Prosobranchia, Mollusca). *Hydrobiologia*, **193**, 99–108.
27. Bohonak, A.J. (1999) Dispersal, gene flow, and population structure. *Quarterly Review of Biology*, **74** (1), 21–43.
28. Palumbi, S.R. (1995) Using genetics as an indirect estimator of larval dispersal. In: *Ecology of Marine Invertebrate Larvae* (ed. L. McEdward), pp. 369–387. CRC Press, Boca Raton, FL.
29. Kinlan, B.P. & Gaines, S.D. (2003) Propagule dispersal in marine and terrestrial environments: a community perspective. *Ecology*, **84** (8), 2007–2020.
30. Richards, S.A, Possingham, H.P. & Noye, B.J. (1995) Larval dispersion along a straight coast with tidal currents: complex distribution patterns from a simple model. *Marine Ecology Progress Series*, **122**, 59–71.
31. Miller, K.J. & Mundy, C.N. (2005) *In situ* fertilisation success in the scleractinian coral *Goniastrea favulus*. *Coral Reefs*, **24**, 313–317.
32. Miller, K. & Mundy, C. (2003) Rapid settlement in broadcast spawning corals: implications for larval dispersal. *Coral Reefs*, **22**, 99–106.
33. Miller, K.J. & Ayre, D.J. (2008) Population structure is not a simple function of reproductive mode and larval type: insights from tropical corals. *Journal of Animal Ecology*, **77**, 713–724.
34. Marshall, D.J. (2002) *In situ* measures of spawning synchrony and fertilization success in an intertidal, free-spawning invertebrate. *Marine Ecology Progress Series*, **236**, 113–119.
35. Gordon, R. & Brawley, S.H. (2004) Effects of water motion on propagule release from algae with complex life histories. *Marine Biology*, **145**, 21–29.
36. Almany, G.R., Berumen, M.L., Thorrold, S.R., Planes, S. & Jones, G.F. (2007) Local replenishment of coral reef fish populations in a marine reserve. *Science*, **316**, 742–744.
37. Sponaugle, S., Cowen, R.K., Shanks, A. *et al.* (2002) Predicting self-recruitment in marine populations: biophysical correlates and mechanisms. *Bulletin of Marine Science*, **70** (Suppl.), 341–375.
38. Queiroga, H., Blanton, J., Southward, A.J., Tyler P.A., Young C.M. & Fuiman I.A. (2005) Interactions between behaviour and physical forcing in the control of horizontal transport of decapod crustacean larvae. *Advances in Marine Biology*, **47**, 107–214.
39. Metaxas, A. (2001) Behaviour in flow: perspectives on the distribution and dispersion of meroplanktonic larvae in the water column. *Canadian Journal of Fisheries and Aquatic Science*, **58** (1), 86–98.
40. Shanks, A.L. & Brink, L. (2005) Upwelling, downwelling, and cross-shelf transport of bivalve larvae: test of a hypothesis. *Marine Ecology Progress Series*, **302**, 1–12.
41. Metaxas, A. & Young, C.M. (1998) Responses of echinoid larvae to food patches of different algal densities. *Marine Biology*, **130**, 433–445.
42. Metaxas, A. & Young, C.M. (1998) Behaviour of echinoid larvae around sharp haloclines: effects of the salinity gradient and dietary conditioning. *Marine Biology*, **131**, 443–459.
43. Mundy, C.N. & Babcock, R.C. (1998) Role of light intensity and spectral quality in coral settlement: implications for depth-dependent settlement? *Journal of Experimental Marine Biology and Ecology*, **223**, 235–255.
44. McCarthy, D.A., Forward, R.B., Jr & Young, C.M. (2002) Ontogeny of phototaxis and geotaxis during larval development of the sabellariid polychaete *Phragmatopoma lapidosa*. *Marine Ecology Progress Series*, **241**, 215–220.
45. Jonsson, P.R, Andre, C. & Lindegarth, M. (1991) Swimming behaviour of marine bivalve larvae in a flume boundary-layer flow: evidence for near-bottom confinement. *Marine Ecology Progress Series*, **79** (1–2), 67–76.

46. Manriquez, P.H. & Castilla, J.C. (2007) Roles of larval behaviour and microhabitat traits in determining spatial aggregations in the ascidian *Pyura chilensis*. *Marine Ecology Progress Series*, **332**, 155–165.

47. Young, C.M. (1995) Behaviour and locomotion during the dispersal phase of larval life. In: *Ecology of Marine Invertebrate Larvae* (ed. L. McEdward), pp. 249–278. CRC Press, Boca Raton, FL.

48. Kingsford, M.J., Leis, J.M., Shanks, A.L., Lindeman, K.C., Morgan, S.G. & Pineda, J. (2002) Sensory environments, larval abilities and local self-recruitment. *Bulletin of Marine Science*, **70** (Suppl.), 309–340.

49. Patel, P., Callow, M.E., Joint, I. & Callow, J.A. (2003) Specificity in the settlement – modifying response of bacterial biofilms towards zoospores of the marine alga *Enteromorpha*. *Environmental Microbiology*, **5** (5), 338–349.

50. Mora, C. & Sale, P.F. (2002) Are populations of coral reef fish open or closed? *Trends in Ecology and Evolution*, **17** (9), 422–428.

51. Sale, P.F. (2004) Connectivity, recruitment variation, and the structure of reef fish communities. *Integrative and Comparative Biology*, **44**, 390–399.

52. Fisher, R. (2005) Swimming speeds of larval coral reef fishes: impacts on self-recruitment and dispersal. *Marine Ecology Progress Series*, **285**, 223–232.

53. Cowen, R.K., Lwiza, K.M.M., Sponaugle, S., Paris, C.B. & Olson, D.B. (2000) Connectivity of marine populations: open or closed? *Science*, **287**, 857–859.

54. Paris, C.B. & Cowen, R.K. (2004) Direct evidence of a biophysical retention mechanism for coral reef fish larvae. *Limnology and Oceanography*, **49** (6), 1964–1979.

55. Cowen, R.K., Paris, C.B. & Srinivasan, A. (2006) Scaling of connectivity in marine populations. *Science*, **311**, 522–527.

56. Roughan, M., Mace, A.J., Largier, J.L., Morgan, S.G., Fisher, J.L. & Carter, M.L. (2005) Subsurface recirculation and larval retention in the lee of a small headland: a variation on the upwelling shadow theme. *Journal of Geophysical Research: Oceans*, **110**, C10027.

57. Mace, A.J. & Morgan, S.G. (2006) Larval accumulation in the lee of a small headland: implications for the design of marine reserves. *Marine Ecology Progress Series*, **318**, 19–29.

58. Bryan, S.E., Cook, A., Evans, J.P. *et al.* (2004) Pumice rafting and faunal dispersion during 2001–2002 in the Southwest Pacific: record of a dacitic submarine explosive eruption from Tonga. *Earth and Planetary Science Letters*, **227**, 135–154.

59. Pettengill, J.B., Wendt, D.E., Schug, M.D. & Hadfield, M.G. (2007) Biofouling likely serves as a major mode of dispersal for the polychaete tubeworm *Hydroides elegans* as inferred from microsatellite loci. *Biofouling*, **23** (3/4), 161–169.

60. Kendrick, G.A. & Walker, D.I. (1995) Dispersal of propagules of *Sargassum* spp. (Sargassaceae: Phaeophyta): observations of local patterns of dispersal and consequences for recruitment and population structure. *Journal of Experimental Marine Biology and Ecology*, **192** (2), 273–288.

61. Olson, R.R. & McPherson, R. (1987) Potential *vs.* realised dispersal: fish predation on larvae of the ascidian *Lissoclinum patella* (Gottschaldt). *Journal of Experimental Marine Biology and Ecology*, **110**, 245–256.

62. Allen, J.D. & McAlister, J.S. (2007) Testing rates of planktonic *versus* benthic predation in the field. *Journal of Experimental Marine Biology and Ecology*, **347** (1–2), 77–87.

63. Hellberg, M.E., Burton, R.S., Neigel, J.E. & Palumbi, S.R. (2002) Genetic assessment of connectivity among marine populations. *Bulletin of Marine Science*, **70** (Suppl.), 273–290.

64. Nielsen, R. & Wakeley, J. (2001) Distinguishing migration from isolation: a Markov chain Monte Carlo approach. *Genetics*, **158** (2), 885–896.

65. Waples, R.S. & Gaggiotti, O. (2006) What is a population? An empirical evaluation of some genetic methods for identifying the number of gene pools and their degree of connectivity. *Molecular Ecology*, **15**, 1419–1439.

66. Thorrold, S.R., Jones, G.P., Hellberg, M.E. *et al.* (2002) Quantifying larval retention and connectivity in marine populations with artificial and natural markers. *Bulletin of Marine Science*, **70** (Suppl. 1), 291–308.

67. Davis, A.R. & Butler, A.J. (1989) Direct observations of larval dispersal in the colonial ascidian *Podoclavella moluccensis* Sluiter: evidence for closed populations. *Journal of Experimental Marine Biology and Ecology*, **127** (2), 189–203.

68. Oliver, J.K. & Willis, B.L. (1987) Coral-spawn slicks in the Great Barrier Reef: preliminary observations. *Marine Biology*, **94** (4), 521–529.

69. Garland, E.D. & Zimmer, C.A. (2002) Techniques for the identification of bivalve larvae. *Marine Ecology Progress Series*, **225**, 299–310.

70. Pradillon, F., Schmidt, A., Peplies, J. & Dubilier, N. (2007) Species identification of marine invertebrate early stages by whole-larvae *in situ* hybridisation of 18S ribosomal RNA. *Marine Ecology Progress Series*, **333**, 103–116.

71. McQuaid, C.D. & Phillips, T.E. (2000) Limited wind-driven dispersal of intertidal mussel larvae: *in situ* evidence from the plankton and the spread of the invasive species *Mytilus galloprovincialis* in South Africa. *Marine Ecology Progress Series*, **201**, 211–220.

72. Hovestadt, T., Messner, S. & Poethke, H.J. (2001) Evolution of reduced dispersal mortality and 'fat-tailed' dispersal kernels in autocorrelated landscapes. *Proceedings of the Royal Society of London, Part B*, **268**, 385–391.

73. Levin, L.A. (1990) A review of methods for labelling and tracking marine invertebrate larvae. *Ophelia*, **32** (1–2), 115–144.

74. Gaines, S.D. & Bertness, M.D. (1992) Dispersal of juveniles and variable recruitment in sessile marine species. *Nature*, **360**, 579–580.

75. Zacherl, D.C. (2005) Spatial and temporal variation in statolith and protoconch trace elements as natural tags to track larval dispersal. *Marine Ecology Progress Series*, **290**, 145–163.

76. Becker, B.J., Fodrie, F.J., McMillan, P.A. & Levin, L.A. (2005) Spatial and temporal variation in trace elemental fingerprints of mytilid mussel shells: a precursor to larval tracking. *Limnology and Oceanography*, **50**, 48–61.

77. Killingley, J.S. & Rex, M.A. (1985) Mode of larval development in some deep-sea gastropods indicated by oxygen-18 values of their carbonate shells. *Deep-Sea Research*, **32** (8), 809–818.

78. Zumholz, K., Klugel, A., Hansteen, T. & Piatkowski, U. (2007) Statolith microchemistry traces the environmental history of the boreoatlantic armhook squid *Gonatus fabricii*. *Marine Ecology Progress Series*, **333**, 195–204.

79. Strasser C.A, Thorrold S.R., Starczak V.R. & Mullineaux L.S. (2007) Laser ablation ICP-MS analysis of larval shell in softshell clams (*Mya arenaria*) poses challenges for natural tag studies. *Limnology and Oceanography Methods*, **5**, 241–249.

80. Levin, L.A., Huggett, D., Myers, P., Bridges, T. & Weaver, J. (1993) Rare-earth tagging methods for the study of larval dispersal by marine invertebrates. *Limnology and Oceanography*, **38** (2), 346–360.

81. Anastasia, J.R., Morgan, S.G. & Fisher, N.S. (1998) Tagging crustacean larvae: assimilation and retention of trace elements. *Limnology and Oceanography*, **43** (2), 362–368.

82. Jones, G.P., Planeds, S. & Thorrold, S.R. (2005) Coral reef fish larvae settle close to home. *Current Biology*, **15**, 1314–1318.

83. Moran, A.L. & Marko, P.B. (2005) A simple technique for physical marking of larvae of marine bivalves. *Journal of Shellfish Research*, **24** (2), 567–571.

84. Johnson, M.P., Allcock, A.L., Pye, S.E,, Chambers, S.J. & Fitton, D.M. (2001) The effects of dispersal mode on the spatial distribution patterns of intertidal molluscs. *Journal of Animal Ecology*, **70**, 641–649.

85. Knowlton, N. & Jackson, J.B.C. (1993) Inbreeding and outbreeding in marine invertebrates. In: *The Natural History of Inbreeding and Outbreeding* (ed. N. Thornhill), pp. 200–249. University of Chicago Press, Chicago.

86. Yund, P.O. (2000) How severe is sperm limitation in natural populations of marine free-spawners? *Trends in Ecology and Evolution*, **15** (1), 10–13.

87. Addison, J.A. & Hart, M.W. (2005) Spawning, copulation and inbreeding coefficients in marine invertebrates. *Biology Letters*, **1**, 450–453.

88. Cowen, R.K., Paris, C.B., Olson, D.B. & Fortuna, J.L. (2002) The role of long distance dispersal *versus* local retention in replenishing marine populations. *Gulf and Caribbean Research Supplement*, **14**, 129–137.

89. Tegner, M.J. (1993) Southern California abalones: can stocks be rebuilt using marine harvest refugia? *Canadian Journal of Fisheries and Aquatic Sciences*, **9**, 2010–2018.

90. Veliz, D., Duchesne, P., Bourget, E. & Bernatchez, L. (2006) Genetic evidence for kin aggregation in the intertidal acorn barnacle (*Semibalanus balanoides*). *Molecular Ecology*, **15** (13), 4193–4202.

91. Ayre, D.J. (1985) Localized adaptation of clones of the sea anemone *Actinia tenebrosa*. *Evolution*, **39**, 1250–1260.

92. Bertness, M.D. & Gaines, S.D. (1993) Larval dispersal and local adaptation in acorn barnacles. *Evolution*, **47** (1), 316–320.

93. Krug, P.J. (2001) Bet-hedging dispersal strategy of a specialist marine herbivore: a settlement dimorphism among sibling larvae of *Alderia modesta*. *Marine Ecology Progress Series*, **213**, 177–192.

94. Marshall, D.J. & Bolton, T.F. (2007) Effects of size on the development time of non-feeding larvae. *Biological Bulletin*, **212** (1), 6–11.

95. Toonen, R.J. & Pawlik, J.R. (2001) Foundations of gregariousness: a dispersal polymorphism among the planktonic larvae of a marine invertebrate. *Evolution*, **55** (12), 2439–2454.

96. Williams, G.C. (1975) *Sex and Evolution*. Princeton University, Princeton.

97. Stoddart, J.A. (1984) Genetical structure within populations of the coral *Pocillopora damicornis*. *Marine Biology*, **81**, 19–30.

98. Ayre, D.J. (1995) Localised adaptation of sea anemone clones: evidence from transplantation over two spatial scales. *Journal of Animal Ecology*, **64**, 186–196.

99. Coffroth, M.A. & Lasker, H.R. (1998) Population structure of a clonal gorgonian coral: the interplay between clonal reproduction and disturbance. *Evolution*, **52** (2), 379–393.

100. Chen, C.A., Wei, N.V. & Dai, C.F. (2002) Genotyping the clonal population structure of a gorgonian coral, *Junceela fragilis* (Anthozoa: Octocorralia: Ellisellidae) from Lanyu, Taiwan, using simple sequence repeats in ribosomal intergenic spacer. *Zoological Studies*, **41** (3), 295–302.

101. Ayre, D.J. & Miller, K.J. (2004) Where do clonal coral larvae go? Adult genotypic diversity conflicts with reproductive effort in the brooding coral *Pocillopora damicornis*. *Marine Ecology Progress Series*, **277**, 95–105.

102. Miller, K.J. & Ayre, D.J. (2004) The role of sexual and asexual reproduction in structuring high latitude populations of the reef coral *Pocillopora damicornis*. *Heredity*, **92**, 557–568.

103. Sherman, C.D.H., Ayre, D.J. & Miller, K.J. (2006) Asexual reproduction does not produce clonal populations of the brooding coral *Pocillopora damicornis* on the Great Barrier Reef, Australia. *Coral Reefs*, **25**, 7–18.

104. Miller, K., Alvarez, B., Battershill, C.N., Northcote, P. & Parthasarathy, H. (2001) Genetic, morphological, and chemical divergence in the sponge genus *Latrunculia* (Porifera: Demospongiae) from New Zealand. *Marine Biology*, **139**, 235–250.

105. Calderon, I., Ortega, N., Duran, S., Becerro, M., Pascual, M. & Turon, X. (2007) Finding the relevant scale: clonality and genetic structure in a marine invertebrate (*Crambe crambe*, Porifera). *Molecular Ecology*, **16** (9), 1799–1810.

106. Porter, J.S., Ryland, J.S. & Carvalho, G.R. (2002) Micro- and macrogeographic genetic structure in bryozoans with different larval strategies. *Journal of Experimental Marine Biology and Ecology*, **272** (2), 119–130.

107. Underwood, A.J. & Fairweather, P.G. (1989) Supply-side ecology and benthic marine assemblages. *Trends in Ecology and Evolution*, **4** (1), 16–20.

Chapter 3
Settlement and Behaviour of Marine Fouling Organisms

Gabrielle S. Prendergast

This chapter defines the term settlement, as it applies to marine biofouling, and outlines the stages and levels of complexity involved in the process. It is beyond the scope of this work to present an exhaustive review of the literature, instead the work aims to outline some broad patterns and provide examples from the wealth of research that exists. The processes of surface encounter and attachment are given particular attention, as well as the role of environmental cues and passive and behavioural responses to them, whilst highlighting the pitfalls associated with inferences made without direct observation. Finally, there is a substantial resource table, summarising the species studied, the cue or influence on settlement, and a brief summary of the findings.

3.1   Introduction

Many fouling species naturally occur in distinct zones (Plate I C), and three possible hypotheses were originally proposed to explain this phenomenon, namely (i) larvae settle randomly and those that do not encounter a suitable site die, (ii) species with a motile adult stage settle randomly then migrate to or are confined to an area based on species-specific biotic or abiotic factors and (iii) larvae may congregate at a particular depth in the water column when they are ready to settle. Since then it has been realised that larvae of many species are capable of sophisticated site-selection behaviours during settlement, as adaptations that maximise the likelihood that a site with an appropriate combination of biotic and abiotic factors (niche) will be arrived at.

Species settlement is not the same phenomenon as recruitment. For the purposes of this chapter, settlement is defined as the process that follows the planktonic phase and precedes metamorphosis, and at the most basic level it involves surface encounter and attachment. Once settlement and metamorphosis has occurred, marine macrofouling organisms are immobile species that grow attached to artificial structures. The most notable of these belong to both algal and invertebrate groups: Algae, Mollusca (e.g. mussels), Crustacea (e.g. barnacles), Bryozoa, Annelida (e.g. tubeworms), Tunicata (colonial and solitary forms), Cnidaria (corals, anemones and hydroids) and Porifera (sponges). As settlement marks the transition from the planktonic to the sessile phase, it has logically become a focus of intense research. From a biofouling perspective, it is of major importance to understand the processes that influence settlement, as

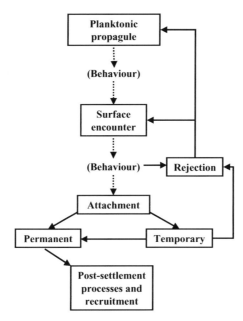

Figure 3.1   The process of settlement can be broken down into a flow chart. It involves the transition of the propagule from the planktonic phase to surface encounter and eventual attachment. Behaviour may influence surface encounter, and may influence attachment. Some, such as inert particles, can bypass the behavioural steps, hence the dotted lines. Behaviours at each point occur in response to the presence or absence of cues, i.e. they receive sensory information that can be physical or chemical, and respond in a complex manner.

this is the very point when fouling of structures and surfaces effectively begins, although it is simply a transitional moment in the life cycle of most species. Prevention may be cheaper and easier than cure in the case of antifouling, and the first step is to find out and understand as much as is possible about why and how fouling species settle where they do. The process of settlement can be broken down into the discrete phases shown in Figure 3.1.

3.2   Cues

There are three fundamental requirements for all life to continue. In the marine environment, there are physical, chemical and biological cues that combine to provide information on these priorities:

(1)   The first priority is, is this a good place to survive? That is, are there any predators, is it too hot or cold, is the location likely to dry out or is the current so strong it will dislodge the settler?

(2)   Assuming condition 1 is met, then is this a good place to grow? That is, is there a good food supply, is the current strong enough to provide enough food (filter feeders), is there enough light for photosynthesis, is there enough space to grow or are there too many others that will compete for space and food?

(3)   Assuming conditions 1 and 2 are met, then is this a good place to reproduce? That is, does this spot provide opportunities for fertilisation?

Of course, what makes up an ideal response to these questions is different for every species, and defines its unique niche. Some of the cues used by fouling species are presented in Table 3.1.

3.3   Planktonic propagules

Most propagules of fouling organisms are planktonic dispersers, a stage that is addressed in Chapters 1 and 2. It was previously thought that all propagules were inert and that their adult distributions were governed entirely by larval supply followed by post-settlement mortality. However, most settlement studies show non-random distributions, and even algal zoospores can show complex responses to cues. From the planktonic state, the first priority is to encounter a substratum on which to attach, otherwise the propagule will eventually perish. There are three ways in which a surface may be encountered:

(1) Basic model – only hydrodynamics and surface properties entrain inert planktonic propagules, presumably mechanical contact triggers settlement reaction.
(2) More complex – propagule involuntarily responds to environmental stimuli, e.g. chemical triggers stimulate the settlement response.
(3) Most complex – propagule actively seeks out and chooses the best settlement site using behavioural responses to environmental cues.

Propagules using the first strategy are likely to settle randomly wherever they find themselves, and then post-settlement mortality removes those that settled on an inappropriate location. This strategy is likely to require minimum investment in each individual disperser, yet huge numbers increase the likelihood of success of at least one. With increasing propagule complexity, more sophisticated abilities to respond to appropriate settlement sites increase their likelihood of survival, by non-randomly choosing to settle in an appropriate location.

3.3.1   Basic

The points on the flow chart (Figure 3.1) marked 'Behaviour' may not exist, depending on the species, and surface encounter may be entirely mediated by properties of the propagule and the external environment. For example:

- Does the species need to settle at a certain shore height to survive?
- Is the propagule buoyant at an appropriate depth?
- Does the local flow deliver the propagules to the surface?
- Does the surface roughness trap particles?
- Does surface contact trigger adhesive release?
- Are there any removal forces, like high shear or predators?
- Once settled, are the requirements for survival met?

Ticks in appropriate boxes in each of these cases can lead to patterns of species settlement that appear to reflect behavioural choices, but are actually entirely selected for by the environment, aided by natural selection along the way, providing traits to maximise the propagule's chances (such as appropriate buoyancy). Yet for many species settlement patterns rarely match predictions of models based on passive propagule behaviour. Settlement rates should increase

Table 3.1 Some annotated examples of the breadth of physical and biological cues and other factors that can affect settlement, with particular emphasis on the last 10 years (1998–2007).

Factor and *species*	Notes
Behaviour	
Agaricia humilis [1]	Larval swimming behaviour leads to concentration of settling larvae at the undersides of surfaces and in response to a chemical inducer from crustose red algae.
Balanus amphitrite [2]	Marked differences in settlement according to cyprid behaviour classes (swimmers, tumblers, sliders, stickers, intermediates).
B. amphitrite and *Bugula neritina* [3]	Exploratory behaviour of *B. amphitrite* was affected by flow and fouling, yet settlement was not. *B. neritina* settled only on 2-week fouled surfaces, and crawled upstream in all conditions unless they encountered a filamentous structure.
Crassostrea gigas [4]	'Dive bombing' exhibited by 4% of larvae, sinking 400 times faster than dead or live larvae, so capable of rapid acceleration and control of approach to the bottom under a wide range of conditions.
Chthamalus montagui and *Chthamalus stellatus* [5]	Supply of early-stage larvae correlated with adult abundance in both species, but not that of competent larvae. Late-stage, competent larvae settlement matched adult distributions, so choices at settlement initiated adult distributions.
Elminius covertus [6]	Transplantation of substrata in mangroves to seawards and landwards, and manipulation of supply revealed larval behaviour governed settlement distribution. Settlement consistently higher in seawards-sourced substrata, regardless of target location.
Styela plicata [7]	Forcing larvae to swim, or delay attachment, used up about 22% of their energy reserves after 48 hours, which was yet not enough to significantly affect post-larval growth.
Biofilm	
B. amphitrite [8]	Cyprids preferentially settled on clean, unfilmed substrata, and those that were filmed received less settlement relative to increased microbial abundance in the film.
Bugula flabellata and *Ciona intestinalis* [9]	Settlement of *Bugula flabellata* inhibited by biofilms of all ages; settlement of *C. intestinalis* positively correlated with biofilm age. Some *C. intestinalis* were attracted to the biofilm and others were trapped by it, but all could proceed to metamorphosis.
Hydroides elegans [10]	Monospecific bacterial biofilms had variable settlement inductive capacity. Inductive properties removed by heat, UV or chemically killing bacteria. Range of inductive bacteria suggests either multiple inductive cues or cue is common to many bacterial species.
B. amphitrite [11]	Larval preference for different substrata disappeared with biofilm age.
B. amphitrite [12]	Three bacterial monospecies films all inhibited settlement, even when formaldehyde or UV-killed, regardless of bacterial cell size. Effect not reversed with addition of settlement promoter, which was the case in natural, multispecies biofilms.
B. amphitrite, Balanus trigonus and *H. elegans* [13]	*B. amphitrite* and *B. trigonus* settlement induced by biofilms grown at high temperatures, but not affected (*B. amphitrite*) or inhibited (*B. trigonus*) by biofilms grown at low temperatures. Salinity during biofilm growth slightly affected *H. elegans* settlement.

(Continued)

Table 3.1   (*Continued*)

Factor and *species*	Notes
B. neritina [14]	Settlement inhibited by bacteria *Pseudoalteromonas* and diatom *Nitzschia frustulum* biofilms, induced by diatom *Achnanthes* sp., *Amphora cofeaeformis*, *A. tenerrima*, *N. constricta* and a 5-day-old natural biofilm, unaffected by alpha-proteobacterium and *Vibrio* sp. biofilms.
C. gigas [15]	Both bacterial inducer and L-DOPA-induced settlement behaviour – larvae extend the foot while swimming, and then crawl on the surface. Some proceeded to settlement and metamorphosis – more did this in older larvae.
H. elegans [16]	The AHLs (acylated homoserine lactone derivatives) used in bacterial quorum sensing induced some initial larval settlement behaviours like reduced swimming speed and crawling on the bottom, but did not effectively induce settlement.
H. elegans [17]	UV killed biofilm's inductive abilities were significantly reduced in both multi- and monospecies films, therefore their inductive abilities for this species require them to be live.
H. elegans [18]	Bacterial exopolymers induced first stage of larval surface exploration. When exopolymers related to UV irradiated, metabolically inactive biofilms complete settlement induced, indicating biofilm conveys information about substratum suitability.
H. elegans [19]	Inductive properties of diatom biofilms were stable even after they were heat killed. Coverage of as little as 1.8% of attractive diatoms was still enough to induce significantly more settlement than filtered sea water control.
H. elegans [20]	The biofilm cue that triggers settlement and metamorphosis of this species is likely to be an insoluble, surface-bound material that is found in many bacterial taxa.
H. elegans [21]	Inductive properties of bacterial metabolites are surface associated, not waterborne.
H. elegans [22]	Inductive bacterial strains, both in a biofilm and in suspension, induced larval settlement. When bacteria killed, the induction was lost. So cue present in bacterial extracellular matrix and a component of metabolic pathway, thus biofilm viability is necessary to promote settlement.
Mytilus galloprovincialis [23]	Larvae stored at different temperatures (ambient and refrigerated), both had settlement and metamorphosis facilitated by biofilm and epinephrine, and inhibited by phentolamine.
M. galloprovincialis [24]	Living biofilms induced more settlement with age, which interacted with immersion month – i.e. films of same age from different season induced differential settlement. Dead biofilms lost their abilities and conditioned water had no effect either.
Multispecies assemblages [25]	Biofilm density was more important than its origin in determining settlement.
Multispecies [25]	Different age and source biofilms, reciprocally transplanted, had variable effects on settlement. Two of the three sites had no effect of source on settlement, at third site local film most heavily settled. This was also the densest film, so heavily filmed surfaces more attractive.

Table 3.1   (*Continued*)

Factor and *species*	Notes
Pomatoceros lamarkii [26]	Larvae did not settle on unfilmed plates, or on air-dried biofilms for up to 5 days after a single drying event. Drying also removed inductive properties of conspecific adults. Freeze drying had same effect, but not produced with biofilm exposure to hyper-saline water.
Semibalanus balanoides [27]	In the laboratory, cyprids preferentially settled on the oldest biofilms, and preferred films grown in the mid-intertidal.
Spirorbis spirorbis, S. tridentatus and *Flustrellidra hispida* [28]	When biofilms on slate and pieces of *Fucus serratus* were air dried for 1 hour at 20°C, to mimic a tidal emersion, their usual settlement inducing properties were lost.
Ulva intestinalis [29]	Ulva zoospores drastically reduced their swimming speed in the presence of acyl-homoserine lactone (AHL, a quorum sensing system used by favourable bacterial biofilms), concentrating them around biofilms with high bacterial densities.
Conspecifics *B. amphitrite* and *S. balanoides* [30]	Larvae of both species were offered settlement-inducing protein complex (SIPC) coated membrane vs. uncoated. All preferred the SIPC region. In addition, both preferred conspecific to allospecific SIPC.
B. amphitrite [2]	Settlement factor only promoted settlement in younger cyprids (0–9 days old).
B. amphitrite [31]	In the presence of only five cyprids, effects of settlement promoter were magnified ten times and effects of settlement inhibitor were lost.
B. amphitrite [32]	Larval attachment and metamorphosis unaffected by cGMP-dependent PDE inhibitors and inhibited by calcium–calmodulin-dependent PDE inhibitors. Calmodium inhibitors also inhibited settlement behaviour when involved with MLCK or CaM-II. Larval attachment and metamorphosis promoted by cAMP-dependent PDE inhibitors.
B. amphitrite [33]	Suggest that SIPC is involved in both adult–larva and larva–larva interactions, as it is also found to be present in the cypris 'footprint' antennular secretions.
B. amphitrite [34]	Adult SIPC-induced settlement can be inhibited by lentil lectin (LCA). Pre-treatment of cyprids with LCA did not affect settlement, but pre-treatment of the SIPC inhibited settlement.
B. amphitrite and *Balanus improvisus* [35]	For both species, there were significant gregarious effects on settlement with as few as three or more cyprids per well. For *B. amphitrite*, this sensitivity to the gregarious cue was not affected by cyprid age.
Balanus (= Semibalanus) balanoides [36]	Force of adhesion increases in the presence of settlement factor.
B. improvisus [37]	Larvae intensified their exploration on surfaces treated with crude conspecific settlement factor, but their propensity to settle or to leave the surface was unaffected.
Chthamalus fragilis [38]	Cyprids will settle in unfavourable environments if conspecifics already present.

(*Continued*)

Table 3.1   (*Continued*)

Factor and *species*	Notes
Chamaesipho tasmanica [39]	The presence of incumbent new settlers and recent recruits did not affect settlement, but the presence of older recruits did promote settlement.
Elminius modestus and *E. covertus* [40]	Found that settlement was driven by the interaction of conspecific chemical cues and flow (as altered by the structure of conspecifics).
Multispecies assemblages [41]	Early settlers had little impact on subsequent settlement processes, compared to biofilms.
S. balanoides [42]	Larvae exhibited reduced specificity for attractive surfaces (settlement factor) halfway through one season, but not another.
S. balanoides [43]	At low densities (<8 per cm^2), cyprids aggregated.
S. balanoides [44]	On a small scale, the presence of conspecifics influenced settlement more than substratum heterogeneity.
S. balanoides [45]	Cyprids altered their exploratory behaviour around chemically treated pits, and settled more than in untreated pits.
S. balanoides [46]	Settlement positively related to number of adults in a cluster on cleared patch of rock, except at highest density, where settler priorities switched from the need to find a mate at low densities, to the need to avoid intraspecific competition for space at high densities.
S. balanoides [47]	There was less settlement on patches with conspecific adults than on those that had been cleared.
S. balanoides [27]	The presence of conspecifics had no effect in the laboratory but promoted settlement more than biofilms in the field.
S. balanoides [45]	Cyprids settled rapidly in both field (around 40% in 10 minutes) and laboratory (mean 24.9 seconds after entering 40×40 mm field of view) when presented with pits pre-treated with conspecific settlement factor. Exploratory tracks not related to settlement likelihood, and cyprids in the vicinity of treated pits moved around four times slower.
Day/night	
C. montagui and *C. stellatus* [48]	More chthamalid barnacles settled during daytime than night-time immersions; however, this effect was spatially variable (significant in Portugal but not Ireland).
S. balanoides [49]	Larvae significantly altered their surface exploratory behaviour between day and night, in terms of gross and net distance travelled, speed and heading angle.
Depth	
Balanus glandula and *Balanus crenatus* [50]	Two species had non-overlapping vertical distributions of larvae in the water column, which matched both the adult and recruit distributions.
B. amphitrite [8]	Tidal height significantly affected larval settlement abundance during the first experiment with abundant larval supply, but not during subsequent experiments with low supply.
C. tasmanica [51]	More cyprids were consistently caught in traps low on the shore, suggesting zonation of larvae in the water column.

Table 3.1 (*Continued*)

Factor and *species*	Notes
E. covertus [52]	At large spatial scale, more larvae settled on seawards side of mangrove forest, matching adult distributions, yet more larvae settled on bottom 5 cm of mangrove pneumatophores, nearest sediment, unlike adults that were nearest the top 5 cm.
S. balanoides [47]	Despite 60% of planktonic cyprids in the surface 4 m, 90% of settlement was in the lower intertidal. Reciprocal transplants of substrata showed settlers behaviourally selected lower intertidal. Settlers had more TAG:CHOL with decreasing shore height, less in plankton with time but not with depth.
Flow velocity	
B. amphitrite [53]	Settlement strongly affected by flow rate and interaction with surface type.
B. improvisus [54]	Larvae would not settle in flows >5–10 cm/s, larval choice adaptively connected to fact that juveniles cannot feed well at these speeds, as cirral fan deforms and cannot retain food or feed for long. So flow environment was sub-optimal for a later life history stage.
B. improvisus [55]	Larvae were more likely to reject settlement sites in a flume with increasing flow velocity. Increasing flow velocity decreased the contact rate of cyprids with a surface, and the contact rate was highly correlated with the settlement rate.
B. neritina, H. elegans and *Balanus sp.* [56]	Species-specific settlement response to velocity gradients in pipe flows. *Balanus* optimum velocity gradient was 50–120 per second, with 8–20 per second for *H. elegans* and 8–25 per second for *B. neritina*. These values thought to be related to larval morphology, behaviour and swimming ability.
Perna canaliculus [57]	More flow led to more settlement and less mortality, and vice versa. Counts at intervals suggested that settlers were releasing and re-settling, implying settlement behaviour, in the low and medium flow environments. Higher O_2 concentrations also enhanced settlement.
Food availability	
Multispecies assemblages [58]	Settlement on protruding bodies is hydrodynamically limited but advantageous due to increased food availability for recruits.
Genetic/maternal influence	
B. amphitrite [59]	Larval attachment strength on different artificial settlement inducers and a control treatment were strongly related to family, i.e. a genetic or maternal influence. This effect was lost with cyprid age on the control treatment.
Hydrodynamics	
Balanus spp., *Mytilus trossulus, Pseudochitinopoma occidentalis, Tubulipora sp.* and *Eupolymnia heterobranchia* [60]	Patterns of settlement on pipe walls consistent with expected larval delivery to walls according to varied speed, turbulence intensity and wall shear stress. However, post-contact, there were effects of pipe wall orientation on settlement that were not related to larval delivery.
Barnacle cyprids and particulates [61]	Settlement onto cylinders whose diameter showed no effect, patterns of barnacle fouling in flow reflected the delivery of particles.

(Continued)

Table 3.1 (*Continued*)

Factor and *species*	Notes
Multispecies assemblages [58]	Settlement on protruding bodies is hydrodynamically limited, favouring species with adhesive threads.
Phragmatopoma lapidosa californica [62]	Interactive effects of flow speed and behaviour led to more individuals present in intermediate flow speeds.
S. balanoides and *B. crenatus* [63]	Correlations between settlement and wind direction/wave turbulence.
S. balanoides [64]	Near shore hydrodynamics deliver planktonic larvae and thereby determine local settlement patterns.
Hydrostatic pressure *B. amphitrite* [65]	Most cyprids settled (60%) at pressures below 10 000 kPa, all died at 40 000 kPa.
B. glandula and *B. crenatus* [50]	Thought that *B. glandula* may respond to pressure minima to retain a surface position in the water column.
Interspecific chemical cue 2 species of *Acropora* [66]	Of five crustose coralline algae (CCA) species presented, some were favoured and had enhanced coral survival, and vice versa. Larvae most sensitive to, settled more on, and survived more on their favourite CCA.
B. amphitrite [67]	Extract of barnacle species, *Megabalanus rosa* and *Balanus eburneus*, was able to induce settlement of *B. amphitrite* larvae, although not as potently as the conspecific extract.
Barnacles [68]	Pedal mucus from the limpet *Patella vulgata* increased barnacle settlement in both the laboratory and field, by enmeshing them and consequently increasing the amount of time spent exploring the surface.
Coronula diadema [69]	Cyprids would not settle in sea water unless whale skin was present in the Petri dish.
Herdmania curvata [70]	In field, non-geniculate coralline algae (NCA) induced settlement of many species, but in laboratory, culture and exposure to three different NCAs inhibited settlement and metamorphosis, in irreversible manner not found in larvae that were not cultured or pre-exposed to NCA.
H. elegans [71]	Larvae preferentially settled on branches of *B. neritina* in field and laboratory. *B. neritina* leachate induced strong settlement and metamorphosis response in *H. elegans*, thought to be predation avoidance mechanism, as *B. neritina*'s metabolites deter local fishes.
M. galloprovincialis [72]	Mussel larvae settled on fibrous macroalgae, whose inductive abilities were decreased or lost with formalin, ethanol or heat, but not lost with antibiotic treatment. No induction with conditioned water, therefore contact dependent chemical settlement cue.
Multiple coral planulae larvae [73]	Corals preferentially settled in grooves and on panels with grazers that removed the filamentous algae that inhibited settlement.
Mytilus edulis [74]	Macroalgae *Cladophora rupestris* attracted larvae compared to *Laminaria saccharina* and *Fucus vesiculosis*. Exudates and biofilms isolated and had similar effects in terms of settlement attraction or repellence, except in *Γ. vesiculosis*, which had no effect.
P. canaliculus [75]	The presence of an algal chemical cue significantly increased the settlement of the mussel larvae onto artificial substrata.

Table 3.1 (*Continued*)

Factor and *species*	Notes
Larval size	
Watersipora subtorquata, *B. neritina* and *Diplosoma listerianum* [76]	Larger larvae delayed settlement longer in absence of positive cues. *B. neritina* settlement accelerated in presence of positive cues regardless of larval size. In field, larger *W. subtorquata* more discriminating and delayed longer. Implications for larval dispersal.
C. gigas [4]	Larvae within interaction zone (one to two body lengths from seafloor, in flume boundary layer flow) better able to control their settlement location and cue response. Larger size classes more likely to enter this zone with time.
C. intestinalis [77]	Larger settlers survived more (up to 1 week) than smaller settlers. Increased density of settlement decreased the survival rate, most markedly in the smaller size class of settlers.
Larval supply	
B. amphitrite [8]	Larval supply was not related to settlement, with fourfold settlement variability in areas of similar supply, low settlement in areas of high supply.
C. tasmanica [51]	Variations in larval supply explained distribution of juveniles on the substratum.
S. balanoides [63]	Clear relationship between supply and settlement.
S. balanoides [64]	Strong correlation between larval supply and settler density.
S. balanoides [78]	Larval supply explained 65 and 96% of the variation in cyprid settlement at two sites.
S. balanoides [79]	Settlement was consistently lower beneath fucoid canopies, whose presence limited supply of larvae to the substratum.
Light	
B. amphitrite [65]	Chose white over black surfaces in the light, no difference in the dark.
B. glandula [50]	Thought that vertical zonation in the water column (cyprids always found at the surface) may be a phototactic response.
C. fragilis [38]	Cyprids settle so that the adult eyes point downwards to the shaded side.
Dreissena polymorpha [80]	Post-veliger mussels settled more on shaded than on sunlit surfaces.
Hincksia irregularis [81]	Spores exhibited positive and negative phototaxis according to the region they had been isolated from (North Carolina positive, Florida negative).
Multispecies assemblages [82]	More barnacles settled on horizontal undersides.
Porites asteroides [83]	In the laboratory, they were observed in choice chamber to avoid swimming in UV exposed regions, in the field they were observed to settle in avoidance of damaging UVR regions of choice chambers.
S. balanoides [84]	On vertical tiles there were significantly more settlers on the top than the bottom.
Various scleractinian coral species [85]	In the laboratory, light quality or quantity affected settlement of five out of six species' planulae larvae, corresponding to the vertical depth distributions of these species in the field.

(Continued)

Table 3.1 (*Continued*)

Factor and *species*	Notes
Physiological condition	
B. amphitrite [2]	Age-dependent settlement on some surfaces, not on others. Settlement factor had no effect in older cyprids.
B. amphitrite [86]	Interaction between cyprid age and lipid content in propensity to settle. Strong positive relationship between lipid content and attachment in young cyprids.
B. (= *Semibalanus*) *balanoides* [87]	The series of releasers involved in instinctive behaviour may all show time dependence.
B. amphitrite [88]	Cyprid age altered their response to biofilms, such that young larvae were inhibited and older larvae were induced to settle, irrespective of biofilm age.
B. amphitrite and *B. improvisus* [35]	*B. improvisus* settlement decreased with larval age, and increased with experiment duration. In *B. amphitrite*, increased larval age and duration of experiment interacted with the presence of a conspecific cue to increase cyprid settlement.
Barnacles [89]	Only those larvae that settled within a recruitment window, excluding first and last three cohorts and regardless of size, survived to reproduction, all others contribute null to the population survival.
D. listerianum [90]	Swimming is more energetically costly than just extending the larval period. Even short periods can significantly deplete reserves that would be used in post-metamorphic growth.
Hydroides dianthus [91]	For 2 days after competency reached, larvae settled equally in response to biofilms and conspecifics. Then, contrary to desperate larvae hypothesis, larvae more specifically attracted to presence of conspecifics (although larval period had been artificially extended).
H. dianthus [92]	Tested desperate larvae hypothesis for feeding (planktotrophic) larvae. Substratum specificity unaffected, but time to achieve competence extended in those exposed to lower food concentrations. Subsequent larval size differences did not influence substratum specificity.
W. subtorquata and *B. neritina* [93]	Settlement inhibition strong in young larvae of both species, but much reduced in older larvae of both species, not to the same degree between species though. Age in field likely to be variable too.
Predators	
H. elegans [94]	Presence of the predator *Tisbe japonica* prevented a significant proportion of *H. elegans* from settling on otherwise suitable substrata. Comparable results were found in the field. Significant differences in settlement and in mortality rate.
Infaunal bivalve [95]	Adult density increases drastically decreased the survival of settling larvae, and increased flow velocity caused only a slightly higher predation risk.
Semibalanus cariosus [96]	In this species, established adult conspecifics act as predators on the substratum and reduce settlement by between 65 and 100%, except during exceptionally high settlement years when their feeding apparatus are inundated.
Resource competitors	
B. amphitrite [97]	Barnacle settlement was inversely correlated with settlement of bryozoan *B. neritina*.

Table 3.1 (*Continued*)

Factor and *species*	Notes
C. intestinalis [77]	Larger larvae survived more. Increased density decreased survival, but this effect was more pronounced in the smaller settlers, that had less capacity to feed and less energetic reserves.
S. balanoides [43]	Cyprids avoided settling at high densities (>8 per cm^2), perhaps to avoid intraspecific competition.
Salinity	
B. amphitrite [65]	More than 50% of cyprids settled at salinities between 22 and 35 PSU.
Shear strength	
Model larvae [98]	The ratio of larval fall velocity to shear velocity affects settlement density.
S. balanoides [99]	*S. balanoides* larvae settle in high velocity/energy environments.
S. balanoides and *C. fragilis* [100]	Replicate surfaces deployed in situ received remarkably similar settlement patterns in microsites with low shear (as evinced by sand deposition in a flume) among broader scale areas with higher shear.
Skin friction (drag)	
B. amphitrite and *S. balanoides* [98]	Drag force necessary to detach cyprids positively correlated with time of attachment for *B. amphitrite*, and an order of magnitude higher for *S. balanoides*.
B. improvisus [101]	Microtextured surfaces, oriented parallel to flow, reduced surface skin friction and led to behavioural rejection of the substratum.
Surface area	
Chthamalus spp. and *Policipes polymerus* [102]	Relatively smaller available surface area increases the density of new settlers, in the decreased-substratum settlement intensification hypothesis.
S. balanoides [84]	Larger tiles led to greater settler density, as explorers were less likely to encounter an edge and leave.
Surface bound chemicals	
B. neritina [103]	K^+ and NH_4^+ induced attachment, while Mg^{2+} inhibited it, and Ca^{2+} toxic to larvae. Gamma-aminobutyric acid induced attachment. Acetylcholine, l-3-l-alanine and dopamine induced settlement but inhibited attachment by restricting mobility during surface exploration. Serotonin inhibited settlement and attachment by inducing swimming behaviour.
Surface colour	
Barnacles [104]	More barnacles settled on black glass panels than on white or clear.
B. amphitrite [65]	Chose white over black surfaces in the light, no difference in the dark.
Surface energy/ wettability	
B. amphitrite (laboratory) and other *Balanus* spp. (field) [53]	Negative correlation between contact angle and percentage settlement.
B. amphitrite [97]	Barnacles settled most on surfaces with high surface wettability.
B. improvisus [105]	High wettability completely inhibited settlement, and vice versa; intermediate levels reduced settlement by 38%.

(*Continued*)

Table 3.1 (*Continued*)

Factor and *species*	Notes
Ectocarpus siliculosus [106]	Spores settled more on hydrophobic than on either positively or negatively charged hydrophilic surfaces, irrespective of the presence of a nutrient mixture or gradient.
Surface heterogeneity	
Multispecies [107]	Intermediate orders of surface heterogeneity and complexity led to the greatest total abundance of settlers.
S. balanoides [44]	Cyprids selected cryptic habitats in one site and horizontal surfaces in another.
S. balanoides [108]	Strong correlation between barnacle settlement abundance and fractal dimension (F) and the number of potential settlement sites (PSS) of the surfaces. After altering F but not PSS, it was found that larvae do not settle in response to F, but to PSS.
Surface orientation	
D. polymorpha [80]	Post-veliger mussels settled more on upper than on lower horizontal surfaces.
Multispecies assemblages [82]	More barnacles settled on horizontal undersides.
S. balanoides [84]	On vertical tiles there were significantly more settlers on the top than the bottom.
Surface texture	
B. improvisus [109]	Cyprids prefer smooth surfaces and behaviourally reject microtextured surfaces.
B. improvisus [37]	Larvae preferentially settled on smooth surfaces compared to ribbed microtextures, and more likely to leave the microtextured surfaces.
B. improvisus [110]	Natural microtopographies of *Mytilus edulis* shells and eggcases of the catshark *Scyliorhinus canicula* temporarily prevented cyprid settlement.
Crambe crambe and *Scopalina lophyropoda* [111]	Found that more larvae settled in grooved microrefuges, and in particular in their shaded portions, suggesting they are detected phototactically. In the laboratory, survival of sponge larvae in grooves was enhanced after exposure to bulldozing by urchins.
D. polymorpha [80]	Post-veliger mussels settled more on textured than on smooth surfaces.
M. edulis, *Polydora ciliata*, *B. improvisus*, diatoms, hydrozoa, bryozoa and several ciliates [112]	Most species preferred to settle in pits, and some on peaks, with species-specific preferences for different scales of surface roughness.
S. balanoides [113]	Larvae selected settlement sites with greater microheterogeneity than surrounding, unselected sites, within 300 μm range.
S. balanoides [114]	Cyprids settled on panels of different surface roughness, in the order of fine>medium>coarse>smooth (fine = <0.5 mm, medium = 0.5–2 mm + fine, coarse = 2–4 mm + fine + medium). They were more likely to cluster on smooth panels.
S. balanoides [115]	Resin replicas of mussel shells (*Mytilus edulis* and *Perna perna*) were settled more when roughened than on natural microtextures or smooth controls.

Table 3.1 (*Continued*)

Factor and *species*	Notes
Tetraclita stailactifera [116]	Larvae settled ten times more on rough than smooth in any given week, differences between weeks accounted for by larval supply.
Ulva linza zoospores [117]	When blasted with a water jet, more spores were removed from the roughest (100 μm) and smoothest (1 μm) topographies than the 25 μm structures, implicating the need to consider the dimensions of the microtextured antifouling surface.
Surface type	
Two species of coral planulae larvae [118]	Aggregating species swam for less than 10 minutes and there was no effect of substratum type. Non-aggregating species swam for longer and showed clear surface type discrimination.
B. amphitrite (laboratory) and other *Balanus* spp. (field) [53]	Highest settlement on glass, lowest on Teflon.
B. amphitrite [2]	Age-dependent settlement on polystyrene, but not on glass.
B. amphitrite [11]	Settlement was significantly lower on glass than on quartz.
B. amphitrite [119]	Non-solid surfaces (hydrogels) inhibited larval settlement compared to a solid-surface control, in a way that did not affect the competence of larvae to settle when offered an alternative solid surface.
D. polymorpha [80]	Post-veliger mussels settled more on PVC than on Plexiglas surfaces, more on PVC and Plexiglas than on glass surfaces, and strongly avoided galvanised steel. They showed no discrimination between wood, fibreglass, limestone, concrete, aluminium and raw steel.
Hexaminius sp, *B. amphitrite* and *Balanus variegatus* [120]	More barnacle recruitment on concrete and plywood than on fibreglass or aluminium.
Multispecies [121]	Found different groups of species, sessile and mobile, attracted to settle among mussel beds from different underlying substrata.
S. balanoides [42]	Larvae exhibited reduced specificity for attractive surfaces halfway through one season, but not another.
S. balanoides [122]	Cyprids preferred slate, quartz and marble significantly to millstone grit sandstone and granitic gneiss, perhaps due to variable ability of rocks to sequester/partition solutes.
S. balanoides [123]	Cyprids preferred to settle on blackstone oil shale, despite smooth surfaces, due to extractable settlement inducing factor.
S. balanoides [124]	Cyprids demonstrated predictably different exploratory behaviour on three different coatings.
Temperature	
S. balanoides [125]	Larvae that had been frozen in sea ice settled and metamorphosed more than those that had not been frozen, although the mean and maximum time to metamorphosis was longer for frozen cyprids.
T. stalactifera and *Chthamalus bisinuatus* [119]	More barnacles settled during winter than summer, summer associated with upwelling that cools the water from 24 to 15°C. Higher winter water temperatures may promote settlement, as well as winter storms that may deliver more larvae to the shore.

(*Continued*)

Table 3.1 (*Continued*)

Factor and *species*	Notes
Tidal height	
Perna perna and *M. galloprovincialis* [126]	Both mussel species settled at appropriate tidal heights, irrespective of gregarious cues.
Various coral larvae [127]	Panels conditioned at 2 and 12 m, and then put in an aquarium as choice assay for various species. Those with shallow adult distributions settled significantly more on 2 m panels, vice versa for deeper adult distributions, and no difference in those with wide adult depth distributions.
Wind velocity and direction	
S. balanoides [128]	Daily larval settlement, recruitment patterns and interannual spatial patterns strongly correlated with local wind patterns.

inversely with surface roughness [129], yet overall levels of biofouling have been found to increase linearly with increasing levels of surface roughness [130].

3.3.2   More complex

Similarly, the propagules may act in a relatively passive manner while the cues act as environmental triggers. For example, for some species the presence of chemicals on a surface (such as settlement-inducing protein complex [SIPC] or the neurotransmitter gamma-aminobutyric acid [GABA] [131]) can trigger adhesive release, in a contact-dependent manner. This is an important example for many reasons: in the absence of surface contact, the chemical encounter will not trigger adhesive release, as there is nothing to adhere to. If a surface is encountered without the chemicals that signify the presence of conspecifics, the adhesive will not trigger, as the site either is not suited to the requirements of survival, or at least does not provide opportunities to reproduce in the future. Thus the cues may interact in a manner that is sometimes clearly adaptive to us, as in this case, whereas in other cases we can only speculate as to the explanation. These complex cue responses may appear to reflect choice, but may be entirely involuntary, triggered responses. There is a great deal of difficulty in discerning the difference between passive/inert responses and behavioural choice, and to compound this issue have been the difficulties in observing these tiny organisms. Thus, in the third case that follows, there is very little observational, experimental evidence to demonstrate that cues can elicit a behavioural response. In many more instances, behaviour has been merely inferred from observed settlement patterns and other indirect indicators.

3.3.3   Most complex

It seems incomprehensible that larvae could overcome the relatively enormous forces of currents and waves, as larvae and spores are invariably much smaller than a millimetre in length, and often only tens of micrometres. Yet larvae may exploit these forces, rather than struggle against them. For instance, young larvae of the intertidal sponges *Cliona celata* and *Microciona prolifera* have been shown to swim towards the water surface, perhaps as a phototactic, geotactic or barokinetic response [132]. By maintaining their position in the surface layers, they can increase their chances of encountering an intertidal surface when the current or waves assist. As technology has advanced, laboratory-based flumes with high-resolution video imaging have revealed many simple yet effective methods used to increase their rate of encounter with suitable substrata. For example, studies of the oyster, *Crassostrea gigas*, have revealed that once larvae are within interaction range with the surface (defined as one to two body lengths), they may begin 'dive-bombing' [4]. This involves rapid acceleration rates of more than 400 times that of dead sinking larvae, which allows them to sample a surface after a short burst of effort. If the surface is unfavourable, the larvae continue to be carried in the flow, presumably reserving their energies for further dives.

Even more complex flume studies of non-fouling gastropod larvae swimming over a surface in turbulent flows have revealed a simple yet effective method of chemically detecting and tracking an appropriate settlement site, without the need to fight the currents: The larvae respond to what are effectively strands of an odour plume emanating from its prey species, a coral. In the absence of this olfactory cue, the larvae keep swimming, as the host must not be present. When the larvae passes through an odour trail, it simply ceases swimming and sinks, in an on–off fashion as trails are detected and lost, until the gastropod eventually lands on its appropriate target [133]. Indeed, it seems that whichever species is observed, surprisingly simple or sophisticated solutions to the surface encounter problem are revealed.

3.4   Inert surface encounter, followed by (active?) attachment or rejection

Research suggests that algal spores are delivered to substrata in a passive and inert manner, apart from when they are very close to the surface, i.e. within the boundary layer, and under conditions of very low flow. However, many species (although never red algae) then switch to motility on surface contact through propulsive flagellae that emerge from the spore body. Some algal spores exhibit thigmotaxis, a preference for settlement sites with a degree of surface roughness, which may be lost over time (see below) or may be species specific. Roughness can affect sedimentation rates, reduce evaporation, protect from some grazers and provide shelter from hydrodynamic shear, as well as from brushing off by fronds of other algae [134]. However, simply counting settlement on these different surfaces and assuming a preference may be a mistaken assumption that results from the textures setting up the conditions in which the algae are not removed. This was demonstrated recently when spores of the green algae *Ulva linza* that settled on surfaces of different roughness were blasted with a water jet. More were removed from the smoothest and roughest surfaces, with maximum spore retention in depressions of the intermediate roughness that was closest to the size of the spore itself [117]. Settling zoospores also exhibit a negatively phototactic response, as swimming down is thought to increase the chances of surface encounter. Chemicals associated with the surface may also present attractive or repellent forces to the settling zoospore. For instance, apart from sunlight and an attachment point, these photosynthesisers also require certain nutrients; and it has been suggested that spores can move along the chemical concentration gradient of certain nutrients, within the boundary layer, to positions that maximise nutrient availability [135, 136]. Other chemical cues to settle for algae can come from the biofilm, whose characteristics may impart indirect environmental information including shore height and light availability.

Once contact has been established in these ways, the process of attachment begins. This may be facilitated through the sticky nature of the flagellae, or through production of, or attachment to, mucilage. In this way, the spore body can be correctly positioned with respect to the surface. If the wettability of the surface ('free energy') is inappropriate, it is less likely that this initial attachment will be possible. Again, some have presumed a behavioural preference for particular levels of surface wettability, while it may simply reflect the chemical nature of the adhesives in question. Once the spore is held in position, permanent attachment begins. At the point where the spore meets the surface, adhesive is released to form a thick pad that rapidly cures to form a strong bond in 2–36 hours, depending on the species. Indeed, the speed of the curing process may have implications for the settlement distribution of the different species, i.e. those with slow curing times may require a more sheltered environment (or those in a sheltered environment may be able to afford to take their time, a circular argument). The settled spore then proceeds to the algal equivalent of metamorphosis, where it forms a cell wall and germinates.

3.5   Behavioural surface encounter, exploration and attachment or rejection

Marine invertebrate larvae frequently exhibit more complex behaviours. As exampled below, there are problems associated with the measurement of behaviour in the laboratory versus the field, and in making inferences versus direct observation. In the former case, these problems are associated with the inherently variable nature of cues in the field, as well as in their complex interactions. Attempts to faithfully reproduce these conditions in a laboratory setting are highly complex, limited and expensive. In the field, the opposite is the case; the lack of control of factors plus the addition of untold extraneous variables makes results 'noisy' at best. Experiments tend to be far 'neater' and more repeatable in the laboratory, whereas in the field attempts to repeat experiments often oppose or confound original findings. There may be temporal and spatial dependence to many factors that cannot be reproduced in the laboratory. In addition, concerns like larval physiological condition and mixed cohorts of unknown age and provenance can increase the variability of response. Settlement on virgin substrata or into monospecies communities can yield entirely different results under otherwise comparable conditions to those on filmed substrata or on multispecies assemblages. Some examples are outlined below and are organised by, but not exclusive to, taxonomic group, for convenience.

3.5.1   *Crustacea*

A great deal of work has focused on barnacle settlement, as they are a common hard fouling organism that is particularly difficult to remove. They have been shown to respond in complex ways to a variety of cues, and in particular to the gregarious chemical cues of their own species. A laboratory-based study found that settlement of *Balanus amphitrite* was affected by the presence of only five other cyprids, such that the effects of a settlement promoter were magnified ten times, and of a settlement inhibitor were lost, highlighting the need to consider cue interactions [31]. A recent field-based study found that significantly more cyprids arrived on surfaces of different textures when they were treated with a gregarious chemical cue, when filmed simultaneously with an untreated control, indicating a level of behavioural control over site-encounter rates [137]. However, many studies have presumed contact rates to be passive and have measured them with, for instance, sticky surfaces. Yet if the nature of the surface itself can behaviourally influence the contact rate, these studies may require reassessment.

Most direct behavioural observation of cyprid settlement has taken place under laboratory conditions and has revealed that after surface encounter these larvae continue to exhibit complex behaviours. When *Semibalanus balanoides* cyprid larvae locate a surface, they may engage in three phases of exploratory behaviour, namely broad (in the range of metre to centimetre), close (in the range of millimetre) and fine scale exploration (inspection, in the range of around 300 μm), involving 'walking' across the surface on their antennules that secrete a finite supply of temporary adhesive [113]. The antennules house mechanoreceptors that respond to the pull required to remove the antennule from the surface as a measure of the force of adhesion achieved on the surface [138]. The antennules probably also contain chemotactic receptors [139], providing further information about the biotic nature of the site. Many studies suggest that close inspection is an indicator behaviour that the surface is favourable for settlement, while straight line exploratory paths indicate a less favourable site. Cyprids may choose to attach temporarily, perhaps to sample the site for a whole tidal cycle before detaching and restarting the process, or they may attach permanently and proceed to metamorphosis. If they choose to detach and re-enter the planktonic phase, there are consequences for their physiological reserves. This can lead to the phenomenon of larval desperation, which is outlined below.

Even within the subclass Cirripedia, different species may sample and evaluate the substratum properties in very different ways according to their particular niche criteria; for instance, *Balanus improvisus* and *B. amphitrite* prefer hydrophobic and hydrophilic surfaces, respectively [97, 105]. *B. improvisus* prefers relatively low flow speeds, presumed to be due to removal of exploring cyprids by increasing drag and lift forces [55], yet *S. balanoides* prefer high flows and *B. amphitrite* select medium flow rates [53]. Again, this could reflect different adhesive properties or indirectly provide other information about the nature of the settlement site, as each barnacle species has its own niche requirements.

3.5.2 *Bryozoa*

Much of the work on bryozoan larvae has focussed on biofilms and flow as cues to settlement, and on the phenomenon of larval desperation. *Bugula simplex*, *Bugula stolonifera*, *Bugula turrita* and *Bugula neritina* all prefer to settle on filmed substrata when offered a choice [76, 140], yet *Bugula flabellata* settlement is inhibited by biofilms. The latter was presumed to be an active rejection choice as the biofilm is sticky [141].

Both *B. neritina* and *B. stolonifera* swim up as a response to gravity (they are not buoyant), even though the former lacks a statocyst [142]. Laboratory-based observations of *B. neritina* surface exploration found that in still water the larvae kept contacting, exploring then swimming away from the test surfaces, but never exited any surface in the presence of flow [3]. Thus, at least in conditions of no flow, larvae of this species are able to control their own surface encounter rate. They crawled upstream during surface exploration, and they only ever settled on surfaces with a 2-week old biofilm.

Although larvae of this species have been shown to respond to several settlement cues (they are typically photonegative and geopositive to find shaded undersides [103]), it has repeatedly been shown that they become less discerning as they age. This has been linked to the fact that, as the larvae do not feed, their physiological reserves must diminish with time, and so they become more and more desperate to settle. They reach a situation where if they do not settle they will die, so the presence of desirable cues becomes relatively less of a priority, and settlement choice patterns become less clear. In terms of the three priorities/questions outlined above, as a larva's reserves run out, the question may go from 'is this a good place to survive?' to 'is this a better place to survive than where I am now?' As time progresses, the answer may inevitably become 'yes'. This hypothesis has been named 'larval desperation', and evidence both for and against it has been gathered for several species from disparate phyla, both in the laboratory and in the field [93]. Larval physiological condition has been linked to larval size: presumably larger larvae have more reserves and so can delay settlement, in the absence of attractive cues, longer than their smaller counterparts. In the case of the bryozoan *Watersipora subtorquata*, this is the case, yet for *B. neritina* it is not [76].

3.5.3   Tunicata

Larger larvae of the colonial ascidian *Diplosoma listerianum* also delay settlement longer in the absence of appropriate cues, inferring larval desperation among smaller individuals that presumably have less energetic reserves [76].

Many ascidians naturally inhabit shaded, down-facing surfaces, perhaps as they lack sedimentation. The cues that they may use to behaviourally locate these surfaces include light/shade and gravity. The colonial ascidians *Didemnum candidum* and *D. listerianum* both use light as a cue [143], and the solitary ascidian *Ascidia mentula* avoids very low light levels and is negatively geotactic [144].

It seems that *Ciona intestinalis* is able to encounter surfaces passively or actively. Settlement of this solitary ascidian is facilitated or attracted by biofilms, and generally increases with increasing biofilm age. This study is interesting in its way of deducing site selection behaviour from observations of settlers: Active choice in this species would be reflected in tadpole larvae attaching at their head/anterior end, while passive entrapment by the sticky biofilm would result in the larvae sticking sideways on. Larvae of each type were recorded, yet both were able to proceed to metamorphosis [141], so perhaps the attachment response of this species is involuntarily triggered by contact with a surface or with surface bound biochemical properties.

3.5.4   Cnidaria

Within the phylum Cnidaria, there is large variation in response to light cues. Planula larvae of the hydrozoan *Clava multicornis* are positively phototactic [145], while those of the scyphozoan *Cyanea capillata* are negatively phototactic [144], and those of the coral *Porites asteroides*, which requires light for its symbiotic photosynthesising zooxanthellae, avoid UV exposed regions in both laboratory and field tests [83]. These responses are all indicative of selective cue responses. In the first case, the ideal settlement site is often on algae [146], the second prefers undersides to avoid sediment clogging and the third can be damaged by certain UV wavelengths. In fact, even among the coral species that require light, several species have been shown to settle in response to different light qualities and quantities, creating their own vertical zonation at settlement that is consistent with that of the adults [85].

However, not all larvae encounter a surface from the plankton; some offspring are delivered directly to the surface by the parent organism. From this beginning, they crawl to a new area before attaching. Still, evidence shows that even these larvae respond to cues, even though they are seemingly already delivered to a suitable substratum. For instance, in the presence of favourable substrata and flow speeds, 90% of the crawling planula larvae of the cup coral *Balanophyllia elegans* settled in 3 days, compared to only 11% when either or both factors were absent. Similarly, the cues of natural rock and fast water velocity interacted to reveal that these larvae are behaviourally discerning [147]. Other crawling planulae, such as the octocoral *Alcyonium siderium*, respond positively to the presence of different crustose algae and negatively to encrusting invertebrate colonies, perhaps to avoid competition for resources [148]. Crawling planulae of the intertidal hydroid *Dynamena pumila* are positively phototactic. Interestingly, adhesion in this species is triggered by temporary exposure to air, when the tide goes out [149]. This could be a cue indicating that an appropriate intertidal settlement site has been arrived at, or a strategy to avoid the problems associated with achieving adhesion in a hydrophilic environment.

3.5.5 Annelida

However, for the spirorbid polychaetes *Spirorbis spirorbis* and *Spirorbis tridentatus*, as well as the bryozoan *Flustrellidra hispida*, the air-drying of biofilmed slates and pieces of algae, both of which usually induce settlement in these species at low tide, removed their inductive properties [28]. Larvae of *Pomatoceros lamarkii* are found in the laboratory to be induced to settle in response to biofilms in a positive density dependant manner. Yet when the film was dried to mimic a tidal emersion (1–2 hours at 20°C), it lost its inductive properties for 5 days [26].

Many fouling polychaetes are induced to settle via cue reception, which can be gregarious attraction to members of their own [150, 151 or other 71] species, appropriate habitat type (often mediated by the biofilm [21]) or the presence of food (presumably mediated by flow detection) [152]. Typical settlement behaviour to contact and explore a potential settlement site, in response to appropriate cue reception in *Hydroides elegans*, is for the swimming larva to slow down and secrete a mucous thread that sticks to the surface, which the larva then crawls over [18].

Just as some groups of corals avoid interspecific competition by settling in different light intensity habitats, the polychaete *H. elegans*, a barnacle (*Balanus* sp.) and a bryozoan (*Bugula* sp.) species have been shown to settle in areas with different, non-overlapping velocity gradients, which could represent micro-niche differentiation [56].

3.5.6 Porifera

Sponges seem to be among the least discriminating or responsive to cues. Sponge larvae are all non-feeding, short-lived and short-dispersing. Most studies show that sponge larvae typically avoid light, and are strong swimmers [153]. Swimming is usually by means of cilia on the larval surface. In *Reneira*, the cilia at the larva's rear end are pigmented and respond to changes in light intensity, such that in the presence of light the cell's cilia become rigidly straight and in the presence of shade this effect is removed, effectively steering them in favour of shaded areas [154]. *Cliona colata* larvae have been shown to swim upwards for around 30 hours, then switch to downwards swimming, irrespective of light or chemical or temperature gradients [132]. This may reflect the transition to competence to settle. Beyond photo responsiveness, there seems to be little evidence of cue response, presumably as they are generalists as adults, so can tolerate a wide range of habitats [155], and are clonal, so can spread out to cover more than just the initial point of attachment [111]. As the sponge lacks adhesive glands, when adhering to a surface, its ectodermal cells secrete an adhesive. In this way, they do not need to attach at any particular point. They then spread out and flatten to increase the area and strength of attachment [156].

3.5.7 *Mollusca*

The molluscs are perhaps an exception among fouling species, as mussels are capable of moving to a different site even after they have settled and metamorphosed, therefore their settlement behaviour has been described in both larvae and in post-metamorphic juveniles. Among post-metamorphic zebra mussel *Dreissena polymorpha* juveniles, cues to settlement include differential attraction to light and dark areas, to black and white substrata, depth and chemically mediated gregarious attraction to conspecifics, leading to their ability to detect areas that are deep, dark, sheltered and protected from predators [80,157]. In the blue mussel, *Mytilus edulis* L., rough, biofilmed, hydrophobic surfaces are preferable for settlement [158]. From our understanding, this presumably reflects shelter from shear, site immersion information and compatibility with adhesives, respectively. They select artificial over living substrata, supposedly as an interspecific competition avoidance mechanism [159]. It has also been noted that young or small juveniles will select different sites than their older or larger counterparts, gradually migrating on water currents to join the adult group as they grow [160]. This could indicate different niche requirements of smaller/younger individuals, or may be their strategy to avoid intraspecific competition with superior individuals until they are better equipped.

Mussel veliger larvae secrete a long, sticky, mucous thread, and many researchers have concluded that they are passive settlers whose thread is more likely to become entangled on rough or filamentous surfaces, which could lead to an erroneous assumption of larval behavioural preference for these surfaces. Yet Dobretsov and Wahl [159] placed identical filamentous surfaces in the vicinity of macroalgae with different chemically attractive or repellent qualities, and found distinctly different settlement on the filamentous surfaces, accordingly. This is a good example of deductive inference of behaviour, in the absence of direct observation in the field, which is not always feasible. The larvae must have either altered their approach to unfavourable/favourable surfaces, or altered their adhesion rates after contacting the surfaces. The latter could, nevertheless, have been either behaviourally or chemically mediated, highlighting the pitfalls associated with this kind of assumption. However, direct video observations of post-metamorphic juvenile *M. edulis* has revealed a clear behavioural preference for areas associated with favourable chemical treatments. On unfavourable surfaces, mussels exhibited a straighter line of exploration, i.e. their gross and net distances travelled were similar. However, on chemically favourable surfaces, the turning rate was high, influencing the index of straightness of exploratory path such that they moved around and around on the area of interest; their gross distance was high, while their net distance travelled was lower [161]. This kind of behavioural analysis has been promoted as a rapid assay for assessing the attractiveness of antifouling coatings.

Post-metamorphic mussels secrete byssus threads (known colloquially as 'beards') that have adhesive pads or 'plaques' at their tips. Several proteins have been described as present in the plaques, and 3,4-dihydroxyphenyl-L-alanine (DOPA) containing proteins are charged as being primarily responsible for their strong adhesion. It appears that one of these forms a structurally strong adhesive while another forms a surrounding protective coat [162]. Many man-made adhesives fail in the marine environment due to their inability to remove water molecules from hydrophilic surfaces, yet these adhesives apparently avoid addressing this problem by binding to hydrophobic surfaces.

There are examples in the literature that tend to concentrate on particular species from each group, i.e. most polychaete work looks at *H. elegans*. However, care should be taken when making inferences applicable to the rest of the group. For example, the preferences for different surface wettabilities, flow speeds, surface textures and shore heights are marked among different species of barnacle, and so we may presume this to be the case among different species of polychaete, bryozoan, etc. Thus, changing the nature of the substratum to repel a particular fouling community may result in attraction of others. The resulting community could be functionally identical yet composed of entirely different representative species from the same groups.

3.6   Conclusions

- The process of settlement marks the transition from the planktonic (in most cases) to the sessile phase.
- In order to secure a location appropriate to the survival, growth and reproduction requirements of the individual, larvae may respond to physical and chemical environmental cues.
- The scale of cue response ranges from inert to active behaviour, yet drawing the distinction requires careful experimental design.
- Clear cue effects found in laboratory studies may be lost in the field due to complex interactions with other cues, and variability in space and time.

References

1. Raimondi, P.T. & Morse, A.N.C. (2000) The consequences of complex larval behavior in a coral. *Ecology*, **81** (11), 3193–3211.
2. Rittschof, D., Branscomb, E.S. & Costlow, J.D. (1984) Settlement and behaviour in relation to flow and surface in larval barnacles, *Balanus amphitrite* Darwin. *Journal of Experimental Marine Biology and Ecology*, **82**, 131–146.
3. Walters, L.J., Miron, G. & Bourget, E. (1999) Endoscopic observations of invertebrate larval substratum exploration and settlement. *Marine Ecology Progress Series*, **182**, 95–108.
4. Finelli, C. & Wethey, D. (2003) Behavior of oyster (*Crassostrea virginica*) larvae in flume boundary layer flows. *Marine Biology*, **143** (4), 703–711.
5. Jenkins, S.R. (2005) Larval habitat selection, not larval supply, determines settlement patterns and adult distribution in two chthamalid barnacles. *Journal of Animal Ecology*, **74** (5), 893–904.
6. Satumanatpan, S. & Keough, M.J. (2001) Roles of larval supply and behavior in determining settlement of barnacles in a temperate mangrove forest. *Journal of Experimental Marine Biology and Ecology*, **260** (2), 133–153.

7. Thiyagarajan, V. & Qian P.Y. (2003) Effect of temperature, salinity and delayed attachment on development of the solitary ascidian *Styela plicata* (Lesueur). *Journal of Experimental Marine Biology and Ecology*, **290** (1), 133–146.

8. Olivier, F., Tremblay, R., Bourget, E. & Rittschof, D. (2000) Barnacle settlement: field experiments on the influence of larval supply, tidal level, biofilm quality and age on *Balanus amphitrite* cyprids. *Marine Ecology Progress Series*, **199**, 185–204.

9. Wieczorek, S.K. & Todd, C.D. (1998) Inhibition and facilitation of settlement of epifaunal marine invertebrate larvae by microbial cues. *Biofouling*, **12** (1–3), 81–118.

10. Unabia, C.R.C. & Hadfield, M.G. (1999) Role of bacteria in larval settlement and metamorphosis of the polychaete *Hydroides elegans*. *Marine Biology*, **133** (1), 55–64.

11. Faimali, M., Garaventa, F., Terlizzi, A., Chiantore, A.M. & Cattaneo-Vietti, R. (2004) The interplay of substrate nature and biofilm formation in regulating *Balanus amphitrite* (Darwin, 1854) larval settlement. *Journal of Experimental Marine Biology and Ecology*, **306**, 37–50.

12. Lau, S.C.K., Thiyagarajan, V. & Qian, P.Y. (2003) The bioactivity of bacterial isolates in Hong Kong waters for the inhibition of barnacle (*Balanus amphitrite* Darwin) settlement. *Journal of Experimental Marine Biology and Ecology*, **282** (1–2), 43–60.

13. Lau, S.C.K., Thiyagarajan, V., Cheung, S.C.K. & Qian, P.-Y. (2005) Roles of bacterial community composition in biofilms as a mediator for larval settlement of three marine invertebrates. *Aquatic Microbial Ecology*, **38** (1), 41–51.

14. Dahms, H.U., Dobretsov, S. & Qian, P.Y. (2004) The effect of bacterial and diatom biofilms on the settlement of the bryozoan *Bugula neritina*. *Journal of Experimental Marine Biology and Ecology*, **313** (1), 191–209.

15. Fitt, W.K., Coon, S.L., Walch, M., Weiner, R.M., Colwell, R.R. & Bonar, D.B. (2005) Settlement behavior and metamorphosis of oyster larvae (*Crassostrea gigas*) in response to bacterial supernatants. *Marine Biology*, **106** (3), 389–394.

16. Huang, Y.-L., Dobretsov, S., Ki, J.-S., Yang, L.-H. & Qian, P.-Y. (2007) Presence of acyl-homoserine lactone in subtidal biofilm and the implication in larval behavioral response in the polychaete *Hydroides elegans*. *Microbial Ecology*, **54** (2), 384–392.

17. Hung, O.S., Thiyagarajan, V., Wu, R.S.S. & Qian, P.-Y. (2005) Effect of ultraviolet radiation on biofilms and subsequent larval settlement of *Hydroides elegans*. *Marine Ecology Progress Series*, **304**, 155–166.

18. Lau, S., Harder, T. & Qian, P.Y. (2003) Induction of larval settlement in the serpulid polychaete *Hydroides elegans* (Haswell): role of bacterial extracellular polymers. *Biofouling*, **19** (3), 197–204.

19. Lam, C., Harder, T. & Qian, P.-Y. (2003) Induction of larval settlement in the polychaete *Hydroides elegans* by surface-associated settlement cues of marine benthic diatoms. *Marine Ecology Progress Series*, **263**, 83–92.

20. Huang, S. & Hadfield, M.G. (2003) Composition and density of bacterial biofilms determine larval settlement of the polychaete *Hydroides elegans*. *Marine Ecology Progress Series*, **260**, 161–173.

21. Harder, T., Lau, S.C.K., Dahms, H.-U. & Qian, P.-Y. (2002) Isolation of bacterial metabolites as natural inducers for larval settlement in the marine polychaete *Hydroides elegans* (Haswell). *Journal of Chemical Ecology*, **28** (10), 2029–2043.

22. Lau, S.C.K. & Qian, P.-Y. (2001) Larval settlement in the serpulid polychaete *Hydroides elegans* in response to bacterial films: an investigation of the nature of putative larval settlement cue. *Marine Biology*, **138** (2), 321–328.

23. Satuito, C.G., Bao, W., Yang, J. & Kitamura, H. (2005) Survival, growth, settlement and metamorphosis of refrigerated larvae of the mussel *Mytilus galloprovincialis* Lamarck and their use in settlement and antifouling bioassays. *Biofouling*, **21** (3), 217–225.

24. Bao, W.-Y., Satuito, C., Yang, J.-L. & Kitamura, H. (2007) Larval settlement and metamorphosis of the mussel *Mytilus galloprovincialis* in response to biofilms. *Marine Biology*, **150** (4), 565–574.

25. Keough, M.J. & Raimondi, P.T. (1996) Responses of settling invertebrate larvae to bio-organic films: effects of large-scale variation in films. *Journal of Experimental Marine Biology and Ecology*, **207**, 59–78.

26. Hamer, J.P., Walker, G. & Latchford, J.W. (2001) Settlement of *Pomatoceros lamarkii* (Serpulidae) larvae on biofilmed surfaces and the effect of aerial drying. *Journal of Experimental Marine Biology and Ecology*, **260** (1), 113–131.

27. Thompson, R.C., Norton, T.A. & Hawkins, S.J. (1998) The influence of epilithic microbial films on the settlement of *Semibalanus balanoides* cyprids – a comparison between laboratory and field experiments. *Hydrobiologia*, **375/376**, 203–216.

28. Hamer, J.P. & Walker, G. (2001) Avoidance of dried biofilms on slate and algal surfaces by certain spirorbid and bryozoan larvae. *Journal of the Marine Biological Association of the UK*, **81**, 167–168.

29. Wheeler, G.L., Tait, K., Taylor, A., Brownlee, C. & Joint, I.A.N. (2006) Acyl-homoserine lactones modulate the settlement rate of zoospores of the marine alga Ulva intestinalis via a novel chemokinetic mechanism. *Plant, Cell & Environment*, **29** (4), 608–618.

30. Matsumura, K., Hills, J.M., Thomason, P.O., Thomason, J.C. & Clare A.S. (2000) Discrimination at settlement in barnacles: laboratory and field experiments on settlement behaviour in response to settlement-inducing protein complexes. *Biofouling*, **16** (2–4), 181–190.

31. Head, R.M., Overbeke, K., Klijnstra, J., Biersteker, R. & Thomason, J.C. (2003) The effect of gregariousness in cyprid settlement assays. *Biofouling*, **19** (4), 269–278.

32. Yamamoto, H., Tachibana, A., Saikawa, W., Nagano, M., Matsumura, K. & Fusetani, N. (1998) Effects of calmodulin inhibitors on cyprid larvae of the barnacle, Balanus amphitrite. *Journal of Experimental Zoology*, **280** (1), 8–17.

33. Matsumura, K., Nagano, M., Kato-Yoshinaga, Y., Yamazaki, M., Clare, A.S. & Fusetani, N. (1998) Immunological studies on the settlement-inducing protein complex (SIPC) of the barnacle Balanus amphitrite and its possible involvement in larva-larva interactions. *Proceedings of the Royal Society of London Series B – Biological Sciences*, **265** (1408), 1825–1830.

34. Matsumura, K., Mori, S., Nagano, M. & Fusetani, N. (1998) Lentil lectin inhibits adult extract-induced settlement of the barnacle, Balanus amphitrite. *Journal of Experimental Zoology*, **280** (3), 213–219.

35. Head, R.M., Berntsson, K.M., Dahlstrom, M., Overbeke, K. & Thomason, J.C. (2004) Gregarious settlement in cypris larvae: the effects of cyprid age and assay duration. *Biofouling*, **20** (2), 123–128.

36. Yule, A.B. & Crisp, D.J. (1983) Adhesion of cypris larvae of the barnacle, Balanus balanoides, to clean Arthropodin treated surfaces. *Journal of the Marine Biological Association of the United Kingdom*, **63**, 261–271.

37. Berntsson, K.M., Jonsson, P.R., Larsson, A.I. & Holdt, S. (2004) Rejection of unsuitable substrata as a potential driver of aggregated settlement in the barnacle Balanus improvisus. *Marine Ecology Progress Series*, **275**, 199–210.

38. Crisp, D.J. (1990) Field experiments on the settlement, orientation and habitat choice of Chthalamus fragilis (Darwin). *Biofouling*, **2**, 131–136.

39. Jeffery, C.J. (2002) New settlers and recruits do not enhance settlement of a gregarious intertidal barnacle in New South Wales. *Journal of Experimental Marine Biology and Ecology*, **275** (2), 131–146.

40. Wright, J.R. & Boxshall, A.J. (1999) The influence of small-scale flow and chemical cues on the settlement of two congeneric barnacle species. *Marine Ecology Progress Series*, **183**, 179–187.

41. Keough, M.J. (1998) Responses of settling invertebrate larvae to the presence of established recruits. *Journal of Experimental Marine Biology and Ecology*, **231** (1), 1–19.

42. Jarrett, J.N. (1997) Temporal variation in substratum specificity of Semibalanus balanoides (Linnaeus) cyprids. *Journal Of Experimental Marine Biology and Ecology*, **211** (1), 103–114.

43. Hills, J.M. & Thomason, J.C. (1996) A multi-scale analysis of settlement density and pattern dynamics of the barnacle Semibalanus balanoides. *Marine Ecology Progress Series*, **138**, 103–115.

44. Chabot, R. & Bourget, E. (1988) Influence of substratum heterogeneity and settled barnacle density on the settlement of cypris larvae. *Marine Biology*, **97** (1), 45–56.

45. Hills, J.M., Thomason, J.C., Milligan, J.L. & Richardson, T. (1998) Do barnacle larvae respond to multiple settlement cues over a range of spatial scales? *Hydrobiologia*, **376**, 101–111.

46. Kent, A., Hawkins, S.J. & Doncaster, C.P. (2003) Population consequences of mutual attraction between settling and adult barnacles. *Journal of Animal Ecology*, **72** (6), 941–952.
47. Miron, G., Boudreau, B. & Bourget, E. (1999) Intertidal barnacle distribution: a case study using multiple working hypotheses. *Marine Ecology Progress Series*, **189**, 205–219.
48. Cruz, T., Castro, J.J., Delany, J., et al. (2005) Tidal rates of settlement of the intertidal barnacles Chthamalus stellatus and Chthamalus montagui in western Europe: the influence of the night/day cycle. *Journal of Experimental Marine Biology and Ecology*, **318** (1), 51–60.
49. Hills, J.M., Thomason, J.C., Davis, H., Kohler, J. & Millett, E. (2000) Exploratory behaviour of barnacle larvae in field conditions. *Biofouling*, **16** (2–4), 171–179.
50. Grosberg, R.K. (1982) Intertidal zonation of barnacles: the influence of planktonic zonation of larvae on vertical distribution of adults. *Ecology*, **63** (4), 894–899.
51. Jeffery, C.J. & Underwood, A.J. (2000) Consistent spatial patterns of arrival of larvae of the honeycomb barnacle Chamaesipho tasmanica Foster and Anderson in New South Wales. *Journal of Experimental Marine Biology and Ecology*, **252** (1), 109–127.
52. Satumanatpan, S., Keough, M.J. & Watson, G.F. (1999) Role of settlement in determining the distribution and abundance of barnacles in a temperate mangrove forest. *Journal of Experimental Marine Biology and Ecology*, **241** (1), 45–66.
53. Qian, P.-Y., Rittschof, D. & Sreedhar, B. (2000) Macrofouling in unidirectional flow: miniature pipes as experimental models for studying the interaction of flow and surface characteristics on the attachment of barnacle, bryozoan and polychaete larvae. *Marine Ecology Progress Series*, **207**, 109–121.
54. Larsson, A.I. & Jonsson, P.R. (2006) Barnacle larvae actively select flow environments supporting post settlement growth and survival. *Ecology*, **87** (8), 1960–1966.
55. Jonsson, P.R., Berntsson, K.M. & Larsson, A.I. (2004) Linking larval supply to recruitment: flow-mediated control of initial adhesion of barnacle larvae. *Ecology*, **85** (10), 2850–2859.
56. Qian, P.-Y., Rittschof, D., Sreedhar, B. & Chia, F.S. (1999) Macrofouling in unidirectional flow: miniature pipes as experimental models for studying the effects of hydrodynamics on invertebrate larval settlement. *Marine Ecology Progress Series*, **191**, 141–151.
57. Alfaro, A.C. (2005) Effect of water flow and oxygen concentration on early settlement of the New Zealand green-lipped mussel, Perna canaliculus. *Aquaculture*, **246** (1–4), 285–294.
58. Abelson, A., Weihs, D. & Loya, Y. (1994) Hydrodynamic impediments to settlement of marine propagules, and adhesive-filament solution. *Limnology and Oceanography*, **39** (1), 164–169.
59. Holm, E.R., McClary, M. & Rittschof, D. (2000) Variation in attachment of the barnacle Balanus amphitrite: sensation or something else? *Marine Ecology Progress Series*, **202**, 153–162.
60. Eckman, J.E. & Duggins, D.O. (1998) Larval settlement in turbulent pipe flows. *Journal of Marine Research*, **56** (6), 1285–1312.
61. Rittschof, D., Sin, T.M., Teo, S.L.M. & Coutinho, R. (2007) Fouling in natural flows: cylinders and panels as collectors of particles and barnacle larvae. *Journal of Experimental Marine Biology and Ecology*, **348** (1–2), 85–96.
62. Pawlik, J.R. & Butman, C.A. (1993) Settlement of a marine tube worm as a function of current velocity: interacting effects of hydrodynamics and behavior. *Limnology and Oceanography*, **38** (8), 1730–1740.
63. Todd, C.D., Phelan, P.J.C., Weinmann, B.E., et al. (2006) Improvements to a passive trap for quantifying barnacle larval supply to semi-exposed rocky shores. *Journal of Experimental Marine Biology and Ecology*, **332** (2), 135–150.
64. Minchinton, T.E. & Scheibling, R.E. (1993) Free space availability and larval substratum selection as determinants of barnacle population structure in a developing rocky intertidal community. *Marine Ecology Progress Series*, **95**, 233–244.
65. Kon-Ya, K. & Miki, W. (1994) Effects of environmental factors on larval settlement of the barnacle Balanus amphitrite reared in the laboratory. *Fisheries Science*, **60** (5), 563–565.
66. Harrington, L., Fabricius, K., De'ath, G. & Negri, A. (2004) Recognition and selection of settlement substrata determine post-settlement survival in corals. *Ecology*, **85** (12), 3428–3437.

67. Kato-Yoshinaga, Y., Nagano, M., Mori, S., Clare, A.S., Fusetani & N., Matsumura, K. (2000) Species specificity of barnacle settlement-inducing proteins. *Comparative Biochemistry and Physiology. Part A, Molecular and Integrative Physiology*, **125** (4), 511–516.

68. Holmes, S.P. (2002) The effect of pedal mucus on barnacle cyprid settlement: a source for indirect interactions in the rocky intertidal? *Journal of the Marine Biological Association of the United Kingdom*, **82** (1), 117–129.

69. Nogata, Y. & Matsumura, K. (2006) Larval development and settlement of a whale barnacle. *Biology Letters*, **2** (1), 92–93.

70. Degnan, B.M. & Johnson, C.R. (1999) Inhibition of settlement and metamorphosis of the ascidian Herdmania curvata by non-geniculate coralline algae. *Biological Bulletin*, **197** (3), 332–340.

71. Bryan, P.J., Kreider, J.L. & Qian, P.-Y. (1998) Settlement of the serpulid polychaete Hydroides elegans (Haswell) on the arborescent bryozoan Bugula neritina (L.): evidence of a chemically mediated relationship. *Journal of Experimental Marine Biology and Ecology*, **220** (2), 171–190.

72. Yang, J.-L., Satuito, C., Bao, W.-Y. & Kitamura, H. (2007) Larval settlement and metamorphosis of the mussel Mytilus galloprovincialis on different macroalgae. *Marine Biology*, **152** (5), 1121–1132.

73. Petersen, D., Laterveer, M. & Schuhmacher, H. (2005) Spatial and temporal variation in larval settlement of reefbuilding corals in mariculture. *Aquaculture*, **249** (1–4), 317–327.

74. Dobretsov, S.V. (1999) Effects of macroalgae and biofilm on settlement of blue mussel (Mytilus edulis L.) larvae. *Biofouling*, **14** (2), 153–165.

75. Alfaro, A., Copp, B., Appleton, D., Kelly, S. & Jeffs, A. (2006) Chemical cues promote settlement in larvae of the green-lipped mussel, Perna canaliculus. *Aquaculture International*, **14** (4), 405–412.

76. Marshall, D.J. & Keough, M.J. (2003) Variation in the dispersal potential of non-feeding invertebrate larvae: the desperate larva hypothesis and larval size. *Marine Ecology – Progress Series*, **255**, 145–153.

77. Marshall, D.J. & Keough, M.J. (2003) Effects of settler size and density on early post-settlement survival of Ciona intestinalis in the field. *Marine Ecology Progress Series*, **259**, 139–144.

78. Todd, C.D. (2003) Assessment of a trap for measuring larval supply of intertidal barnacles on wave-swept, semi-exposed shores. *Journal of Experimental Marine Biology and Ecology*, **290** (2), 247–269.

79. Jenkins, S.R. & Hawkins, S.J. (2005) Barnacle larval supply to sheltered rocky shores: a limiting factor? *Hydrobiologia*, **503** (1), 143–151.

80. Marsden, J. & Lansky, D. (2000) Substrate selection by settling zebra mussels, Dreissena polymorpha, relative to material, texture, orientation, and sunlight. *Canadian Journal of Zoology*, **78** (5), 787–793.

81. Greer, S.P. & Amsler, C.D. (2004) Clonal variation in phototaxis and settlement behaviours of Hincksia irregularis (Phaeophyceae) spores. *Journal of Phycology*, **40** (1), 44–53.

82. Glasby, T.M. (2000) Surface composition and orientation interact to affect subtidal epibiota. *Journal of Experimental Marine Biology and Ecology*, **248** (2), 177–190.

83. Gleason, D., Edmunds, P. & Gates, R. (2006) Ultraviolet radiation effects on the behavior and recruitment of larvae from the reef coral Porites astreoides. *Marine Biology*, **148** (3), 503–512.

84. Hills, J.M. & Thomason, J.C. (1998) On the effect of tile size and surface texture on recruitment pattern and density of the barnacle, Semibalanus balanoides. *Biofouling*, **13** (1), 31–50.

85. Mundy, C.N. & Babcock, R.C. (1998) Role of light intensity and spectral quality in coral settlement: implications for depth-dependent settlement? *Journal of Experimental Marine Biology and Ecology*, **223** (2), 235–255.

86. Harder, T., Thiyagarajan, V. & Qian, P.-Y. (2001) Combined effect of cyprid age and lipid content on larval attachment and metamorphosis of Balanus amphitrite Darwin. *Biofouling*, **17** (4), 257–262.

87. Crisp, D.J. & Meadows, P.S. (1963) Adsorbed layers: the stimulus to settlement in barnacles. *Proceedings of the Royal Society London B*, **158**, 364–387.

88. Harder, T.N., Thiyagarajan, V. & Qian, P.-Y. (2001) Effect of cyprid age on the settlement of Balanus amphitrite Darwin in response to natural biofilms. *Biofouling*, **17** (3), 211–219.

89. Pineda, J. & Starczak, V. (2005) Recruitment windows: implications for 'effective' connectivity in benthic populations. *EOS, Transactions of the American Geophysical Union, Ocean Sciences Meeting Supplement*, **87** (36), abstract Os46E-09.

90. Bennett, C.E. & Marshall, D.J. (2005) The relative energetic costs of the larval period, larval swimming and metamorphosis for the ascidian Diplosoma listerianum. *Marine and Freshwater Behaviour and Physiology*, **38** (1), 21–29.

91. Toonen, R.J. & Pawlik, J.R. (2001) Settlement of the gregarious tube worm Hydroides dianthus (Polychaeta: Serpulidae). I. Gregarious and nongregarious settlement. *Marine Ecology Progress Series*, **224**, 103–114.

92. Toonen, R.J. & Pawlik, J.R. (2001) Settlement of the gregarious tube worm Hydroides dianthus (Polychaeta: Serpulidae). II. Testing the desperate larva hypothesis. *Marine Ecology Progress Series*, **224**, 115–131.

93. Gribben, P.E., Marshall, D.J. & Steinberg, P.D. (2006) Less inhibited with age? Larval age modifies responses to natural settlement inhibitors. *Biofouling*, **22** (2), 101–116.

94. Dahms, H.U., Harder, T. & Qian, P.-Y. (2004) Effect of meiofauna on macrofauna recruitment: settlement inhibition of the polychaete Hydroides elegans by the harpacticoid copepod Tisbe japonica. *Journal of Experimental Marine Biology and Ecology*, **311** (1), 47–61.

95. Andre, C., Jonsson, P.R. & Lindegarth, M. (1993) Predation on settling bivalve larvae by benthic suspension feeders: the role of hydrodynamics and larval behaviour. *Marine Ecology Progress Series*, **97**, 183–192.

96. Navarrete, S.A. & Wieters, E.A. (2000) Variation in barnacle recruitment over small scales: larval predation by adults and maintenance of community pattern. *Journal of Experimental Marine Biology and Ecology*, **253** (2), 131–148.

97. Rittschof, D. & Costlow, J.D. (1989) Bryozoan and barnacle settlement in relation to initial surface wettability: a comparison of laboratory and field studies. *Scientia Marina*, **53** (2–3), 411–416.

98. Eckman, J.E.S., Savidge, W.B. & Gross, T.F. (1990) Relationship between duration of cyprid attachment and drag forces associated with detachment of Balanus amphitrite cyprids. *Marine Biology*, **107** (1), 111–118.

99. Crisp, D.J. (1955) The behaviour of barnacle cyprids in relation to water movement over a surface. *Journal of Experimental Biology*, **32**, 569–590.

100. Wethey, D.S. (1986) Ranking of settlement cues by barnacle larvae: influence of surface contour. *Bulletin of Marine Sciences*, **39** (2), 393–400.

101. Berntsson, K.M., Andreasson, H., Jonsson, P.R., et al. (2000) Reduction of barnacle recruitment on micro-textured surfaces: analysis of effective topographic characteristics and evaluation of skin friction. *Biofouling*, **16** (2–4), 245–261.

102. Pineda, J. & Caswell, H. (1997) Dependence of settlement rate on suitable substrate area. *Marine Biology*, **129**, 541–548.

103. Yu, X., Yan, Y. & Gu, J.-D. (2007) Attachment of the biofouling bryozoan Bugula neritina larvae affected by inorganic and organic chemical cues. *International Biodeterioration & Biodegradation*, **60** (2), 81–89.

104. Pomerat, J.D. & Reiner, E.R. (1942) The influence of surface angle and of light on the attachment of barnacles and their sedentary organisms. *Biological Bulletin*, **82**, 14–25.

105. Dahlstrom, M., Jonsson, H., Jonsson, P.R. & Elwing, H. (2004) Surface wettability as a determinant in the settlement of the barnacle Balanus improvisus (Darwin). *Journal of Experimental Marine Biology and Ecology*, **305** (2), 223–232.

106. Amsler, C.D., Shelton, K.L., Britton, C.J., Spencer, N.Y. & Greer, S.P. (1999) Nutrients do not influence swimming behaviour or settlement rates of Ectocarpus siliculosus (Phaeophyceae) spores. *Journal of Phycology*, **35**, 239–244.

107. Pech, D., Ardisson, P.L. & Bourget, E. (2002) Settlement of a tropical marine epibenthic assemblage on artificial panels: influence of substratum heterogeneity and complexity scales. *Estuarine Coastal and Shelf Science*, **55** (5), 743–750.

108. Hills, J.M., Thomason, J.C. & Muhl, J. (1999) Settlement of barnacle larvae is governed by Euclidean and not fractal surface characteristics. *Functional Ecology*, **13** (6), 868–875.
109. Berntsson, K.M., Jonsson, P.R., Lejhall, M. & Gatenholm, P. (2000) Analysis of behavioural rejection of micro-textured surfaces and implications for recruitment by the barnacle Balanus improvisus. *Journal of Experimental Marine Biology and Ecology*, **251** (1), 59–83.
110. Bers, A.V. & Wahl, M. (2004) The influence of natural surface microtopographies on fouling. *Biofouling*, **20** (1), 43–51.
111. Maldonado, M. & Uriz, M.J. (1998) Microrefuge exploitation by subtidal encrusting sponges: patterns of settlement and post-settlement survival. *Marine Ecology Progress Series*, **174**, 141–150.
112. Köhler, J., Hansen, P.D. & Wahl, M. (1999) Colonization patterns at the substratum-water interface: how does surface microtopography influence recruitment patterns of sessile organisms? *Biofouling*, **14** (3), 237–248.
113. Le Tourneux, F. & Bourget, E. (1988) Importance of physical and biological settlement cues used at different spatial scales by the larvae of Semibalanus balanoides. *Marine Biology*, **97**, 57–66.
114. Hills, J.M. & Thomason, J.C. (1998) The effect of scales of surface roughness on the settlement of barnacle (Semibalanus balanoides) cyprids. *Biofouling*, **12** (1–3), 57–69.
115. Bers, A.V., Prendergast, G.S., Zurn, C.M., Hansson, L.J., Head, R.M. & Thomason, J.C. (2006) A comparative study of the anti-settlement properties of mytilid shells. *Biology Letters*, **2**, 88–91.
116. Skinner, L.F. & Coutinho, R. (2005) Effect of microhabitat distribution and substrate roughness on barnacle Tetraclita stalactifera (Lamarck, 1818) settlement. *Brazilian Archives of Biology and Technology*, **48** (1), 109–113.
117. Granhag, L.M., Finlay, J.A., Jonsson, P.R., Callow, J.A. & Callow, M.E. (2004) Roughness-dependent removal of settled spores of the green alga Ulva (syn. Enteromorpha) exposed to hydrodynamic forces from a water jet. *Biofouling*, **20** (2), 117–122.
118. Carlon, D. & Olson, R. (1993) Larval dispersal distance as an explanation for adult spatial pattern in two Caribbean reef corals. *Journal of Experimental Marine Biology and Ecology*, **173** (2), 247–263.
119. Rasmussen, K., Willemsen, P.R. & Ostgaard, K. (2002) Barnacle settlement on hydrogels. *Biofouling*, **18** (3), 177–191.
120. Anderson, M.J. & Underwood, A.J. (1994) Effects of substratum on the recruitment and development of an intertidal estuarine fouling assemblage. *Journal of Experimental Marine Biology and Ecology*, **184**, 217–236.
121. People, J. (2006) Mussel beds on different types of structures support different macroinvertebrate assemblages. *Austral Ecology*, **31** (2), 271–281.
122. Holmes, S.P., Sturgess, C.J. & Davies, M.S. (1997) The effect of rock-type on the settlement of Balanus balanoides (L) cyprids. *Biofouling*, **11** (2), 137–147.
123. Huxley, R., Holland, D.L., Crisp, D.J. & Smith, R.S.L. (1984) Influence of oil-shale on intertidal organisms – effect of oil-shale surface-roughness on settlement of the barnacle Balanus balanoides (L). *Journal of Experimental Marine Biology and Ecology*, **82** (2–3), 231–237.
124. Thomason, J.C., Hills, J.M. & Thomason, P.O. (2002) Field-based behavioural bioassays for testing the efficacy of antifouling coatings. *Biofouling*, **18** (4), 285–292.
125. Pineda, J., DiBacco, C. & Starczak, V. (2005) Barnacle larvae in ice: survival, reproduction, and time to postsettlement metamorphosis. *Limnology and Oceanography*, **50** (5), 1520–1528.
126. Porri, F., Zardi, G., McQuaid, C. & Radloff, S. (2007) Tidal height, rather than habitat selection for conspecifics, controls settlement in mussels. *Marine Biology*, **152** (3), 631–637.
127. Baird, A.H, Babcock, R.C. & Mundy, C.P. (2003) Habitat selection by larvae influences the depth distribution of six common coral species. *Marine Ecology Progress Series*, **252**, 289–293.
128. Bertness, M.D., Gaines, S.D. & Wahle, R.A. (1996) Wind-driven settlement patterns in the acorn barnacle Semibalanus balanoides. *Marine Ecology Progress Series*, **137**, 103–110.
129. Eckman, J.E. (1990) A model of passive settlement by planktonic larvae onto bottoms of differing roughness. *Limnology and Oceanography*, **35** (4), 887–901.

130. Kerr, A., Beveridge, C.M., Cowling, M.J., Hodgkiess T., Parr, A.C.S. & Smith, M.J. (1999) Some physical factors affecting the accumulation of biofouling. *Journal of the Marine Biological Association of the United Kingdom*, **79** (2), 357–359.

131. Fusetani, N. (2004) Biofouling and antifouling. *Natural Product Reports*, **21**, 94–104.

132. Warburton, F.E. (1966) The behavior of sponge larvae. *Ecology*, **47** (4), 672–674.

133. Koehl, M.A.R., Strother, J.A., Reidenbach, M.A., Koseff, J.R. & Hadfield, M.G. (2007) Individual-based model of larval transport to coral reefs in turbulent, wave-driven flow: behavioral responses to dissolved settlement inducer. *Marine Ecology Progress Series*, **335**, 1–18.

134. Callow, M.E. (2002) Marine biofouling: a sticky problem. *Biologist*, **49** (1), 1–5.

135. Amsler, C.D. & Neushul, M. (1990) Nutrient stimulation of spore settlement in the kelps Pterygophora californica and Macrocystis pyrifera. *Marine Biology*, **107** (2), 297–304.

136. Callow, M.E. & Callow, J.A. (2000) Substratum location and zoospore behaviour in the fouling alga Enteromorpha. *Biofouling*, **15** (1–3), 49–56.

137. Prendergast, G.S., Zurn, C.M., Bers, A.V., Head, R.M., Hansson, L.J. & Thomason J.C. (2008) Field-based video observations of the surface exploratory behaviour of wild Semibalanus balanoides (L.) cypris larvae: effects of texture and gregariousness. *Biofouling*, **24** (6), 449–459.

138. Crisp, D.J. (1975) Surface chemistry and life in the sea. *Chemical Industry Developments*, **5**, 187–193.

139. Nott, J.A. & Foster, B.A. (1969) On the structure of the antennular attachment organ of the cypris larva of Balanus balanoides (L.). *Philosophical Transactions of the Royal Society of London Series B, Biological Sciences*, **256** (803), 115–134.

140. Brancato, M.S. & Woollacott, R.M. (1982) Effect of microbial films on settlement of bryozoan larvae (Bugula simplex, B. stolonifera and B. turrita). *Marine Biology*, **71** (1), 51–56.

141. Wieczorek, S.K. & Todd, C.D. (1997) Inhibition and facilitation of bryozoan and ascidian settlement by natural multi-species biofilms: effects of film age and the roles of active and passive larval attachment. *Marine Biology*, **128**, 463–473.

142. Pires, A. & Woollacott, R.M. (1983) A direct and active influence of gravity on the behavior of a marine invertebrate larva. *Science*, **220** (4598), 731–733.

143. Hurlbut, C.J. (1993) The adaptive value of larval behavior of a colonial ascidian. *Marine Biology*, **115**, 253–262.

144. Svane, I. & Dolmer, P. (1995) Perception of light at settlement: a comparative study of two invertebrate larvae, a scyphozoan planula and a simple ascidian tadpole. *Journal of Experimental Marine Biology and Ecology*, **187**, 51–61.

145. Orlov, D. (1996) Observations on the settling behaviour of planulae of Clava multicornis Forskål (Hydroidea, Athecata). *Scientia Marina*, **60** (1), 121–128.

146. Gili, J.M., Hughes, R.G. & Rossi, S. (2000) The effects of exposure to wave action on the distribution and morphology of the epiphytic hydrozoans Clava multicornis and Dynamena pumila. *Scientia Marina*, **64** (1), 135–140.

147. Altieri, A.H. (2003) Settlement cues in the locally dispersing temperate cup coral Balanophyllia elegans. *Biological Bulletin*, **204** (3), 241–245.

148. Sebens, K.P. (1983) The larval and juvenile ecology of the temperate octocoral Alcyonium siderium Verrill. I. Substratum selection by benthic Larvae. *Journal of Experimental Marine Biology and Ecology*, **71** (1), 73–89.

149. Orlov, D.V. (1997) The role of larval settling behaviour in determination of the specific habitat of the hydrozoan Dynamena pumila (L.) larval settlement in Dynamena pumila (L.). *Journal of Experimental Marine Biology and Ecology*, **208** (1–2), 73–85.

150. Scheltema, R.S., Williams, I.P., Shaw, M.A. & Loudon, C. (1981) Gregarious settlement by the larvae of Hydroides dianthus (Polychaeta: Serpulidae). *Marine Ecology Progress Series*, **5**, 69–74.

151. Jensen, R.A. & Morse, D.E. (1990) Chemically induced metamorphosis of polychaete larvae in both the laboratory and ocean environment. *Journal of Chemical Ecology*, **16** (3), 911–930.

152. Qian, P.-Y. (1999) Larval settlement of polychaetes. *Hydrobiologia*, **402**, 239–253.

153. Maldonado, M. & Young, C.M. (1996) Effects of physical factors on larval behavior, settlement and recruitment of four tropical demosponges. *Marine Ecology Progress Series*, **138**, 169–180.
154. Leys, S.P. & Degnan, B.M. (2001) Cytological basis of photoresponsive behavior in a sponge larva. *Biological Bulletin*, **201** (3), 323–338.
155. Ilan, M. & Loya, Y. (1990) Sexual reproduction and settlement of the coral reef sponge Chalinula sp. from the Red Sea. *Marine Biology*, **105** (1), 25–31.
156. Chia, F.S. & Bickell, L.R. (1978) Mechanisms of larval attachment and the induction of settlement and metamorphosis in coelenterates: a review. In: *Settlement and Metamorphosis of Marine Invertebrate Larvae* (eds F.S. Chia & M.E. Rice), pp. 1–12. Elsevier, New York.
157. Kobak, J. (2001) Light, gravity and conspecifics as cues to site selection and attachment behaviour of juvenile and adult Dreissena polymorpha Pallas, 1771. *Journal of Molluscan Studies*, **67** (2), 183–189.
158. Dobretsov, S.V. & Railkin, A.I. (1996) Effects of substrate features on settling and attachment of larvae in blue mussel Mytilus edulis (Mollusca, Filibranchia). *Zoologichesky Zhurnal*, **75** (4), 499–506.
159. Dobretsov, S. & Wahl, M. (2001) Recruitment preferences of blue mussel spat (Mytilus edulis) for different substrata and microhabitats in the White Sea (Russia). *Hydrobiologia*, **445** (1), 27–35.
160. Bayne, B.L. (1964) Primary and secondary settlement in Mytilus edulis L. (Mollusca). *Journal of Animal Ecology*, **33** (3), 513–523.
161. Stagg, M. (2003) Behavioural bioassays for non-biocidal coatings. PhD thesis, Newcastle University.
162. Lin, Q., Gourdon, D., Sun, C., et al. (2007) Adhesion mechanisms of the mussel foot proteins mfp-1 and mfp-3. *Proceedings of the National Academy of Sciences*, **104** (10), 3782–3786.

Chapter 4
Succession on Hard Substrata

Stuart R. Jenkins and Gustavo M. Martins

The aim of this chapter is to provide an overview of the processes and interactions which occur after organisms have settled onto hard substrata. Following settlement, marine benthic organisms must cope with a range of stresses imposed by the novel benthic environment. We examine how complex ecological communities develop on hard substrata within the context of various models of succession. We review these models and assess the role of disturbance, life history characteristics of settling organisms and seasonality in determining the pattern and trajectory of succession. We use the large body of work which has used marine hard substrata to understand such interactions in order to gain insight into succession on fouled surfaces.

4.1 Introduction

The concept of succession has a central place in ecology. It refers to the sequence of species replacements and the change in the composition or structure of ecological communities following the provision of new unoccupied habitat. Such provision may occur through some form of physical or biological disturbance to existing communities or after the creation of entirely new habitat, owing to, for example a lava flow, severe landslide or anthropogenic activity. Whilst early work on succession, conducted in terrestrial habitats, emphasised a series of orderly species replacements, culminating in a single climax community (e.g. [1, 2]), it is now recognised that the successional process is not so predictable (see reviews [3, 4]). The order of colonisation may change from place to place and time to time, there may be several different endpoints and owing to a range of physical and biological disturbance events, succession never really stops. Much of the experimental work over the past decades which has elucidated models and mechanisms of succession has taken place on marine hard substrata and hence is directly applicable to the concept of biofouling.

4.2 Succession and the role of disturbance

Succession may be characterised as either primary or secondary, depending on the extent to which surfaces are already colonised. Primary succession refers to the colonisation of entirely virgin substrata. In the marine environment, this can occur through volcanic activity producing subtidal lava flows, fracturing of rock surfaces, for example in exposed intertidal and shallow subtidal environments, and through a whole range of anthropogenic activities introducing submerged structures to the sea. Secondary succession refers to colonisation of a disturbed habitat where established communities are partially removed. Here physical or biological disturbance creates new resources (often space), but unlike primary succession, colonisation occurs within an environment where other organisms are already established.

In considering fouling of hard surfaces in the marine environment, both primary and secondary successions are relevant. Introduction of submerged structures such as breakwaters, oil and gas platforms and the hulls of ships to the sea provides unfouled surfaces which are rapidly colonised through primary succession. However, secondary succession is equally relevant, since fouled surfaces, like any other hard surfaces in shallow waters, are subjected to a range of physical and biological disturbance regimes which open up free unoccupied space. Open space is a key resource for most sessile marine organisms, providing a site for attachment as well as potential access to key resources of light and food supply.

The important role of disturbance in structuring marine benthic communities was highlighted by the pioneering work of Dayton [5] on the Pacific coast of North America. He demonstrated the effects of wave action and specifically the physical effects of drifting logs battering the intertidal shoreline, in providing new space for colonisation and influencing the outcome of interspecific competition. His work and hundreds of subsequent scientific studies show that disturbance provides the trigger for ecological succession. In a wider context, disturbance also plays a key role in maintaining species diversity (see [6] for review). Pickett and White [4] define disturbance as any discrete event in time or space that changes the structure of populations, communities or ecosystems, and changes the resources available or the physical environment. Disturbance can be the result of both physical (i.e. landslides, wave action, ice scouring) and biological (i.e. predation, grazing) processes and, although the former is most often associated with the term, both sources of disturbance are generally thought to have equivalent effects on the structure of communities [3]. Disturbance is generally non-selective, although various forms of natural disturbance (for example where species sensitivities vary) and anthropogenic disturbance (for example exploitation of key marine invertebrates [7]) may result in proportionally greater loss of certain taxa. Where disturbance selectively removes species with strong interactions (e.g. key herbivores or predators), the consequences for successional change may be disproportionate to the level of physical disturbance.

The extent of resources made available by disturbance is very much a function of the intensity of the disturbance itself. Extreme disturbances (usually physical), such as storms, earthquakes, ice scouring events and eruptions, can make available a large amount of resources [8,9]. However, such large-scale disturbances are typically infrequent and stochastic, such that there are few opportunities to study their effects [10]. More frequently, small-scale disturbances create gaps of space that are readily colonised. One area of disturbance ecology which has received relatively little attention is the kind of non-catastrophic disturbance which only partially removes small portions of the established community [11]. This may actually form the most common type of disturbance in natural marine assemblages. Airoldi [12] contrasted the effects of differing disturbance intensity on succession in marine subtidal communities. She showed that colonisation in completed denuded areas was slow and allowed species other than the previous residents to become established. In contrast, in areas that were abraded (but organisms were not completely removed) recovery of the resident members was rapid. Space was quickly regained by disturbance-resistant species, which prevented the colonisation of other species through space pre-emption. Other intertidal studies [9] confirm this general prediction of more rapid recovery to a pre-disturbed state in plots which are only partially disturbed.

4.3   Models of succession

The influential paper of Connell and Slatyer [13] in the late 1970s defines three alternative models of community succession: facilitation, tolerance and inhibition, which differ in the mechanisms that are proposed to drive species replacements. The facilitation model assumes that, of those species arriving at available substrata, only certain 'early successional' species are able to colonise owing to their particular life history characteristics. These early colonisers modify the environment so that it is more suitable for other 'late successional' species. They therefore facilitate the colonisation of other species. The 'tolerance' and 'inhibition' models differ from that of facilitation in that any species arriving at a disturbed or new area is capable of colonising. There is no requirement for previous settlers to modify conditions. In the tolerance model, environmental modifications, brought about by early colonisers, neither promote nor retard colonisation by later arriving taxa. Here the sequence of species is determined solely by their life history characteristics. Later colonising species are generally more efficient at utilising resources and hence ultimately out-compete early colonisers, by depriving them of resources. For example, shade-tolerant algae may grow up within the shade cast by early colonisers, but ultimately out-compete them through depriving them of light (e.g. [14]). The third model, inhibition, proposes that early colonisers inhibit the settlement and growth of later arrivals. Here succession proceeds through the replacement of early colonisers when they are damaged or die. In general, late colonising species will gradually accumulate in the community through outliving early species.

It is worth noting that under all scenarios Connell and Slatyer [13] consider there are certain species that will usually appear because they have evolved characteristics such as abundant, widely dispersing propagules, rapid growth and ability to withstand variable environmental stresses (see below). Because such species are typically not adapted to germinating and growing in densely occupied sites, then under all models, as succession proceeds, the environment is modified such that it becomes unsuitable for further recruitment of these early successional species.

A variety of manipulative experiments on hard substrata have been undertaken to investigate the validity of these models (e.g. [15–19]). All models have been supported to a certain extent acting either independently or in conjunction with another model within any particular successional sequence. Numerous examples of inhibition have been demonstrated, including many intertidal studies demonstrating inhibition of late arriving algae by early algal settlers (see [20] for review). Subtidally, Sutherland and Karlson [21] showed that the identity of early colonisers of settlement plates depended entirely on when plates were submerged and succession could only proceed when initial colonisers were removed, suggesting inhibitory effects of all species on each other. There are fewer clear examples of facilitation, possibly because of the greater experimental difficulties in demonstrating the need for certain species to be present to allow colonisation by another. Facilitation is most likely to occur in stressful environments where stress tolerant early colonisers can ameliorate environmental conditions [22]. In primary succession on virgin substrata, the biochemical conditioning of surfaces, bacterial colonisation and then colonisation by unicellular organisms such as diatoms is frequently assumed a requirement before settlement of macroorganisms [23]. However, there are limited experimental data to support this assumption and although invertebrate larvae will preferentially settle on 'filmed' surfaces, they will settle on totally clean surfaces when no other option is available. Facilitation by macroorganisms probably occurs in mussel assemblages; this taxon rarely occurs early in colonisation and it is generally perceived that mussel larvae require some filamentous or complex structure to allow attachment [24].

The models of Connell and Slatyer [13] have been criticised for their oversimplified nature, lacking appreciation of seasonal and spatial variation in recruitment, growth and mortality, of the density-dependent nature of species interactions and the prevalence of indirect interactions. Sousa and Connell [20], however, point out that a general model of succession can never be expected to incorporate all sources of variation. There is no doubt about the influence of this work on subsequent experimental studies of succession. One very useful advance on the general model was put forward by Farrell [25] who considered the effect of consumers on the rate of succession in each of Connel and Slatyer's three alternative models [13]. Consumers may target early or late successional species or alternatively show no preference. Consumption of late successional species is predicted to retard succession, whatever the underlying model (facilitation, tolerance or inhibition). However preferential consumption of early successional species will advance species replacements in the case of the inhibition model, but retard this process when these species are required to facilitate the arrival of later species.

Experimental evidence of the positive effects of consumers on the rate of succession was presented by Sousa [15] on an intertidal boulder shore. Here, disturbed lower shore boulders were rapidly colonised by *Ulva*, which was a superior competitor for space in this system. As long as this alga remained healthy, it prevented the colonisation of later species (inhibition model). However, consumption of *Ulva* by the herbivorous crab, *Pachygrapsus crassipes*, breaks this inhibitory effect and accelerates succession, allowing establishment of long-lived red algae. Similar results were recorded by Lubchenco [26] on the mid-shore rocky intertidal area, where succession was accelerated by grazing periwinkles that removed early successional ephemerals, thus promoting the establishment of the late successional *Fucus*.

4.4 Extension of General Models

More recent work has shown how a variety of factors, both internal and external to the community, can modify the outcomes based on these general models. The extent to which the process of succession is variable and context-dependent has been addressed by a number of authors. For example, Berlow [27] addressed the question of how historical events affect the predictability of successional patterns in mussel beds. Succession showed complex patterns based on the interaction of different historical factors including the timing of disturbance and the presence or absence of key competitors and predators. However, some consistent successional trends were identified as a consequence of 'noise-dampening' forces, which included life history trade-offs in early and late successional species (see below) and strong predictable biotic interactions. Benedetti-Cecchi [28] used simple interaction webs based on individual species life history traits. He built up a series of qualitative models based on the outcomes of competition among key species and whose predictions could easily be tested. He put forward this approach as a means of improving the predictive capabilities of general models of succession.

A well-used example demonstrating the influence of species interactions on the successional development of communities is that of the three-way interaction between algae, barnacles and limpets on intertidal rocky shores [17, 19, 29]. In the absence of limpets, algae promptly monopolise space and prevent the recruitment and survival of barnacles. However, where present, limpets significantly decrease algal cover, indirectly leading to a barnacle-dominated assemblage. However, as barnacles grow larger, there is a negative effect on the foraging activity of limpets, which indirectly create escapes for algae. Succession in this assemblage is therefore mediated by a set of direct and indirect interactions among the organisms present and suggests that indirect interactions may not always feed back positively for a species. Such examples, and countless others, present increasing evidence of the ecological importance of indirect effects during succession and hence suggest that understanding of successional processes during biofouling will be based on a solid knowledge of the system and species life history traits [30].

4.5   Life history characteristics

The models of Connell and Slatyer [13] are based on the premise that a subset of species in the pool of available colonists is most likely to colonise newly disturbed or created space. These species, labelled 'early successional', arrive first because of certain key life history traits. They generally have abundant and freely available propagules and rapid growth rates and can be considered 'opportunistic' species. In contrast, mid- and late-successional species tend to be better overall competitors. They rely on efficient utilisation of resources to out-compete early colonists. Such differences in life history characteristics are probably the single most important factor driving predictable patterns of succession. Given finite resources, it is difficult for any particular species to be a 'master of all trades' and typical life history traits are based on trade-offs between different characteristics such as dispersal and competitive ability [27].

The extensive literature from experimental studies on rocky intertidal substrata provides a huge resource of information on species life history traits which can be utilised to understand mechanisms of succession on fouled surfaces. Early colonists of disturbed rocky substrata tend to be opportunistic and fast-growing species which are available for colonisation throughout the year. This group of species typically include the microalgae of biofilms (e.g. diatoms, cyanobacteria) and foliose and filamentous algae belonging to genera such as *Ulva* (syn. *Enteromorpha* [31]), *Polysiphonia*, *Ceramium*, *Porphyra* and *Ectocarpus*. Taxa such as *Ulva* release spores all year round and may have very large dispersal shadows [32]; hence they tend to be available as colonists whenever opportunity arises. Later colonists tend to be perennial algae with more complex, often upright morphologies, and common invertebrates, such as barnacles and mussels (e.g. [17, 19, 26, 29]) which may eventually dominate space. Such taxa may have distinct reproductive periods with propagules available only for a short period of time. Additionally, many perennial algal species have a very limited range of dispersal [33] and hence their role as potential colonists is limited to areas where adults are abundant. Such characteristics do not dictate that these species cannot arrive as early colonisers, merely that on average they are likely to arrive later in a colonising sequence (if at all).

The sequence of early to late colonising species follows the general pattern of increasing body size, and increasing competitive ability. The possession of distinct differences in life history strategies within colonising species defines a range of probable trajectories which form only a small subset of what would be possible, hence ensuring some level of consistency (or canalisation) on patterns of succession [27].

4.6   Patch characteristics

Natural disturbed patches vary considerably in size and shape [34] and the influence of these factors on the rate and pattern of succession has received considerable attention in a variety of marine systems including boulder fields [35, 36], seagrass beds [37], the rocky intertidal [34, 38] and rock pools [39, 40]. The area of a disturbed patch or new substrata will dictate the number and identity of colonising propagules since a larger area will generally 'sample' a greater number of propagules per unit time than a smaller one [41]. In this sense, a new man-made structure introduced to the sea represents a huge 'patch' and may be expected to be colonised by a greater diversity of biota than a small patch produced through a discrete disturbance event. Differing identity and diversity of colonisers will undoubtedly influence the course of succession, although the extent to which generalities can be made is probably limited.

Patch size is also an important determinant of the mode of re-colonisation (see below). With increasing patch size, there is a corresponding increase in the importance of propagule supply and settlement from the plankton relative to vegetative re-growth from the surrounding community. Essentially patch size determines the importance of edge effects [34, 38]. Small patches (and those with complex shapes) are likely to be colonised via vegetative growth from the adjacent intact community, hence succession may be dominated by organisms abundant in the surrounding community. In larger patches, however, recruitment from the plankton can be the predominant mechanism of re-colonisation [42]. The extreme of this continuum of patch size is reached on entirely virgin substrata through introduction of surfaces into the sea. Thus initial colonisation of new (or recently de-fouled) ships will occur entirely through planktonic means.

Differences in patch size may affect succession by alteration of the internal physical environment. This may be particularly true in patchy habitats such as crevices, rock pools or ponds. For instance, Martins *et al.* [40] found significant effects of rock pool size on the composition and structure at both early and late stages in succession, which was related to increasing stability in the physical environment with pool size [43]. Large pools generally supported greater species diversity which may potentially lead to greater complexity of competitive interactions [44].

Other unexpected consequences of patch size on succession can be mediated by grazers. Sousa [35] showed experimentally that the size of clearances within mussel beds had a strong indirect influence on succession. Small patches supported higher densities of grazers, especially limpets. As a result, the algal assemblage in small patches consisted of grazer-resistant species (*Analipus*, *Endocladia*, *Cladophora*) which were competitively inferior to grazer vulnerable species (ulvoids, fucoids) which occupied larger clearings. In effect small patches appeared to offer refuges from competition for grazer-resistant species. Such concentration of grazers in small patches has also been recorded by other investigators (e.g. [38, 45]) and is thought to be related to the increased protection from physical (desiccation, wave action) and biological (predation) stresses provided by the surrounding intact community.

4.7 Mode of colonisation

Different modes of colonisation of new or disturbed habitat will have an important effect on the rate and trajectory of succession. Depending on the extent to which virgin substrata is exposed and the proximity of established organisms, colonisation may occur by vegetative re-growth of survivors within a patch (e.g. [12]), germination and growth of propagules within a patch that survive the disturbance (e.g. [46]), lateral encroachment from the surrounding undisturbed assemblage by vegetative growth (e.g. [47]) or recruitment from planktonic propagules (e.g. [48]).

In those situations in which recruitment from the plankton may represent the main form of colonisation, the pattern of succession will firstly be dictated solely by those taxa with a planktonic stage. Secondly the timing of disturbance or creation of new substrata in relation to the reproductive periods of colonists will clearly have an important influence on the successional trajectory [49, 50]. Thirdly the spatial pattern of recruitment of potential colonists is also of obvious importance, particularly where dispersal distances are low. The influential role of spatial and temporal variability in recruitment in modifying successional trajectories is well demonstrated by the work of Hawkins and Hartnoll [29, 50–52] on exposed rocky shores of the Isle of Man in the Irish Sea. A conceptual model of the generation and maintenance of patchiness was developed, centred on the importance of grazing by the limpet *Patella vulgata*. Transition between various states (limpet, barnacle or fucoid dominated) is accelerated or slowed by the recruitment inputs of key species. For example, enhanced recruitment and/or survival of *Patella* can retard the development of *Fucus* populations, whereas high *Fucus* recruitment can swamp the ability of grazing patellids to control algal growth.

The ability of certain taxa to become established on recently created or cleared areas may depend on its local abundance and ability to disperse. Sousa [35] noted that in the absence of grazers, the recruitment of late successional species was highly correlated with its proximity and local abundance, suggesting that the majority of their propagules are dispersed relatively short distances (see [32, 33]). This suggests that succession of virgin substrata can be delayed depending on a nearby source of colonists.

Whilst colonisation through recruitment from planktonic propagules is frequently emphasised in fouling of hard surfaces, vegetative re-growth of survivors, lateral inward encroachment by juveniles or adults from the margins of disturbed areas and recruitment of propagules that survive the disturbance may play a more important role [33]. Vegetative re-growth is common in many algal assemblages and those dominated by colonial organisms. However, many sessile invertebrates (e.g. barnacles, mussels) are unable to vegetatively regenerate. Lateral encroachment of juvenile and adult mussels is a common feature of dynamic mosaics of mussel beds [53], while the growth and development of propagules that survive a disturbance event may be the dominant mechanism of regeneration in many algal assemblages [54].

The relative contribution of different modes of colonisation will clearly depend on numerous complex interacting factors including the extent and type of disturbance, the nature of the surrounding community, time of year and the nature of environmental stress. Bulleri and Benedetti-Cecchi [55] examined the balance between differing forms of re-colonisation in rocky intertidal areas. They found that local processes (within patch, e.g. edge effects) prevailed over larger scale processes (among patches, e.g. regional pool of propagules) in determining early patterns of colonisation of space and that patches at the boundary of different habitats were relatively more stable and resilient than patches well within each habitat.

4.8   Seasonality

The timing of disturbance or creation of new substrata can have an enormous impact on succession because of seasonality in organisms' reproductive patterns and/or growth and seasonal variation in environmental conditions. A vast body of experimental evidence has shown that the timing of disturbance can have a dramatic effect on the trajectory and rate of succession [16,45,56–59]. For example, Hawkins [29] initiated algal colonisation on intertidal rocky shores at different times of the year through exclusion of grazers. The sequence of colonisation varied between seasons: *Fucus* sporelings grew directly on primary substratum in summer, but in autumn *Fucus* was preceded by diatoms and in winter and spring by diatoms, and then green algae. In none of the situations was there evidence of facilitation of early colonisers on the recruitment of *Fucus*. Differences in the sequence of colonisation were considered to be due to the availability of propagules and environmental conditions. This illustrates the effect of life history traits on the seasonal abundance of propagules and hence on the sequence of colonisation.

Experimental studies have often shown that time of clearing has an effect on the rate or trajectory of succession but not on the endpoint (see below), with the stronger competitor eventually monopolising space, despite initial differences in early succession. However, in some cases, there is more than one possible endpoint and time of disturbance can determine the dominance at late succession through priority effects. Benedetti-Cecchi [60] showed that on the west coast of Italy, littoral rock pools can be dominated by either canopy or turf-forming algae and that dominance is set by the timing of disturbance. Canopy dominated rock-pools only occurred where disturbances coincided with the main peak of recruitment of these algae. Outside this period, turfs were able to monopolise the space and prevent the recruitment of canopy algae. Such pre-emption of space by turfing algae can have long-lasting effects. This suggests that timing of clearance may become increasingly important in areas where more than one species with strong competitive abilities occur.

Seasonal variation in environmental conditions can affect the establishment, survival and growth of numerous taxa and are therefore thought to have an influence on the composition and structure of communities, and hence on succession. Kaehler and Williams [61] showed that on a tropical shore the composition of algae developing on bare space depended on season (cool or hot), probably because of the differing ability among taxa to withstand periods of adverse environmental conditions.

4.9   Variable endpoints of succession

The occurrence of distinct communities within the same habitat and under similar environmental conditions has always intrigued ecologists, in general, and marine ecologists, in particular, who have endeavoured to understand the mechanisms and processes that drive community divergence. The distribution and composition of communities may be strongly dependent on events occurring early in succession [27,62]. In other words, history matters in community development. However, much experimental work suggests that varying successional trajectories ultimately converge towards the local community. Hence different communities within the same environment are merely transitory points along a successional continuum. Recent work questions the extent to which such conclusions are biased, through small-scale experimental work, and suggests a scale dependency to such observations. In the Gulf of Maine, USA, Petraitis and co-workers [48, 63, 64] hypothesise that macroalgal stands (*Ascophyllum nodosum*) and barnacle/mussel beds may represent alternate stable states. They experimentally demonstrated that convergence late in succession in this system is scale-dependent and that variability in the outcomes of succession increases with disturbance intensity. Experimental clearings of varying size, mimicking the effects of ice scour were made within dense stands of *A. nodosum*. In the smaller clearings, post-settlement predation and edge effects (e.g. sweeping of fronds) were more intense and precluded the recruitment of mussels and *Fucus* and the assemblage converged towards the surrounding community. In contrast, in the larger clearings, recruitment of *A. nodosum*, a poor disperser, was low and assemblages diverged towards mussel or *Fucus*-dominated as a function of variable recruitment and biological interactions. The proposal of alternate stable states in this environment has proved contentious [65, 66] but has re-invigorated the examination of physically induced differences in succession and community development.

Alternative states may also arise due to biological disturbances through dramatic changes in the abundance of species with strong feedbacks on the community. The proliferation of no-take marine reserves has generally shown that worldwide depletion of commercially important species can have strong cascading effects through the community. For instance, exploitation of top-predator fish in both temperate and tropical regions [67, 68] has resulted in dramatic effects on successional trajectories leading to community divergence. One such example is that presented by Guidetti [67] in the Adriatic Sea, Italy. The greater abundance of predatory fish inside marine reserves significantly decreased the abundance of sea urchins, a key grazer in the Mediterranean Sea. As a consequence, two functionally different algal assemblages occurred within and outside marine reserves: one dominated by upright turf-forming algae (within) and the other dominated by the grazing-resistant encrusting coralline algae, 'barrens' (outside).

The question as to whether such divergent communities represent alternate stable states has been the focus of lively debate [20, 49, 69, 70]. In South Africa, exploitation is believed to have shifted the balance of community structure from a community regulated by strong top-down control by the rock lobster (*Jasus lalandii*) to a community regulated by a predatory mollusc (*Burnupena papyracea*). By contrast, over-harvesting of intertidal fauna in South America has led to significant but reversible changes in community structure. When harvesting was restricted along parts of the Chilean coastline, a dramatic change occurred in intertidal community structure over the following 5 years [71, 72]. Hence shifts in community composition may at times represent an alternate stable state, while in other regions divergent assemblages are only maintained while disturbance persists with the system returning to its original state when disturbance declines.

4.10   Conclusions

- Experimental investigations of succession on marine hard substrata have contributed considerably to understanding of the biofouling process.
- Any introduction of hard surfaces to the sea will be followed by series of species replacements which will change the composition and structure of the fouling assemblage through time.

- The successional process is in part predictable, with early opportunistic species being replaced by more competitive longer lived taxa. However, an array of direct and indirect species interactions, stochastic recruitment events and physical–environmental change dictates that the fouling assemblage generated will follow a range of successional trajectories with potentially variable endpoints.
- The highly variable disturbance regime to which hard substrata in shallow marine environments are subjected ensures that composition of fouling communities will show continual change through the generation of a mosaic of patches at differing stages in succession.

References

1. Clements, F. (1928) *Plant Succession and Indicators*. H.W. Wilson, New York.
2. Clements, F. (1936) Nature and the structure of the climax. *Journal of Ecology*, **24**, 252–284.
3. Sousa, W.P. (1984) The role of disturbance in natural communities. *Annual Review of Ecology and Systematics*, **15**, 353–391.
4. Pickett, S.T.A. & White, P.S. (1985) *The Ecology of Natural Disturbance and Patch Dynamics*. Academic Press, New York.
5. Dayton, P.K. (1971) Competition, disturbance and community organisation: the provision and subsequent utilization of space in a rocky intertidal community. *Ecological Monographs*, **41**, 351–389.
6. Petraitis, P.S., Latham, R.E. & Niesenbaum, R.A. (1989) The maintenance of species diversity by disturbance. *The Quarterly Review of Biology*, **64**, 393–418.
7. Castilla, J.C. (1999) Coastal marine communities: trends and perspectives from human-exclusion experiments. *Trends in Ecology & Evolution*, **14**, 280–283.
8. McCook, L.J. & Chapman, A.R.O. (1997) Patterns and variations in natural succession following massive ice-scour of a rocky intertidal seashore. *Journal of Experimental Marine Biology and Ecology*, **214**, 121–147.
9. Underwood, A.J. (1998) Grazing and disturbance: an experimental analysis of patchiness in recovery from a severe storm by the intertidal alga *Hormosira banksii* on rocky shores in New South Wales. *Journal of Experimental Marine Biology and Ecology*, **231**, 291–306.
10. Turner, M., Baker, W., Peterson, C. & Peet, R. (1998) Factors influencing succession: lessons from large, infrequent natural disturbances. *Ecosystems*, **1**, 511–523.
11. Platt, W.J. & Connell, J.H. (2003) Natural disturbances and directional replacement of species. *Ecological Monographs*, **73**, 507–522.
12. Airoldi, L. (1998) Roles of disturbance, sediment stress, and substratum retention on spatial dominance in algal turf. *Ecology*, **79**, 2759–2770.
13. Connell, J.H. & Slayter, R.O. (1977) Mechanisms of succession in natural communities and their role in community stability and organization. *American Naturalist*, **111**, 1119–1144.
14. Keser, M. & Larson, B.R. (1984) Colonization and growth of *Ascophyllum nodosum* (Phaeophyta) in Maine. *Journal of Phycology*, **20**, 83–87.
15. Sousa, W.P. (1979) Experimental investigations of disturbance and ecological succession in a rocky intertidal algal community. *Ecological Monographs*, **49**, 227–254.
16. Breitburg, D.L. (1985) Development of a subtidal epibenthic community – factors affecting species composition and the mechanisms of succession. *Oecologia*, **65**, 173–184.
17. van Tamelen, P.G. (1987) Early successional mechanisms in the rocky intertidal: the role of direct and indirect interactions. *Journal of Experimental Marine Biology and Ecology*, **112**, 39–48.
18. Benedetti-Cecchi L. & Cinelli, F. (1996) Patterns of disturbance and recovery in littoral rock pools: nonhierarchical competition and spatial variability in secondary succession. *Marine Ecology Progress Series*, **135**, 145–161.
19. Kim, J.H. (1997) The role of herbivory, and direct and indirect interactions, in algal succession. *Journal of Experimental Marine Biology and Ecology*, **217**, 119–135.

20. Sousa, W.P. & Connell, J.H. (1985) Further comments on the evidence for multiple stable points in natural communities. *American Naturalist*, **4**, 612–615.

21. Sutherland, J. & Karlson, R. (1977) Development and stability of the fouling community at Beaufort North Carolina. *Ecological Monographs*, **47**, 425–446.

22. Bertness, M.D. & Callaway, R.M. (1994) Positive interactions in communities. *Trends in Ecology & Evolution*, **9**, 191–193.

23. Wahl, M. (1989) Marine epibiosis. I. Fouling and antifouling: some basic aspects. *Marine Ecology Progress Series*, **58**, 175–189.

24. Seed, R. & Suchanek, T.H. (1992) Population and community ecology of *Mytilus*. In: *The Mussel Mytilus: Ecology, Physiology, Genetics and Culture* (ed. E. Gosling), pp. 87–169. Elsevier, Amsterdam.

25. Farrell, T.M. (1991) Models and mechanisms of succession: an example from a rocky intertidal community. *Ecological Monographs*, **6**, 95–113.

26. Lubchenco, J. (1983) *Littorina* and *Fucus*: effects of herbivores, substratum heterogeneity, and plant escapes during succession. *Ecology*, **64**, 1116–1123.

27. Berlow, E.L. (1997) From canalization to contingency: historical effects in a successional rocky intertidal community. *Ecological Monographs*, **67**, 435–460.

28. Benedetti-Cecchi, L. (2000) Predicting direct and indirect interactions during succession in a midlittoral rocky shore assemblage. *Ecological Monographs*, **70**, 45–72.

29. Hawkins, S.J. (1981) The influence of season and barnacles on algal colonization of *Patella vulgata* (L.) exclusion areas. *Journal of the Marine Biological Association of the United Kingdom*, **61**, 1–15.

30. Menge, B.A. (1995) Indirect effects in marine rocky intertidal interaction webs: patterns and importance. *Ecological Monographs*, **65**, 21–74.

31. Hayden, H.S., Blomster, J., Maggs, C.A., Silva, P.C., Stanhope, M.J. & Waaland, J.R. (2003) Linnaeus was right all along: *Ulva* and *Enteromorpha* are not distinct genera. *European Journal of Phycology*, **38**, 277–294.

32. Norton, T.A. (1992) Dispersal by macroalgae. *British Phycological Journal*, **27**, 293–301.

33. Santelices, B. (1990) Patterns of reproduction, dispersal and recruitment in seaweeds. *Oceanography and Marine Biology Annual Review*, **28**, 177–276.

34. Airoldi, L. (2003) Effects of patch shape in intertidal algal mosaics: roles of area, perimeter and distance from edge. *Marine Biology*, **143**, 639–650.

35. Sousa, W.P. (1984) Intertidal mosaics: patch size, propagule availability, and spatially variable patterns of succession. *Ecology*, **65**, 1918–1935.

36. McGuinness, K.A. (1987) Disturbance and organisms on boulders. 2. Causes of patterns in diversity and abundance. *Oecologia*, **71**, 420–430.

37. Reed, B.J. & Hovel, K.A. (2006) Seagrass habitat disturbance: how loss and fragmentation of eelgrass *Zostera marina* influences epifaunal abundance and diversity. *Marine Ecology Progress Series*, **326**, 133–143.

38. Farrell, T.M. (1989) Succession in a rocky intertidal community: the importance of disturbance size and position within a patch. *Journal of Experimental Marine Biology and Ecology*, **128**, 57–73.

39. Underwood, A.J. & Chapman, M.G. (1998) A method for analysing spatial scales of variation in composition of assemblages. *Oecologia*, **117**, 570–578.

40. Martins, G.M., Hawkins, S.J., Thompson, R.C. & Jenkins, S.R. (2007) Community structure and functioning in intertidal rock pools: effects of pool size and shore height at different successional stages. *Marine Ecology Progress Series*, **329**, 43–55.

41. Sousa, W.P. (2001) Natural disturbance and the dynamics of marine benthic communities. In: *Marine Community Ecology* (eds M. Bertness, S. Gaines & M. Hay), pp. 85–130. Sinauer Associates, Sunderland, Massachusetts.

42. Whitlatch, R.B., Lohrer, A.M., Thrush, S.F., *et al.* (1998) Scale-dependent benthic recolonization dynamics: life stage-based dispersal and demographic consequences. *Hydrobiologia*, **375/376**, 217–226.

43. Metaxas, A., Hunt, H.L. & Scheibling, R.E. (1994) Spatial and temporal variability of macrobenthic communities in tidepools on a rocky shore in Nova-Scotia, Canada. *Marine Ecology Progress Series*, **105**, 89–103.

44. Keough, M.J. (1984) Effects of patch size on the abundance of sessile marine invertebrates. *Ecology*, **65**, 423–437.

45. Benedetti-Cecchi, L. & Cinelli, F. (1993) Early patterns of algal succession in a mid-littoral community of the Mediterranean Sea: a multifactorial experiment. *Journal of Experimental Marine Biology and Ecology*, **169**, 15–31.

46. Worm, B., Lotze, H., Bostrom, C., Engkvist, R., Labanauskas, V. & Sommer, U. (1999) Marine diversity shift linked to interactions among grazers, nutrients and propagule banks. *Marine Ecology Progress Series*, **185**, 309–314.

47. Dethier, M.N. (1984) Disturbance and recovery in intertidal pools: maintenance of mosaic patterns. *Ecological Monographs*, **54**, 99–118.

48. Dudgeon, S.R. & Petraitis, P.S. (2001) Scale-dependent recruitment and divergence of intertidal communities. *Ecology*, **82**, 991–1006.

49. Sutherland, J.P. (1974) Multiple stable points in natural communities. *American Naturalist*, **108**, 859–873.

50. Hawkins, S.J. (1981) The influence of *Patella* grazing on the fucoid-barnacle mosaic on moderately exposed rocky shores. *Kieler Meeresforschung*, **5**, 537–543.

51. Hawkins, S.J. (1983) Interaction of *Patella* and macroalgae with settling *Semibalanus balanoides* (L.). *Journal of Experimental Marine Biology and Ecology*, **71**, 55–72.

52. Hartnoll, R.G. & Hawkins, S.J. (1985) Patchiness and fluctuations on moderately exposed rocky shores. *Ophelia*, **24**, 53–64.

53. Tanaka, M.O. & Magalhaes, C.A. (2002) Edge effects and succession dynamics in Brachidontes mussel beds. *Marine Ecology Progress Series*, **237**, 151–158.

54. Hoffmann, A.J. & Santelices, B. (1991) Banks of algal microscopic forms: hypotheses on their functioning and comparisons with seed banks. *Marine Ecology Progress Series*, **79**, 1–2.

55. Bulleri, F. & Benedetti-Cecchi, L. (2006) Mechanisms of recovery and resilience of different components of mosaics of habitats on shallow rocky reefs. *Oecologia*, **149**, 482–492.

56. Dayton, P.K., Currie, V., Gerrodette, T., Keller, B.D., Rosenthal, R. & Ven Tresca, D. (1984) Patch dynamics and stability of some California kelp communities. *Ecological Monographs*, **54**, 253–289.

57. Sousa, W.P. (1985) Disturbance and patch dynamics on rocky intertidal shores. In: *The Ecology of Natural Disturbance and Patch Dynamics* (eds S.T.A. Pickett & P.S. White), pp. 101–124. Academic Press, New York.

58. Jenkins, S.R., Coleman, R.A., Della Santina, P., Hawkins, S.J., Burrows, M.T. & Hartnoll, R.G. (2005) Regional scale differences in the determinism of grazing effects in the rocky intertidal. *Marine Ecology Progress Series*, **287**, 77–86.

59. Underwood, A.J. & Chapman, M.G. (2006) Early development of subtidal macrofaunal assemblages: relationships to period and timing of colonization. *Journal of Experimental Marine Biology and Ecology*, **330**, 221–233.

60. Benedetti-Cecchi, L. (2000) Priority effects, taxonomic resolution, and the prediction of variable patterns of colonisation of algae in littoral rock pools. *Oecologia*, **123**, 265–274.

61. Kaehler, S. & Williams, G.A. (1998) Early development of algal assemblages under different regimes of physical and biotic factors on a seasonal tropical rocky shore. *Marine Ecology Progress Series*, **172**, 61–71.

62. Chapman, M.G. & Underwood, A.J. (1998) Inconsistency and variation in the development of rocky intertidal algal assemblages. *Journal of Experimental Marine Biology and Ecology*, **224**, 265–289.

63. Petraitis, P.S. & Latham, R.E. (1999) The importance of scale in testing the origins of alternative community states. *Ecology*, **80**, 429–442.

64. Petraitis, P.S. & Dudgeon, S.R. (2005) Divergent succession and implications for alternative states on rocky intertidal shores. *Journal of Experimental Marine Biology and Ecology*, **326**, 14–26.

65. Bertness, M.D., Trussell, G.C., Ewanchuk, P.J. & Silliman, B.R. (2004) Do alternate stable community states exist in the Gulf of Maine rocky intertidal zone? Reply. *Ecology*, **85**, 1165–1167.

66. Petraitis, P.S. & Dudgeon, S.R. (2004) Do alternate stable community states exist in the Gulf of Maine rocky intertidal zone? Comment. *Ecology*, **85**, 1160–1165.

67. Guidetti, P. (2006) Marine reserves re-establish lost predatory interactions and cause wide-community changes in rocky reefs. *Ecological Applications*, **16**, 963–976.

68. Mumby, P.J., Harborne, A.R., Williams, J., et al. (2007) Trophic cascade facilitates coral recruitment in a marine reserve. *Proceedings for the National Academy of Science of the USA*, **104**, 8362–8367.

69. Connell, J.H. & Sousa, W.P. (1983) On the evidence needed to judge ecological stability or persistence. *American Naturalist*, **121**, 789–824.

70. Sutherland, J.P. (1990) Perturbations, resistance and alternative views of the existence of multiple stable points in nature. *American Naturalist*, **136**, 270–275.

71. Branch, G.M. & Moreno, C.A. (1994) Intertidal and subtidal grazers. In: *Rocky Shores: Exploitation in Chile and South Africa* (ed. R. Siegfried), pp. 75–100. Springer-Verlag, Berlin.

72. Thompson, R.C., Crowe, T.P. & Hawkins, S.J. (2002) Rocky intertidal communities: past environmental changes, present status and predictions for the next 25 years. *Environmental Conservation*, **29**, 168–191.

Chapter 5
Patterns of Fouling on a Global Scale

João Canning-Clode and Martin Wahl

The aim of this chapter is to seek and explain a possible trend in the distribution of biofouling assemblages at a global scale. For this, we have selected various case studies, one of them new, to provide an overview in plausible interpretations of the spatial variation of biofouling diversity. What types of biofoulers would settle on artificial panels deployed close to the tropics and those in higher latitudes? How fast is biomass accrual in these different locations? How do biofouling communities differ between regions with regard to taxonomic and functional properties? In an attempt to answer these questions, here we review the present understanding of diversity in biofouling communities at a global scale with special emphasis on the latitudinal gradient and on the relationship between region and local diversity in these assemblages.

5.1 Background

Global-scale patterns in biodiversity have challenged biologists since the eighteenth and nineteenth centuries [1]. In fact, large-scale patterns are one of the most studied trends in ecology [2]. The observation that the tropics hold more species than higher latitudes is the oldest paradigm about a large-scale ecological pattern [3,4]. Although not as well investigated as the latitudinal gradient, longitudinal gradients can also be recognised in some marine and terrestrial environments [5–7].

   A congregation of theories that aim to explain the spatial variation in biodiversity has been investigated [2]. For example, the species–area hypothesis is a relationship recognised since the beginning of the twentieth century in ecology [8,9]. It proposes that larger areas hold more species, since their larger populations and more stable conditions reduce the risk of extinction and because they contain more barriers and heterogeneous habitat which promote speciation [9,10]. The relationship between local and regional richness can be illustrated by plotting local species diversity against regional species diversity [11,12]. A linear relationship between local and regional diversity is expected if communities are unsaturated, and a curvilinear relationship if communities are saturated [13,14] (Figure 5.1). Other patterns of spatial variation in species richness were explored along environmental gradients (depth, exposure, emersion, pollution, peninsulas, bays, isolation, productivity/energy) [2]. Nonetheless, in the marine environment, there is still a lack of global studies as compared to the terrestrial environment [15]. While in recent decades patterns on a large scale have been acknowledged for some marine taxa [16–20], major trends in marine biodiversity are still poorly understood [21–23].

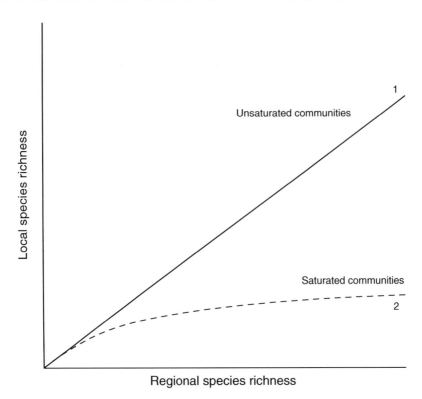

Figure 5.1   The local and regional diversity relationship. If local assemblages are unsaturated, community diversity will continue to rise with regional diversity in a linear way (solid line – line 1). If saturated with species, local communities will reach a plateau or maximum species richness (dashed line – line 2).

Biofouling communities are distributed worldwide, especially in shallow and exposed marine environments. They are among the most diverse and productive assemblages of macroorganisms. Their ecological importance and the fact that their abundance and distribution are currently affected by human activities (aquaculture, artificial reefs, ship hull transport) and climate change make them suitable for the study of global patterns.

5.2   Is there a latitudinal cline in fouling organisms?

Among the spatial patterns of biological diversity, the latitudinal gradient of species richness, which states that maximum species richness occurs in the tropics and decreases towards higher latitudes, is considered one of the most reliable patterns in ecology [9]. When compared to the terrestrial realm, however, this pattern appears to be less consistent in the sea [17, 22, 24]. Clear gradients have been reported, e.g. in bryozoans in the North Atlantic [25], molluscs in the western Atlantic and eastern Pacific oceans [19, 26, 27], deep-sea isopods, gastropods and bivalves in the North Atlantic [28], and in the intertidal sessile assemblages along the northwestern Pacific coast [29]. In contrast, studies on polychaetes and nematodes [17, 30, 31], macroalgae [32] and on marine soft sediments [33, 34] did not find a latitudinal gradient of diversity.

Hillebrand [13] performed a meta-analysis on approximately 600 published articles and concluded that generally marine organisms demonstrate a decrease in species richness towards the poles but the strength and slope of the gradient depend on regional, habitat and organism characteristics. Sessile organisms, i.e. the major constituents of biofouling communities, showed a weak relationship to latitude at both local and regional scales [13].

In a recent comparison of benthic marine algae, Kerswell [35] performed a meta-analysis based on 191 species lists from primary literature. Data were compiled at the genus level for the classes Rhodophyceae, Phaeophyceae and Chlorophyceae and at species level for all algae belonging to the order Bryopsidales. This latter order was treated with higher resolution because it is a group well studied and taxonomically stable. The author concluded that algal genera display an inverse latitudinal gradient with highest diversity in temperate regions. In contrast, number of species of the order Bryopsidales was maximum in the tropics and decreases towards the poles [35]. The genus richness was highest in southern Australia and Japan and moderate in the Indo-Australian Archipelago and southern Indian Ocean. Algal richness in the Atlantic Ocean was higher in the eastern coastline with a hotspot situated at the European coast. Lowest diversity in algal genera was described for the polar regions [35]. We cannot exclude that some research bias contributes to certain features of this distributional pattern (high diversity on European shores, low diversity in Polar regions). A common explanation for a decrease in algal richness in the tropics is its competition with corals and elevated herbivore pressure at low latitudes. According to Kerswell [35], global patterns of algal richness can partially be explained by the species–area hypothesis. Bryopsidalean species richness is closely related to corals and reef fish diversity, implying a common regulatory mechanism. The mesoscale location of algal-richness hotspots may be determined by major ocean currents through propagule dispersal and changes in oceanic conditions [35].

A recent review [36] discussed the different hypotheses around the commonly observed latitudinal diversity pattern. They conclude that (i) the tropical diversity maximum is geologically very old and (ii) speciation rates are higher in the tropics because (iia) molecular evolution rates in ectotherms is higher in warmer climates, (iib) larger biome areas enhance the chance for geological or ecological isolation of sub-populations and (iic) more biotic interactions – in a positive feedback – drive specialisation and speciation. As a result, tropical diversification rates are higher due to faster speciation and slower extinction. Species originating in the tropics tend to disperse to higher latitudes while retaining their presence in the tropics [36].

Recently, Witman *et al.* [37] conducted an investigation on epifaunal invertebrate communities encrusting subtidal vertical rock wall habitats using high resolution photos. The authors compared 12 biogeographic regions from 62° S to 63° N latitude to test the effects of latitude and richness of the regional species pool on local species richness. In their study, local diversity at each site was estimated as the asymptote of the species accumulation curve, using both the number of species observed (S_{obs}) and Chao2 [38] as a local species richness estimator. Regional diversity was assembled from published species lists and by consulting taxonomic experts in each of the 12 different biogeographic regions. They reported that the sessile communities were from 10 different invertebrate phyla, where sponges, cnidarians and ascidians were the dominant groups [37]. They also found a clear latitudinal pattern with maximum species richness at low latitudes at both regional and local scales (Figure 5.2). However, this pattern was more robust at a regional scale [37].

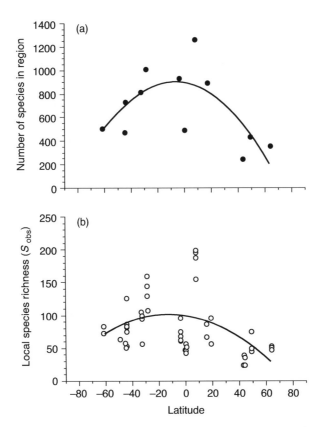

Figure 5.2 Species richness as a function of latitude. (a) Regional species richness; (b) local species richness as number of species observed (S_{obs}). Lines represent significant, best fits to second-order polynomial equations. (Reprinted from Reference [37]. Copyright (2007) National Academy of Sciences, USA.)

5.3 The pattern revealed by a global modular fouling experiment

The core of a Global Approach by Modular Experiment (GAME) project [39] conducted by our group was to achieve a broad knowledge in community ecology by conducting identical experiments in different coastal shallow water hard-bottom ecosystems of both hemispheres. Identical experimental designs allow consistent comparisons and global interpretations. Here we analyse 23 GAME studies performed between 2003 and 2007 in 16 biogeographic regions from 42° S to 59° N latitude (Table 5.1). All studies were focused on a benthic ecology topic, e.g. the intermediate disturbance hypothesis in biofouling communities [41, 42, 46, 47, 49, 53, 55, 59], temporal variability of disturbances [60–43] and diversity and stability of marine communities [40, 44, 48, 50, 51, 54, 57, 61, 62]. In all regions, 15 × 15 cm polyvinylchloride (PVC) panels were submerged for colonisation at approximately 0.5 m depth during a period between 5 and 8 months. For the purpose of this analysis, only common sessile species (mean abundance >1% cover) on untreated control panels ($n = 6$) were taken into consideration.

Table 5.1   Dominant biofoulers on the PVC panels recorded from the 23 studies analysed from the GAME project [39] between 2003 and 2007.

Location	Latitude	Longitude	Total species	Dominant algae	Dominant invertebrates	References
Finland	59° N	23° E	9	*Cladophora rupestris*, *Ectocarpus siliculosus*	*Balanus improvisus*, *Electra crustulenta*	[40]
Sweden	58° N	11° E	39	*Ceramium rubrum*, *Polysiphonia fucoides*	*Ciona intestinalis*, *Ascidiella aspersa*, *Laomedea flexuosa*	[41]
England	54° N	1° W	13	—	*Ascidiella aspersa*	[42]
			12	*E. siliculosus*	*Ascidiella aspersa*, *Botrylloides leachi*	[43]
			25	*E. siliculosus*	*Ascidiella aspersa*, *Didemnum* spp.	[44]
Poland	54° N	18° E	11	*Enteromorpha ahlneriana*, *C. rupestris*	*Mytilus edulis*, *B. improvisus*	[45]
Italy	43° N	10° E	21	*Cladophora* sp., *Ceramium* sp.	*M. edulis*, *Serpula vermicularis*	[46]
Japan	38° N	141° E	51	*Ceramium kondoi*, *Ulva arasakii*, *Sargassum horneri*	*Tricellaria occidentalis*, *Watersipora subovoidea*, *Mytilus galloprovincialis*	[47]
	35° N	139° E	48	*Enteromorpha intestinalis*	*Molgula manhattensis*, *Balanus amphitrite*, *M. galloprovincialis*	[48]
Portugal	32° N	16° W	36	*Lithophyllum incrustans*, *Polysiphonia* sp.	*Diplosoma* sp., *S. vermicularis*	[49]
			32	*Cladophora coelothrix*, *Ceramium virgatum*	Family Serpuliadae	[50]
Malaysia	5° N	102° E	23	*Padina australis*, *Lobophora variegata*	*B. amphitrite*	[51]
			21	Rodophyta	*Balanus* sp.	[52]
Brazil	22° S	43° W	25	*Bryopsis pennata*, *Codium taylorii*	*B. neritina*, *Schizoporella errata*	[53]

(Continued)

Table 5.1 (*Continued*)

Location	Latitude	Longitude	Total species	Dominant algae	Dominant invertebrates	References
			32	*Enteromorpha* sp., *Ulva* sp.	*B. neritina*, *Styela* sp.	[54]
	23° S	42° W	27	Phaeophyceae	*B. trigonus*, *Megabalanus* sp.	[47]
Chile	29° S	71° W	32	*Ulva* sp. *Polysiphonia* sp.	*Pyura chilensis*, *Diplosoma* sp.	[55]
			40	*Ulva* sp.	*Austromegabalanus psittacus*, *P. chilensis*	[56]
			25	*Ectocarpus* sp., *Polysiphonia* sp.	*P. chilensis*, *Diplosoma* sp., *B. neritina*	[57]
Australia	34° S	150° E	25	—	*Hydroides elegans*, *B. leachi*, *Pyura stolonifera*	[58]
Australia	35° S	138° E	30	Encrusting coralline algae, *Ulva* sp., *Mychodea* sp.	*Galeolaria* sp.	[59]
New Zealand	36° S	174° E	33	brown filamentous algae; *Ulvella sp.*, *Acrochaetium* sp.	*B. trigonus*, *Elminius modestus* *Elminius modestus*	[60]
			28	*Ralfsia* sp., *Ulvella* sp.	*B. trigonus, E. modestus*	[61]
Tasmania Australia	42° S	147° E	26	*Scytosiphon, lomentaria Asperococcus bullosus*	*E. modestus*, *Watersipora subtorquata*, *Cryptosula pallasiana Cryptosula pallasiana*	[62]

Note: Site coordinates are indicated. Total species refer to the total number of species observed during the course of the study in each location.

According to the GAME results, the most common biofoulers worldwide were ascidians, bryozoans, barnacles and mussels. High abundances of the ascidians *Ascidella aspersa* and *Ciona intestinales* were observed in several sites in the northern hemisphere. In contrast, the presence of the genera *Styela* and *Pyura* were more frequent in the southern hemisphere. Furthermore, species like *Botrylloides leachi* and *Diplosoma* sp. were present in both hemispheres (Table 5.1). In what is concerned to bryozoans, the genus *Watersipora* was dominant in two sites from different hemispheres (Tasmania and Japan). Considerable abundances of *Bugula neritina* were verified in Brazil and Chile. Barnacles belonging to the genus *Balanus* had a very broad distribution. The species *Balanus improvisus* and *Balanus amphitrite* were more

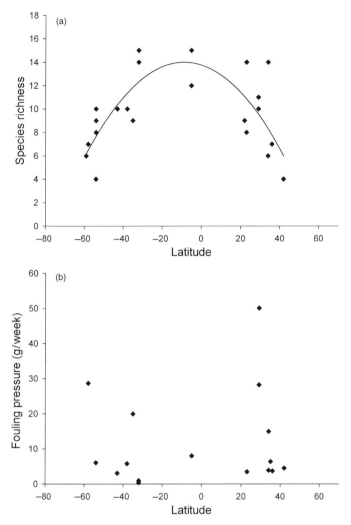

Figure 5.3 The GAME analysis: relationship between macrofouling species richness and latitude (a) and biofouling pressure across latitude assessed as wet weight (g) accrual per time (weeks) (b).

common in the north (e.g. Finland, Poland, Japan and Malaysia), while *Balanus trigonus* was regularly seen on panels deployed in the southern hemisphere (e.g. Brazil, New Zealand). The mussels *Mytilus edulis* and *Mytilus galloprovincialis* were also prevailing in various locations above 32° N. In addition, macroalgae had a vast distribution with the genera *Ulva*, *Ceramium* and *Polysiphonia* being common in several sites from both hemispheres (Table 5.1). A latitudinal diversity cline was found. Species diversity was significantly greater in tropical areas than at higher latitudes ($R^2 = 0.52$; $p = 0.0003$; Figure 5.3(a)). Within this general pattern, regional differences are evident. High species richness was found on Malaysia (5° N) and Madeira Island, Portugal (32° N), while minimum species richness was registered in Poland and Tasmania. Since the biofouling communities on the 5–8 months old panels in many cases cannot be considered 'mature', these results represent more a sampling of the diversity of the different coloniser pools.

The speed of community assemblage is likely to differ between regions. In order to investigate this, we calculated the biofouling pressure in 16 of the 23 GAME studies, as the quantity of biomass (wet weight) accrual (representing recruitment plus growth) per week and panel (Figure 5.3(b)). Biofouling pressure varied from 3 to 50 g wet weight of biomass per week. Latitude does not explain this large variability. Therefore, other factors than those linked to the sum parameter 'latitude' (as a surrogate for warm temperature, environmental constancy and high diversity) like regional salinity, productivity or species identity may override any global scale trend. Biofouling pressure was highest in Sweden and Chile where ascidians were the dominant organisms.

5.4   Are biofouling communities saturated? The relationship between local and regional diversity

Diversity in local communities is regulated by local and regional processes. Local processes include predation, competition and disturbance, while migration, speciation, extinction and historical events are regional processes [14, 63]. Moreover, local and regional richness are differentiated by spatial scale. Local diversity is measured on a scale in which all the species in the community potentially interact with each other in some unit of ecological time, typically a generation [12]. In contrast, regional diversity refers to a larger spatial scale like the biogeographic distribution of colonists in a location [64].

One means to separate evolutionary and ecological patterns consists in testing which proportion of regionally available species is represented in a local community. Ecological limitation (i.e. saturation) means that with increasing number of available species in the regional pool (i.e. along the latitudinal gradient) or with invasion events, local richness does not increase beyond an intrinsically determined maximum [64]. Therefore, based on ecological models that rely on species interactions, a saturation pattern is predicted when species are interacting and not expected when species are not or only weakly interacting [65, 66]. The relationship between regional and local richness can be demonstrated by plotting local species diversity against regional species diversity. Here, communities in which local diversity is linearly dependent on regional diversity over the entire range of regional diversities are referred to as 'unsaturated' or 'Type I' communities [67, 68], i.e. the number of coexisting species does not seem limited – at least during the period of observation. Alternatively, as regional richness increases, local diversity might reach a ceiling above which it does note rise despite further increases in regional diversity. In this case local communities are said to be 'saturated' with species and are referred to as 'Type II' communities [67, 68].

In the previously mentioned study conducted by Witman *et al.* [37], the exploration of the relationship between regional and local diversity in epifaunal invertebrate communities was another objective. To test saturation in epifaunal communities, they used the slopes in a log–log analysis. A steady rise of local species richness with increasing regional species richness is represented by a slope of 1 in a log–log plot, while local saturation (an asymptote) is represented by a slope significantly smaller than 1 [37]. They found that the slopes did not differ from 1 for both S_{obs} and Chao2. This indicates that the number of species coexisting in local communities is not limited within the size of the regional species pool ('Type I' communities, Figure 5.4). Witman *et al.* [37] further emphasised that the strong effect of

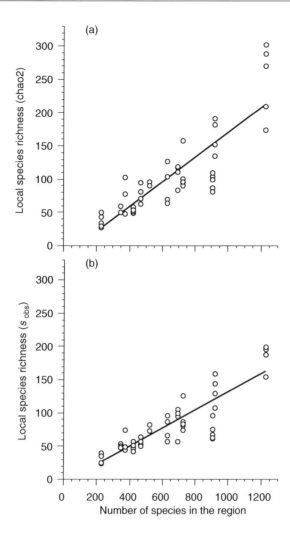

Figure 5.4 Plots of regional versus local species richness based on the Chao2 [38] (a) and S_{obs} (b) estimates of species richness. Lines represent significant, best fits by least-squares linear regression. (Reprinted from Reference [37]. Copyright (2007) National Academy of Sciences, USA.)

regional diversity on local communities involves that patterns of local diversity in marine benthic assemblages at a global scale are influenced by processes of speciation, migration, extinction and geologic events.

5.5   Discussion and future perspectives

The examples presented in this overview have shown a clear relationship between latitude and biofouling species richness. The study performed by Witman *et al.* [37] as well as the analysis we presented here with the GAME data both demonstrated that the tropical regions hold more biofouling species when compared to areas at higher latitudes. Our findings are, therefore, in accordance to Hillebrand's meta-analysis [13] where from 600 published gradients the author found a significant average tendency of decreasing marine biodiversity with increasing latitude. This global diversity pattern reflects that the regions closer to the tropics are richer in biofouling species than at higher latitudes, and highlights the need of protection for certain species in these areas. Factors such as predation, age, competition, climate, productivity and regional species pool have been proposed as possible explanations for the latitudinal gradient [1, 4, 9]. More recently, interest in evolutionary and historical explanations for the gradient is increasing with the availability of molecular, phylogenetic and palaeontological data [36]. Furthermore, in their interpretation of the relationship between local and regional diversity, Witman *et al.* [37] found a 'Type I' curve which indicates that the number of coexisting species is unsaturated.

The most striking outcome of this review is that, while large scale patterns have been recognised for some taxonomic groups [16–20, 35] and habitats, e.g. intertidal [15, 29, 69], in the past, there is not a satisfying understanding about global patterns and their causes in biofouling communities. The relative simplicity by which such assemblages can be studied and manipulated on artificial substrata should facilitate further latitudinal and regional comparisons.

The two investigations described here have found the latitudinal pattern at different depths, community age and substrata. In contrast to the GAME investigation, Witman *et al.* [37] reported that sponges were dominant organisms in their survey which might have been influenced by depth (see Chapter 6 for the influence of depth in biofouling patterns). To avoid this heterogeneity, standardisation in the methodology of future biofouling studies with regard to depth, age and substrata would facilitate global comparisons.

In the GAME analysis, only one study was carried out in the tropics proper (Malaysia) leaving a sampling gap. More knowledge in biofouling assemblages in areas between 30° N and 20° S latitude might have created a different pattern than the one we found. Additionally and in contrast to the study conducted by Witman *et al.* [37], our analysis did not include a study site in the Antarctic region. Although little is known concerning biofouling assemblages in Antarctica (but see [70]), for a more adequate and accurate global perspective on biofouling organisms this area should be included in future surveys.

A relevant fact that requires further investigation is the presence of non-native biofouling species in regional diversity and its influence in the global diversity pattern. The introduction of non-native species in the marine realm is considered a relevant threat for the biodiversity of the marine system [71]. Recent studies have revealed ship traffic as the main vector for marine non-native species. Species transported in the ballast water of vessels as well as sessile and sedentary organisms growing on ship hulls are considered a major source of invaders [72]. Common biofoulers like barnacles, bryozoans and mussels present life forms that make possible their survival on ship hulls (see Chapter 24). Moreover, the availability of open space has been demonstrated as a limiting resource in marine hard-bottom assemblages (e.g. [73]). In a study to assess the effects of diversity on the invasion of sea squirts in a subtidal epifaunal community, Stachowicz *et al* [74] observed that space availability decreased and was more consistent over time, in more diverse communities. In their study, they have demonstrated that the survival and percent cover of the invaders was reduced with diversity [74]. This may indicate that locations close to the tropics, i.e. that hold more different species, are more resistant to the invasion process.

Finally, the question whether community richness is ecologically limited or relates directly to regional richness is not only essential for our understanding of an ecosystem. In addition, in the context of the dramatic decrease of species numbers on both global and regional scales, we must be able to evaluate at which point of the process the richness and functionality of local communities will be affected. Furthermore, the relationship between local and regional richness may be affected by other parameters such as facilitation and abiotic stress [75]. It is certainly of importance which section of the regional richness gradient is examined.

5.6   Conclusions

- The latitudinal gradient and the relationship between local and regional species richness are common spatial patterns that seek to explain biological diversity.
- A clear relationship between latitude and biofouling diversity was reported in two different investigations. Moreover, one of the studies showed that local diversity was dependent on regional diversity in a linear way.
- A poor understanding in biofouling global trends still persists. This fact highlights the need for additional large-scale studies focusing on these assemblages.

Acknowledgements

We thank Dr. Sergey Dobretsov for his constructive comments on this manuscript. We further thank all students involved in the GAME project between 2003 and 2007. J. Canning-Clode studies were supported by a scholarship from the German Academic Exchange Service (DAAD).

References

1. Ricklefs, R.E. (2004) A comprehensive framework for global patterns in biodiversity. *Ecology Letters*, **7** (1), 1–15.
2. Gaston, K.J. (2000) Global patterns in biodiversity. *Nature*, **405** (6783), 220–227.
3. Hillebrand, H. (2004) On the generality of the latitudinal diversity gradient. *American Naturalist*, **163** (2), 192–211.
4. Willig, M.R., Kaufman, D.M. & Stevens, R.D. (2003) Latitudinal gradients of biodiversity: pattern, process, scale, and synthesis. *Annual Review of Ecology, Evolution and Systematics*, **34**, 273–309.
5. Callisto, M. & Goulart, M. (2005) Invertebrate drift along a longitudinal gradient in a neotropical stream in Serra do Cipo National Park, Brazil. *Hydrobiologia*, **539**, 47–56.
6. Jetz, W. & Rahbek, C. (2001) Geometric constraints explain much of the species richness pattern in African birds. *Proceedings of the National Academy of Sciences of the United States of America*, **98** (10), 5661–5666.
7. Roberts, C.M., McClean, C.J., Veron, J.E.N., *et al.* (2002) Marine biodiversity hotspots and conservation priorities for tropical reefs. *Science*, **295** (5558), 1280–1284.
8. Arrhenius, Q. (1921) Species and area. *Journal of Ecology*, **9**, 95–99.
9. Rosenzweig, M.L. (1995) *Species Diversity in Space and Time*. Cambridge University Press, Cambridge.
10. Chown, S.L. & Gaston, K.J. (2000) Areas, cradles and museums: the latitudinal gradient in species richness. *Trends in Ecology & Evolution*, **15** (8), 311–315.
11. Caley, M.J. & Schluter, D. (1997) The relationship between local and regional diversity. *Ecology*, **78** (1), 70–80.
12. Krebs, C.J. (2001) *Ecology, The Experimental Analysis of Distribution and Abundance*, 5th edn. Benjamin Cummings, San Francisco.

13. Hillebrand, H. (2004) Strength, slope and variability of marine latitudinal gradients. *Marine Ecology Progress Series*, **273**, 251–267.

14. Hillebrand, H. & Blenckner, T. (2002) Regional and local impact on species diversity – from pattern to processes. *Oecologia*, **132** (4), 479–491.

15. Kuklinski, P., Barnes, D.K.A. & Taylor, P.D. (2006) Latitudinal patterns of diversity and abundance in North Atlantic intertidal boulder-fields. *Marine Biology*, **149** (6), 1577–1583.

16. Giangrande, A. & Licciano, M. (2004) Factors influencing latitudinal pattern of biodiversity: an example using Sabellidae (Annelida, Polychaeta). *Biodiversity and Conservation*, **13** (9), 1633–1646.

17. Gobin, J.F. & Warwick, R.M. (2006) Geographical variation in species diversity: a comparison of marine polychaetes and nematodes. *Journal of Experimental Marine Biology and Ecology*, **330** (1), 234–244.

18. Holmes, N.J., Harriott, V.J. & Banks, S.A. (1997) Latitudinal variation in patterns of colonisation of cryptic calcareous marine organisms. *Marine Ecology Progress Series*, **155**, 103–113.

19. Roy, K., Jablonski, D., Valentine, J.W. & Rosenberg, G. (1998) Marine latitudinal diversity gradients: tests of causal hypotheses. *Proceedings of the National Academy of Sciences of the United States of America*, **95** (7), 3699–3702.

20. Stehli, F.G. & Wells, J.W. (1971) Diversity and age patterns in hermatypic corals. *Systematic Zoology*, **20**, 115–126.

21. Clarke, A. (1992) Is there a latitudinal diversity cline in the sea? *Trends in Ecology & Evolution*, **7** (9), 286–287.

22. Clarke, A. & Crame, J.A. (1997) Diversity, latitude and time: patterns in the shallow sea. In: *Marine Biodiversity: Patterns and Processes* (eds R.F.G. Ormond, J.D. Gaje, & M.V. Angel), pp. 122–147. Cambridge University Press, Cambridge.

23. Gray, J.S. (2001) Antarctic marine benthic biodiversity in a world-wide latitudinal context. *Polar Biology*, **24** (9), 633–641.

24. Rivadeneira, M.M., Fernandez, M. & Navarrete, S.A. (2002) Latitudinal trends of species diversity in rocky intertidal herbivore assemblages: spatial scale and the relationship between local and regional species richness. *Marine Ecology Progress Series*, **245**, 123–131.

25. Clarke, A. & Lidgard, S. (2000) Spatial patterns of diversity in the sea: bryozoan species richness in the North Atlantic. *Journal of Animal Ecology*, **69** (5), 799–814.

26. Roy, K., Jablonski, D. & Valentine, J.W. (1994) Eastern Pacific molluscan provinces and latitudinal diversity gradient – no evidence for rapoports rule. *Proceedings of the National Academy of Sciences of the United States of America*, **91** (19), 8871–8874.

27. Roy, K., Jablonski, D. & Valentine, J.W. (2000) Dissecting latitudinal diversity gradients: functional groups and clades of marine bivalves. *Proceedings of the Royal Society of London Series B – Biological Sciences*, **267** (1440), 293–299.

28. Rex, M.A., Stuart, C.T. & Coyne, G. (2000) Latitudinal gradients of species richness in the deep-sea benthos of the North Atlantic. *Proceedings of the National Academy of Sciences of the United States of America*, **97** (8), 4082–4085.

29. Okuda, T., Noda, T., Yamamoto, T., Ito, N. & Nakaoka, M. (2004) Latitudinal gradient of species diversity: multi-scale variability in rocky intertidal sessile assemblages along the North-western Pacific coast. *Population Ecology*, **46** (2), 159–170.

30. Boucher, G. (1990) Pattern of nematode species-diversity in temperate and tropical subtidal sediments. *Pubblicazioni Della Stazione Zoologica Di Napoli I: Marine Ecology*, **11** (2), 133–146.

31. Mackie, A.S.Y., Oliver, P.G., Darbyshire, T. & Mortimer, K. (2005) Shallow marine benthic invertebrates of the Seychelles Plateau: high diversity in a tropical oligotrophic environment. *Philosophical Transactions of the Royal Society of London Series A*, **363** (1826), 203–227.

32. Santelices, B. & Marquet, P.A. (1998) Seaweeds, latitudinal diversity patterns, and Rapoport's rule. *Diversity and Distributions*, **4**, 71–75.

33. Ellingsen, K.E. & Gray, J.S. (2002) Spatial patterns of benthic diversity: is there a latitudinal gradient along the Norwegian continental shelf? *Journal of Animal Ecology*, **71** (3), 373–389.

34. Gray, J.S. (2002) Species richness of marine soft sediments. *Marine Ecology Progress Series*, **244**, 285–297.
35. Kerswell, A.P. (2006) Global biodiversity patterns of benthic marine algae. *Ecology*, **87** (10), 2479–2488.
36. Mittelbach, G.G., Schemske, D.W., Cornell, H.V., *et al.* (2007) Evolution and the latitudinal diversity gradient: speciation, extinction and biogeography. *Ecology Letters*, **10** (4), 315–331.
37. Witman, J.D., Etter, R.J. & Smith, F. (2004) The relationship between regional and local species diversity in marine benthic communities: A global perspective. *Proceedings of the National Academy of Sciences of the United States of America*, **111** (44), 15664–15669.
38. Colwell, R.K. & Coddington, J.A. (1994) Estimating terrestrial biodiversity through extrapolation. *Philosophical Transactions of the Royal Society of London Series B – Biological Sciences*, **345** (1311), 101–118.
39. GAME (2002–2009) Global Approach by Modular Experiments (GAME). Available from http://www.ifm-geomar.de/index.php?id=1325&L=1. Accessed 1 July 2009.
40. Stockhausen, B. (2007) *The role of community structure for invasion dynamics in marine fouling communities in the Baltic Sea.* MSc thesis, University of Cologne.
41. Svensson, J.R., Lindegarth, M., Siccha, M. *et al.* (2007) Maximum species richness at intermediate frequencies of disturbance: consistency among levels of productivity. *Ecology*, **88** (4), 830–838.
42. Sugden, H., Lenz, M., Molis, M., Wahl, M. & Thomason, J. (2008) The interaction between nutrient availability and disturbance frequency on the diversity of benthic marine communities on the north-east coast of England. *Journal of Animal Ecology*, **77** (1), 24–31.
43. Sugden, H., Panusch, R., Lenz, M., Wahl, M. & Thomason, J.C. (2007) Temporal variability of disturbances: is this important for diversity and structure of marine fouling assemblages? *Marine Ecology: An Evolutionary Perspective*, **28** (3), 368–376.
44. Lauterbach, L. (2007) *The role of community structure for invasion dynamics in marine fouling communities in the North Sea.* MSc thesis, University Stuttgart.
45. Schröder, S.E. (2006) *Effects of temporally variable disturbance regimes on structure and diversity of marine macrobenthic communities in the southern Baltic Sea.* MSc thesis, University of Cologne.
46. Spindler, G. (2005) *Composition of hard-bottom communities in a frequency-gradient of mechanical disturbance.* MSc thesis, Christian-Albrechts-University of Kiel.
47. Miethe, T. (2005) *Interactive effects of productivity and disturbance on the diversity of hard-bottom communities in Japan.* MSc thesis, Humboldt University.
48. Link, H. (2007) *Factors determining the stability of marine fouling communities in Tokyo Bay, Japan.* MSc thesis, University of Heidelberg.
49. Canning-Clode, J., Kaufmann, M., Wahl, M., Molis, M. & Lenz, M. (2008) Influence of disturbance and nutrient enrichment on early successional fouling communities in an oligotrophic marine system. *Marine Ecology: an Evolutionary Perspective*, **29** (1), 115–124.
50. Jochimsen, M.C. (2007) *Role of community structure for invasion dynamics.* MSc thesis, University of Osnabrück.
51. Weseloh, A. (2007) *The role of community structure for invasion dynamics in marine fouling communities in the South China Sea.* MSc thesis, University of Bremen.
52. Herbon, C. (2005) *Der Einfluss der zeitlichen Variabilität von Störung auf die Struktur mariner makrobenthischer Gemeinschaften.* MSc thesis, Christian-Albrechts-University of Kiel.
53. Jara, V.C., Miyamoto, J.H.S., da Gama, B.A.P., Molis, M., Wahl, M. & Pereira, R.C. (2006) Limited evidence of interactive disturbance and nutrient effects on the diversity of macrobenthic assemblages. *Marine Ecology Progress Series*, **308**, 37–48.
54. Wunderer, L. (2006) *Beeinflusst das Alter von Aufwuchsgemeinschaften ihre Resistenz gegenüber veränderten Umweltbedingungen? – Ein Transplantationsexperiment an der Südostküste Brasiliens.* MSc thesis, Ludwig-Maximilians-University München.
55. Valdivia, N., Heidemann, A., Thiel, M., Molis, M. & Wahl, M. (2005) Effects of disturbance on the diversity of hard-bottom macrobenthic communities on the coast of Chile. *Marine Ecology Progress Series*, **299**, 45–54.

56. Cifuentes, M., Kamlah, C., Thiel, M., Lenz, M. & Wahl, M. (2007) Effects of temporal variability of disturbance on the succession in marine fouling communities in northern-central Chile. *Journal of Experimental Marine Biology and Ecology*, **352** (2), 280–294.

57. Krüger, I. (2007) *Succession of fouling communities from the Chilean coast*. MSc thesis, Westfälische Wilhelms-University Münster.

58. Rich, C. (2005) *The effect of disturbance and its temporal variability on the structure and composition of a marine fouling community*. BSc thesis, University of Wollongong.

59. Valdivia, N., Stehbens, J.D., Hermelink, B., Connell, S.D., Molis, M. & Wahl, M. (2008) Disturbance mediates the effects of nutrients on developing assemblages of epibiota. *Austral Ecology*, **33** (8), 951–962.

60. Atalah, J., Otto, S.A., Anderson, M.J., Costello, M.J., Lenz, M. & Wahl, M. (2007) Temporal variance of disturbance did not affect diversity and structure of a marine fouling community in north-eastern New Zealand. *Marine Biology*, **153**, 199–211.

61. Wiesmann, L. (2006) *Stabilität mariner Aufwuchsgemeinschaften unter veränderten Umweltbedingungen im Nordosten Neuseelands*. MSc thesis, Westfälische Wilhelms-University Münster.

62. Keller, S. (2006) *Eine Untersuchung zum invasiven Potential frühsukzessionaler mariner Aufwuchsgemeinschaften in Abhängigkeit von sukzessionalem Stadium und Diversität*. MSc thesis, Carl von Ossietzky University.

63. He, F.L., Gaston, K.J., Connor, E.F. & Srivastava, D.S. (2005) The local-regional relationship: immigration, extinction, and scale. *Ecology*, **86** (2), 360–365.

64. Srivastava, D.S. (1999) Using local-regional richness plots to test for species saturation: pitfalls and potential. *Journal of Animal Ecology*, **68**, 1–16.

65. Loreau, M. (2000) Are communities saturated? On the relationship between alpha, beta and gamma diversity. *Ecology Letters*, **3** (2), 73–76.

66. Shurin, J.B. & Allen, E.G. (2001) Effects of competition, predation, and dispersal on species richness at local and regional scales. *American Naturalist*, **158** (6), 624–637.

67. Cornell, H.V. (1985) Local and regional richness of cynipine gall wasps on California oaks. *Ecology*, **66**, 1247–1260.

68. Cornell, H.V. & Lawton, J.H. (1992) Species interactions, local and regional processes, and limits to the richness of ecological communities: a theoretical perspective. *Journal of Animal Ecology*, **61**, 1–12.

69. Leonard, G.H. (2000) Latitudinal variation in species interactions: a test in the New England rocky intertidal zone. *Ecology*, **81** (4), 1015–1030.

70. Bowden, D.A., Clarke, A., Peck, L.S. & Barnes, D.K.A. (2006) Antarctic sessile marine benthos: colonisation and growth on artificial substrata over three years. *Marine Ecology Progress Series*, **316**, 1–16.

71. Hewitt, C.L. & Campbell, M.L. (2007) Mechanisms for the prevention of marine bioinvasions for better biosecurity. *Marine Pollution Bulletin*, **55** (7–9), 395–401.

72. Minchin, D. & Gollasch, S. (2003) Fouling and ships' hulls: how changing circumstances and spawning events may result in the spread of exotic species. *Biofouling*, **19**, 111–122.

73. Stachowicz, J.J. & Byrnes, J.E. (2006) Species diversity, invasion success, and ecosystem functioning: disentangling the influence of resource competition, facilitation, and extrinsic factors. *Marine Ecology Progress Series*, **311**, 251–262.

74. Stachowicz, J.J., Fried, H., Osman, R.W. & Whitlatch, R.B. (2002) Biodiversity, invasion resistance, and marine ecosystem function: Reconciling pattern and process. *Ecology*, **83** (9), 2575–2590.

75. Russell, R., Wood, S.A., Allison, G. & Menge, B.A. (2006) Scale, environment, and trophic status: the context dependency of community saturation in rocky intertidal communities. *American Naturalist*, **167** (6), 158–170.

Chapter 6
Biofouling Patterns with Depth

Phillip R. Cowie

The aim of this chapter is to outline the mechanisms which cause vertical zonation in subtidal epifaunal communities on natural substratum and describe how these mechanisms also relate to the zonation patterns displayed by biofouling communities. Examples have been selected to illustrate the historical development of subtidal studies and the schemes used to describe vertical community changes with depth, the main biotic and abiotic factors responsible for causing vertical zonation and the patterns of zonation displayed by biofouling communities. Deficiencies in our current knowledge of vertical zonation patterns at greater depths (>100 m) are also identified.

6.1 Introduction

Within the marine environment, there are a variety of hard surfaces, which are suitable for colonisation by epifaunal organisms, these surfaces can be classified as being either natural or man-made. Natural surfaces include primary substrata such as underwater rocks, reefs, hard ground, clastic rocks, stones and manganese nodules [1, 2] and secondary hard substrata formed by bioconstructors [3]. Sublittoral man-made hard materials in the marine environment include metals, plastics, modified concrete, pipelines and cables. Traditionally, research on the epifaunal communities of natural hard substratum has focussed on the development of assemblages within the intertidal zone and the abiotic and biotic factors that influence intertidal zonation patterns. These studies have shown that intertidal zonation is influenced by a variety of factors, including wave exposure, substratum orientation and composition, temperature, tidal dynamics and biological interactions [4–9].

In contrast, studies of the forcing factors, which influence the development of sessile communities on natural and man-made hard substrata in the subtidal, lagged behind those on the intertidal. This was due to the logistical and scientific difficulties associated with carrying out experiments in the sublittoral zone. Historically, much of the information on hard substratum epifaunal and biofouling communities was obtained from the use of opportunistic grab and dredge sampling, primitive diving equipment and the retrieval of panels or man-made materials from the marine environment [10–12]. The commercial availability of self-contained underwater breathing apparatus (SCUBA) to scientists in the 1950s enabled studies of subtidal epifaunal assemblages to be conducted in situ. Early SCUBA studies focussed on providing basic descriptions of epibenthic communities on different hard substrata [13–15]. It was clear

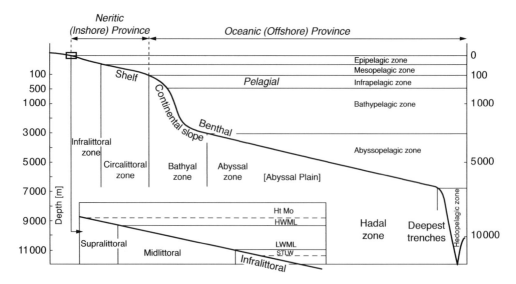

Figure 6.1 Main vertical and horizontal divisions of the marine environment. HtMo, highest level of moistening by high water spring tide or waves and sprays; HWML, high water mean level; LWML, low water mean level; STLW, spring tide low water. (Reprinted from Pérès [20]. Copyright O. Kinne.)

from these and subsequent studies that sublittoral communities were more complex than in the intertidal and that marked changes in community assemblages were apparent at different depths [16]. In view of the different characteristics of these zones and the different habitats that they offer, researchers developed schemes to categorise subtidal vertical zonation based on biological criteria where vertical zones are based on the plants and animals that characterise them [17–19]. Instead of measurements of environmental factors, these schemes use general features of the epibenthic assemblages which result from the cumulative, interferent or antagonistic influences of environmental factors on the organisms and the assemblages they constitute [18, 20]. Figure 6.1 shows the main vertical and horizontal divisions of the marine environment in the scheme developed by Pérès [20]. The precise limits and subdivisions of these zones vary between different geographical locations of the world [18, 19, 21]. This discussion will focus on zonation of subtidal communities in the infralittoral zone and lower.

Surveys and studies conducted during the last few decades on the subtidal regions of oil and gas platforms using advanced SCUBA techniques and remotely operated vehicles (ROVs) to satisfy regulatory authorities have also enabled scientists to obtain a greater understanding of the changes in community composition that occur with increased depth on man-made structures and the forcing factors involved in their development [22]. The combination of studies undertaken on both natural and man-made hard substrata has shown that a variety of abiotic and biotic factors contribute towards differences in the community structure of sessile assemblages that occurs with increasing depth. These factors include light levels, temperature, pressure, food and nutrient levels, substratum composition and availability, and the biology and physiology of individual species. Several of these factors are functions of another (e.g. interdependent), for example light penetration, salinity, temperature and food resources often change with depth [23].

6.2　Major forcing factors determining subtidal sessile assemblages

6.2.1　Light levels

Light availability is a primary factor in determining the proportion of photosynthetic algae within sessile and biofouling communities [24]. In the upper surface waters (within the photic zone), seaweeds may form a considerable component of biofouling communities, due to the availability of high light levels [25]. With increasing depth, there is an exponential decrease in the available light levels for photosynthesis as light is absorbed, scattered and refracted (Figure 6.2). Different wavelengths of light are absorbed and scattered at different rates in water and have different extinction coefficients with red light being attenuated most rapidly and blue light penetrating deepest in clear water [27]. Investigations of the effects of the reduction of light irradiance on benthic assemblages have traditionally focused on plants that being constrained by photosynthetic requirements are strongly influenced by light intensity [28]. With increasing depth, algal cover becomes sparser until even the most receptive plants cannot gain enough light. At approximately 150 m (in clear water), the levels of photosynthetically active radiation usable by photosynthetic organisms are about 1% of the light levels in the surface waters. At the lowest light levels, the crustose coralline algae are the only conspicuous algal growth form and one of the deepest recorded records of algal life was of a red crustose coralline alga at 268 m near the Bahamas [29]. Light levels can also influence epifauna, and sessile invertebrates tend to be more abundant on shaded surfaces [30]. Marine organisms and consequently the assemblages they constitute may be classified in relation to their light

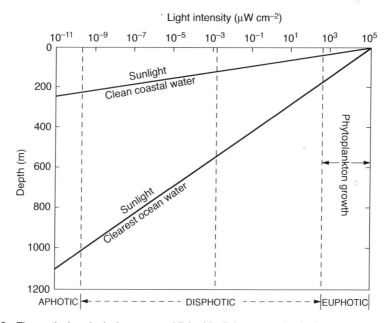

Figure 6.2　The vertical ecological zones established by light penetration in the sea. Note that the light intensity is logarithmic with depth. (Adapted from figure published in Reference [26], reproduced by permission of Elsevier, Copyright Elsevier.)

requirements or tolerances into photophilic and sciaphilic forms. Numerous intermediate levels of requirement and tolerance may exist between these two groups [20]. Variations in sea water turbidity can greatly alter the bathymetric distribution of photophilic and sciaphilic epibenthic assemblages by altering the extent of light penetration – either increasing or decreasing their typical levels [31]. In deeper waters, photosynthetic organisms are absent from biofouling communities. The deep-sea environment is dark below a depth of 300–900 m – depending on latitude, season and differences in suspended matter content – and light levels do not directly influence the vertical distribution of benthic epifauna which inhabit deeper water [32].

6.2.2   Temperature and Pressure

Similar to the pattern with light levels, with increasing depth there is a decrease in sea water temperature. The upper layers of oceans in the mid and low latitudes are affected by turbulent mixing (caused by winds and waves) which transfers heat from the surface waters to a few metres or several hundred metres deep. Below the mixed layer at depths of 200–300 m, the temperature decreases rapidly down to 1000 m and below [26]. The importance of temperature in the growth, reproduction and development of marine organisms is well known [33]. Where thermoclines do occur, distinctly different communities at different depths can occur as a result of the temperature regime above and below the thermocline. Above the thermocline, waters are subject to large temperature changes throughout the year, whilst below it such changes are slight. In coastal regions, the depth of the thermocline, which separates the two zones, is determined by both hydrodynamic conditions and the local topography. In calm waters, the thermocline might rise to 40 m, but generally the colder waters are limited to depths below 70–80 m [18]. In most oceanic areas, the water temperature at 2000–3000 m never rises above 4°C regardless of latitude. At greater depths, the temperature declines to between about 0°C and 3°C. The temperature of deep water at the equator is within a few degrees of that of deep water in polar regions [32].

In contrast with light and temperature, there is an increase in water pressure with depth. Pressure increases at the rate of 10^2 kPa every 10 m [34]. This means that the organisms living at great depths are subjected to very high pressures. In the deepest ocean basins, organisms exist at pressures exceeding 10^5 kPa. The combination of low temperatures and increased pressure means that species occupying lower bathyal, abyssal and hadal zones possess biochemical adaptations to allow them to cope with the testing conditions. Alterations in hydrostatic pressure have been found to affect gene expression, protein synthesis and the extent of lipid saturation in deep-sea bacteria [35] and the biochemistry and physiology of invertebrates and fish [36, 37]. In general terms, the species and communities found at great depths are highly specialised and are typically slower growing than shallow-water species and communities.

6.2.3   Food supplies and water currents

Biofouling communities in shallow waters can be described as being fuelled by photosynthetic organisms – many biofouling animals are suspension- and filter-feeders (e.g. barnacles, sponges, hydroids, bryozoa, tunicates and bivalve molluscs) which directly extract photosynthetic phytoplankton or their zooplankton grazers from the water column [23]. The availability of adequate supplies of microphagous food is crucial in the growth and development of biofouling assemblages. In shallow, coastal waters there is normally an adequate supply of food, in deeper waters (below the compensation depth) high levels of living phytoplankton are not present because of the absence of light. In the lower half of the circalittoral zone, the nutrition of sessile species depends in part on exogenous organic matter and in waters below the continental shelf edge (200 m depth) this dependence becomes total. The input of exogenous organic material stems from three sources: (1) planktonic organisms and their detritus which sink to the bottom from upper water layers, (2) detritus transport from land or shallow benthos and (3) the circadian migration of organisms [38]. In the case of the first two sources, the distribution of such exogenous food resources depends on water currents.

Epifaunal communities in both shallow coastal and deep offshore waters rely on a sufficient external flow of water to carry nutrient particles to them. Within limits, the more rapid the flow of water, the more food becomes available to the organism. In coastal waters, sessile organisms can benefit from tide, wind and residual currents. In the deep-sea, suspension-feeders were not expected to thrive because of low current flow speeds, low organic particle concentration and poor advective food flux [39]. However, photographs of hard substrata in deep-sea areas show that dense communities of suspension-feeding organisms can occupy hard substrata in some regions [40, 41]. Locally abundant epifaunal communities in some deep-sea habitats can occur as a result of topographically induced strong currents such as those which occur around seamounts and where suspension-feeders can dominate communities in some bathymetric zones [40, 42]. Sessile/suspension-feeders in the deep sea can also take advantage of benthic nepheloid layers (BNL) containing high loads of continental slope and locally derived suspended material and which can extend several hundred meters above the seafloor [43].

Currents are also important to sessile assemblages for two other main reasons: (1) they maintain the amounts of hard substrata available and (2) they influence the distribution and recruitment of planktonic larvae. Currents influence the nature of the substratum available for colonisation by epifauna: strong currents prevent sedimentation and leave bare hard substrata or coarse sediments, whereas weak currents result in high sedimentation rates and lead to silty or muddy substrata [20, 42]. Sessile organisms that produce planktonic larvae also rely on currents to disperse their larvae and populations of these species are maintained by the recruitment of competent larvae from the water column. In offshore locations, in the absence of large resident breeding populations in the vicinity, the settlement of some organisms on hard substrata is dependent on currents bringing larvae from coastal brood sites. Over a wide depth range, the watermasses influencing the larval and food supply of sessile communities may be different at the top and bottom of the depth range.

All the above factors combine with others such as competition and predation [44–46] and the biology and physiology of individual species to influence the bathymetric distribution of epifaunal species and communities, producing vertical zones on natural and man-made hard substrata. In ecological terms, a vertical zone can be defined as 'the depth interval of the benthic domain where the ecological conditions related to the main environmental factors are homogeneous or exhibit a gradient between two critical levels which corresponds to the boundaries of the zone'. Usually, the boundary between two adjacent vertical zones corresponds to a sharp change in the composition of the living assemblage [38]. However, a more gradual continuum of change in community composition may also be seen between different zones [47].

6.3   Patterns of vertical zonation on natural substratum subtidal communities

Vertical zonation has been recorded in many marine subtidal communities around the world. The main feature of vertical zonation is the arrangement of communities in belts with distinctive species compositions and dominance. Community changes with depth concern not only the composition, but also the organisation of the communities [48].

One of the longest histories of research on vertical zonation has been conducted on the hard bottom zoobenthos of the Mediterranean. Early studies described the complex communities thriving on hard substrata [17, 49]. These researchers identified two main zones in the Mediterranean between surface waters and the edge of the continental shelf: the infralittoral and circalittoral zones. The infralittoral zone is characterised by an upper limit corresponding to the highest level inhabited by species which cannot endure prolonged emergence and its lower limit corresponds to the maximal depth consistent with the survival of photophilic algae. Multicellular, photophilic soft algae typically dominate assemblages within the infralittoral. The circalittoral zone extends from the lowest boundary of the infralittoral zone down to the maximum depth where photoautotrophic multicellular algae can survive [20]. This zone is characterised by the presence of the sciaphilic algae community and increases in the abundance and diversity of epifaunal suspension-feeders (e.g. sponges, cnidarians, bryozoans, tunicates) at the expense of algal cover [49, 50]. Later studies also confirmed the importance of light as a major factor influencing the occurrence and depth distribution of shallow coralligenous assemblages in the Mediterranean Sea and the spatio-temporal variability of these communities [51]. The potential importance of other factors such as water currents, bottom topography and sedimentation on the distribution of growth forms and trophic guilds of epibenthic communities were also studied [31, 52].

Studies of subtidal communities in other regions of the world, e.g. Canada [16, 53], Norway [54], Greenland [55] and the United Kingdom [10, 14, 15, 56] have also examined the factors determining community composition with depth. These researchers, and others, have affirmed the importance of light gradients in determining community composition and other factors such as nutrient inputs, habitat heterogeneity, substratum composition, topography and water velocity.

Studies undertaken in other regions of the world have also shown that the vertical zonation schemes initially developed for Mediterranean coastal waters have general applicability worldwide (with modifications) to take into account the geographical differences which occur in subtidal zonation as a result of local interactions between the main forcing factors. In general, predicting patterns of subtidal epifaunal and epifloral zonation, even on a local scale, is greatly complicated by the spatial and temporal heterogeneity of these communities where differences among depths in assemblage structure can vary interactively with time and across horizontal scales. Overall, subtidal assemblages appear as a mosaic of patches superimposed on the depth gradient [3, 51].

In contrast to coastal and slope epifaunal communities, studies of assemblages on hard substrata in the deep-sea are rarely studied with the exceptions of seamounts [40, 42, 57], hydrothermal vents [58] and deep-sea corals [59, 60]. Even fewer studies have examined the vertical zonation of deep-sea epifaunal species and communities. Most research on deep-sea benthic community structure has focussed on the diversity and distribution of fauna living in and on the sediments and the distribution of fish [34]. With increasing depth and in offshore environments, rocky substrata become scarce and are replaced by soft substratum. However, where hard substrata are present in the form of seamounts, abyssal hills, pinnacles, boulders and metallic nodules and crusts, these surfaces can provide an oasis of hard substratum in the deep oceans of the world. Seamounts are undersea mountains that rise steeply from the sea bottom to below the sea level and have a minimum elevation of 1000 m [57]. They are associated with increases in the speed of sea currents, upwellings, turbulence, Taylor cones and eddies in the zones where the seamounts interact with ocean currents [42]. Hard substrata atypical of the deep-sea environment are common on seamounts and may take a variety of forms including pinnacles, knobs, rocks and cobbles. There are potentially more than 100 000 seamounts in the world's oceans that can represent a large area of hard substratum [40]. Because seamounts are subject to very vigorous currents and are associated with hard substrata and high levels of productivity in their upper regions, they are host to very distinctive biological communities. Studies conducted using epibenthic sledge surveys and manned submersibles on deeper seamounts have shown that distinct vertical distributions of suspension-feeders such as stony corals, gorgonian corals, black corals, sea anemones, sea pens, hydroids, sponges, sea squirts (Ascidiacea) and crinoids can occur [42, 57]. On the Patton Seamount, deep-sea dives were conducted using the research vehicle Alvin at depths between 151 and 3375 m. They recorded three distinct faunal associations based on depth: (1) a shallow-water community (151–950 m) consisting of rockfishes, flatfishes, seastars and large numbers of attached suspension-feeders, (2) a mid-depth community (400–1500 m) also consisting of numerous attached suspension-feeding organisms such as corals, sponges, crinoids, sea anemones and sea cucumbers and fish, and (3) a deep-water community (500–3375 m) consisting of fewer attached suspension-feeders [57]. The high concentrations of organisms in the upper 1500 m were related to organisms taking advantage of the productive photic zone and the complex habitat created by heterogeneous hard substrata in this zone. In this study, reduction in sessile suspension-feeders with depth was related to the reduction in hard substrata and the increase in sedimented material. Other researchers [61, 62] have also recorded a strong decrease in the abundance of suspension-feeding and other organisms with increasing depth from the summits of Jasper and Cross seamounts. Decreases in filter-feeders were related to decreased current velocities and food supplies at greater depths with a related increase in sedimentation.

On a smaller-scale, seamount researchers have also recorded the importance of localised, topographically induced current acceleration (associated with the rims of summits, vertical walls of calderas, knobs and pinnacles) on the distribution and abundance of epifaunal suspension-feeders on seamounts. Increased abundances of suspension-feeders (gorgonians, black corals, antipatharians, sponges, ascidians and hydroids) are found in regions of increased mean current speeds which bring increased levels of food, prevent sedimentation both on the substratum and on the suspension-feeding organisms themselves and increased larval supply and recruitment [61–63].

Seamounts are not the only natural hard substratum present in the deep-sea. Studies have shown that epifaunal communities may also develop on metallic nodules and crusts – a very common hard substratum found worldwide in the abyssal deep-sea [1, 64]. The distribution of organisms on small manganese nodules at 4500 m depth in the North Pacific was found to follow a vertical zonation with suspension-feeders located at nodule summits and deposit-feeders at the base [64]. Vertical distribution was related to vertical gradients in boundary shear stress, horizontal particulate flux and particle contact rates. Small-scale vertical zonations have also been found in the distribution of epifaunal species associated with biogenic structures in the deep-sea (4100 m depth) such as the stalks of glass sponges where larger suspension-feeders were typically found at the top of stalks and zonation on stalks appeared to be mainly controlled by biological interactions among species [39].

6.4   Patterns of vertical zonation in biofouling communities on man-made structures

Man-made structures associated with the oil and gas industries, when located on soft bottoms or in offshore regions, can provide surfaces for attachment in regions where there are few such naturally occurring substrata. The vertical zonation of biofouling communities on oil and gas platforms is well documented and many organisms found on these platforms are restricted to a particular depth range [65]. It is tempting to use studies of community zonation on natural substratum as models for the community development and vertical distribution of species that occur on man-made substrata. However, the physical and biological environment at offshore platforms differs from natural coastal rocky habitats and there are limitations in directly extrapolating from the natural situation. Platform habitats are comparatively homogeneous when compared with complex, heterogeneous natural habitats. Habitat complexity on platforms is created by the presence of diagonal and horizontal cross-members and associated pipelines. Platforms may harbour an abbreviated list of hard bottom species compared to similarly complex natural substrata, and communities are typically dominated by filter-feeders with pelagic larvae capable of being carried to the platforms from the nearest breeding population. Species which can successfully populate platforms often expand their depth range due to the paucity of normal predators and/or competitors. The same organisms on natural substrata may occupy a more limited depth range than on man-made structures, because they are commonly exposed to predators and competitors not necessarily present on the platforms [66, 67]. Platform structures typically descend semi-vertically towards the seabed, sometimes to depths of hundreds of metres in oceanic waters. The sharp changes in physical–chemical properties of the sea water encountered may be reflected in sharp delineations between biofouling assemblages in comparison with the gradual gradient of change typically seen on natural substrata.

Despite the differences in habitat composition between the two hard substratum types and the epifaunal communities that may eventually develop, the forcing factors that determine vertical zonation on platforms are similar. In temperate waters, light levels are important in determining the lower limit of the zone where macroalgae dominate the biofouling community. In the North Sea, this lower limit varies according to platform location and local turbidity. In the shallow, turbid waters of the southern North Sea algae are confined to areas shallower than 5 m depth, whereas in the central and northern sector plants have been found as deep as 40 m depth [25]. Plates II A and B show differences in biofouling community composition on a mooring chain at 5 m (A) and 10 m depth (B) in Scapa Flow, Scotland. At 5 m depth the extensive biofouling community is dominated by red and green algae, erect hydroids, scallops, mussels and encrusting barnacles and tubeworms. This community is greatly reduced at 10 m depth. Green algae are virtually absent, red algal cover is reduced and erect colonial hydroids dominate the community.

Below the photic zone invertebrates typically dominate. The dominant invertebrate taxa found on North Sea platforms include *Mytilus edulis* (mussel), *Tubularia larynx* (hydroid),

Metridium senile (anemone), *Alcyonium digitatum* (soft coral), *Filograna implexa* (colonial tubeworm), *Balanus hameri* (deep water barnacle), *Ascidiella scabra* (sea squirt) and bryozoans [23, 68]. The vertical distribution of these organisms is complex, zonation patterns are not consistent throughout the North Sea and can vary from platform to platform in the same locality [69]. Zonation patterns vary according to the depth of the actual platform, season, length of time of platform deployment, geographical location, distance from shore, the influence of prevailing water currents and masses on the supply of nutrients and larvae and complex biological interactions [68–70]. Interactions between biotic and abiotic factors can result in the development of distinct biofouling regions within a geographical location, where the biofouling communities which develop on platforms in these areas and their vertical distributions are substantially different from one another. This is seen in the distinct biofouling regions which occur in the North Sea and within the Gulf of Mexico [23, 67, 71, 72].

There is an extensive amount of information on the biofouling of man-made structures to depths of 60 m; however, information on the biofouling of artificial substrata at great depths is limited. There is some limited information available on biofouling in deeper waters from early studies conducted on technical materials [73, 74] and microbial biofouling research conducted on the feasibility of ocean thermal energy conversion (OTEC) heat exchangers [75, 76]. Information obtained from these studies was typically non-quantitative and limited from an ecological point of view. Later research has examined the development of biofilms on artificial substrata, e.g. panels [77, 78] and instruments [79]. A recent study by Kogana *et al.* [41] of the Acoustic Thermometry of Ocean Climate (ATOC)/Pioneer telecommunications cable highlights the potential for biofouling to occur on man-made substrata at depth. Actinarians (*Metridium farcimen*) were found to colonise the cable at various depths (Plate II C) and hard substratum communities were shown in proximity to the cable at 1160 m water depth (Plate II D).

6.5   Conclusions

- Depth-related vertical zonation in epifaunal assemblages has been found in intertidal, sublittoral, bathyal and abyssal marine communities on natural hard substrata. Studies have shown that community forcing factors with depth on hard substrata include light levels, temperature, pressure, food and nutrient levels, substratum composition and availability and the biology and physiology of individual species.
- Studies of natural substratum assemblages have focussed on zonation patterns in comparatively shallow waters (0–100 m depth); studies at greater depths are comparatively rare and there is a lack of information on both the composition and structure of communities with great depth.
- Diver and ROV inspections of oil and gas platforms have accumulated decades of information regarding depth-related changes in biofouling communities around the world. This information shows that the forcing factors which structure biofouling communities on man-made structures are similar to those on natural substrata, but the relative importance of these factors varies.
- Systematic studies of the biofouling on man-made structures are commonly restricted to those from surface waters to 150 m depth. There is a worldwide lack of information on

the biofouling patterns with depth on both natural and man-made structures and a need for more research in this area.

- Research may be driven by close collaborations between the oil and gas industries and scientific institutes – such as project SERPENT (Scientific and Environmental ROV Partnership using Existing Industrial Technology) [80] and the development and availability to scientists of the new generation of deep-sea submersibles which are being built.

References

1. Mullineaux, L.S. (1988) The role of settlement in structuring a hard-substratum in the deep sea. *Journal of Experimental Marine Biology and Ecology*, **120**, 247–261.
2. Railkin, A.I. (2004) *Marine Biofouling: Colonization Processes and Defences*. CRC Press, London.
3. Terlizzi, A., Anderson, M.J., Fraschetti, S. & Benedetti-Cecchi, L. (2007) Scales of spatial variation in Mediterranean subtidal sessile assemblages at different depths. *Marine Ecology Progress Series*, **332**, 25–29.
4. Evans, R.G. (1947) The intertidal ecology of selected localities in the Plymouth neighbourhood. *Journal of the Marine Biological Association of the United Kingdom*, **27**, 173–218.
5. Connell, J.H. (1972) Community interactions on marine rocky intertidal shores. *Annual Review of Ecology and Systematics*, **3**, 169–192.
6. Wethey, D.S. (1983) Geographic limits and local zonation: the barnacles *Semibalanus* (Balalzus) and *Chthamalus* in New England. *Biological Bulletin*, **165**, 330–341.
7. Raimondi, I.T. (1988) Rock type affects settlement, recruitment and zonation of the barnacle *Chthurnalus anisopoma* Pilsbury J. *Journal of Experimental Marine Biology and Ecology*, **123**, 253–267.
8. Anderson, M.J. & Underwood, A.J. (1994) Effects of substratum on the recruitment and development of an intertidal estuarine fouling assemblage. *Journal of Experimental Marine Biology and Ecology*, **184**, 217–236.
9. Jenkins, R.J., Norton, T.A. & Hawkins, S.J. (1999) Settlement and post-settlement interactions between *Semibalanus balanoides* (L.) (Crustacea: Cirripedia) and three species of fucoid canopy algae. *Journal of Experimental Marine Biology and Ecology*, **236**, 49–67.
10. Kitching, J.A., Macan, T.T. & Gilson, H.C. (1934) Studies in sublittoral ecology. I. A submarine gully in Wembury Bay, South Devon. *Journal of the Marine Biological Association of the United Kingdom*, **19**, 677–706.
11. Fuller, J.L. (1946) Season of attachment and growth of sedentary marine organisms at Lamoine, Maine. *Ecology*, **27** (2), 150–158.
12. Jones, N.S. (1951) The bottom fauna off the south of the Isle of Man. *The Journal of Animal Ecology*, **20** (1), 132–144.
13. Norton, T.A. & Milburn, J.A. (1972) Direct observations on the sublittoral marine algae of Argyll, Scotland. *Hydrobiologia*, **40** (1), 55–68.
14. Hiscock, K. & Hoare, R. (1975) The ecology of sublittoral communities at Abereiddy quarry. Pembrokeshire. *Journal of the Marine Biological Association of the United Kingdom*, **55**, 833–864.
15. Knight-Jones, E.W. & Nelson-Smith, A. (1976) Sublittoral transects in the Menai Straits and Milford Haven. In: *Biology of Benthic Organisms. 11th European Symposium on Marine Biology* (eds B.F. Keegan, Ó.P. Céidigh & P.J.S. Boaden), pp. 379–389. Pergamon Press, Oxford.
16. Logan, A., Page, F.H. & Thomas, M.L.H. (1984) Depth zonation of epibenthos on sublittoral hard substrates off Deer Island, Bay of Fundy, Canada. *Estuarine, Coastal and Shelf Science*, **18**, 571–592.
17. Pérès, J.M. & Molinier, R. (1957) Compte/rendu du colloque tenue à Gênes par la comité du benthos de la Commission Internationale pour l'Exploration Scientifique de la Mer Mediterranée. *Recueil des travaux de la Station marine d'Endoume, Marseille*, **22** (Bull. 12), 5–15.

18. Hiscock, K. & Mitchell, R. (1980) The description and classification of sublittoral epibenthic ecosystems. In: *The Shore Environment Volume 2: Ecosystems* (eds J.H. Price, D.E.G. Irvine & W.F. Farnham), pp. 323–370. Academic Press, London.

19. Connor, D.W., Brazier, D.P., Hill, T.O. & Northen, K.O. (1997) *Marine Nature Conservation Review: marine biotope classification, classification for Britain and Ireland. Volume 2. Sublittoral biotopes.* Report no. 229, Joint Nature Conservation Committee, Peterborough.

20. Pérès, J.M. (1982) Zonations. In: *Ocean Management* (ed. O. Kinne), pp. 9–45. John Wiley & Sons, Chichester.

21. Hedgpeth, J.W. (1957) Classification of marine environments. *Memoirs of the Geological Society of America*, **67**, 17–28.

22. Whomersley, P. & Picken, G.B. (2003) Long-term dynamics of fouling communities found on offshore installations in the North Sea. *Journal of the Marine Biological Association of the United Kingdom*, **83**, 897–902.

23. Kingsbury, R.W.S.M. (1981) Marine fouling of North Sea installations. In: *Marine Fouling of Offshore Structures* (ed. D.J. Crisp), pp. 4–31. Society for Underwater Technology, London.

24. Irving, A.D. & Connell, S.D. (2002) Sedimentation and light penetration interact to maintain heterogeneity of subtidal habitats: algal versus invertebrate dominated assemblages. *Marine Ecology Progress Series*, **245**, 83–91.

25. Terry, L.A. & Picken, G.B. (1986) Algal fouling in the North Sea. In: *Algal Biofouling* (eds L.V. Evans & K.D. Hoagland), pp. 179–192. Elsevier, Amsterdam.

26. Herring, P. (2002) *The Biology of the Deep Ocean.* Oxford University Press, Oxford.

27. Lüning, K. (1990) *Seaweeds: Their Environment, Biogeography, and Ecophysiology.* John Wiley & Sons, New York.

28. Dring, M.J. (1981) Chromatic adaptation of photosynthesis in benthic marine algae: an examination of its ecological significance using a theoretical model. *Limnological Oceanography*, **26**, 271–284.

29. Littler, M.M., Littler, D.S., Blair, S.M. & Norris, J.N. (1985) Deepest known plant life discovered on an uncharted seamount. *Science*, **227**, 57–69.

30. Glasby, T.M. & Connell, S.D. (2001) Orientation and position of substrata have large effects on epibiotic assemblages. *Marine Ecology Progress Series*, **214**, 127–135.

31. Cocito, S., Bedulli, D. & Sgorbini, S. (2002) Distribution patterns of the sublittoral epibenthic assemblages on a rocky shoal in the Ligurian Sea (NW Mediterranean). *Scientia Marina*, **66** (2), 175–181.

32. Lalli, C.M. & Parsons, R.T. (1993) *Biological Oceanography: an Introduction.* Pergamon Press, Oxford.

33. Newell, R.C. & Branch, G.M. (1980) The influence of temperature on the maintenance of metabolic energy balance in marine invertebrates. *Advances in Marine Biology*, **17**, 329–396.

34. Carney, R.S. (2005) Zonation of deep biota on continental margins. *Oceanography and Marine Biology: an Annual Review*, **43**, 211–278.

35. Fang, L., Barcelona, M.J., Nogi, Y. & Kato, C. (2000) Biochemical implications and geochemical significance of novel phospholipids of the extremely barophilic bacteria from the Marianas Trench at 11,000 m. *Deep Sea Research I*, **47**, 1173–1182.

36. Gillet, M.B., Suko, J.R., Santoso, F.O. & Yancey, P.H. (1997) Elevated levels of trimethylamine oxide in muscles of deep-sea gadiform teleosts: a high pressure adaptation. *Journal of Experimental Zoology*, **279**, 386–391.

37. Somero, G.N. (1998) Adaptations to cold and depth: contrasts between polar and deep-sea animals. In: *Society for Experimental Biology Seminar Series* (eds H.O. Pörtner & R.C. Playle), pp. 33–57. Cambridge University Press, Cambridge.

38. Pérès, J.M. (1982) Major benthic assemblages. In: *Ocean Management* (ed. O. Kinne), pp. 373–521. John Wiley & Sons, Chichester.

39. Beaulieu, S.E. (2001) Life on glass houses: sponge stalk communities in the deep sea. *Marine Biology*, **138**, 803–817.

40. Rogers, A.D. (2004) *The Biology, Ecology and Vulnerability of Seamount Communities*. International Union for Conservation of Nature & Natural Resources, Gland.

41. Kogana, I., Paull, C.K., Kuhnz, L.A., *et al.* (2006) ATOC/Pioneer Seamount cable after 8 years on the sea floor: observations, environmental impact. *Continental Shelf Research*, **26**, 771–787.

42. Rogers, A.D. (1994) The biology of seamounts. In: *Advances in Marine Biology* (eds J.H.S. Blaxter & A.J. Southward), pp. 305–350. Academic Press, Oxford.

43. Thomsen, L., Graf, G., Juterzenka, K.V. & Witte, U. (1995) An *in situ* experiment to investigate the depletion of seston above an interface feeder field on the continental slope of the western Barents Sea. *Marine Ecology Progress Series*, **123**, 295–300.

44. Paine, R.T. (1974) Intertidal community structure. Experimental studies on the relationship between a dominant competitor and its principal predator. *Oecologia*, **15**, 93–120.

45. Russ, G.R. (1982) Overgrowth in a marine epifaunal community: competitive hierarchies and competitive networks. *Oecologia*, **53**, 12–19.

46. Sebens, K.P. (1986) Spatial relationships among encrusting marine organisms in the New England subtidal zone. *Ecological Monographs*, **56** (1), 73–96.

47. Boudouresque, C.F. (1971) Contribution à L'étude phytosociologique des peuplements algaux des côtes varoises. *Vegetatio*, **22** (1–3), 83–104.

48. Garrabou, J., Ballesteros, E. & Zabala, M. (2002) Structure and dynamics of North-western Mediterranean rocky benthic communities along a depth gradient. *Estuarine, Coastal and Shelf Science*, **55**, 493–508.

49. Pérès, J.M. (1967) The Mediterranean benthos. *Oceanography and Marine Biology: An Annual Review*, **5**, 449–533.

50. Zabala, M. & Ballesteros, E. (1989) Surface-dependent strategies and energy flux in benthic marine communities or, why corals do not exist in the Mediterranean. *Scientia Marina*, **53**, 3–17.

51. Balata, D., Acunto, S. & Cinelli, F. (2006) Spatio-temporal variability and vertical distribution of a low rocky subtidal assemblage in the north-west Mediterranean. *Estuarine, Coastal and Shelf Science*, **67**, 553–561.

52. Gili, J.M. & Ballesteros, E. (1991) Structure of cnidarian populations in Mediterranean sublittoral benthic communities as a result of adaptations to different environmental conditions. *Ocealogia Aquatica*, **10**, 243–354.

53. Logan, A. (1988) A sublittoral hard substrate epibenthic community below 30 m in Head Harbour Passage, New Brunswick, Canada. *Estuarine, Coastal and Shelf Science*, **27**, 445–449.

54. Gulliksen, B. (1978) Rocky bottom fauna in a submarine gulley at Loppkalven, Finnmark, Northern Norway. *Estuarine and Coastal Marine Science*, **7**, 361–372.

55. Thorbjørn, L. & Petersen, G.H. (2003) The epifauna on the carbonate reefs in the Arctic Ikka Fjord, SW Greenland. *Ophelia*, **57** (3), 177–202.

56. Hiscock, K. (2005) A re-assessment of rocky sublittoral biota at Hilsea Point Rock after fifty years. *Journal of the Marine Biological Association of the United Kingdom*, **85**, 1009–1010.

57. Hoff, G.R. & Stevens, B. (2005) Faunal assemblage structure on the Patton Seamount (Gulf of Alaska, USA). *Alaska Fishery Research Bulletin*, **11** (1), 27–36.

58. Shank, T.M., Fornari, D.J., Von Damm, K.L., Lilley, M.D., Haymon, R.M. & Lutz, R.A. (1998) Temporal and spatial patterns of biological community development at nascent deep-sea hydrothermal vents (9°50′ N, East Pacific Rise). *Deep-Sea Research II*, **45**, 465–515.

59. Krieger, K.J. & Wing, B.L. (2002) Megafauna associations with deepwater corals (*Primnoa* spp.) in the Gulf of Alaska. *Hydrobiologia*, **471**, 82–90.

60. Etnoyer, P. & Morgan, L. (2003) *Occurrences of Habitat-Forming deep Sea Corals in the Northeast Pacific Ocean: A Report to NOAA's Office of Habitat Conservation*. Marine Conservation Biology Institute, Redmond.

61. Genin, A., Dayton, P.K., Lonsdale, P.F. & Speiss, F.N. (1986) Corals on seamount peaks provide evidence of current acceleration over deep-sea topography. *Nature*, **322**, 59–61.

62. Grigg, R.W., Malahoff, A., Chave, E.H. & Landahl, J. (1987) Seamount benthic ecology and potential environmental impact from manganese crust mining in Hawaii. In: *Seamounts, Islands and Atolls*

(eds B. Keating, P. Fryer, R. Batiza & G. Boehlert), pp. 379–390. American Geophysical Union, Washington.

63. Tunnicliffe, V., Juniper, S.K. & de Burgh, M.E. (1985) The hydrothermal vent community on Axial Seamount, Juan de Fuca Ridge. *Bulletin of the Biological Society of Washington*, **6**, 453–464.

64. Mullineaux, L.S. (1989) Vertical distributions of the epifauna on manganese nodules: implications for settlement and feeding. *Limnological Oceanography*, **34** (7), 1247–1262.

65. Dokken, Q.R., Withers, K., Childs, S. & Rigs, T. (2000) *Characterization and Comparison of Platform Reef Communities off the Texas Coast*. Texas Parks and Wildlife Department, Centre for Coastal Studies, Texas A & M University-Corpus Christi.

66. Wolfson, A., Van Blaricom, G., Davis, N. & Lewbel, G.S. (1979) The marine life of an offshore oil platform. *Marine Ecology Progress Series*, **1**, 81–89.

67. Lewbel, G.S., Howard, R.L. & Gallaway, B.J. (1987) Zonation of dominant fouling organisms on northern Gulf of Mexico petroleum platforms. *Marine Environmental Research*, **21**, 199–224.

68. Picken, G.B. (1986) Moray Firth marine fouling communities. *Proceedings of the Royal Society of Edinburgh*, **91B**, 213–220.

69. Sell, D. (1992) Marine fouling. *Proceedings of the Royal Society of Edinburgh*, **100B**, 169–184.

70. Forteath, G.N.R., Picken, G.B., Ralph, R. & Williams, J. (1982) Marine growth studies on the North Sea oil platform Montrose Alpha. *Marine Ecology Progress Series*, **8**, 61–68.

71. Gallaway, B.J., Johnson, M.F., Martin, L.R., *et al.* (1981) *Vol. II – The artificial Reef Studies. Ecological Investigations of Petroleum Production Platforms in the central Gulf of Mexico*. Bureau of Land Management, New Orleans.

72. Gallaway, B.L. & Lewbel, G.S. (1982) *The ecology of petroleum platforms in the northwestern Gulf of Mexico: A community profile*. Open-File report 82-03, U.S. Fish and Wildlife Service, Washington, DC.

73. Muraoka, J.S. (1966) *Deep-ocean biodeterioration of materials. Part IV. One year at 6800 feet*. Technical Report R-456, U.S. Naval Civil Engineering Laboratory, California.

74. Muraoka, J.S. (1966) *Deep-ocean biodeterioration of materials. Part V. Two years at 5640 feet*. Technical Report R-495, U.S. Naval Civil Engineering Laboratory, California.

75. Sasscer, D.S., Morgan, T.O., Rivera, C., *et al.* (1981) OTEC biofouling, corrosion and materials study from a moored platform at Punta Tuna, Puerto Rico. I. Fouling resistance. *Ocean Science Engineering*, **6**, 499–532.

76. Berger, L.R. & Berger, J.A. (1986) Countermeasures to microbiofouling in simulated ocean thermal energy conversion heat exchangers with surface and deep ocean waters in Hawaii. *Applied and Environmental Microbiology*, **51** (6), 1186–1198.

77. Guezennec, J., Ortega-Morales, O., Raguenes, G. & Geesey, G. (1998) Bacterial colonization of artificial substrate in the vicinity of deep-sea hydrothermal vents. *FEMS Microbiology Ecology*, **26**, 89–99.

78. Head, R.M., Davenport, J.A. & Thomason, J.C. (2004) The effect of depth on the accrual of marine biofilms on glass substrata deployed in the Clyde Sea, Scotland. *Biofouling*, **20** (3), 177–180.

79. Amram, P., Anghinolfi, M., Anvar, S., *et al.* (2003) Sedimentation and fouling of optical surfaces at the ANTARES site. *Astroparticle Physics*, **19**, 253–267.

80. Hudson, I.R., Jones, D.O.B. & Wigham, B.D. (2005) A review of the uses of work-class ROVs for the benefits of science: Lessons learned from the SERPENT project. *Underwater Technology. International Journal of the Society for Underwater Technology*, **26** (3), 7.

Chapter 7
Epibiosis

Martin Wahl

In this chapter I aim to demonstrate how the physicochemical properties of the water medium favour a sessile mode of life of consumers and, consequently, the commonest epibioses in aquatic communities. The possible consequences of an epibiotic association for epi- and basibiont are described. Finally, the chapter attempts to provide an overview of the distributional patterns of epibiosis at different scales.

7.1 Sessile mode of life

Water is much denser and substantially more viscous than air. Consequently, the net weight of particles is reduced and the drag on them increased. In addition, water is a suitable solvent for numerous ions and molecules. These properties make water a good vector for particulate and dissolved organic matter. Organisms may exploit this rich energy source in three ways: absorption of dissolved matter, collection of deposited particles at interfaces, or filtering of suspended particle. The first process may take place through any thin-skinned body surface, the second necessitates still water and the third requires a relative movement between the medium carrying the particles and the filtering apparatus. This filtering current is given in a water flow but can also be generated by the swimming of a suspension feeder or by the biogenic creation of a filtering current through or around a stationary organism.

The availability of dissolved and particulate energy in the surrounding fluid allows a stationary mode of life even for heterotrophic organisms. Additionally, the reduced weight and increased drag in the aqueous environment require some form of attachment for an organism to stay in place. These selective drivers have produced a large number of sessile life forms in the aquatic, specifically the marine, ecosystem. Most marine phyla are well represented in the guilds of sessile suspension or deposit feeders. A sessile mode of life, at least during some ontogenetic phase, is led by all sponges, most cnidarians, many mollusca (bivalves and gastropods), some rotifers, most tentaculata, the tube-building polychaetes, some echinoderms, a few crustaceans and most tunicates. Diversification within these guilds has led to the coexistence of hundreds or thousands of sessile species in many regions. While niche segregation is possible with regard to some resources (e.g. type and size of particles, current regime, irradiation), all of the sessile animal and plant species require a settlement substratum. This may easily become a limiting resource in an energy-rich and low-disturbance environment. Competition for hard substratum for settlement may have driven the evolution of a life style which is typical for the marine benthos: epibiosis.

'Epibiosis' is the spatial association between a substratum organism ('basibiont') and a sessile organism ('epibiont') attached to the basibiont's outer surface without trophically depending on it. These restrictions allow the differentiation of epibiosis from ectoparasitism or trophic symbiosis or from the ephemeral relationship engaged by a 'visiting' motile animal. Epizoans and epiphytes are animal and alga epibionts, respectively. These definitions are derived from an earlier version [1]. There are no established terms for epibiotic prokaryotes or fungi, though epibacteria and epifungi might represent a logical solution.

Epibiotic communities on living substrata and biofouling communities on non-living hard substrata share a major proportion of species (see below). Consequently, the establishment of epibiotic and biofouling communities follow very similar rules and are treated in Chapters 3 and 4.

7.2   Consequences of epibioses

For most aquatic organisms, major portions of their body surface serve the exchange of a variety of substances with the environment. Those substances include nutrients, small ions, dissolved micro- and macromolecular compounds, organic particles, info-chemicals, defensive metabolites, waste products or gases. Other properties of the outer body surface influence various interactions of an organism with its biotic and abiotic surroundings. Roughness, wettability and texture influence both drag and microhydrodynamics, which in turn may affect the (passive) settlement of colonisers. Colour and shape are important for mate or predator–prey recognition. Light is absorbed through translucent surfaces. Osmoregulation is surface-bound.

Any kind of epibiosis, from the first adsorbed macromolecules to the larvae of multicellular sessile animals, will modify some or all of these properties of the basibiont. As a consequence, epibiosis modulates the basibiont's interactions. These effects may be small or large, beneficial, neutral or detrimental. In many instances, the sign and strength of epibiosis effects are context-specific [2, 3].

Passively adsorbed sugars or proteins may affect the physicochemical properties of the surface (electrical charge, wettability), but will not infer with other functions. Epibacterial films may constitute a physiological filter affecting the flux of compounds into or out off the basibiont, or metabolically processing these compounds during their passage [4]. A biofilm composed of bacterial consortia, diatoms, fungi and protozoans will alter the chemical conditions at the living interface through its metabolic activities (concentrations of O_2, CO_2, nutrients, H^+) [1, 2]. When a biofilm reaches a certain thickness and/or when it is opaque due to the presence of pigmented components (e.g. diatoms, detritus), it can be expected to insulate a basibiotic alga's surface from the vital resource light [3]. The insulating effect of thick biofilms or larger epibionts may be both chemical and optical. One common aspect of this 'insulation' is camouflaging the basibiont from optical, tactile or chemical detection. The basibiont benefits from this masking when the deterred interactors are consumers (see below), detrimental epibionts or pathogens [4]. It is detrimental when recognition by symbionts or mates is hindered.

How the biotic interactions may be affected by the presence, quantity and identity of epibionts or the combination of particular epibiotic species is particularly well studied for predator–prey relationships. Well-defended epibionts like some bacteria, sponges, hydrozoans, actinians, ascidians and certain algae may reduce the predation on their basibiont by gastropods

[5–7], urchins [8], starfish [7, 9, 10], crabs [8, 11], fishes [12] or birds [13]. This protection conferred by one partner of an association to the other is called 'associational resistance'. In contrast, impending danger may also be enhanced by epibionts ('shared doom'). Borrowing bivalves are usually protected from optical detection to some degree. However, epibiotic hydrozoa (Plate II E) may facilitate their detection by crab predators [12]. One advantage of the fusiform shape of mussel shells may be that crab claws slide off easily. Barnacles on mussel shells (Plate II F) improve the grip and thus predation pressure by the predator crab *Carcinus maenas* [14]. Fast-growing mussels may escape crab predation when their shells reach a resistance to breakage surpassing the strength of crab claws. Some epibiotic hydrozoans and boring polychaetes weaken the shell of molluscs and thus enhance the success of crushing predators [15,16]. Basibiont molluscs bearing these epibionts will leave the window of vulnerability at a larger size, i.e. later in life, than unfouled conspecifics. This effect may also be expected from other boring epibionts including certain sponges, green algae, phoronids and actinians. Apparency of an otherwise inconspicuous basibiont (translucent, dull coloured or small) may be enhanced by colourful and/or large epibionts. Epibiont-enhanced apparency, the opposite of camouflage, may increase predation [17].

The effects of epibionts on basibionts may depend on the identities of the epibiont, the basibiont and the third interactor species involved. A given epibiont species may have opposite effects on its host, i.e. increase and reduce predation, depending on which basibiont species it grows on or which predator species is around. When an epibiont is more palatable to urchins than the basibiont species it grows on, consumption on the latter is enhanced relative to an unfouled conspecific. The inverse effect is observed, when the epibiont is less attractive than the basibiont [8]. An epibiotic hydroid protects its clam basibiont against fish predation but facilitates detection and consumption by several crustacean predators [12]. The context specificity of epibiont impact may be extended to abiotic interactions. Epibionts on swimming bivalves may have an overall beneficial effect by protecting their hosts against predators. However, when predators are absent, their presence is detrimental because they increase drag and weight of their swimming host [18]. Multiple effects of epibionts may be additive. Epibiotic barnacles on the snail *Littorina littorea* increase drag and consequently reduce growth [19] of the basibiont and consequently keep the snails for longer in the size window of vulnerability [20] with regard to crab predation and further enhance predatory success by improving the handling of the basibiont prey by the crab [14]. Additionally, the presence of epibiotic barnacles facilitates the settlement of shell-boring polychaetes [21] which weaken the shell and extent the time the slow-growing snails remain vulnerable to crab predation. Finally, barnacle epibionts mechanically hinder the mating and, consequently, the reproduction of the basibiont snails [20] so that the enhanced mortality by extended predation may not be compensated for.

Filamentous epiphytes on seagrass represent a further example for complex and sometimes contrasting effects of an epibiont guild on their basibiont (personal observation). Epiphytic filamentous algae may increase drag (detrimental), compete with the host for light (detrimental) and nutrients (detrimental), deflect grazing pressure being more palatable to some isopod grazers than their host (beneficial), offer shelter from UV radiation (beneficial) and desiccation during emersion (beneficial) and attract mussel recruits who may be responsible for a fatal increase in weight (detrimental). Whether the net effect of these epiphytes is positive or negative depends on the ecological context, i.e. current regime, depth, nutrient availability, growth rate, defence level and isopod abundance, habitat and season.

In contrast to predator–prey relationships where the consequences for the prey are always detrimental, epibiosis may be detrimental, tolerable or favourable depending on which epibiont species are involved, which epibiotic coverage is reached or which third interactors are present. The evolutionary incentive to develop defences against epibionts may be more complex or ambiguous than in predator–prey relationships. We would expect that many antifouling defences are species-specific and tuned to need (inducible and reducible).

For the epibiont partner, the consequences of an epibiotic association seem to be more straightforward reviewed in [22]. A first and immediate advantage of settling on a living surface is the acquisition of a solid substratum for attachment. Many bacteria and diatoms grow better at such a solid/liquid interface where nutrients are concentrated thermodynamically or by sedimentation and where the relative velocity between water and the – now sessile – organism is higher than in a planktonic life stage. Most propagules of algae and sessile animals require attachment for metamorphosis. In sessile species, without successful settlement of the propagules further development generally is jeopardised. Relative to settlement on firm rock, the elevated position on a basibiont allows escaping the boundary layer near the bottom which is often stagnating and/or shaded. At the same time, temporarily strong water velocity may not constitute a serious problem when the basibiont has a flexible built, or reacts by body contraction or motility to stressful drag. While epibionts, according to the definition given in the beginning, do not trophically depend on their host, they may in a facultative manner benefit from, e.g., sloppy feeding manners of their host or from exuded metabolites, mucus and wastes.

Via associational resistance, epibionts may enjoy the shelter offered by a well-defended basibiont. On the other hand, epibiosis is associated with a certain number of risks. For one thing, the epibiont has to cope with or adapt to a naturally limited life expectancy of the substratum. One requirement for a given epibiosis to be considered successful from the epibiont's point of view is that reproduction is achieved before the basibiont perishes, renews its outer surface (mucus secretion, sloughing, moulting) or simply burrows to pass the winter in the sediment like some crabs do. Physiological processes of the basibiont sometimes create quite particular chemical conditions at the living surface. These may be the result of transcutaneous exchanges linked to the basibiont's primary metabolism or of the production and excretion of secondary defence metabolites. Epibionts on motile basibionts may be exposed to drastic changes of environmental variables when the host switches between epi- and endobenthic, or between aquatic and aerial habitats. Also, epibionts may accidentally fall victim to predators of their basibiont – an associational accident called 'shared doom'.

Thus, the consequences of an epibiotic association are multiple, complex and variable for either partner. The life history of a species determines whether the net effect of this association tends to be beneficial or detrimental. Potential epibionts may avoid unsuitable basibionts behaviourally, e.g. cueing on exudates from indicator species of the biofilm, from conspecifics or from the basibiont itself. Potential basibionts may limit undesirable degrees of epibiosis or harmful epibiont species by sporadic, regulated or constitutive defences.

7.3   Distributional patterns of epibioses

As the effects of epibiosis vary among species, ecological context and season, it is not surprising to realise that the prevalence of epibionts and basibionts varies among species and that the quality and quantity of epibiontic colonisation varies among body parts of a given basibiont, between habitats, and in time.

Those portions of the body surface which are engaged in transcutaneous exchange of nutrients, gases or light may not tolerate extensive biofouling. Likewise, suspension-feeding devices like the intake pores of sponges and colonial ascidians, or the tentacles of bryozoans and sessile polychaetes would lose their functionality when fouled. Sensory organs depend on an unobstructed access to the optical, mechanical or chemical stimuli they were evolved to register. Joints of body appendages require a certain freedom of movement. Consequently, only on body parts not serving any of these functions or during periods where these functions are not vital some degree of epibiosis may be tolerable. Organisms which depend on flexibility or light weight should avoid being colonised by rigid or heavy epibionts.

On the epibiont side, a successful coloniser should reach reproductive maturity before the basibiont surface sloughs or the basibiont dies. In general, the epibiont must be able to cope with all aspects of the basibiont's life style and surface properties. Obviously, it would be disadvantageous if the epibiont itself contributed to a premature mortality of the basibiont by interfering with some of its vital functions by excessive size or weight.

Because successful epibionts and basibionts are characterised by certain suites of properties (described in more detail in [22], it is not surprising to state that these functional groups are not homogeneously represented in all phyla (Figure 7.1) as found when over 2000 epibiont–basibiont pairings were analysed [23]. Larger hard-shelled molluscs and crustaceans seem to tolerate epibiosis (i.e. feature as basibionts) particularly well. Epibionts are common among small filamentous algae or hydrozoa, encrusting algae or bryozoa, and unicellular algae (bacteria were not recorded).

Within a given basibiont species, epibiosis varies at a number of scales. Different ontogenetic stages and different parts of the same organism may exhibit very different degrees of biofouling. Also, epibiotic cover may substantially vary between seasons or among years. Often, stressed individuals and old or decaying parts of an organism become particularly prone to biofouling (e.g. [6, 24–26]). The body surface of one basibiont may represent a number of different microhabitats occupied by different species of epibionts [27,28]. Functionally similar epibiotic hydroids may differentiate trophically (phenotypic plasticity) in response to certain species combinations in a given epibiotic assemblage [29]. Also, males and females of the same basibiont species may exhibit differences in the composition and quantity of epibioses [24,28]. Populations of a given basibiont species sampled in different habitats or regions often feature different epibiont species or different degrees of epibiotic coverage [6, 26, 30, 31]. Within the same habitat and exposed to the same coloniser pool, epibiotic communities often differ among co-occurring basibiont species [32, 33]. Seasonal or between-year variability of epibiotic communities is common [6, 26, 32, 33]. One should expect that the susceptibility to epibiosis should also vary among basibiont genotypes and with stress. Genetic causes for variability in epibiosis could reside in different capacities of antifouling defence. Single or suites of stressors weakening the basibiont could jeopardise antifouling defences. However, too little data are published on these two aspects to draw any conclusions.

Epibiosis is common in the marine environment and thousands of epibiotic pairings have been described [23]. In view of the particular conditions on a basibiont's surface and the apparent dominance of benefits for epibionts, we would expect a high degree of co-evolutive adaptation and specialisation at least on the side of the epibionts. Host specificity, however, between basibiont and epibionts seems to be the exception. The nature of epibiosis is more

Species per phylum described as

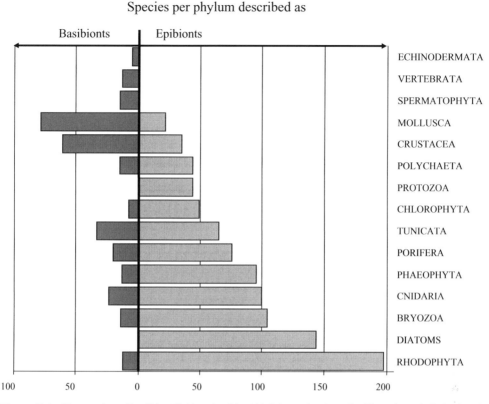

Figure 7.1   The number of basibiont (left) and epibiont (right) species described in various phyla (ordered by commonness of epibionts). The qualifications for either functional group tend to be inversely related among the phyla.

random and facultative. In most surveys, less than 20% of the epibionts attach exclusively to living substrata, and less than 5% are restricted to one basibiont species [23, 34–36]. This generalisation may apply less well to microepibionts. New technologies for the identification of non-cultivable strains have allowed an increasing number of papers which often report on host-specific assemblages of epibacteria [37–39].

The conspicuous variability in quantity and quality of epibiosis at a number of scales has two main causes. The composition of the pool of locally available potential coloniser species differs regionally for historical reasons, and temporally with season, currents regimes and dynamics within the plankton. At small spatial scale, depth, light and the exudates of macroorganisms may affect the behaviour of the colonisers. Which of the available potential epibionts will successfully settle and grow when a substratum becomes available depends on the properties of the basibiont's surface and the behavioural response of the coloniser to it. These properties comprise its consistency, texture, wettability, the deployment of mechanical, chemical or behavioural defences, and by the presence of previous settlers (e.g. biofilms).

Since defences are treated in detail in Chapter 8, it suffices to say here that typical antifouling protection in the most thoroughly studied species is assured by highly complex defence systems composed of multiple single compounds, or by mechanisms targeting the multitude of potential colonisers. The defence components may act complementarily but often overlap with regard to single coloniser species. The result is that in well-defended species all potential colonisers face more than one defence mechanism. The deployment of these defence adaptations varies at all scales: among organs of an individual, conspecifics, species, habitats, times. From the point of view of a potential epibiont, this results in a highly variable, 'flickering' defence landscape represented by the members of a basibiont population expressing at any given time different defences to different degrees. The difficulty to adapt to such a multiple-stress situation might explain the scarcity of obligate and specialised epibionts.

7.4   Conclusions

- Epibiosis is common in the marine environment. In view of the constant and sometimes impressively strong biofouling pressure, the 'null-condition' of any living surface would be 'fouled'.
- In nature we observe all states of biofouling, from sterile surfaces to those bearing the same communities as on the adjacent rock.
- Epibiosis modulates biotic and abiotic interactions, and contributes an often-underestimated dimension to biofouling.
- The composition and quantity of epibiotic communities vary between host species, between different parts of the host, between different physiological states of the host and between seasons, regions and depths. However, many living surfaces constitute a suitable substratum for many species of biofoulers.
- In some soft bottom regions, biogenic or living surfaces constitute the major portion of colonisable substratum, thus contributing to the amount and diversity of hard bottom communities. In addition, epibiosis on motile organisms facilitates dispersal of sessile organisms. Epibiosis on forms resistant to antifouling protection, such as some bryozoans, enables many species to settle on ship hulls and be dispersed over long distances.
- By opening access to additional substratum and by modulating substratum properties, epibiosis has an enormous impact in marine biofouling phenomena worldwide.

References

1. Araujo Jorge, T.C., Coutinho, C.M.L.M. & Aguiar, L.E.V. (1992) Sulphate-reducing bacteria associated with biocorrosion – a review. *Memórias do Instituto Oswaldo Cruz*, **87** (3), 329–337.
2. Thevanathan, R., Nirmala, N., Manoharan, A., *et al.* (2000) On the occurrence of nitrogen fixing bacteria as epibacterial flora of some marine green algae. *Seaweed Research Utiln*, **22**, 189–197.
3. Costerton, J.W., Cheng, K.J., Geesey, G.G., *et al.* (1987) Bacterial biofilms in nature and disease. *Annual Review of Microbiology*, **41**, 435–464.
4. Turnes, G. Turnes, M.S. & Fenical, W. (1992) Embryos of *Homarus americanus* are protected by epibiotic bacteria. *Biological Bulletin*, **182** (1), 105–108.
5. Cerrano, C., Puce, S., Chiantore, M., Bavestrello, G. & Cattaneo-Vietti, R. (2001) The influence of the epizoic hydroid *Hydractinia angusta* on the recruitment of the Antarctic scallop *Adamussium colbecki*. *Polar Biology*, **24** (8), 577–581.
6. Dougherty, J.R. & Russell, M.P. (2005) The association between the coquina clam *Donax fossor* Say and its epibiotic hydroid *Lovenella gracilis* Clarke. *Journal of Shellfish Research*, **24** (1), 35–46.

7. Marin, A. & Belluga, M.D.L. (2005) Sponge coating decreases predation on the bivalve *Arca noae*. *Journal of Molluscan Studies*, **71**, 1–6.

8. Wahl, M. & Hay, M.E. (1995) Associational resistance and shared doom: effects of epibiosis on herbivory. *Oecologia*, **102**, 329–340.

9. Laudien, J. & Wahl, M. (1999) Indirect effects of epibiosis on host mortality: starfish predation on differently fouled mussels. *Marine Ecology-Pubblicazioni Della Stazione Zoologica Di Napoli I*, **20** (1), 35–47.

10. Laudien, J. & Wahl, M. (2004) Associational resistance of fouled blue mussels (*Mytilus edulis*) against starfish (*Asterias rubens*) predation: relative importance of structural and chemical properties of the epibionts. *Helgoland Marine Research*, **58** (3), 162–167.

11. Wahl, M., Hay, M.E. & Enderlein, P. (1997) Effects of epibiosis on consumer-prey interactions. *Hydrobiologia*, **355**, 49–59.

12. Manning, L.M. & Lindquist, N. (2003) Helpful habitant or pernicious passenger: interactions between an infaunal bivalve, an epifaunal hydroid and three potential predators. *Oecologia*, **134** (3), 415–422.

13. Prescott, R.C. (1990) Sources of predatory mortality in the bay scallop *Argopecten irradiance* (Lamarck): interactions with seagrass and epibiotic coverage. *Journal of Experimental Marine Biology and Ecology*, **144**, 63–83.

14. Enderlein, P., Moorthi, S., Rohrscheidt, H. & Wahl, M. (2003) Optimal foraging *versus* shared doom effects: interactive influence of mussel size and epibiosis on predator preference. *Journal of Experimental Marine Biology and Ecology*, **292** (2), 231–242.

15. Bach, C.E., Hazlett, B.A. & Rittschof, D. (2006) Sex-specific differences and the role of predation in the interaction between the hermit crab, *Pagurus longicarpus*, and its epibiont, *Hydractinia symbiolongicarpus*. *Journal of Experimental Marine Biology and Ecology*, **333** (2), 181–189.

16. Buschbaum, C., Buschbaum, G., Schrey, I. & Thieltges, D.W. (2006) Shell-boring polychaetes affect gastropod shell strength and crab predation. *Marine Ecology Progress Series*, **329**, 123–130.

17. Threlkeld, S.T. & Willey, R.L. (1993) Colonization, interaction, and organization of cladoceran epibiont communities. *Limnology and Oceanography*, **38** (3), 584–591.

18. Forester, A.J. (1979) The association between the sponge *Halichondria panicea* (Pallas) and the scallop *Chlamys varia* (L.): a commensal-protective mutualism. *Journal of Experimental Marine Biology and Ecology*, **36**, 1–10.

19. Wahl, M. (1997) Increased drag reduces growth of snails: comparison of flume and *in-situ* experiments. *Marine Ecology Progress Series*, **151**, 291–293.

20. Buschbaum, C. & Reise, K. (1999) Effects of barnacle epibionts on the periwinkle *Littorina littorea* (L.). *Helgoland Marine Research*, **53**, 56–61.

21. Thieltges, D.W. & Buschbaum, C. (2007) Vicious circle in the intertidal: facilitation between barnacle epibionts, a shell boring polychaete and trematode parasites in the periwinkle *Littorina littorea*. *Journal of Experimental Marine Biology and Ecology*, **340** (1), 90–95.

22. Wahl, M. (1997) Living attached: aufwuchs, fouling, epibiosis. In: *Fouling Organisms of the Indian Ocean: Biology and Control Technology* (eds R. Nagabhushanam & M.F. Thompson), pp. 31–83. Oxford & IBH Publishing Company, New Delhi.

23. Wahl, M. & Mark, O. (1999) The predominantly facultative nature of epibiosis: experimental and observational evidence. *Marine Ecology Progress Series*, **187**, 59–66.

24. Maldonado, M. & Uriz, M.J. (1992) Relationship between sponges and crabs: patterns of epibiosis on *Inachus aguiarii* (Decapoda: Majidae). *Marine Biology*, **113**, 281–286.

25. Warner, G.F. (1997) Occurrence of epifauna on the periwinkle, *Littorina littorea* (L.), and interactions with the polychaete *Polydora ciliata* (Johnston). *Hydrobiologia*, **355**, 41–47.

26. Fernandez, L., Parapar, J., Gonzalez-Gurriaran, E. & Muino, R. (1998) Epibiosis and ornamental cover patterns of the spider crab *Maja squinado* on the Galician coast, northwestern Spain: Influence of behavioral and ecological characteristics of the host. *Journal of Crustacean Biology*, **18** (4), 728–737.

27. Gili, J.M., Abello, P. & Villanueva, R. (1993) Epibionts and intermoult duration in the crab *Bathynectes piperitus*. *Marine Ecology Progress Series*, **98**, 107–113.

28. Patil, J.S. & Anil, A.C. (2000) Epibiotic community of the horseshoe crab *Tachypleus gigas*. *Marine Biology*, **136** (4), 699–713.

29. Orlov, D.V. & Marfenin, N.N. (1995) Behavior and settling of actinulae of *Tubularia larynx* (Leptolida, Tubulariidae). *Zoologichesky Zhurnal*, **73** (9), 5–11.

30. Key, M.M., Jeffries, W.B., Voris, H.K. & Yang, C.M. (1996) Epizoic bryozoans, horseshoe crabs, and other mobile benthic substrates. *Bulletin of Marine Science*, **58** (2), 368–384.

31. Reiss, H., Knauper, S. & Kroncke, I. (2003) Invertebrate associations with gastropod shells inhabited by *Pagurus bernhardus* (Paguridae) – secondary hard substrate increasing biodiversity in North Sea soft-bottom communities. *Sarsia*, **88**, 404–414.

32. Chiavelli, D.A., Mills, E.L. & Threlkeld, S.T. (1993) Host preference, seasonality, and community interactions of zooplankton epibionts. *Limnology and Oceanography*, **38** (3), 574–583.

33. Davis, A.R. & White, G.A. (1994) Epibiosis in a guild of sessile subtidal invertebrates in southeastern Australia: a quantitative survey. *Journal of Experimental Marine Biology and Ecology*, **177** (1), 1–14.

34. Barnes, D.K.A. (1995) Sublittoral epifaunal communities at Signy Island, Antarctica. 2. Below the ice-foot zone. *Marine Biology*, **121** (3), 565–572.

35. Cook, J.A., Chubb, J.C. & Veltkamp, C.J. (1998) Epibionts of *Asellus aquaticus* (L.) (Crustacea, Isopoda): an SEM study. *Freshwater Biology*, **39** (3), 423–438.

36. Gutt, J. & Schickan, T. (1998) Epibiotic relationships in the Antarctic benthos. *Antarctic Science*, **10** (4), 398–405.

37. Harder, T., Lau, S.C.K., Dobretsov, S., Fang, T.K. & Qian, P.Y. (2003) A distinctive epibiotic bacterial community on the soft coral *Dendronephthya* sp. and antibacterial activity of coral tissue extracts suggest a chemical mechanism against bacterial epibiosis. *FEMS Microbiology Ecology*, **43** (3), 337–347.

38. Lee, O.O. & Qian, P.Y. (2004) Potential control of bacterial epibiosis on the surface of the sponge *Mycale adhaerens*. *Aquatic Microbial Ecology*, **34** (1), 11–21.

39. Dobretsov, S. & Qian, P.Y. (2006) Facilitation and inhibition of larval attachment of the bryozoan *Bugula neritina* in association with mono-species and multi-species biofilms. *Journal of Experimental Marine Biology and Ecology*, **333** (2), 263–274.

Chapter 8
Natural Control of Fouling

Rocky de Nys, Jana Guenther and Maria J. Uriz

The aim of this chapter is to highlight the diversity and complexity of the natural defence mechanisms that control biofouling in marine organisms. We have selected examples and case studies to provide an overview of the impacts of biofouling on organisms and the effectiveness of a range of defence mechanisms that prevent or reduce biofouling. The focus of this chapter is on the natural control of macrofouling organisms given this reflects the majority of work on the natural control of biofouling to date, acknowledging the growing body of literature on defences against bacteria, biofilms and pathogens. We also identify, where we can, the potential to further our understanding of the complexity of defence mechanisms and their potential synergies.

8.1 Biofouling and its biological consequences

Biofouling on a living surface, more commonly termed epibiosis, is the non-symbiotic, often facultative, association between the biofouling organism (the epibiont) and the living surface (the basibiont) [1, 2] (see also Chapter 7). Epibiosis can have significant consequences for the epibiont, the basibiont or both organisms. The colonisation of biofouling organisms may be beneficial because epibionts have a protective role for some basibionts. For example, the hydroid *Hydractinia* sp. on the shells of the hermit crab *Pagurus pollicaris* [3], as well as a range of epibionts on the carapace of the spider crab *Maja squinado* [4] and the mussel *Mytilus edulis* [5], protect the basibionts from predation. Epiphytes on the intertidal seagrass *Zostera marina* also protect the seagrass from desiccation during low tide [6], while hydroid colonies on the giant kelp *Macrocystis pyrifera* enhance frond growth during periods of low concentrations of inorganic nitrogen in sea water due to the provision of ammonium excreted by the hydroid colonies [7, 8].

In contrast to these positive impacts, epibiosis may also have major negative impacts on marine organisms. For marine plants, two of the most significant impacts of biofouling are those on photosynthesis and growth. Epiphytes attenuate light reaching the surface of the seagrass *Z. marina* and reduce the photosynthesis of the seagrasses *Thalassia testudinum* and *Z. marina* [9, 10]. Encrusting bryozoans also reduce the photosynthesis of *Fucus serratus* and *Gelidium rex* [11,12]. Contrary to the positive effects of hydroids on *M. pyrifera* during periods of low nitrogen concentrations [7], encrusting bryozoans can also reduce growth rates of this kelp [13] and cause tissue damage resulting in lower pigment concentration [8]. Furthermore, the epiphyte *Polysiphonia lanosa* may also interfere with the reproductive output of the brown alga *Ascophyllum nodosum* by reducing the receptacle biomass of its basibiont [14].

A broad range of marine animals may also be negatively affected by epibionts. For bivalves, epibionts may negatively affect the shell condition, growth rate and survival of the pearl oysters *Pinctada margaritifera* [15] and *Pinctada maxima* [16], the scallop *Euvola ziczac* [17] and the mussel *Perna perna* [18]. The endolithic boring sponge *Cliona* sp. and the polychaete *Polydora ciliata* also weaken the shells of the gastropod *Littorina littorea* making them more prone to predation [19, 20]. These boring species together with the barnacle *Balanus crenatus* and tissue-invading trematodes further reduce the fecundity, growth and survival of this gastropod [20]. Epibionts also increase mass and drag [21] and restrict the mobility and lower the crawling speed of *L. littorea* [22] and *Batillaria zonalis* [23].

Some epibionts may affect the biological function of a basibiont's appendages. For example, mussels attached to the branchial appendages of the horseshoe crab *Limulus polyphemus* can impair aeration of the crab's gills [24], while bryozoans on the eyes and antennae of the crab *Carcinus maenas* can lead to the loss of function of these organs [25]. Another unusual example is epizoic barnacles, in particular the little known barnacle *Platylepas ophiophilus*, which impair the ability of sea snakes, such as *Aipysurus laevis* and *Lapemis hardwickii*, to shed their skin [26].

Finally, basibionts may be damaged by predators of epibionts. Kelp blades of *M. pyrifera* encrusted with the bryozoan *Membranipora membranacea* and the barnacle *Lepas pacifica* are more readily damaged by the labrid fish *Oxyjulis californica* [27]. The protozoan *Colacium vesiculosum* on the pelagic crustacean zooplankton *Daphnia* may also make it more susceptible to planktivorous fish predation due to increased visibility [28]. In summary, given the overwhelmingly negative ecological effects associated with epibiosis, it is not surprising that many marine plants and animals have evolved intricate defence mechanisms to prevent or reduce biofouling and its associated disadvantages and costs.

8.2   Defence mechanisms against biofouling

Defence mechanisms that prevent or reduce biofouling can be broadly defined into four categories: behavioural, mechanical, physical and chemical. However, these mechanisms are not mutually exclusive, and as we begin to better understand the nature of defence mechanisms against biofouling it is becoming clearer that there is a complexity of mechanisms. These mechanisms are often employed together, and potentially synergistically, to provide optimum multiple defence strategies.

8.2.1   Behavioural defence mechanisms

A number of behavioural mechanisms have been identified that deter or control the settlement, growth and survival of biofouling organisms. Some benthic motile species, such as isopods, crabs and gastropods, burrow into soft bottom substrata or hide in rock crevices, thereby abrading the surface and minimising their exposure to biofouling organisms [29, 30]. The nocturnal activity of some crabs also restricts algal growth on their carapace [31], while intertidal exposure to air leads to desiccation stress of biofouling organisms on crabs and gastropods [30, 31].

In high-density populations of the gastropod *L. littorea*, individuals often move over one another, grazing and secreting mucus onto each other's shells, which inhibits biofouling [32]. Similarly, the gastropod *Calliostoma zizyphinum* regularly wipes the surface of its shell with its foot, which has an important role for both feeding and as a defence against biofouling [33]. When specimens of *C. zizyphinum* were prevented from wiping their shells, the biofouling cover on their shells was nine times higher than on wiped shells [33].

While behavioural mechanisms against biofouling organisms are important, they may not have evolved specifically as a defence mechanism. For example, crustaceans gain additional benefits from burrowing, such as reduced predation [34], which may be of even greater importance than biofouling [31]. This makes identifying specific behaviours as defence strategies against biofouling and quantifying their importance difficult, given that burrowing, nocturnal activity and exposure are common among many marine organisms in a diversity of environments.

8.2.2   *Mechanical defence mechanisms*

Mechanical defence mechanisms also reduce biofouling on basibionts, primarily through the removal of epibionts with the shedding of surface layers [35] (Plate III A). For example, the encrusting coralline alga *Spongites yengoi* sloughs off deep layers within the thallus to remove associated micro- and macrofouling organisms [35], whereas *Sporolithon ptychoides*, *Neogoniolithon fosliei* and *Hydrolithon onkodes* only slough off surface layers [36], and the algae *A. nodosum*, *Chondrus crispus* and *Dilsea carnosa* slough off the cuticle [37, 38]. Sponges have a similar mechanical mechanism and this appears to be widespread. The best example is the Mediterranean sponge *Crambe crambe* which has a post-reproduction resting period in autumn when it rearranges internal structures and produces a non-cellular, external cuticle, which is colonised by microfoulers. After a few weeks, the sponge casts off the cuticle with the attached microfoulers and renews its cellular ectosome [39]. A similar process to this sloughing of a 'glassy' cuticle also occurs in sponges of the orders Dictyoceratida, Verongida, Poecilosclerida and Chondrosida [40, 41], and has been reported in colonial ascidians [42] and cnidarians [43]. However, the frequency at which the epidermis is renewed is species dependent. For the sponge *Halichondria panicea* sloughing of the complete exterior tissue layer occurs every 3 weeks, and the removed tissue is covered by a microfouling community that is not present on the surface of freshly sloughed sponges [44].

An alternative, or additional mechanical strategy, to sloughing is the removal of epibionts by renewing of surfaces through continuous or periodical mucus secretion [45, 46]. Glycoproteins in the mucus of the sea star *Marthasterias glacialis* inhibit the adhesion of bacteria by causing bacterial clumpings, which are subsequently washed away [46]. Given the number of marine plants and animals that secrete mucus onto their surface, this mechanism may be utilised much more broadly across a diversity of taxa, and may also be linked with the retention, or release of chemical defences.

The mechanical removal of epibionts can also be achieved by friction between the surface of the basibiont and the sediment for burrowing species [47] or between the surface and water for fast-swimming species [1]. Additionally, surface structures, such as spines of the bryozoans *Microciona atrasanguinea* and *Electra pilosa* [48], and cleaning of the surface, such as the gills of the shrimps *Nihonotrypaea japonica* and *Upogebia major*, by active scraping with specialised appendages may reduce epibionts [49].

Figure 8.1 Pedicellaria of the crown-of-thorns sea star *Acanthaster planci*. Although a mechanical defence against fouling had been proposed, these pedicellariae are ineffective in deterring the settlement of fouling organisms on the surface of the sea stars.

An unusual proposed mechanical defence is the use of pedicellariae (Figure 8.1), which are forcep- or pincer-like appendages made of calcareous ossicles, found in sea urchins and some sea stars [50]. Pedicellariae have reported functions in deterring predators and competitors [51], capturing mobile prey [52] and deterring settling larvae of biofouling organisms [50]. Mobile pedicellariae and spines of the sea urchin *Echinus esculentus* do prevent the settlement of cypris larvae of the barnacle *Balanus balanoides* [50], and pedicellariae are also proposed to protect the surface of sea stars from settling biofouling organisms [53]. However, in the only study explicitly investigating this role, pedicellariae were singularly ineffective as a defence against biofouling [54]. This does not exclude a defensive role given the range of pedicellariae types and densities, but it does exclude a universal role as a mechanical defence in echinoderms.

Finally, mechanical defences against biofouling are not restricted to invertebrates, with higher organisms also effectively removing epibionts through mechanical defences. For example, the loggerhead sea turtle *Caretta caretta* sheds portions of their scutes, which remove some sessile organisms [55]. Again, given the scale of epidermal shedding by marine mammals, this mechanism may be widespread and effective across marine fish and mammals.

8.2.3 Physical defence mechanisms

Research on physical defence mechanisms against the settlement of biofouling organisms has focused on the wettability and microtopography of natural surfaces. Surface wettability and surface tension influence the attachment and strength of attachment of biofouling organisms to solid surfaces. While the effects of the surface wettability of artificial surfaces on the settlement of marine organisms are well documented [56], contact angle measurements of marine organisms are scarce [57, 58], and the effects of natural surface energies in situ are not well known. In one of the few studies in the field low surface energies of 23–27 mNm^{-1} were

(a)   (b)   (c)

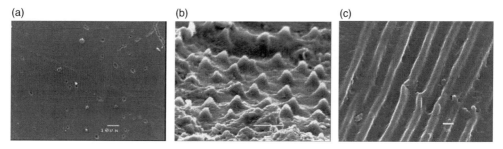

Figure 8.2 SEM images of the exterior surface of the shells highlighting the diversity of surface micro-topographies present on bivalve molluscs ranging from featureless, to complex peak and ridge profiles. Shells are (a) *Amusium balloti* (scale bar 10 μm), (b) *Tellina plicata* (scale bar 5 μm) and (c) *Mytilus galloprovincialis* (scale bar 1 μm). (Reproduced with permission from References [66,67].)

measured for the gorgonians *Pseudopteragorgia americana* and *Pseudopteragorgia acerose* [57], and this range is correlated with the reduced settlement of microfouling organisms [59,60].

Physical surface microtopographies, and their potential biofouling deterrent properties, have become a focus of biofouling research given they offer the potential to develop non-toxic biomimetic antifouling technologies [61]. One of the best examples of a natural surface with a physical defence mechanism is the skin of the pilot whale *Globicephala melas*, where 0.1–1.2 μm^2 pores enclosed in a network of nanoridges reduce the surface available for adhesion and attachment to the pore margins and tips of the nanoridges [62]. Physical defence mechanisms of the periostracum of shell-bearing molluscs, which is a thin, flexible and sclerotinised protein layer covering calcified shells [63], have also been investigated because the periostracum has a secondary function in deterring epibionts [45,64,65]. Studies on the surface microto-pographies of the periostracum show that micro-ripples (Figure 8.2) with wavelengths of 1.8–1.9 μm on *Mytilus galloprovincialis* [65,66], 1.0–2.0 μm on *Mytilus edulis* [68,69], 1.5–2.0 μm on *P. perna* [69] and 0.8 μm on *Pteria penguin* [70] correlate with low biofouling cover.

Similarly, parallel ridges with irregular distances of 15–115 μm on the eggcase of the dogfish *Scyliorhinus canicula*, and evenly distributed knobbed surface structures, 10 μm in diameter, on the aboral skeleton plates of the brittle star *Ophiura texturata* had repellent effects on the ciliates *Zoothamnium commune* and *Vorticella* sp. [68]. Evenly distributed circular elevations, 200 μm in diameter, and spicule-like structures, 2–2.5 μm in length, between the elevations of the carapace of the crab *Cancer pagurus* repelled the barnacle *Balanus improvisus* [68].

An area of potential is the biofouling deterrent role of the surface micro- and macrostructures of sponges (Plate III A), where there are unique obstacles to deter organisms from settling. Most sponge surface skeletons consist of specialised spicules that protrude to various degrees from the sponge body. Sponge microtopography may reduce or impede settlement by reducing the area for settlement and decreasing its stability since spicules are continuously shed [71]. For instance, a dense palisade of needles (oxea) protrudes 50–100 μm from the surface of Polymastidae and Suberitidae species. Long oxea with a spacing of 1 mm extend several millimetres beyond the surface of *Raspailia* species, and divergent bundles of thin, shorter styloids surround each oxea [72]. Sponges from the genus *Pheronema* form dense mazes of several long spicules which hold diatoms and small crustaceans away from the sponge surface and prevent the biofouling of ostia [73]. The sponge genera *Spirastrella, Diplastrella*,

Corticium, and many calcareous sponges, have multi-actinated microscleres within the ectosomal cells, and the 5–20 μm long actines (rays) of these microscleres protrude from the sponge surface producing microtopographies similar to those reported to inhibit biofouling on artificial [74] and natural surfaces [66]. Additionally, smooth sponge surfaces, particularly those of keratose sponges (the orders Verongida, Dictyoceratida and Dendroceratida), have characteristic conules and depressions resulting in a continuous change of the surface tension, with maximal tension at the conule apex, and tensile stresses in the dermal membrane [75], which may prevent biofouling in a similar manner to gorgonian surfaces [57].

While surface topography and structure can be effective in deterring the settlement and growth of biofouling organisms, and may be important in reducing biofouling through facilitating weak adhesion and the release of biofouling, the deterrent effects of surface topography are by no means universal. For example, the surfaces of sea stars that are not fouled [76] were investigated for their physical effects against biofouling, and were ineffective [77].

There does, however, appear to be a common scale of microtopography from nanometres to microns that is most effective across a range of natural surfaces, and this scale is similar to scale of the propagules and larvae of many biofouling organisms. Surface structures within this range can effectively reduce the surface available for the settlement and growth of biofouling organism, which many biofouling organisms avoid [67, 78–80]. It also provides a template for the development of biomimetic antifouling surfaces [74].

8.2.4   Chemical defence mechanisms

Many studies have concentrated on the antifouling activity of surface-associated or exuded natural products extracted from marine algae and invertebrates [1, 81–83]. It has been repeatedly reported that benthic micro- and macroorganisms produce bioactive molecules that can kill, impair or inhibit the settlement of bacteria, spores or larvae in laboratory assays [84]. However, to unequivocally determine whether marine organisms use natural products to keep their surfaces free of biofouling organisms, several criteria need to be fulfilled [82]. Firstly, the marine organisms need to be generally free of biofouling organisms in their natural environment. Secondly, the presence of putative compounds on or near the surface of the organism needs to be verified and their natural concentrations quantified. Thirdly, ecologically relevant concentrations of these compounds need to be tested in bioassays with ecologically relevant biofouling species to determine whether these compounds have any antifouling effects [81, 85–88]. This is a challenge given the difficulty in quantifying metabolites in low concentrations in sea water and testing them against biofouling organisms, and has restricted the identification of 'natural antifoulants' to a small subset of the array of biologically active metabolites characterised from marine organisms. Most studies on the ecological role of surface-associated or exuded compounds have targeted marine algae [87–92]. To date, only terpenoids from the brown alga *Dictyota menstrualis* [85], furanones from the red alga *Delisea pulchra* [87, 90], phlorotannins from the brown alga *Fucus vesiculosus* [92] and triterpene glycosides of the sponges *Erylus formosus* and *Ectyoplasia ferox* [93] have been localised at or near the surface at biologically effective concentrations to deter biofouling organisms. Given the presence of these compounds at or near the surface, there is also the potential that they may affect post-settlement processes in settled organisms, and subsequently reduce biofouling through decreasing post-settlement survival [94]. This is an area of chemical defences we need to develop given that there is a constant loss of metabolites, either primary or secondary, from marine organisms into the environment, and these have the potential to affect post-settlement survival, either independent of, or in addition to, affecting the settlement process.

To determine the role of natural products as a defence against biofouling, we also need to be able to characterise the cellular, biochemical and molecular processes involved in the production, storage and release of metabolites. A lack of knowledge on how and where bioactive molecules are produced and stored within an organism, and how these are released, either as water-borne cues or surface-associated compounds, makes it difficult to design realistic (ecologically relevant) experiments in the field. Several studies demonstrate that potential antifouling compounds are localised in specific cells. Monitoring the fate of these cells provides important information on the biological process of defence against biofouling. The sponge *C. crambe* (Plate III A) provides an example where biologically active compounds are localised in spherulous cells that accumulate close to the sponge periphery and along the exhalant channels, and, once released, remain attached to the external sponge surface [95]. The challenge remains to link the processes of localisation, internal transport, release, external accumulation or diffusion, and biological efficacy (either pre- or post-settlement), to provide realistic models for chemical defence mechanisms against biofouling.

8.2.5   Multiple defence strategies

Natural antifouling mechanisms are highly complex and throughout this chapter we describe the diversity of effective strategies used to prevent or reduce biofouling. An optimised defence against biofouling may employ a combination of behavioural, mechanical, physical and/or chemical defence mechanisms to deter the settlement of the widest range of biofouling organisms. For example, the gorgonian *Pseudopterogorgia americana* uses both mechanical and physical defence mechanisms against biofouling by secreting large amounts of mucus and having low surface energies of 23–27 mN m^{-1} [57,96]. Another gorgonian *Leptogorgia virgulata* uses mechanical and chemical defences to deter biofouling organisms by shedding thin layers of spicule-containing material and possessing secondary metabolites [97,98]. The colonial ascidian *Polysyncraton lacazei* combines mechanical and chemical defences against a broad range of biofouling organisms by sloughing surface layers and producing secondary metabolites. They also have surface-associated copepod and mite species grazing on its surface [99]. The mussel *Mytilus edulis* and the pilot whale *G. melas* use both physical and chemical defence mechanisms. The periostracum of the mussel *M. edulis* has micro-ripples with wavelengths of 1–2 μm [67,68] and contains compounds, which deter the settlement of biofouling organisms [100]. Similarly, the skin of the pilot whale *G. melas* has 0.1–1.2 μm^2 pores enclosed in a network of nanoridges, and a zymogel with hydrolytic activities, both of which enhance the self-cleaning abilities of the skin [62,101]. Sponges also provide examples of multiple defences with *C. crambe* as a model where an external glassy pellicle forms a physical defence, which when lost (with any biofouling organisms) is replaced by a chemical defence localised within spherulous cells in the ectosome [39]. Ectosomal skeletons of mega- and microscleres also produce surface macro-and microtopographies, while physical surface tension variation also contributes to the effective control of biofouling.

8.3   Conclusions

- The natural control of biofouling is diverse, complex and functional.
- Understanding natural defence strategies provides not only an understanding of specialised biological and ecological processes, but also a template for the development of new non-toxic antifouling solutions.
- Non-release technologies using biomimetic physical, mechanical and chemical principles provide an opportunity to maximise biofouling deterrence and/or release, while minimising or alleviating environmental effects.

References

1. Wahl, M. (1989) Marine epibiosis. I. Fouling and antifouling: some basic aspects. *Marine Ecology Progress Series*, **58**, 175–189.
2. Wahl, M. & Mark, O. (1999) The predominantly facultative nature of epibiosis: experimental and observational evidence. *Marine Ecology Progress Series*, **187**, 59–66.
3. Brooks, W.R. & Mariscal, R.N. (1985) Protection of the hermit crab *Pagurus pollicaris* Say from predators by hydroid-colonized shells. *Journal of Experimental Marine Biology and Ecology*, **87**, 111–118.
4. Parapar, J., Fernandez L., Gonzalez-Guirriaran, E. & Muino, R. (1997) Epibiosis and masking material in the spider crab *Maja squinado* (Decapoda: Majidae) in the Ria de Arousa (Galicia, NW Spain). *Cahiers de Biologie Marine*, **38**, 221–234.
5. Thieltges, D.W. (2005) Benefit from an invader: American slipper limpet *Crepidula fornicata* reduces star fish predation on basibiont European mussels. *Hydrobiologia*, **541**, 241–244.
6. Penhale, P.A. & Smith, W.O. (1977) Excretion of dissolved organic carbon by eelgrass (*Zostera marina*) and its epiphytes. *Limnology and Oceanography*, **22**, 400–407.
7. Hepburn, C.D. & Hurd, C.L. (2005) Conditional mutualism between the giant kelp *Macrocystis pyrifera* and colonial epifauna. *Marine Ecology Progress Series*, **302**, 37–42.
8. Hepburn, C.D., Hurd, C.L. & Frew, R.D. (2006) Colony structure and seasonal differences in light and nitrogen modify the impact of sessile epifauna on the giant kelp *Macrocystis pyrifera* (L.) C Agardh. *Hydrobiologia*, **560**, 373–384.
9. Brush, M.J. & Nixon, S.W. (2002) Direct measurements of light attenuation by epiphytes on eelgrass *Zostera marina*. *Marine Ecology Progress Series*, **238**, 73–79.
10. Drake, L.A., Dobbs, F.C. & Zimmerman, R.C. (2003) Effects of epiphyte load on optical properties and photosynthetic potential of the seagrasses *Thalassia testudinum* Banks ex Konig and *Zostera marina* L. *Limnology and Oceanography*, **48**, 456–463.
11. Oswald, R.C., Telford, N., Seed, R. & Happey-Wood, C.M. (1984) The effects of encrusting bryozoans on the photosynthetic activity of *Fucus serratus* L. *Estuarine, Coastal and Shelf Science*, **19**, 697–702.
12. Cancino, J.M., Munoz, J., Munoz, M. & Orellana, M.C. (1987) Effects of the bryozoan *Membranipora tuberculata* (Bosc) on the photosynthesis and growth of *Gelidium rex* Santilices et Abbott. *Journal of Experimental Marine Biology and Ecology*, **113**, 105–112.
13. Dixon, J., Schroeter, S.C. & Kastendiek, J. (1981) Effects of the encrusting bryozoan *Membranipora membranacea* on the loss of blades and fronds by the giant kelp *Macrocystis pyrifera* (Laminariales). *Journal of Phycology*, **17**, 341–345.
14. Kraberg, A.C. & Norton, A.T. (2007) Effect of epiphytism on reproductive and vegetative lateral formation in the brown, intertidal seaweed *Ascophyllum nodosum* (Phaeophyceae). *Phycological Research*, **55**, 17–24.

15. Mao Che, L., Le Campion-Alsumard, T., Boury-Esnault, N., Payri, C., Golubic, S. & Bezac, C. (1996) Biodegradation of shells of the black pearl oyster, *Pinctada margaritifera var. cumingii*, by microborers and sponges of French Polynesia. *Marine Biology*, **126**, 509–519.

16. Taylor, J.J., Southgate, P.C. & Rose, R.A. (1997) Fouling animals and their effect on the growth of silver-lip pearl oysters, *Pinctada maxima* (Jameson) in suspended culture. *Aquaculture*, **153**, 31–40.

17. Lodeiros, C.J.M. & Himmelman, J.H. (1996) Influence of fouling on the growth and survival of the tropical scallop, *Euvola* (*Pecten*) *ziczac* (L. 1758) in suspended culture. *Aquaculture Research*, **27**, 749–756.

18. Kaehler, S. & McQuaid, C.D. (1999) Lethal and sub-lethal effects of phototrophic endoliths attacking the shell of the intertidal mussel *Perna perna*. *Marine Biology*, **135**, 497–503.

19. Stefaniak, L.M., McAtee, J. & Shulman, M.J. (2005) The costs of being bored: effects of a clinoid sponge on the gastropod *Littorina littorea* (L.). *Journal of Experimental Marine Biology and Ecology*, **327**, 103–114.

20. Thieltges, D.W. & Buschbaum, C. (2007) Vicious circle in the intertidal: facilitation between barnacle epibionts, a shell boring polychaete and trematode parasites in the periwinkle *Littorina littorea*. *Journal of Experimental Marine Biology and Ecology*, **340**, 90–95.

21. Wahl, M. (1997) Increased drag reduces growth of snails: comparison of flume and *in situ* experiments. *Marine Ecology Progress Series*, **151**, 291–293.

22. Buschbaum, C. & Reise, K. (1999) Effects of barnacle epibionts on the periwinkle *Littorina littorea* (L.). *Helgoland Marine Research*, **53**, 56–61.

23. Chan, D.H.L. & Chan, B.K.K. (2005) Effect of epibiosis on the fitness of the sandy shore snail *Batillaria zonalis* in Hong Kong. *Marine Biology*, **146**, 695–705.

24. Botton, M.L. (1981) The gill book of the horseshoe crab (*Limulus polyphemus*) as a substrate for the blue mussel (*Mytilus edulis*). *Bulletin New Jersey Academy of Science*, **26**, 26–28.

25. Cadee, G.C. (1991) Carapaces of the shore crab *Carcinus maenas* as a substrate for encrusting organisms. In: *Bryozoa: Living and Fossil* (ed. F.P. Bigey), pp. 71–79. Société des Sciences Naturelles de l'Ouest de la France, Nantes.

26. Zann, L.P., Cuffey, R.J. & Kropach, C. (1975) Fouling organisms and parasites associated with the skin of sea snakes. In: *The Biology of Sea Snakes* (ed. W.A. Dunson), pp. 251–265. University Park Press, Baltimore.

27. Bernstein, B.B. & Jung, N. (1979) Selective pressures and coevolution in a kelp canopy community in southern California. *Ecological Monographs*, **49**, 335–355.

28. Chiavelli, D.A., Mills, E.L. & Threlkeld, S.T. (1993) Host preference, seasonality, and community interactions of zooplankton epibiont. *Limnology and Oceanography*, **38**, 574–583.

29. Olafsdottir, S.H. & Svavarsson, J. (2002) Ciliate (Protozoa) epibionts of deep-water asselote isopods (Crustacea): patterns and diversity. *Journal of Crustacean Biology*, **22**, 607–618.

30. Vasconcelos, P., Cúrdia, J., Castro, M. & Gaspar, M.B. (2007) The shell of *Hexaplex* (*Truncular-iopsis*) *trunculus* (Gastropoda: Muricidae) as a mobile hard substratum for epibiotic polychaetes (Annelida: Polychaeta) in the Ria Formosa (Algarve coast – southern Portugal). *Hydrobiologia*, **575**, 161–172.

31. Becker, K. & Wahl, M. (1996) Behaviour patterns as natural antifouling mechanisms of tropical marine crabs. *Journal of Experimental Marine Biology and Ecology*, **203**, 245–258.

32. Wahl, M. & Sönnichsen, H. (1992) Marine epibiosis. IV. The periwinkle *Littorina littorea* lacks typical antifouling defences – why are some populations so little fouled? *Marine Ecology Progress Series*, **88**, 225–235.

33. Holmes, S.P., Sturgess, C.J., Cherrill, A. & Davies, M.S. (2001) Shell wiping in *Calliostoma zizyphinum*: the use of pedal mucus as a provendering agent and its contribution to daily energetic requirements. *Marine Ecology Progress Series*, **212**, 171–181.

34. Kuhlmann, M.L. (1992) Behavioural avoidance of predation in an intertidal hermit crab. *Journal of Experimental Marine Biology and Ecology*, **157**, 143–158.

35. Keats, D.W., Groener, A. & Chamberlain, Y.M. (1993) Cell sloughing in the littoral zone coralline alga, *Spongites yengoi* (Foslie) Chamberlain (Corallinales, Rhodophyta). *Phycologia*, **32**, 143–150.

36. Keats, D.W., Knight, M.A. & Pueschel, C.M. (1997) Antifouling effects of epithallial shedding in three crustose coralline algae (Rhodophyta, Coralinales) on a coral reef. *Journal of Experimental Marine Biology and Ecology*, **213**, 281–293.

37. Sieburth, J. & Tootle, J.L. (1981) Seasonality of microbial fouling on *Ascophyllum nodosum* (L.) Lejol., *Fucus vesiculosus* L., *Polysiphonia lanosa* (L.) Tandy and *Chondrus crispus* Stackh. *Journal of Phycology*, **17**, 57–64.

38. Nylund, G.M. & Pavia, H. (2005) Chemical *versus* mechanical inhibition of fouling in the red alga *Dilsea carnosa*. *Marine Ecology Progress Series*, **299**, 111–121.

39. Turon, X., Uriz, M.J. & Willenz, P. (1999) Cuticular linings and remodelisation processes in *Crambe crambe* (Demospongiae: Poecilosclerida). *Memoirs of the Queensland Museum*, **44**, 617–625.

40. Carballo, J.L. (1994) *Taxonomia, zoogeografia y autoecologia de los poriferos del estrecho de Gibraltar*. PhD thesis, University of Sevilla.

41. Vacelet, J. & Perez, T. (1998) Two new genera and species of sponges (Porifera: Demospongiae) without skeleton from a Mediterranean cave. *Zoosystema*, **20**, 5–22.

42. Turon, X. (1992) Periods of on-feeding in *Polysyncrator lacazei* (Ascidiacea: Didemnidae): a rejuvenative process? *Marine Biology*, **112**, 647–655.

43. Garrabou, J. (1997) *Structure and dynamics of north-western Mediterranean rocky benthic communities along a depth gradient: a geographical information system (GIS) approach*. PhD thesis, University of Barcelona.

44. Barthel, D. & Wolfrath, B. (1989) Tissue sloughing in the sponge *Halochondria panicea*: a fouling organism prevents being fouled. *Oecologia*, **78**, 357–360.

45. Wahl, M., Kröger, K. & Lenz, M. (1998) Non-toxic protection against epibiosis. *Biofouling*, **12**, 205–226.

46. Bavington, C.D., Lever, R., Mulloy, B. *et al.* (2004) Anti-adhesive glycoproteins in echinoderm mucus secretions. *Comparative Biochemistry and Physiology. Part B: Biochemistry & Molecular Biology*, **139**, 607–617.

47. Svavarsson, J. & Davidsdottir, B. (1994) Foraminiferan (Protozoa) epizoites on Arctic isopods (Crustacea) as indicators of isopod behavior. *Marine Biology*, **118**, 239–246.

48. Dyrynda, P.E.J. (1986) Defensive strategies of modular organisms. *Philosophical Transactions of the Royal Society of London B: Biological Sciences*, **313**, 227–243.

49. Batang, S.B. & Suzuki, H. (2003) Gill-cleaning mechanisms of the burrowing thalassinidean shrimps *Nihonotrypaea japonica* and *Upogebia major* (Crustacea: Decapoda). *Journal of Zoology*, **261**, 69–77.

50. Campbell, A.C. & Rainbow, P.S. (1977) The role of pedicellariae in preventing barnacle settlement on the sea urchin test. *Marine Behaviour and Physiology*, **4**, 253–260.

51. van Veldhuizen, H.D. & Oakes, V.J. (1981) Behavioural responses of 7 species of asteroids to the asteroid predator *Solaster dawsoni*. *Oecologia*, **48**, 214–220.

52. Lauerman, L.M.L. (1998) Diet and feeding behaviour of the deep-water sea star *Rathbunaster californicus* (Fisher) in the Monterey Submarine Canyon. *Bulletin of Marine Science*, **63**, 523–530.

53. Ruppert, E.E. & Barnes, R.D. (1994) *Invertebrate Zoology*. Saunders College Publishing, Fort Worth.

54. Guenther, J., Heimann, K. & de Nys, R. (2007) Pedicellariae of the crown-of-thorns sea star *Acanthaster planci* are not an effective defence against fouling. *Marine Ecology Progress Series*, **340**, 101–108.

55. Caine, E.A. (1986) Carapace epibionts of nesting loggerhead sea turtles: Atlantic coast of U.S.A. *Journal of Experimental Marine Biology and Ecology*, **95**, 15–26.

56. Aldred, N., Ista, L.K., Callow, M.E., Callow, J.A., Lopez, G.P. & Clare, A.S. (2006) Mussel (*Mytilus edulis*) byssus deposition in response to variations in surface wettability. *Journal of the Royal Society Interface*, **3**, 37–43.

57. Vrolijk, N.H., Targett, N.M., Baier, R.E. & Meyer, A.E. (1990) Surface characterisation of two gorgonian coral species: implications for a natural antifouling defence. *Biofouling*, **2**, 39–54.

58. Becker, K., Hormchong, T. & Wahl, M. (2000) Relevance of crustacean carapace wettability for fouling. *Hydrobiologia*, **426**, 193–201.

59. Baier, R.E. (1970) Surface properties influencing biological adhesion. In: *Adhesion in Biological Systems* (ed. R.S. Manly), pp. 15–48. Academic Press, New York.

60. Genzer, J. & Efimenko, K. (2006) Recent developments in superhydrophobic surfaces and their relevance to marine fouling: a review. *Biofouling*, **22**, 339–360.

61. de Nys, R. & Steinberg, P.D. (2002) Linking marine biology and biotechnology. *Current Opinion in Biotechnology*, **13**, 244–248.

62. Baum, C., Meyer, W., Stelzer, R., Fleischer, L.G. & Siebers, D. (2002) Average nanorough skin surface of the pilot whale (*Globicephala melas*, Delphinidae): considerations on the self-cleaning abilities based on nanoroughness. *Marine Biology*, **140**, 653–657.

63. Harper, E.M. (1997) The molluscan periostracum: an important constraint in bivalve evolution. *Palaeontology*, **40**, 71–97.

64. Harper, E.M. & Skelton, P.W. (1993) A defensive value of the thickened periostracum in the Mytiloidea. *Veliger*, **36**, 36–42.

65. Scardino, A.J., de Nys, R., Ison, O., O'Connor, W. & Steinberg, P. (2003) Microtopography and antifouling properties of the shell surface of the bivalve molluscs *Mytilus galloprovincialis* and *Pinctada imbricata*. *Biofouling*, **19**, 221–230.

66. Scardino, A.J. & de Nys, R. (2004) Fouling deterrence on the bivalve shell *Mytilus galloprovincialis*: a physical phenomenon? *Biofouling*, **20**, 249–257.

67. Scardino, A.J., Harvey, E. & de Nys, R. (2006) Testing attachment point theory: diatom attachment on microtextured polyimide biomimics. *Biofouling*, **20**, 249–257.

68. Bers, A.V. & Wahl, M. (2004) The influence of natural surface microtopographies on fouling. *Biofouling*, **20**, 43–51.

69. Bers, A.V., Prendergast, G.S., Zürn, C.M., Hansson, L., Head, R.M. & Thomason, J.C. (2006) A comparative study of the anti-settlement properties of mytilid shells. *Biology Letters*, **2**, 88–91.

70. Guenther, J. & de Nys, R. (2006) Differential community development of fouling species on the pearl oysters *Pinctada fucata*, *Pteria penguin* and *Pteria chinensis* (Bivalvia, Pteriidae). *Biofouling*, **22**, 163–171.

71. Maldonado, M. & Uriz, M.J. (1998) Microrefuge exploitation by subtidal encrusting sponges: patterns of settlement and post-settlement survival. *Marine Ecology Progress Series*, **174**, 141–150.

72. Uriz, M.J. (1988) Deep-water sponges from the continental shelf and slope off Namibia (South-West Africa): Classes Hexactinellida and Demospongia. *Monografías de Zoología Marina*, **3**, 9–157.

73. Uriz, M.J. (2006) Mineral skeletogenesis of sponges. *Canadian Journal of Zoology*, **84**, 322–356.

74. Schumacher, J.F., Aldred, N., Callow, M.E., *et al.* (2007) Species-specific engineered antifouling topographies: correlations between the settlement of algal zoospores and barnacle cyprids. *Biofouling*, **23**, 307–317.

75. Taragawa, C.K. (1990) Mechanical function and regulation of the skeletal network in Dysidea. In: *New Perspectives in Sponge Biology* (ed. K. Rützler), pp. 252–258. Smithsonian Institute Press, Washington, DC.

76. Guenther, J., Walker-Smith, G., Warén, A. & de Nys, R. (2007) Fouling-resistant surface of tropical sea stars. *Biofouling*, **23**, 413–418.

77. Guenther, J. & de Nys, R. (2007) Surface microtopographies of tropical sea stars: lack of an effective defence mechanism against fouling. *Biofouling*, **23**, 419–429.

78. Hills, J.M., Thomason, J.C. & Muhl, J. (1999) Settlement of barnacle larvae is governed by Euclidean and not fractal surface characteristics. *Functional Ecology*, **13**, 868–875.

79. Callow, M.E., Jennings, A.R., Brennan, A.B., *et al.* (2002) Microtopographic cues for settlement of zoospores of the green fouling alga *Enteromorpha*. *Biofouling*, **18**, 237–245.

80. Scardino, A.J., Guenther, J. & de Nys, R. (2008) Attachment point theory revisited: the fouling response to a microtextured matrix. *Biofouling*, **21** (1), 45–53.

81. Hay, M.E. (1996) Marine chemical ecology: what's known and what's next? *Journal of Experimental Marine Biology and Ecology*, **200**, 103–134.

82. Steinberg, P.D., de Nys, R. & Kjelleberg, S. (2001) Chemical mediation of surface colonization. In: *Marine Chemical Ecology* (eds J.B. McClintock & B.J. Baker), pp. 355–387. CRC Press, Boca Raton.

83. Steinberg, P.D., de Nys, R. & Kjelleberg, S. (2002) Chemical cues for surface colonization. *Journal of Chemical Ecology*, **28**, 1935–1951.

84. Fusetani, N. (2004) Biofouling and antifouling. *Natural Product Reports*, **21**, 94–204.

85. Schmitt, T.M., Hay, M.E. & Lindquist, N. (1995) Constraints on chemically mediated coevolution: multiple functions for seaweed secondary metabolites. *Ecology*, **76**, 107–123.

86. Nylund, G.M. & Pavia, H. (2003) Inhibitory effects of red algal extracts on larval extracts settlement of the barnacle *Balanus improvisus*. *Marine Biology*, **143**, 875–882.

87. Dworjanyn, S.A., de Nys, R. & Steinberg, P.D. (2006) Chemically mediated antifouling in the red alga *Delisea pulchra*. *Marine Ecology Progress Series*, **318**, 153–163.

88. Nylund, G.M., Gribben, P.E., de Nys, R., Steinberg, P.D. & Pavia, H. (2007) Surface chemistry *versus* whole-cell extracts: antifouling tests with seaweed metabolites. *Marine Ecology Progress Series*, **329**, 73–84.

89. de Nys, R., Dworjanyn, S.A. & Steinberg, P.D. (1998) A new method for determining surface concentrations of marine natural products on seaweeds. *Marine Ecology Progress Series*, **162**, 79–87.

90. Dworjanyn, S.A., de Nys, R. & Steinberg, P.D. (1999) Localisation and surface quantification of secondary metabolites in the red alga *Delisea pulchra*. *Marine Biology*, **133**, 727–736.

91. Dobretsov, S., Dahms, H.U., Harder, T. & Qian, P.Y. (2006) Allelochemical defense against epibiosis in the macroalga *Caulerpa racemosa* var. *turbinata*. *Marine Ecology Progress Series*, **318**, 165–175.

92. Brock, E., Nylund, G.M. & Pavia, H. (2007) Chemical inhibition of barnacle settlement by the brown alga *Fucus vesiculosus*. *Marine Ecology Progress Series*, **337**, 165–174.

93. Kubanek, J., Whalen, K.E., Engel, S., *et al.* (2002) Multiple defensive roles for triterpene glycosides from two Caribbean sponges. *Oecologia*, **131**, 125–136.

94. Wikström, S.A. & Pavia, H. (2004) Chemical settlement inhibition *versus* post-settlement mortality as an explanation for differential fouling of two congeneric seaweeds. *Oecologia*, **138**, 223–230.

95. Uriz, M.J., Becerro, M.A., Tur, J.M. & Turon, X. (1996) Location of toxicity within the Mediterranean sponge *Crambe crambe* (Demospongiae: Poecilosclerida). *Marine Biology*, **124**, 583–590.

96. Ciereszko, L.S., Mizelle, J.W. & Schmidt, R.W. (1973) Occurrence of taurobetaine in coelenterates and of polysaccharide sulfate in the gorgonian *Pseudopteragorgia americana*. In: *Food Drugs from the Sea: Proceedings, 1972* (ed. L.R. Worthen), pp. 177–180. Marine Technology Society, Washington, DC.

97. Patton, W.K. (1972) Studies on the animal symbionts of the gorgonian coral, *Leptogorgia virgulata* (Lamarck). *Bulletin of Marine Science*, **22**, 419–431.

98. Targett, N.M., Bishop, S.S., McConnell, O.J. & Yoder, J.A. (1983) Antifouling agents against the benthic marine diatom, *Navicula salinicola* homarine from the gorgonians *Leptogorgia virgulata* and *L. setacea* and analogs. *Journal of Chemical Ecology*, **9**, 817–829.

99. Wahl, M. & Banaigs, B. (1991) Marine epibiosis. III. Possible antifouling defense adaptations in *Polysyncraton lacazei* (Giard) (Didemnidae, Ascidiacea). *Journal of Experimental Marine Biology and Ecology*, **145**, 49–63.

100. Bers, A.V., D'Souza, F., Klijnstra, J.W., Willemsen, P.R. & Wahl, M. (2006) Chemical defence in mussels: antifouling effect of crude extracts of the periostracum of the blue mussel *Mytilus edulis*. *Biofouling*, **22**, 251–259.

101. Baum, C., Meyer, W., Roessner, D., Siebers, D. & Fleischer, L.G. (2001) A zymogel enhances the self-cleaning abilities of the skin of the pilot whale (*Globicephala melas*). *Comparative Biochemistry and Physiology. Part A: Molecular & Integrative Physiology*, **130**, 835–847.

Introduction to Microbial Fouling

Sergey Dobretsov, Anna M. Romaní, David A. Spratt,
Derren Ready and Jonathan Pratten

Biofilms are structured microbial communities that develop on solid surfaces within an aqueous phase. Although the first detailed description of microbial attachment to surfaces appeared more than 50 years ago [1], it was not until the late 1970s that the term 'biofilm' made its first appearance in the scientific literature. Biofilm development is a multi-step process dependent on the properties of a substratum, environmental conditions and composition of biofilms. It includes adhesion of microbes, formation of biofilms, biofilm growth and, finally, biofilm sloughing and cell detachment [2].

Bacterial adhesion is the first and most important step in biofilm formation. At 50 nm, reversible, low-specificity (e.g. van der Waals) forces are important before short-range forces (e.g. hydrogen bonds) create a stronger bond. Additionally, hydrophobic molecules as well as pili and fimbria predispose to bacterial adhesion. Once adhered, the microorganisms will usually form a biofilm. Biofilms are matrix-encased communities which are specialised for surface persistence. This community of microorganisms becomes irreversibly attached to a surface and exhibits distinctive phenotypic properties [3].

Biofilms have multiple impacts on the Earth's resources: they can be essential (i.e. sewage processing and self-depuration mechanisms in aquatic habitats), useful (i.e. food digestion and biodegradation) but also problematic (i.e. metal corrosion and biofouling, growth in water conduits, dental plaque, medical infections) [4]. Although biofilms are complex and different from environment to environment, they can be characterised by three common features [2]. Firstly, biofilms develop at phase interfaces, either solid–liquid or air–liquid. Secondly, a biofilm contains a number of microorganisms from one or several species. Finally, the sessile microorganisms produce an extracellular polymer matrix within which they are embedded. Contact with a solid surface triggers the expression of a panel of microbial enzymes, which catalyse the formation of sticky polysaccharides that promote microbial colonisation and protection. Molecular techniques have now identified genes which are up- and down-regulated both between biofilm grown organisms and those grown in suspension, and within different regions of the biofilm. Additionally, bacteria in biofilms produce small chemical signals that can coordinate their adhesion, growth and virulence by a process called 'quorum sensing'. This has important implications in assessing the behaviour of organisms in terms of virulence, susceptibility and response to host defence mechanisms.

In this section, marine biofilms (Chapter 9), freshwater biofilms (Chapter 10) and biofilms in medicine (Chapter 11) are considered. Some methodology and scientific approaches are similar between the studies of these biofilms such as the use of the improving imaging technology.

For example, confocal laser scanning microscopy (CLSM), real time image capture, atomic force microscopy (AFM), nuclear magnetic resonance imaging (NMRI) and fluorescent in situ hybridisation (FISH) have been used as powerful tools to aid in our understanding of the processes of adhesion, when considering the three-dimensional structure of biofilms and the relationships between the microorganisms which comprise these structures. Additionally, modern molecular techniques, such as 'community fingerprinting' approaches (restriction length polymorphism [RFLP] and denaturing gel gradient electrophoresis [DGGE] and DNA cloning based on polymerase chain reaction [PCR]), as well as FISH with specific fluorescent probes for microbial RNAs and DNAs, are widely used for identification of both uncultivable and cultivable microorganisms in biofilms. In marine and freshwater biofilms (Chapters 9 and 10), species composition and autotrophic–heterotrophic relationships are further described. However, in all three chapters (Chapters 9, 10 and 11), the combination of different techniques and the experiences from different disciplines underlines the complexity and dynamics of biofilms and the challenges which face us when trying to eradicate these persistent communities.

References

1. Zobell, C.E. (1943) The effect of solid surfaces upon bacterial activity. *Journal of Bacteriology*, **46**, 39–56.
2. Lewandowski, Z. (2000) Structure and function of biofilms. In: *Biofilms: Recent Advances in their Study and Control* (ed. L.V. Evans), pp. 1–17. Harwood Academic Publishers, Amsterdam.
3. Donlan, R.M. & Costerton, J.W. (2002) Biofilms: survival mechanisms of clinically relevant microorganisms. *Clinical Microbiology Reviews*, **15**, 167–193.
4. Jenkinson, H.F. & Lappin-Scott, H.M. (2001) Biofilms adhere to stay. *Trends in Microbiology*, **9**, 9–10.

Chapter 9
Marine Biofilms

Sergey Dobretsov

The aim of this chapter is to highlight the dynamics and complexity of marine biofilms. In this chapter, I mainly focus on the composition of marine biofilms and their prokaryote–eukaryote interactions. Additionally, I provide examples of quorum-sensing signalling in biofilms and demonstrate different mechanisms of quorum-sensing inhibition by metabolites of prokaryotes and eukaryotes. Inhibition and induction of larval settlement by microbial biofilms is also discussed. Finally, I highlight directions of future research that require our close attention.

9.1   Biofilm development

Any natural and man-made substrata in the marine environment are quickly colonised with different species of micro- and macroorganisms [1]. Generally, researchers defined four phases of marine biofouling colonisation: adsorption of dissolved organic molecules (molecular biofouling), colonisation by prokaryotes, colonisation by unicellular eukaryotes (diatoms, flagellates, amoebae and ciliates), and recruitment of invertebrate larvae and algal spores [1, 2] (see Chapter 4). These phases can happen sequentially, overlap or occur in parallel [3, 4]. At the same time, some causality can be demonstrated, as the early phases, i.e. adsorption of molecules and colonisation by prokaryotes, are important for the later phases, i.e. larval and spore settlement [5, 6].

During initial stages of biofouling, not all proteins, glycoproteins, proteoglycans and polysaccharides that occur in the water can absorb equally to the same substratum [3]. Molecular biofouling can inhibit or facilitate microbial attachment and furnish microorganisms with a carbon source for their growth. Microbial adhesion depends on the substratum surface free energy, substratum properties and water turbulence [7–10]. Most microorganisms attach more strongly to hydrophobic materials, such as Teflon^{TM}, than to hydrophilic materials, such as glass [2, 11]. Other physical factors, such as surface roughness and the hydrodynamic regime near the surface, also affect microbial biofouling [12].

The mechanisms of bacterial and diatom adhesion to hard substrata have been reviewed in several publications [2–4, 13] and the reader is directed to these reviews. Adhesion of bacteria and microalgae to surfaces – metals, plastics, organic particles and living tissues – is mediated by secretion of a slimy, glue-like substance of extracellular polymers (EPS), which contain polysaccharides, lipopolysaccharides, proteins and nucleic acids [13]. Additionally, microorganisms may respond to chemical attractants and repellents and swim away or towards the substratum [14, 15]. Adhered organisms can stimulate or suppress the co-adhesion of other

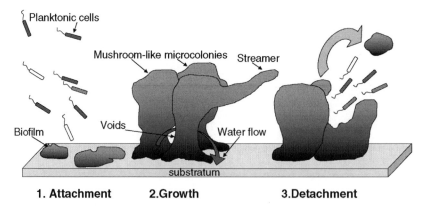

Figure 9.1 A schematic picture illustrating the life cycle of a biofilm. (1) Attachment of microorganisms, (2) growth of the biofilm and formation of three-dimensional structure, (3) detachment and sloughing of the biofilm. (Modified from Reference [22].)

planktonic microorganisms, which, in turn, stimulate formation of multispecies heterogeneous biofilms [16].

Although being small (a hundred microns in thickness), biofilms can cause severe industrial problems by increasing drag force, promoting corrosion and reducing heat transfer efficiency [17]. Biofilms induce corrosion of metals by a number of mechanisms including cathodic and anodic depolarisation, hydrogen production, metal reduction and production of corrosive metabolites such as organic acids and exopolymers [18]. In aquatic environments, biofilms may induce corrosion of stainless steel through a process known as ennoblement, characterised by an increase of the open-circuit potential towards positive (noble) values by metal deposition at the cathode, oxygen removal and acid production [18, 19]. Control and eradication of biofilms are difficult since their resistance towards most antibiotics and biocides is substantially increased compared to planktonic species (see Chapter 11).

All mature biofilms have a three-dimensional structure [20] (Figure 9.1) and the architecture of biofilms is strongly dependent on the species and function of microorganisms within them [21]. A number of investigations have shown that different physical factors (i.e. flow rate, hydrodynamic forces, substrate properties, viscosity), chemical factors (i.e. nutrient availability, EPS production) and biological factors (i.e. competition, predation) affect biofilm architecture both separate and in combination. With improvement in imaging technology and the development of scanning confocal laser microscopy (SCLM), real time image capture, atomic force microscopy (AFM), nuclear magnetic resonance imaging (NMRI) and fluorescent in situ hybridisation (FISH) methods (see Chapters 10, 11 and 22), our understanding of biofilm structure is increasing rapidly.

According to the modern model, biofilms are made of microcolonies separated by interstitial voids [22] (Figure 9.1). The shape of the microcolonies is different within a biofilm and between biofilms. The microcolonies include compact sub-layers filled with EPS, roundly shaped mushroom-like microcolonies and a network of interstitial voids that ensure a free water flow inside the biofilms [10]. At higher flow velocities, microcolonies change their shape, slowly flow under the strain and form elongated streamers extending downstream [22]. Overall, microbial biofilms in aquatic environments are very heterogenic and dynamic structures and these features make them difficult to model and investigate.

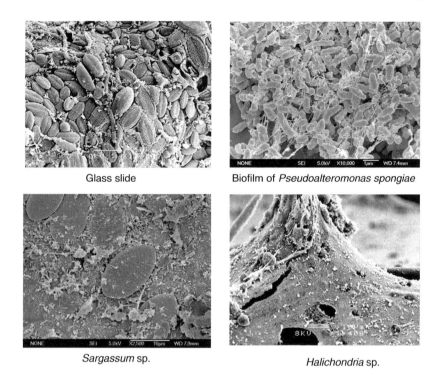

Glass slide

Biofilm of *Pseudoalteromonas spongiae*

Sargassum sp.

Halichondria sp.

Figure 9.2   Scanning electron microphotographs of a monospecies biofilm of the bacterium *Pseudoalteromonas spongiae* (magnification 10 000 ×) and multispecies microbial biofilms developed on the surface of a glass slide (magnification 1900 ×), on the alga *Sargassum* sp. (magnification 2500 ×), and on the sponge *Halichondria* sp. (magnification 1400 ×).

9.2   Composition of microbial biofilms

Compared to freshwater and medical biofilms, marine biofilms consist of different species of bacteria, *Archaea* and unicellular organisms, such as diatoms, and flagellates (compare Figure 9.2 and Figure 10.2). The densities of other microorganisms – fungi, sarcodines and ciliates – in marine biofilms remain quite low [2] (Table 9.1). Experiments performed in the White Sea have shown that rod-shaped bacteria were among the first colonisers of any clean

Table 9.1   The densities of dominant groups of microorganisms in marine biofilms on a hard substratum.

Group of microorganisms	Density of microorganisms	Reference
Viruses	$(10–200) \times 10^7$ virus ml^{-1}	59[a]
Bacteria	3×10^6 to 1.3×10^7 cell cm^{-2}	27, 42, 57
Archaea	1×10^4 to $\times 10^4$ cell cm^{-2}	23
Diatoms	6×10^3 to 3×10^4 cell cm^{-2}	2, 42
Flagellates	3×10^3 cell cm^{-2}	2, 33, 60
Infusoria	Approximately 6 cell cm^{-2}	2, 33
Other protists	Approximately 7 cell cm^{-2}	2

[a] Density in plankton.

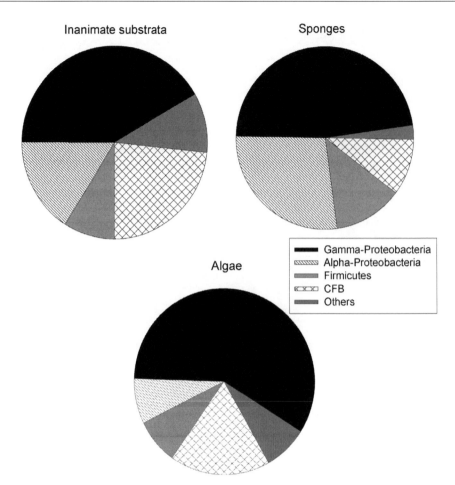

Figure 9.3 The dominant bacterial groups that can be cultivated in biofilms from the surface of different inanimate (PVC panels, rocks) and animate (sponges, algae) substrata from South China Sea, Caribbean and North Sea. CFB, Cytophaga-Flexibacter-Bacteroides group.

substrata [2] (Dobretsov, unpublished data). Following then settle cocci-, vibrios- and spirilli-like bacteria. The last to colonise the substrata are stalked bacteria belonging to the genera *Caulobacter* and *Hyphomicrobium* [2]. In mature marine biofilms, bacteria belonging to α-proteobacteria, γ-proteobacteria and the *Cytophaga–Flavobacterium* group of *Bacteroidetes* are dominating [23–27] (Figure 9.3). In contrast, fresh water biofilms are mainly composed of β-proteobacteria [25] (see Chapter 10). According to our data, most isolates from marine biofilms are affiliated to γ-proteobacteria (41–58%; Figure 9.3). Conversely, studies that used FISH methods showed that α-proteobacteria is the most dominant group in marine biofilms on hard substrata [5, 23]. Differences in these results can be explained by the biases of culture-dependent methods. Analysis of 16S rRNA genes showed that bacterial clones from early biofilms were affiliated to the *Roseobacter* subgroup of α-proteobacteria and *Alteromonas* subgroup of γ-proteobacteria [28]. Similar results were demonstrated for mature biofilms developed on stainless steel [27], but *Cytophaga*-like bacteria dominated in biofilms from an oyster farm [24]. According to our data, most bacterial isolates from biofilms developed in Hong Kong waters belonged to the *Vibrio* and *Pseudoalteromonas* groups of γ-proteobacteria [6, 29, 30].

Species richness of marine benthic diatoms is quite high over geographical scales, while within a particular environment it is usual to find only few taxa [31]. Diatoms belonging to the genera *Achnanthes, Amphora, Navicula* and *Licmophora* are dominant in marine biofilms [3, 31]. Protists (mainly flagellates – *Bodo* spp.; amoebas – *Platyamoeba* spp.; infusoria – *Chlamydonella* spp.) can be found in marine biofilms but in low densities [2, 32, 33] (Dobretsov, unpublished data; Table 9.1). Fungi (mainly *Cladosporium* spp.) [34] were isolated from marine biofilms, while their mycelia have been rarely found in biofilms (Dobretsov, unpublished data). Viruses are possibly very abundant in marine biofilms but their presence has not been confirmed. Marine fungi, protists and viruses in marine biofilms are rarely studied while they are playing an important role as decomposers, grazers and parasites (Figure 9.4). Investigation of these groups in marine biofilms shall be one of the future research directions.

In most laboratory experiments, investigators have been studying monospecies microbial films in static conditions without the addition of nutrients (reviewed by Maki [4] and Dobretsov *et al.* [5]). These studies give only limited information about the structure and the role of biofilms in the field, as natural biofilms are composed of numerous species of microorganisms (Table 9.1), which mostly cannot be cultivated [5]. Turbulent flow around the biofilms in the field determines biofilm structure, composition, biomass and production of chemical compounds by microorganisms [9, 23, 35]. Therefore, marine biofilms should be investigated in the field or in mesocosm experiments under flow conditions.

Marine biofilms can develop even on the surface of antifouling coatings that contain biocides. However, there is only limited information about biofilms composition and development on

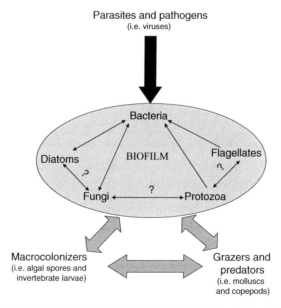

Figure 9.4 Prokaryote–eukaryote interactions in biofilms. Interactions with question marks are proposed but have not yet been demonstrated.

antifouling coatings. It has been proposed that biofilms can increase or decrease release of biocides from coatings and that microorganisms are capable of accumulation and biodegradation of chemical compounds from coatings [17]. Usually, isolates from the surface of antifouling paints include bacteria belonging to gram-positive (*Micrococcus*) and gram-negative (*Pseudomonas*, *Vibrio*) groups [36]. Several studies reported that diatoms belonging to the genera *Amphora*, *Navicula* and *Synedra* are the most dominant on the biofouling release coatings [17, 36], whilst *Achnantes* is very common on tributyltin (TBT) coatings [37]. Investigations of composition and dynamics of biofilms developing on antifouling coatings will be another important future research direction.

9.3   Biofilm dynamics

The species composition of early biofilms depends primarily on two factors: (1) presence of colonisers and (2) the physical and chemical conditions of the substratum and environment. Mature biofilms are subject to constant dynamic changes. Any modifications in abiotic (e.g. climate, depth, light regime, season, water chemistry, nutrient supply, substratum characteristics) and biotic (e.g. availability and physiology of colonising species, competition, predation, grazing, cooperation among species) environmental conditions change the composition of biofilms, biofilm density, productivity, architecture, succession rate and the production of chemical compounds [6]. Biofilm productivity, biomass and structure of biofouling communities showed distinctive seasonal variations: some microorganisms were found all year, while others were present only during particular seasons [38, 39] and even hours [32, 40]. For example, composition of biofilms dominated by naviculoid and nitzschioid taxa or euglenoid species were different during midday, when the diatoms *Gyrosigma balticum* and *Pleurosigma angulatum* were dominant [40]. This phenomenon can be explained by rapid (within minutes) migrations of diatoms that have tendency to adjust their position in a biofilm according to light intensity [32]. Additionally, biofilm sloughing and detachment (Figure 9.1) can open space for re-colonisation by newcomers and existing microorganisms [40].

The unique characteristics of a habitat, like water turbulence, macroalgal canopy cover, grazer activity, temperature, amount of sediments and salinity affect the properties of biofilms [32, 38, 39]. For example, a number of field studies indicated that the diatoms *Fallacia pygmaea* and *Navicula salinarum* predominantly distributed in waters with high nutrient concentrations and organic loads [32]. In other words, biofilms can reflect the key environmental factors of a substratum. This relationship has an important implication for the larval settlement of marine invertebrates [41]. The effect of intertidal and subtidal biofilms on larval attachment of the bryozoan *Bugula neritina* was investigated in our previous study [42]. When biofilms from different tidal regions were combined together, larvae in all cases attached to the subtidal biofilms, which had different microbial composition and thickness. Biofilms developed in the intertidal region and moved to the subtidal region quickly changed their microbial composition and became inductive. On the whole, environmental conditions and interactions between microorganisms can have a significant influence on the composition, structure, physiology and function of marine biofilms.

9.4   Signalling in biofilms

9.4.1   QS signalling in biofilms

In biofilms, bacteria can control their growth and densities, as well as bioluminescence, adhesion and compound production by a regulatory mechanism named quorum sensing (QS). During this process, a small chemical compound (the signal) produced by bacteria releases into the environment [43]. When bacterial density is low, QS signals diffuse away from the cell and quickly degrade. When bacterial density in biofilms is high enough, these signal molecules reach a threshold concentration and start binding to a receptor protein, which leads to transcription of target genes. Usually, single bacterial species use a unique QS signal molecule, while some species can use multiple chemical signals. Based on the properties of bacterial signal receptors, QS signalling can be grouped into two categories. Signal *N*-acyl-L-homoserine lactones (AHLs), LuxR-type signal receptor and LuxR-type I synthase are the major components of gram-negative bacteria [43]. On the other hand, gram-positive bacteria use a two-component QS system, in which QS signals (oligopeptides) are transported outside the cell and detected by a membrane bonded histidine kinase [43, 44].

Bacteria from the marine environment can produce a variety of QS signals. For example, bioluminescence of the marine symbiotic bacterium *Vibrio fisheri* from the light organ of the sepiolid squid *Euprymna scolopes* is regulated by production of AHLs [44]. Similarly, isolates from marine snow (*Roseobacter* spp., *Marinobacter* sp.) produce different AHL signals [45]. At the same time, information about the production of QS by marine biofilms in the natural environment is limited [26]. Our investigation showed that different types of AHLs signals produced in subtidal biofilms of different ages and the concentration of one of these compounds – *N*-dodecanoyl-homoserine lactone – did not exceed 4 mM [46].

A range of microorganisms have adopted QS signals to mediate biosynthesis of compounds and coordinate their basic cellular functions, which give advantages for a population of microorganisms to response to environmental factors. Similarly, the ability of a microorganism to counteract the QS signals of its competitors could also significantly boost the microorganism competitive strength in the ecosystem. For example, gram-positive *Bacillus* species produced AHLs analogs which block QS of gram-negative bacteria [43]. Two groups of QS signal degradation enzymes (AHL-lactonase, AHL-acylase) were identified in several bacterial species [43]. Additionally, QS signals can play a part in interactions between bacteria and eukaryotes, such as the squid *E. scolopes* [43] and the alga *Ulva* (syn. *Enteromorpha*) sp. [47]. Only specific bacterial AHLs induced the settlement of *Ulva* sp. spores [47]. On the other hand, the red alga *Delisea pulchra* excretes halofuranones that interferes with bacterial AHLs and inhibits the growth of gram-negative bacteria [43]. These examples demonstrate that QS signalling may be an important factor that regulates biofouling processes.

9.4.2   Inhibition and induction of larval settlement by microbial biofilms

Numerous studies demonstrated that biofilms can induce or inhibit settlement and metamorphosis of invertebrate larvae and algal spores (see reviews [4–6, 48]). Composition of microbial communities plays an important role in the selection of biofilms by propagules. For example, the barnacles *Balanus amphitrite* and *Balanus improvisus* selected specific biofilms developed intertidally or subtidally correspondingly in choice experiments [6] (see also Chapter 3). Similar results were obtained in our experiments with larvae of *Bugula neritina* [42]. Since microbial community composition in biofilms varies substantially among tidal zones and substrata, differential larval response may allow larvae to evaluate substrata and select the suitable microhabitats.

The amount of inhibitive and inductive bacterial isolates in marine biofilms is approximately equal and properties of a biofilm depend on bacterial densities and interactions [5]. Previously, it was proposed that most of bacterial inhibitive strains belong to the genus *Pseudoalteromonas* [49]. Although subsequently, it has been demonstrated that a wide range of bacterial taxa inhibit larval settlement and there is no relationship between the phylogenetic affiliation of bacteria and their effects on larval or spore settlement (cited by Dobretsov *et al.* [5]). Most of the diatoms isolated from biofilms in Hong Kong waters induced larval settlement of the polychaete *Hydroides elegans*, while there were always low-inductive and non-inductive diatom species [50]. At the same time, the effect of other groups of microorganisms from biofilms on settlement and metamorphosis of propagules has not been well documented [2] and will be an important future direction.

How can bacteria and diatoms affect larval and spores settlement? In most studies, investigators demonstrated that microorganisms are able to produce inductive and inhibitive chemical compounds that affect larval settlement [5]. These compounds can diffuse into the water or remain attached to the biofilm. It is largely unknown how biofilm derived cues are detected by propagules and which signal transduction pathways are involved in this process. Several studies showed that lectin-like receptors on the surface of larvae play an important role in the recognition of marine biofilms [2, 51]. Inhibitive or inductive compounds isolated from marine bacteria and diatoms include polysaccharides, lipids, oligopeptides, glycoconjugates, amino acids and quinines [5, 48]. Only few from these compounds have been fully identified and elucidation of the structure and mode of action of microbial compounds should be an important target of future investigations.

9.5   Prokaryote–eukaryote interactions in biofilms

Microbial biofilms in the sea are complex assemblages in which different microorganisms interact with each other and new colonisers, as well as with eukaryotes (Figure 9.4). These intra- and interspecific interactions include beneficial relationships between microorganisms (mutualism), associations in which one microorganism benefits but another does not (commensalism), interactions beneficial for one species and harmful for another microorganism (predation, grazing, parasitism, disease) and competitive interactions in which both species are adversely affected by the interaction. While some of these interactions have been widely investigated, others remain undiscovered (Figure 9.4).

Beneficial interactions between the diatom *Navicula* sp. and the bacterium *Pseudoalteromonas* sp. have been reported by Wigglesworth-Cooksey and Cooksey [52]. The authors demonstrated that agglutination of diatom cells enhanced with the presence of spent culture medium of the bacterium (Figure 9.5). The investigators suggested that putative bacterial lectins bind to diatoms surface receptors, which causes inhibition of diatom motility and enhances their adhesion. Similarly, the bacterium *Pseudoalteromonas tunicata* requires the

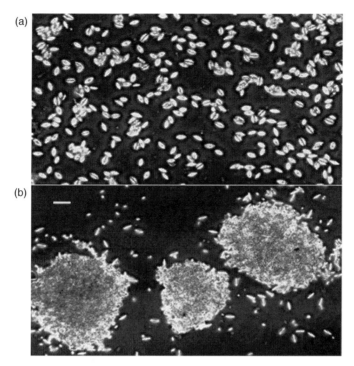

Figure 9.5 Attachment of the diatom *Navicula* sp. to the surface of a plastic plate (a) without and (b) with the presence of the spent media from the bacterium *Pseudoalteromonas* sp. No motile cells were present in this experiment. Bar = 40 μm. (Reproduced from Reference [52] with permission.)

presence of diverse species of bacteria in the inoculum for its effective colonisation of the surface of the alga *Ulva australis* [53]. The effect of bacterial QS that results in simultaneous bacterial attachment and growth can be considered as a result of beneficial interactions between individuals of the same species.

Microorganisms in biofilms always compete for resources and space. The initial colonisers on submerged surfaces are rod-shaped chemotrophic bacteria that appear on surfaces within hours and deplete a large percentage of carbon sources absorbed on the substrate surface [2]. This makes it more difficult for the invading bacteria to sequester themselves in the existing biofilm community [53].

Microorganisms in a biofilm can affect the growth of other microorganisms in the same biofilm by the production of antibiotics, toxins or QS inhibitors. It has been shown that bacterial species that belong to the *Streptomyces*, *Alteromonas*, *Pseudoalteromonas* and *Roseobacter* genera produce a range of antibiotics (cited by Dobretsov *et al.* [5]). In the mixed-species biofilms, the antibiotic produced by the bacterium *P. tunicata* removes the competing bacterial strains unless its competitor is relatively insensitive to an antibacterial protein or produces strong inhibitory activity against the bacterium *P. tunicata* [54]. It was found that antagonistic antimicrobial activity widely exists between common biofouling bacteria and fungi in marine biofilms [34]. In this investigation, 70% of tested fungal species (mostly *Cladosporium*) inhibited growth of marine bacterial isolates from biofilms but only 17% of tested bacterial strains (mostly *Staphylococcus*) inhibited growth of fungi in laboratory experiments. This suggests that the competitive interactions between different components of biofilms may affect biofilm growth, composition and signal production (Figure 9.4) and cannot be neglected in future investigations.

Micrograzers, like heterotrophic flagellates, amoebas and ciliates, may control density and diversity of microorganisms in biofilms (Figure 9.4). Heterotrophic flagellates are among the major regulators of bacterial biomass. It was estimated that ciliates removed approximately 0.2% of bacterial crop per day, amoebae –0.3% and flagellates consumed 2.5% of bacterial carbon per day [33]. The grazing rate of heterotrophic *Bodonid* flagellates was one to ten bacteria per flagellate per hour and the grazing rate increased with increase of bacterial density [55]. The density of bacteria and their behaviour in marine snow biofilms was largely controlled by the grazing pressure from flagellates. Presence of heterotrophic flagellates changed biofilm architecture and shifted the composition of bacterial communities towards less diverse with dominance of grazing-resistant strains. Contrary to the assumption that protozoan feeding involves ingestion of biofilm cells, the study of Joubert and Wolfaardt [56] demonstrated that the ciliates *Tetrahymena* sp. preferentially grazed on the EPS matrix of biofilms, with selective ingestion of yeast and *Pseudomonas* sp. planktonic cells over biofilm-derived cells. Predation of these biofilms increased bacterial EPS production [56]. This example shows that associations within a biofilm very often are complex and cannot be characterised as totally negative or positive.

Large grazers such as limpets can control microbial biomass and biofilm diversity on rocky shores [57]. Reduction of limpet density resulted in a 20% increase of microbial biofilm biomass during a 2-week experiment. Additionally, grazers and predators controlled recruitment of micro- and macrocolonisers by their direct elimination or, indirectly, by modification of biofilms and production of biofilm-derived chemical signals [58] (Figure 9.4). It has been demonstrated earlier (see Section 9.4) that biofilms and chemical cues produced by them induced or inhibit larval and spore settlement. Increased abundance of copepods in biofilms decreased density of bacteria and led to changes in microbial composition due to bioturbation, selective grazing and fertilisation [58]. Finally, changes in biofilms due to the presence of copepods or their chemical compounds (kairomones) reduced larval settlement of *H. elegans* [58]. This study suggests that grazers and predators can have indirect effects on the formation of biofouling communities via the modification of biofilms.

9.6   Conclusions

- Biofilms are very dynamic and complex systems, which include both prokaryote and eukaryote components.
- Environmental changes and modification of parasites and consumer pressure lead to changes in biofilms, which in turn affect recruitment of larvae and algal spores.
- These complex changes are very important in formation of heterogeneous biofouling communities and require considerable future research. Of particular interest are:
  o viruses in marine biofilms;
  o interactions between different components of marine biofilms and propagules;
  o isolation of signal compounds from marine biofilms and investigation of their transduction pathways;
  o marine biofilm dynamics on different substrata from different regions.

- Future collaboration between microbiologists, physicists, chemists and engineers will lead to the development of new non-toxic technological applications which prevent colonisation of targeted species or promote them to settle.

Acknowledgements

I thank Prof M. Wahl and Dr J.C. Thomason for their constructive comments on the manuscript. This study was supported by an Alexander von Humboldt Fellowship.

Dedication

This chapter is dedicated to the memory of my father who loved the sea and passed his passion to me.

References

1. Wahl, M. (1997) Living attached: aufwuchs, fouling, epibiosis. In: *Fouling Organisms of the Indian Ocean: Biology and Control Technology* (eds R. Nagabhushanam & M. Thompson), pp. 31–84. Oxford & IBH Publishing, New Delhi.
2. Railkin, A.I. (2004) *Marine Biofouling: Colonization Processes and Defenses*. CRC Press, Boca Raton.
3. Cooksey, K.E. & Wigglesworth-Cooksey, B. (1995) Adhesion of bacteria and diatoms to surfaces in the sea: a review. *Aquatic Microbial Ecololgy*, **9**, 87–96.
4. Maki, J.S. (2002) Biofouling in the marine environment. In: *Encyclopedia of Environmental Microbiology* (ed. G. Bitton), pp. 610–619. John Wiley & Sons, New York.
5. Dobretsov, S., Dahms, H.-U. & Qian, P.-Y. (2006) Inhibition of biofouling by marine microorganisms and their metabolites. *Biofouling*, **22**, 43–54.
6. Qian, P.-Y., Lau, S.C.K., Dahms, H.-U., Dobretsov, S. & Harder, T. (2007) Marine biofilms as mediators of colonization by marine macroorganisms: implications for antifouling, aquaculture, and conservation management. *Marine Biotechnology*, **9**, 399–410.
7. Zobell, C.E. (1943) The effect of solid surfaces upon bacterial activity. *Journal Bacteriology*, **43**, 39–56.
8. Dexter, S.C., Sullivan, J.D., Williams, J. & Watson, S.W. (1975) Influence of substrate wettability on the attachment of marine bacteria to various surfaces. *Applied Environmental Microbiology*, **30**, 298–308.
9. Characklis, W.G., Marshall, K.C. & McFeters, G.A. (1990) The microbial cell. In: *Biofilm*, Vol. **1** (eds W.G. Characklis & K.C. Marshall), pp. 131–159. John Wiley & Sons, New York.
10. Costeron, J.W., Lewandowski, Z., Caldwell, D.E., Korber, D.R. & Lappin-Scott, H.M. (1995) Microbial biofilms. *Annual Review Microbiology*, **49**, 711–745.
11. Fletcher, M. & Loeb, G. (1979) Influence of substratum characteristics on the attachment of marine pseudomonas to solid surfaces. *Applied Environmental Microbiology*, **37**, 67–72.
12. Cao, R.S. & Alaerts, G.J. (1995) Influence of reactor type and shear stress on aerobic biofilm morphology, population and kinetics. *Water Research*, **29**, 107–118.
13. Flemming, H.-C., Wingender, J., Griebe, T. & Mayer, C. (2001) Physico-chemical properties of biofilms. In: *Biofilms: Recent Advances in their Study and Control* (ed. L.V. Evans), pp. 19–34. Harwood Academic Publishers, Amsterdam.

14. Marshall, K.C., Stout, R. & Mitchell, R. (1971) Mechanism of the initial events in the sorption of marine bacteria to surfaces. *Journal of Genetic Microbiology*, **68**, 337–348.
15. Vandevivere, P. & Kirchman, D.L. (1993) Attachment stimulates exopolysaccharide synthesis by a bacterium. *Applied Environmental Microbiology*, **59**, 3280–3286.
16. McEldowney, S. & Fletcher, M. (1986) Effect of growth conditions and surface characteristics of aquatic bacteria on their attachment to solid surfaces. *Journal of Dental Microbiology*, **132**, 513–523.
17. Yebra, D.M., Søren, K., Weinell, C.E. & Dam-Johansen, K. (2006) Presence and effects of marine microbial biofilms on biocide-based antifouling paints. *Biofouling*, **22**, 33–41.
18. Geesey, G.G., Beech, I., Bremer, P.J., Webster, B.J. & Wells, D.B. (2000) Biocorrosion. In: *Biofilms II: Process Analysis and Applications* (ed. J.D. Bryers), pp. 281–325. Wiley-Liss, New York.
19. Mansfeld, F., Tsai, C.H., Shih, H., Little, B. & Wagner, P.A. (1992) An electrochemical and surface analytical study of stainless steels and titanium exposed to natural seawater. *Corrosion Science*, **33**, 445–456.
20. O'Toole, G., Kaplan, H.B. & Kolter, R. (2000) Biofilm formation as microbial development. *Annual Review Microbiology*, **54**, 49–79.
21. Stoodley, P., Sauer, K., Davies, D.G. & Costerton, J.W. (2002) Biofilms as complex differentiated communities. *Annual Review Microbiology*, **56**, 187–209.
22. Lewandowski, Z. (2000) Structure and function of biofilms. In: *Biofilms: Recent Advances in their Study and Control* (ed. L.V. Evans), pp. 1–17. Harwood Academic Publishers, Amsterdam.
23. Webster, N.S., Smith, L.D., Heyward, A.J., *et al.* (2004) Metamorphosis of a scleractinian coral in response to microbial biofilms. *Applied Environmental Microbiology*, **70**, 1213–1221.
24. Nocker, A., Lepo, J.E. & Snyder, R.A. (2004) Influence of an oyster reed on development of the microbial heterotrophic community on an estuare biofilm. *Applied Environmental Microbiology*, **70**, 6834–6845.
25. Kjellerup, B.V., Thomsen, T.R., Nielsen, J.L., Olensen, B.H., Frølund, O. & Nielsen P.H. (2005) Microbial diversity in biofilms from corroding heating systems. *Biofouling*, **21**, 19–29.
26. Dobretsov, S., Dahms, H.-U., Huang, Y.L., Wahl, M. & Qian, P.-Y. (2007) The effect of quorum sensing blockers on the formation of marine microbial communities and larval attachment. *FEMS Microbiology Ecology*, **60**, 177–188.
27. Jones, P.R., Cottrell, M.T., Kirchman, D.L. & Dexter, SC. (2007) Bacterial community structure of biofilms on artificial surfaces in an estuary. *Microbial Ecology*, **53**, 153–162.
28. Dang, H. & Lovell, C.R. (2000) Bacterial primary colonization and early succession on surfaces in marine waters as analyzed by amplified rRNA gene restriction analysis and sequence analysis of 16S rRNA genes. *Applied Environmental Microbiology*, **66**, 467–475.
29. Dobretsov, S. & Qian, P.Y. (2002) Effect of bacteria associated with the green alga *Ulva reticulata* on marine micro- and macrofouling. *Biofouling*, **18**, 217–228.
30. Dobretsov, S. & Qian, P.Y. (2004) The role of epibotic bacteria from the surface of the soft coral *Dendronephthya* sp. in the inhibition of larval settlement. *Journal of Experimental Marine Biology Ecology*, **299**, 35–50.
31. Thornton, D.C.O, Dong, L.F., Underwood, G.J.C. & Nedwell, D.B. (2002) Factors affecting microphytobenthic biomass species composition and production in the Colne Estuary (UK). *Aquatic Microbial Ecology*, **27**, 285–300.
32. Underwood, G.J.C. (2005) Microalgal (Microphytobenthic) biofilms in shallow coastal waters: how important are species? *Proceedings California Academy of Sciences*, **56**, 162–169.
33. Maybruck, B.T. & Rogerson, A. (2004) Protozoan on the prop roots of the red mangrove tree, *Rhizophora mangle*. *Protistology*, **3**, 265–272.
34. Li, M. & Qian, P.Y. (2005) Antagonistic antimicrobial activity of marine fungi and bacteria isolated from marine biofilm and seawaters of Hong Kong. *Aquatic Microbial Ecology*, **38**, 231–238.
35. Pereira, M.O., Kuehn, M., Wuertz, S., Neu, T. & Melo, L.F. (2002) Effect of flow regime on the architecture of a *Pseudomonas fluorecens* biofilm. *Biotechnology & Bioengineer*, **78**, 164–171.

36. Casse, F. & Swain, G. (2006) The development of microfouling on four commercial antifouling coatings under static and dynamic conditions. *International Biodeterioration and Biodegradation*, **57**, 179–185.

37. Callow, M.E. (1986) Fouling algae from 'in service' ships. *Botanical Marina*, **24**, 351–357.

38. Forster, R.M., Creach, V., Sabbe, K., Vyverman, W. & Stal, L.J. (2006) Biodiversity-ecosystem function relationship in microphytobenthic diatoms of the Westerschelde estuary. *Marine Ecology Progress Series*, **311**, 191–201.

39. Moss, J.A., Nocker, A., Lepo, J.E. & Snyder, R.A. (2006) Stability and change in estuarine biofilm bacterial community diversity. *Applied Environmental Microbiology*, **72**, 5679–5688.

40. Underwood, G.J.C., Perkins, R.G., Consalvey, M.C., *et al.* (2005) Patterns in microphytobenthic primary productivity: species-specific variation in migratory rhythms and photosynthesis in mixed-species biofilms. *Limnology and Oceanography*, **50**, 755–767.

41. Lau, S.C.K., Thiyagarajan, V., Cheung, S.C.K. & Qian, P.-Y. (2005) Roles of bacterial community composition in biofilms as a mediator for larval settlement of three marine invertebrates. *Aquatic Microbial Ecology*, **38**, 41–51.

42. Dobretsov, S. & Qian, P.-Y. (2006) Facilitation and inhibition of larval attachment of the bryozoan *Bugula neritina* in association with mono-species and multi-species biofilms. *Journal Experimental Marine Biology Ecology*, **333**, 263–264.

43. Zhang, L.H. & Dong, Y.H. (2004) Quorum sensing and signal interference: diverse implications. *Molecular Microbiology*, **53**, 1563–1571.

44. Ruby, E.G. & Lee, K.H. (1998) The *Vibrio fischeri* and *Euprymna scolopes* light organ association: current ecological paradigms. *Applied Environmental Microbiology*, **64**, 805–812.

45. Gram, L., Grossart, H.P., Schlingloff, A. & Kiørboe, T. (2002) Possible quorum sensing in marine snow bacteria: Production of acylated homoserine lactones by *Roseobacter* strains isolated from marine snow. *Applied Environmental Microbiology*, **68**, 4111–4116.

46. Huang, Y.-L., Dobretsov, S., Ki, J.-S., Yang, L.-H. & Qian, P.-Y. (2007) Presence of acyl-homoserine lactones in subtidal biofilms and the implication in inducing larval settlement of the polychaete *Hydroides elegans*. *Microbial Ecology*, **54**, 384–392.

47. Tait, K., Joint, I., Daykin, M., Milton, D.L., Williams, P. & Camara, M. (2005) Disruption of quorum sensing in seawater abolishes attraction of zoospores of the green alga *Ulva* to bacterial biofilms. *Environmental Microbiology*, **7**, 229–240.

48. Wieczorek, S.K. & Todd, C.D. (1998) Inhibition and facilitation of the settlement of epifaunal marine invertebrate larvae by microbial biofilm cues. *Biofouling*, **12**, 81–93.

49. Holmström, C. & Kjelleberg, S. (1999) Marine *Pseudoalteromonas* species are associated with higher organisms and produce active extracellular compounds. *FEMS Microbiology Ecology*, **30**, 285–293.

50. Harder, T., Lam, C. & Qian P.-Y. (2002) Induction of larval settlement in the polychaete *Hydroides elegans* by marine biofilms: an investigation of monospecific diatom films as settlement cues. *Marine Ecology Progress Series*, **229**, 105–112.

51. Maki, J.S. & Mitchell, R. (1985) Involvement of lectins in the settlement and metamorphosis of marine invertebrate larvae. *Bulletin Marine Science*, **37**, 675–683.

52. Wigglesworth-Cooksey, B. & Cooksey, K.E. (2005) Use of fluorophore-conjugated lectins to study cell-cell interactions in model marine biofilms. *Applied Environmental Microbiology*, **71**, 428–435.

53. Rao, D., Webb, J. & Kjelleberg, S. (2006) Microbial colonization and competition on the marine alga *Ulva australis*. *Applied Environmental Microbiology*, **72**, 5547–5555.

54. Rao, D., Webb, J.S. & Kjelleberg, S. (2005) Competitive interactions in mixed-species biofilms containing the marine bacterium *Pseudoalteromonas tunicate*. *Applied Environmental Microbiology*, **71**, 1729–1736.

55. Kiørboe, T., Tang, K., Grossart, H.P. & Ploug, H. (2003) Dynamics of microbial communities on marine snow aggregates: colonization, growth, detachment, and grazing mortality of attached bacteria. *Applied Environmental Microbiology*, **69**, 3036–3047.

56. Joubert, L.M. & Wolfaardt, B.A. (2006) Microbial exopolymers link predator and prey in a model yeast biofilm system. *Microbial Ecology*, **52**, 187–197.

57. Thompson, R.C., Moschella, P.S., Jenkins, S.R., Norton, T.A. & Hawkins, S.J. (2002) Differences in photosynthetic marine biofilms between sheltered and moderately exposed rocky shores. *Marine Ecology Progress Series*, **296**, 53–63.

58. Dahms, H.-U. & Qian, P.-Y. (2005) Exposure of biofilms to meiofaunal copepods affects the larval settlement of *Hydroides elegans* (Polychaeta). *Marine Ecology Progress Series*, **297**, 203–214.

59. Weinbauer, M. (2004) Ecology of prokaryotic viruses. *FEMS Microbiology Reviews*, **28**, 127–181.

60. Fenchel, T. (1986) The ecology of heterotrophic microflagellates. *Advances in Microbial Ecology*, **9**, 57–97.

Chapter 10
Freshwater Biofilms

Anna M. Romaní

The aim of this chapter is to highlight the tight coupling between biofilm structure and function which is, at the same time, submitted to a constant dynamism. Firstly, an overview of the knowledge of biofilm structure is given focusing on its three-dimensional structure and the role of extracellular polymeric substances (EPSs). A collection of bacterial and algal biomass data from biofilms found in river systems is included. Relationships between biofilm structure and function are highlighted in two relevant aspects: the biofilm role in the aquatic food web and the biofilm development process. In both aspects, interesting research has been developed during the last years and there are still many open questions to be solved. Amongst these open questions, how microbial interactions within the biofilm determine its metabolism and organic matter use capacity is discussed. Finally, the knowledge of the process of biofilm formation and the physical and chemical factors mostly affecting their evolution is included, which, at the same time, can be very useful for fouling control.

10.1 Introduction

Biofilms in freshwater habitats are consortia of microorganisms (algae, bacteria, cyanobacteria, fungi and protozoa) attached to a surface and embedded in an extracellular matrix of polymeric substances. Naturally, in freshwater ecosystems, biofilms are present in various compartments [1] (e.g. benthic, hyporheic, aquifer sediments) playing a key role in the uptake or retention of inorganic and organic nutrients [2–4]. Because of the high area covered by biofilms, the organisms involved dominate the heterotrophic metabolism in many aquatic ecosystems, being major sites for the uptake and storage of fluvial dissolved organic matter (DOM) [5, 6], and contribute significantly to the carbon cycling in rivers and streams. In aquatic ecology, biofilms are seen as having a positive effect on the ecosystem being responsible for the self-depuration capacity and defined as key elements in the processes of water purification [7].

Freshwater and marine biofilms share the same basic structure and function characteristics and the main differences are due to the physical–chemical environment (salinity, water movement/hydrodynamics, light availability; Table 10.1) and the composition of biofilms. Substrata for biofilm development can also be different between ecosystems and highly diverse, from sand, sediment, rocks and cobbles, wood and leaves to littoral sediment and the surface of submerged plants. Historically, biofilm research in marine and freshwater ecosystems has followed some distinct lines (mostly focused on structure and taxonomy in marine and in metabolism and nutrient cycling in freshwater), although recently other aspects are also

Table 10.1 Differences between biofilms in freshwater and marine ecosystems.

	Freshwater (streams, rivers, lakes) biofilms	Marine (coastal, deep sea) biofilms
Taxonomy	Microorganisms adapted to low salinity	Microorganisms adapted to high salinity
Species composition	Mainly composed of bacteria (β-proteobacteria), cyanobacteria, green algae, diatoms, flagellates and protozoa	Mainly composed of bacteria (α-proteobacteria), *Archaea*, cyanobacteria, green algae, diatoms and flagellates
Biofilm thickness	Thicker than marine biofilm	Thinner than freshwater biofilm
Temperature fluctuations	Biofilms adapted for rapid temperature fluctuations	Biofilms highly sensitive to temperature fluctuations
Light limitation	Riparian canopy cover	Depth limitation
Hydrodynamics, disturbances	Current velocity and floods	Waves, current and tides
Substrata for biofilm development	Rocks, cobbles, sand, leaves and wood	Sand, rocks, submerged plants and animals and marine snow

covered (mostly due to genetics and image technology development) including marine–freshwater interactions.

10.2   Structure and architecture of freshwater biofilms

Biofilms grown on any benthic surface (rocks, cobbles, sand, leaves, wood) are mainly composed of bacteria, algae, cyanobacteria, fungi and protozoa embedded in a polymeric matrix [1]. Figure 10.1 shows a scheme of a theoretical biofilm, including all possible components of a biofilm structure as well as most relevant environmental parameters affecting biofilm structure and function (light, nutrient availability and water flow velocity). Within the biofilm, microbial metabolism is responsible for the uptake of inorganic nutrients and the use and decomposition of organic matter (particulate organic matter [POM] and DOM). Available organic matter for biofilm microorganisms might come from the surrounding water and/or from those produced or entrapped in the main biofilm (including algal exudates and/or EPS). Since most organic molecules available have high molecular weight, the activity of extracellular enzymes is needed before they can be uptaken by the microorganisms [8]. Biofilm extracellular enzymes are linked to microbial cells (outside the cell or in the periplasmic space in gram-negative bacteria [9]) or released into the biofilm media (free in the biofilm matrix or linked to extracellular polymers [10]). Most microorganisms produce extracellular enzymes, especially bacteria and fungi (producing polysaccharidic, proteolytic and ligninolytic enzymes), but algae and cyanobacteria (mainly producing phosphatases) as well as protozoa can also be producers of extracellular enzymes.

The EPS matrix is a crucial structural parameter for biofilm integrity, stability and architecture [1, 11] and provides a refuge for the microbial community against shear stress and protection against desiccation [12]. The polymeric matrix is made of extracellular polymers produced by most microorganisms found in the biofilm by natural secretions, cell lysis and

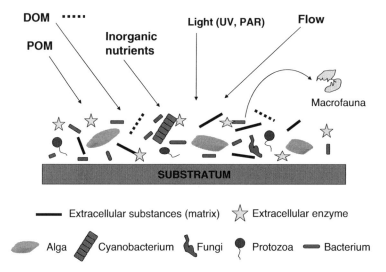

Figure 10.1 Scheme of a theoretical biofilm structure. All possible biofilm components and the main environmental factors affecting biofilm structure and function are included. DOM, dissolved organic matter; POM, particulate organic matter. (Adapted from Reference [2].)

hydrolysis products. Such EPSs are rich in high molecular weight polysaccharides as well as other non-sugar compounds such as proteins, lipids, nucleic acids and various heteropolymers [13, 14]. The biofilm EPS plays a key role in the adhesion to the substratum [15–17] as well as adsorption or entrapment of soluble and particulate matter, and metals and toxic compounds, thereby removing them from the flowing water. The EPS matrix also provides protection for extracellular enzymes and products from hydrolysis [18].

The biofilm thickness determines its three-dimensional structure where layers or patches as well as gradients are described [19–21]. The decrease in nutrient diffusion in thick biofilms results in nutrient gradients [11], such as oxygen, which can lead to anoxic and reducing conditions close to the substratum surface. The confocal laser scanning microscopy (CLSM) is a very powerful tool to study the biofilm as a three-dimensional structure. With this technique, the structural distribution of EPS content (using different dyes), algal content (autofluorescence) and bacterial content (dyeing with Syto9) can be analysed [22]. It has been shown that biofilms are highly hydrated open structures containing a high fraction of EPS and large channels or void spaces between microcolonies [23]. The existence of channels and/or voids might help water flowing through the biofilm structure [24]. The suggested cell-to-cell signalling between biofilm microorganisms can also determine the spatial biofilm structure [25]. Although the development of non-destructive techniques for biofilm research are improving fast, including molecular techniques and the development of biofilm models [26], it is still necessary to combine them together with classical extraction and optical microscopy methods for understanding biofilm structure and functioning [27, 28].

In freshwater environments, biofilms are abundant and occupy diverse inorganic and organic substrata. In lakes, reservoirs and the lower part of large rivers, epiphytic biofilms develop on littoral plant vegetation, while in low-order rivers and streams, epilithic (those developing on rocks and cobbles) and epipsammic or epipelic biofilms (those developing on sand) are the most important [1]. In highly forested rivers, epixylic biofilms (those developing on wood)

(a)   (b)   (c)

Figure 10.2 Scanning electron microscope photographs showing an example of an epiphytic biofilm (a), epipsammic (b) and epilithic (c) biofilm. (a) Corresponds to a colonised *Platanus* leave and fungal hyphae and bacteria are observed. (b) Corresponds to a colonised sand grain where bacteria and scarce diatom frustules are observed. In (c) an epilithic community that developed after 43 days on a clay tile (1 cm^2) immersed in a stream is shown. Mainly diatoms and bacteria can be observed. All three samples were collected from a Mediterranean forested stream (Fuirosos, Spain).

or epiphytic (those developing on decaying leaves) became more relevant. The nature of the substratum determines the composition and structure of the biofilm and, in consequence, its metabolism. Similar to biofilm structure but much more complex (and usually thicker) are the microbial cyanobacterial mats that usually develop in high calcareous ecosystems (and extreme environments), where the microorganisms are included in the inorganic substratum itself conforming different layers. In Figure 10.2, scanning electron microscope (SEM) photographs of an epilithic, epipsammic and epiphytic biofilm developed in a Mediterranean river are shown. The differential accrual of bacteria, algae and fungi can be observed. Throughout this chapter mainly the epilithic biofilm or that developed on inorganic substrata is considered.

10.3   Biofilm biomass

In freshwater biofilms in light conditions, algae comprise the majority of the biofilm biomass together with the EPS, while bacterial biomass is usually lower. Data from river epilithic biofilms showed a proportion of total biofilm carbon of 60–90% of algae, 10–40% of EPS, 1–5% of bacteria and less than 1% of fungi (A. Romaní, unpublished data). However, in close environments (i.e. forested rivers and streams, hyporheic habitats), heterotrophic biomass (bacteria, fungi and protozoa) became more important. In Table 10.2 a collection of data on algae (chlorophyll-*a*) and bacteria (bacterial density) from different freshwater biofilms is shown.

The highest algal biomass is usually reached in eutrophic non-light limiting environments. The positive effect of light on biofilm algal growth has been described elsewhere [40], especially for those sites where light may be limiting primary production [41]. Nutrients are also an important factor for algal growth [42]. The increased level of inorganic nutrients (nitrate and phosphate) usually increases algal biomass accrual in the biofilm [43]. In an oligotrophic river, an increase of inorganic N and P (to a maximum of 400 and 20 µg L^{-1}) caused chlorophyll to increase from less than 1 µg cm^{-2} to about 10 µg cm^{-2} in 15 days [44]. Tank and Dodds [45] collected literature (30 studies) of nutrient effects on epilithic algal biomass and showed that the most common response was the positive effect of N and P together (41% of occasions),

Table 10.2 Algal biomass (chlorophyll-*a*) and bacterial densities in river and stream biofilms from different ecosystems.

Site	Chlorophyll-a (μg cm^{-2})	Bacteria cell ($\times$ 10^8 cm^{-2})	Source
National Capital Region, Québec, epilithic	6.41–8.73	1.04–2.56	[29]
Waterloo region, Québec, epilithic	10.50	3.58	[29]
Bradley Brook, New York, USA, epilithic	3.9–5.4	0.3–0.8	[31]
White Clay Creek, Delaware, USA, slow flow, epilithic	40	16	[32]
White Clay Creek, Delaware, USA, fast flow, epilithic	18	4	[32]
Walker Branch, Tennessee, USA, epilithic	0.3–1.7	0.1	[33]
Ishite, Japan, epilithic	0.7–25.9	0.3–1.4	[34]
Nant Waen, Wales, UK, epilithic	0.46	0.59	[35]
River Clywedog, Wales, UK, epilithic	0.15	0.36	[35]
Driffield Beck, UK, epilithic	2.93	1.73	[36]
Birk Gill, UK, epilithic	0.47	0.73	[36]
Walzbach, Germany, autumn, epilithic	0.84	0.23	Artigas *et al.*, personal observation
Breitenbach, Germany, autumn, epilithic	7.53	1.00	[37]
Riera Major, Spain, epipsammic	1.79	56.1	[30]
Riera Major, Spain, subsurface sand	0.29	11.2	[30]
Riera Major, Spain, epilithic	4.07	266.4	[93]
Fuirosos, Spain, autumn, epilithic	0.32	0.54	Romaní *et al.*, in prep.
La Solana, Spain, cyanobacterial crust	13.33	52.4	[38]
River Ter, Spain, epilithic	27.90	151.6	[39]

Note: Most values come from 1 year's study at each study site. When a specific period was considered, this is included in the site description. The kind of biofilm is also specified (epilithic, epipsammic, cyanobacterial crust, subsurface sand). Values are means and/or ranges.

while only P and only N limitation was reported for 22 and 12% of the studies, respectively, as well as showing any effect of nutrients on algal accrual in shaded sites. As reported by Mosisch *et al.* [46] and Taulbee *et al.* [47], algae respond to nutrient addition in non-limiting light conditions, while their response under limiting light conditions is highly variable [48,49]. Similarly, Ylla *et al.* [50] found a significant effect of light on epilithic primary production and chlorophyll which was increased under high nutrient conditions while the sole effect of nutrients on primary producers was not significant. The response of algal growth and community composition to added nutrients can be also modulated by species-specific requirement in the biofilm community [42] as well as the stoichiometric relationship (C:N:P) between the biofilm structure and that of the flowing water [51]. Water temperature and hydrologic variation are also modulating factors of benthic algal biomass [52]. At low flow conditions, a higher algal biomass is accumulated [32]. Epilithic biofilm chlorophyll-*a* usually ranges from less than 1 $\mu g \, cm^{-2}$ in oligotrophic low-light environments to more than 10 $\mu g \, cm^{-2}$ in eutrophic not-light-limiting environments (Table 10.2).

Biofilm bacterial densities usually measured by DAPI (4′,6′-diamidino-2-phenylindole hydrochloride) staining and counting under epifluorescence microscope [53] range from 10^5 to 10^{10} cells cm^{-2} (Table 10.2). Microbial abundance and activity in aquatic biofilms are influenced by inorganic nutrients and DOM availability [3, 54, 55]. Epilithic bacterial density can be affected indirectly by inorganic nutrients via the influence of nutrients on algal biomass [31, 55]. Variations in DOM and inorganic nutrients also affect the biofilm bacterial community composition [56]. All bacterial cells in the biofilm are not metabolically active. Active bacteria may account for between 20 and 70% of total biofilm bacteria [19, 57]. Inactive, damaged or dead bacteria are usually found in biofilms and recently it has been shown that bacterial cell lysis and the corresponding release of cell content is important for biofilm development processes [58]. Methodologies for considering life/dead, active/inactive proportions of bacteria are developing (see Chapter 22).

The relevance of protozoan biomass in epilithic biofilms is in general lower than that for algae and bacteria and few data exist in the literature, most of them from epipsammic biofilms. In the Fuirosos stream (oligotrophic forested Mediterranean stream), mean protozoa densities in streambed sediments were approximately 210–460 ciliates cm^{-2} and 1000–2700 flagellates cm^{-2} [59]. Similarly, in a second-order gravel stream, 300 ciliates cm^{-2} and 3950 flagellates cm^{-2} were measured [60]. However, Bott and Kaplan [61] reported ciliates densities between 5000 and 36 000 individuals cm^{-2} and between 6×10^5 and 230×10^5 flagellates from a first- and third-order Piedmont stream.

Although in epilithic biofilms fungal biomass is very rare usually reaching less than 1% of total biofilm biomass, they dominate biofilms developed on organic substrata such as leaves and wood material, reaching more than 80% of microbial biomass. In general, high nutrient concentrations enhance fungal biomass and decomposition activities [62], although in heavily nutrient-rich rivers a negative effect can occur [63]. Fungal biomass is usually measured by extracting and analysing ergosterol, a membrane compound of most Eumycota [64]. In a Mediterranean river, biofilm fungal biomass was on average 4.2 μg ergosterol cm^{-2} for epipsammic, 1.9 for epilithic and 14.6 for biofilms colonising leaves during a 12-month study [65].

Trophic interactions between biofilm microorganisms may occur due to the close contact between them, modulating biofilm biomass. Protozoa grazing on bacteria has been described [66, 67], although its negative consequences for the biofilm are not clear. Gücker and Fischer [68] found too low flagellate and cilliate densities in contrast to bacteria to ascertain a grazing control of protozoa on bacteria in the studied river biofilms. Joubert *et al.* [69] showed a preferential grazing on the EPS than on the biofilm bacterial cells by a ciliate as well as a positive effect of protozoa on biofilm biomass and cell viability. Protozoan predation could also promote the evolution of diverse adaptive mechanisms in bacterial biofilms [70]. Herbivorousness of macroinvertebrates can also regulate the benthic biomass [71, 72], also affecting biofilm stoichiometry (decrease in N:P and C:P ratios) [73]. Antagonistic interactions between bacteria and fungi colonising plants have been described, where the lower fungal growth in the presence of bacteria has been related to the efficient bacterial use of simple polysaccharides and peptides provided by fungi after the decomposition of the most recalcitrant compounds [74, 75]. A positive effect of algal biomass on bacterial biomass, probably due to the use of algal exudates by bacteria as well as increasing colonising surface, has been described in many studies and it is generally assumed that algal biomass favours bacterial biomass accrual in the biofilm (this will be considered below).

10.4   Biofilm metabolism and its role on the aquatic food web

The microorganisms within biofilms use, decompose and uptake organic matter both from the surrounding water (allochthonous sources) and from within the main biofilm (autochthonous sources) [6, 76, 77]. Organic matter in flowing water is made up of DOM and POM, DOM being the dominant fraction in streams and rivers [78]. Since a large part of DOM in the flowing water is made up of humic substances and polymeric molecules [78], extracellular enzyme activity for heterotrophic uptake is usually required [9]. The uptake rate of organic compounds is heavily related to their lability, and microorganisms show a faster and preferential use of the most labile and fresh molecules [79]. Bacterial utilisation of DOM is determined by its size and diagenetic state [80]. However, the composition of organic compounds rather than molecular weight can be more important for the availability of dissolved organic carbon (DOC) for the biofilm heterotrophs [81]. Sun *et al.* [82] showed that variations in the H/C and O/C ratios of the DOC determined the bioavailability of DOC for bacteria in fluvial ecosystems. Kaplan and Newbold [83] suggested that labile compounds are quickly recycled within the benthic biofilm, while more refractory substances may be transported further before uptake.

   The biofilm microorganisms also feed on organic molecules from within the biofilm [84]. The use of algal material by microorganisms within the biofilm is favoured by the close contact between the algal and the heterotrophic community in biofilms [85, 86]. Algal accumulation and activity enhances the heterotrophic community's use of organic matter by increasing the amount of substrate available for bacteria [87, 88]. Stimulation of bacterial growth by epilithic algae was also suggested by Sobczak [89]. Although the use of algal exudates is the main internal C-cycling mechanism occurring within the biofilm, the decomposition and use of the main biofilm EPS as sources of proteins and polysaccharides for the microbial community has been also reported [90].

   Most allochthonous and autochthonous organic matter sources need the action of extracellular enzymes before being uptaken by the biofilm heterotrophs. The function of extracellular enzymes for the biofilm microbial community is mainly the acquisition of N and C compounds for their growth and reproduction. In this regard, the most relevant enzymes, as well as those most studied in freshwater biofilms, include proteolytic and polysaccharidic enzymes such as leucine aminopeptidase (involved in peptide decomposition), cellobiohydrolase (involved in the decomposition of cellobiose), β-glucosidase and β-xylosidase (involved in the final step of cellulose and hemicellulose decomposition, respectively) and ligninolytic activities (peroxidase and phenoloxidase). Extracellular enzymes may also be involved in other functions such as lysis of microbial cell walls for microbial growth or protozoa grazing (i.e. bacterivory). β-Glucosaminidase activity, which is involved in the decomposition of peptidoglycan and chitin, might also be found in biofilms, especially when protozoa and fungi are present [91, 92]. Biofilm enzyme activities can be further involved in the acquisition of inorganic P (phosphatases). The biofilm extracellular enzymatic activity can be an indicator of the capacity for biofilm organic matter use, also indicating its quality (or source) and availability [93].

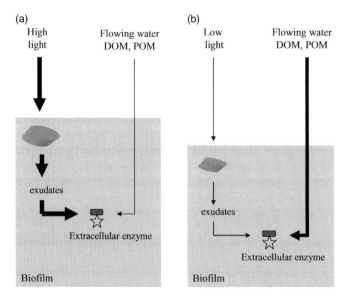

Figure 10.3 Biofilm carbon cycling at low light and high light conditions. In high light conditions (a), algal biomass is developed and thus photosynthetic activity. In this thicker and more structured biofilm, microbes are mainly using organic matter released by algae (high quality fresh molecules) instead of organic molecules from the flowing water. In low light conditions (b), usually a thinner biofilm is developed with lower algal biomass and lower organic matter available within the biofilm and, therefore, microbes are mainly using molecules from the dissolved organic matter (DOM) and particulate organic matter (POM) pool of the surrounding water. Changes in the quality of DOM in the flowing water are not considered in the model, although they could also modulate biofilm C cycling.

Biofilm heterotrophic metabolism is thus highly dependent on the available organic matter without and within the biofilm and this can be very variable and depending on many factors. A simplification of all real possibilities of reuse of biofilm organic matter is shown in Figure 10.3 by considering incident light (and thus, the presence of active algae) the main factor and obtaining two distinct biofilms with two distinct C-cycling dynamics. In high light conditions, algal biomass is developed and thus photosynthetic activity. Algae release extracellular organic molecules during active metabolism or following cell lysis and natural senescence. The release of extracellular organic molecules from living algal cells increases with photosynthetic production [94] and is generally correlated with increases in irradiance [95]. Algal release molecules consist of low-molecular-weight compounds [96, 97] which are utilised rapidly by bacteria and can result in increased bacterial productivity. Increases of bacterial biomass together with algal accrual and photosynthetic activity have been reported in freshwater biofilms studies [87, 36]. A positive effect of algal biomass and photosynthetic activity to bacterial extracellular enzyme activity of β-xylosidase, β-glucosidase and peptidase has been also reported [87, 88, 98]. In high light conditions, a thicker [21], more structured biofilm with a higher C/N ratio is usually built [4, 99]. The higher C/N ratio in such biofilms might be related to a major contribution of the EPS to the total biofilm [4]. As suggested by Freeman and Lock [100], the development of the polymeric matrix could act as a buffer against changing organic substrate supplies, explaining the resilience of structured biofilms to depletion of organic matter from the overlying waters. In these conditions, microbes are mainly using organic matter released by algae (high-quality fresh molecules) instead of organic molecules from the flowing water (Figure 10.3(a)). In contrast, in low light conditions, usually a thinner biofilm is developed, with lower algal biomass and, therefore, microbes are mainly using molecules from the DOM and POM pool (Figure 10.3(b)). In this sense, in an epilithic biofilm grown under dark conditions, extracellular enzyme activities (related to polysaccharides, peptides and lipid decomposition) were positively correlated to biodegradable dissolved organic carbon (BDOC) content in the flowing water while any significant relationship was found between enzyme activities and BDOC for thicker light grown biofilms [4]. This simplified scheme will be obviously modulated by other factors such as inorganic and organic matter (quantity and quality) available and flow velocity.

At the same time, biofilm microorganisms are responsible for inorganic nutrient uptake from the flowing water. Bacteria and algae have been reported to be efficient utilisers of inorganic nutrients [42, 54]. Uptake of ammonium, nitrate and phosphate has been reported for epilithic biofilm communities [101, 102]. In oligotrophic rivers and streams, nutrient uptake rates are also modulated by water velocity [103].

Biofilm metabolism and its role on the aquatic food web is then a complex process where both biofilm structure and complexity and inorganic and organic nutrient availability play a role. The stoichiometric theory might provide a tool to synthesis such relationships [104]. The elemental composition (C, N and P content) of the different compartments can indicate imbalances as well as determine possible relationships between them [105].

10.5 Dynamic structure–function in freshwater biofilms

Biofilm structure is a highly dynamic biological layer which can change its metabolism, thickness, density and composition depending on its own evolution [106, 107]. Throughout biofilm formation, the structure and thus the metabolism of the biofilm change. Processes of substratum colonisation for biofilm formation and detachment usually occur in nature. In rivers and streams, biofilm attachment and detachment processes are frequent. Collapse (dramatic decrease of cells) and recolonisation (start to increase the number of cells) can occur periodically in biofilms [108]. Especially after high flood events, an important detachment occurs due to shear forces decreasing the accumulated biomass to previous colonisation steps, and a new colonisation process begins [109]. An increase in the water level can also induce biofilm formation on the new wetted surfaces. On the other hand, detachment of dying cells can occur after intensive growth [107], as it can also occur during the growing season (spring to summer).

It is generally considered that, after the adsorption of an organic layer on a clean surface, known as the 'conditioning film', bacteria are the first colonisers [110], although other studies state that diatoms can be the first colonisers of a conditioned surface [16]. Throughout studies of biofilm colonisation sequences by using CLSM, a prevalence of EPS has been measured during the early stages, followed by bacterial and algal colonisation after 6 and 12 days [28]. The higher proportion of EPS in young biofilms has been related to the process of early biofilm formation such as cell attachment [16, 17]. The use of the live/dead staining together with the CLSM showed an increase of the dead cells from the early phases of biofilm development (day 7) to 29 days when live bacteria were mainly concentrated in the outermost biofilm layers [19]. The combination of CLSM and FISH showed changes in the distribution of specific EPS glycoconjugates as well as in the β-proteobacteria community composition throughout biofilm formation [111]. Neu *et al.* [111] suggested that changes in algal and

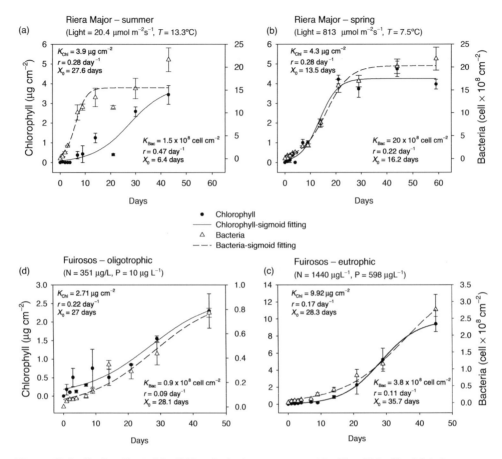

Figure 10.4 Algal and bacterial epilithic colonisation sequences at the Riera Major (Spain) during summer (a) and spring (b) and at the Fuirosos stream's oligotrophic (c) and eutrophic reaches (d). Riera Major is an undisturbed second-order forested stream (average N–NO$_3$ of 260 µg L^{-1} and P–PO$_4$ of 3–8 µg L^{-1}). Colonisation of immersed clay tiles was performed in June to July 1993 [114], a period of high canopy cover by the riparian vegetation (*Alnus glutinosa*) and high temperature of the river water. A second colonisation experiment was performed in April to May 1994 [87], a period with low canopy cover and low water temperature. Fuirosos (Spain) is an undisturbed third-order forested Mediterranean stream (average incident light 50 µmol m^{-2} s^{-1}). Colonisation of immersed clay tiles were performed at two reaches, an oligotorphic reach (not perturbed) and a eutrophic reach (submitted to a continuous addition of phosphate and ammonium) during January to March 2001 [112]. Time sequence of bacterial and algal density were analysed by non-linear regression and fit to a sigmoid curve (three parameters) and the parameters K (carrying capacity), r (growth rate) and X_0 (time when maximal population growth rate is achieved) were estimated.

bacterial biomass are highly influencing the glycoconjugate structure. Changes in C, N and P content and in the density (and consequently diffusion properties) of the biofilm have also been reported. Significant decreases in the C/N ratio after 7 days of epilithic biofilm formation have been related to the accrual of living microbial biomass [112]. Biofilm development in a freshwater environment showed a sharp decrease of the N/P ratio in the biofilm matrix after 4 days of colonisation, this correlated with algal community changes [99]. Changes in algal composition during biofilm formation are also reported, usually characterised by an initial colonisation of diatoms shifting to a major biomass of filamentous forms [113]. Associated with structural changes, autotrophic and heterotrophic metabolic changes have been reported throughout biofilm formation. Specifically, extracellular enzyme activities (such as proteolytic and polysaccharidic enzymes) were positively associated with algal biomass accrual, C, N and EPS content [112]. A higher respiration (electron transport system [ETS]) activity per cell has been measured during the first week of colonisation [114]. During the early phase of biofilm colonisation, microorganisms might be highly active [113]. This is also suggested by the high algal growth and respiration reported for a recolonising initial biofilm after wave disturbance and high flood events [109, 115].

Most colonisation sequences of epilithic biofilms in freshwater environments (followed in the field) show an early phase of colonisation of 3–7 days when increases in biomass is slow, followed by a higher rate of biomass increase (about 4–15 days) and a stabilisation of the biofilm biomass (at about 15–60 days) [19, 32, 33, 48, 87, 99, 112, 114, 116, 117] (Figure 10.4). However, the time for this process can change very much upon environmental conditions and substratum quality. Figure 10.4 shows four examples of algal and bacterial colonisation for the formation of an epilithic biofilm. The mathematical expression of the increase in biomass of a certain group of microorganisms through the biofilm formation can be expressed as a sigmoid curve. The sigmoid curve expresses the classical steps in population growth: lag phase (conditioning of the biofilm and attachment of first colonisers), exponential phase (colonisation and growth) and stationary phase (plateau, equilibrium between attachment and detachment and growth and death). The results show that at the Riera Major during spring the higher incident light made chlorophyll to increase earlier although reaching similar maximum values at the end of the sequence than in summer (low light). In contrast, bacteria were mainly favoured in summer (higher growth rate and lower X_0 – time when maximal population growth rate is achieved) probably due to the higher temperature. In Fuirosos, the effect of nutrient addition is basically affecting maximal carrying capacities of both algae and bacteria (Figure 10.4).

10.6   Conclusions

- Freshwater biofilms are highly structured microbial communities, which affects their functioning. Microorganisms living in the biofilm (algae, bacteria, protozoa, fungi) interact between them (competition, synergism) and with the flowing water as well, determining the metabolic role of the biofilm as a whole for the aquatic habitat. The extracellular enzymes synthesised by biofilm microorganisms are key activities for decomposition and organic matter cycling processes taking place in biofilms.
- The biofilm is highly dynamic and attachment and detachment processes usually occur in natural aquatic ecosystems (flood events, grazing), where detachment is followed by microbial colonisation processes. Physical factors such as nutrient content, incident light and flow velocity are relevant for the process of biofilm development and will determine the mature biofilm properties (thickness, EPS content, microbial biomass).
- The structural–functional complexity and dynamics of biofilms need to be studied at different levels: from general extraction procedures and metabolic measurements to specific three-dimensional and single cell analysis.

- The present and improving knowledge of natural biofilm development process together with structure–function relationships in biofilms might be relevant when developing technologies for biofouling control.

Acknowledgements

The writing of this chapter was supported by the Spanish Ministry of Science and Education (CGL2005-06739-C02-02 and CGL2008-05618-C02-01). I would like to thank Dr Sergey Dobretsov for his comments and suggestions on the comparison between marine and freshwater biofilms.

References

1. Lock, M.A. (1993) Attached microbial communities in rivers. In: *Aquatic Microbiology: An Ecological Approach* (ed. T.E. Ford), pp. 113–138. Blackwell, Oxford.
2. Pusch, M., Fiebig, D., Brettar, I., *et al.* (1998) The role of micro-organisms in the ecological connectivity of running waters. *Freshwater Biology*, **40**, 453–495.
3. Fischer, H., Sachse, A., Steinberg, C.E.W. & Pusch, M. (2002) Differential retention of dissolved organic carbon by bacteria in river sediments. *Limnology and Oceanography*, **47**, 1702–1711.
4. Romaní, A.M., Guasch, H., Muñoz, I., *et al.* (2004) Biofilm structure and function and possible implications for riverine DOC dynamics. *Microbial Ecology*, **47**, 316–328.
5. Kaplan, L.A. & Bott, T.L. (1983) Microbial heterotrophic utilization of dissolved organic matter in a piedmont stream. *Freshwater Biology*, **13**, 363–377.
6. Battin, T.J., Butturini, A. & Sabater, F. (1999) Immobilization and metabolism of dissolved organic carbon by natural biofilms in a Mediterranean and temperate stream. *Aquatic Microbial Ecology*, **19**, 297–305.
7. Cazelles, B., Fontvieille, D. & Chau, N.P. (1991) Self-purification in a lotic ecosystem: a model of dissolved organic carbon and benthic microorganisms dynamics. *Ecological Modelling*, **58**, 91–117.
8. Chróst, R.J. (1991) Environmental control of the synthesis and activity of aquatic microbial ectoenzymes. In: *Microbial Enzymes in Aquatic Environments* (ed. R.J. Chróst), pp. 29–59. Brock/Springer, New York.
9. Chróst, R.J. (1990) Microbial ectoenzymes in aquatic environments. In: *Aquatic Microbial Ecology: Biochemical and Molecular Approaches* (eds J. Overbeck & R.J. Chróst), pp. 47–78. Springer, New York.
10. Thompson, A.J. & Sinsabaugh, R.L. (2000) Matric and particulate phosphatase and aminopeptidase activity in limnetic biofilms. *Aquatic Microbial Ecology*, **21**, 151–159.
11. Hamilton, W.A. (1987) Biofilms: microbial interactions and metabolic activities. In: *Ecology of Microbial Communities. Symposium 41* (eds M. Fletcher, T.R.G. Gray & J.G. Jones), pp. 361–385. Cambridge University Press, Cambridge.
12. Ramasamy, P. & Zhang, X. (2005) Effects of shear stress on the secretion of extracellular polymeric substances in biofilms. *Water Science and Technology*, **52**, 217–223.
13. Christensen, B.E. (1989) The role of extracellular polysaccharides in biofilms. *Journal of Biotechnology*, **10**, 181–202.
14. Lazarova, V. & Manem, J. (1995) Biofilm characterization and activity analysis in water and wastewater treatment. *Water Research*, **29**, 2227–2245.
15. Low, C.S.F. & White, D.C. (1989) Regulation of external polymer production in benthic microbial communities. In: *Microbial Mats: Physiological Ecology of Benthic Microbial Communities*

(eds Y. Cohen & E. Rosenberg), pp. 228–238. American Society for Microbiology, Washington, DC.

16. Cooksey, K.E. & Wigglesworth-Cooksey, B. (1995) Adhesion of bacteria and diatoms to surfaces in the sea: a review. *Aquatic Microbial Ecology*, **9**, 87–96.

17. Costerton, J.W., Lewandowski, Z., Caldwell, D.E., Korber, D.R. & Lappin-Scott, H.M. (1995) Microbial biofilms. *Annual Review of Microbiology*, **49**, 711–745.

18. Lock, M.A., Wallace, R.R., Costerton, J.W., Ventullo, R.M. & Charlton, S.E. (1984) River epilithon: toward a structural-functional model. *OIKOS*, **42**, 10–22.

19. Neu, T.R. & Lawrence, J.R. (1997) Development and structure of microbial biofilms in river water studied by confocal laser scanning microscopy. *FEMS Microbiology Ecology*, **24**, 11–25.

20. DeBeer, D. & Schramm, A. (1999) Micro-environments and mass transfer phenomenon in biofilms studied with microsensors. *Water Science and Technology*, **39**, 173–178.

21. Zippel, B. & Neu, T.R. (2005) Growth and structure off phototrophic biofilms under controlled light conditions. *Water Science and Technology*, **52**, 203–209.

22. Neu, T.R., Woelfl, S. & Lawrence, J.R. (2004) Three-dimensional differentiation of photo-autotrophic biofilm constituents by multi-channel laser scanning microscopy (single-photon and two-photon excitation). *Journal of Microbiological Methods*, **56**, 161–172.

23. Lawrence, J.R., Korber, D.R., Hoyle, B.D., Costerton, J.W. & Caldwell, D.E. (1991) Optical sectioning of microbial biofilms. *Journal of Bacteriology*, **173**, 6558–6567.

24. DeBeer, D., Stoodley, P. & Lewandowski, Z. (1994) Liquid flow in heterogeneous biofilms. *Biotechnology and Bioengineering*, **44**, 636–641.

25. Davies, D.G., Parsek, M.R., Pearson, J.P., Iglewski, B.H., Costerton, J.W. & Greenberg, E.P. (1998) The involvement of cell-to-cell signals in the development of a bacterial biofilm. *Science*, **280**, 295–298.

26. Alpkvist, E., Picioreanu, C., van Loosdrecht, M.C.M. & Heyden, A. (2006) Three-dimensional biofilm model with individual cells and continuum EPS matrix. *Biotechnology and Bioengineering*, **94**, 961–979.

27. Wimpenny, J., Manz, W. & Szewzyk, U. (2000) Heterogeneity in biofilms. *FEMS Microbiology Reviews*, **24**, 661–671.

28. Barranguet, C., van Beusekom, S.A.M., Veuger, B., *et al.* (2004) Studying undisturbed autotrophic biofilms: still a technical challenge. *Aquatic Microbial Ecology*, **34**, 1–9.

29. Carr, G.M., Morin A. & Chambers, P.A. (2005) Bacteria and algae in stream periphyton along a nutrient gradient. *Freshwater Biology*, **50**, 1337–1350.

30. Romaní, A.M. & Sabater, S. (2001) Structure and activity of rock and sand biofilms in a Mediterranean stream. *Ecology*, **82**, 3232–3245.

31. Hepinstall, J.A. & Fuller, R.L. (1994) Periphyton reactions to different light and nutrient levels and the response of bacteria to these manipulations. *Archiv für Hydrobiologie*, **131**, 161–173.

32. Battin, T.J., Kaplan, L.A., Newbold, J.D. & Hansen, C.M.E. (2003) Contributions of microbial biofilms to ecosystem processes in stream mesocosms. *Nature*, **426**, 439–442.

33. Steinman, A.D. & Parker, A.F. (1990) Influence of substrate conditioning on periphytic growth in a heterotrophic woodland stream. *Journal of the North American Benthological Society*, **9**, 170–179.

34. Fukuda, M., Matsuyama J., Katano T., Nakano S. & Dazzo F. (2006) Assessing primary and bacterial production rates in biofilms on pebbles in Ishite Stream, Japan. *Microbial Ecology*, **52**, 1–9.

35. Jones, S.E. & Lock, M.A. (1993) Seasonal determinations of extracellular hydrolytic activities in heterotrophic and mixed heterotrophic/autotrophic biofilms from two contrasting rivers. *Hydrobiologia*, **257**, 1–16.

36. Chapell, K.R. & Goulder, R. (1994) Seasonal variation of extracellular enzyme activity in three diverse headstreams. *Archiv für Hydrobiologie*, **130**, 195–214.

37. Romaní, A.M. & Marxsen, J. (2002) Extracellular enzymatic activities in epilithic biofilms of the Breitenbach: microhabitat differences. *Archiv für Hydrobiologie*, **155**, 541–555.

38. Romaní, A.M. & Sabater, S. (1998) A stromatolitic cyanobacterial crust in a Mediterranean stream optimizes organic matter use. *Aquatic Microbial Ecology*, **16**, 131–141.
39. Romaní, A.M. & Sabater, S. (1999) Epilithic ectoenzyme activity in a nutrient-rich Mediterranean river. *Aquatic Sciences*, **61**, 122–132.
40. Hill, W. (1996) Effects of light. In: *Algal Ecology. Freshwater Benthic Ecosystems* (eds R.J. Stevenson, M.L. Bothwell & R.L. Lowe), pp. 121–148. Academic Press, San Diego.
41. Guasch, H. & Sabater, S. (1994) Primary production of epilithic communities in undisturbed Mediterranean streams. *Verhandlungen der Internationalen Vereinigung für Theoretische und Angewandte Limnologie*, **25**, 1761–1764.
42. Borchardt, M.A. (1996) Nutrients. In: *Algal Ecology. Freshwater Benthic Ecosystems* (eds R.J. Stevenson, M.L. Bothwell & R.L. Lowe), pp. 183–227. Academic Press, San Diego.
43. Dodds, W.K. (2006) Eutrophication and trophic state in rivers and streams. *Limnology and Oceanography*, **51**, 671–680.
44. Perrin, C.J., Bothwell, M.L. & Slaney, P.A. (1987) Experimental enrichment of a coastal stream in British Columbia: effects of organic and inorganic additions on autotrophic periphyton production. *Canadian Journal of Fisheries and Aquatic Sciences*, **44**, 1247–1251.
45. Tank, J.L & Dodds, W.K. (2003) Nutrient limitations of epilithic and epixylic biofilms in 10 North American streams. *Freshwater Biology*, **48**, 1031–1049.
46. Mosisch, T.D., Bunn, S.E. & Davies, P.M. (2001) The relative importance of shading and nutrients on algal production in subtropical streams. *Freshwater Biology*, **46**, 1269–1278.
47. Taulbee, W.K., Cooper, S.D. & Melack, J.M. (2005) Effects of nutrient enrichment on algal biomass across a natural light gradient. *Archiv für Hydrobiologie*, **164**, 449–464.
48. Roberts, S., Sabater, S. & Beardall, J. (2004) Benthic microalgal colonization in streams of differing riparian cover and light availability. *Journal of Phycology*, **40**, 1004–1012.
49. Rier, S.T. & Stevenson, R.J. (2006) Response of periphytic algae to gradients in nitrogen and phosphorus in streamside mesocosms. *Hydrobiologia*, **561**, 131–147.
50. Ylla, I., Romaní, A.M. & Sabater, S. (2007) Differential effects of nutrients and light on the primary production of stream algae and mosses. *Fundamental and Applied Limnology*, **170**, 1–10.
51. Stelzer, R.S. & Lamberti, G.A. (2001) Effects of N:P ratio and total nutrient concentration on stream periphyton community structure, biomass, and elemental composition. *Limnology and Oceanography*, **46**, 356–367.
52. Acuña, V., Giorgi, A., Muñoz, I., Uehlinger, U. & Sabater, S. (2004) Flow extremes and benthic organic matter shape the metabolism of a headwater Mediterranean stream. *Freshwater Biology*, **49**, 960–971.
53. Porter, K.G. & Feig, Y.S. (1980) The use of DAPI for identifying and counting aquatic microflora. *Limnology and Oceanography*, **25**, 943–948.
54. Kirchman, D.L. (1994) The uptake of inorganic nutrients by heterotrophic bacteria. *Microbial Ecology*, **28**, 255–271.
55. Rier, S.T. & Stevenson, R.J. (2002) Effects of light, dissolved organic carbon, and inorganic nutrients on the relationship between algae and heterotrophic bacteria in stream periphyton. *Hydrobiologia*, **489**, 179–184.
56. Olapade, O.A. & Leff, L.G. (2005) Seasonal response of stream biofilm communities to dissolved organic matter and nutrient enrichments. *Applied and Environmental Microbiology*, **71**, 2278–2287.
57. Haglund, A.-L., Törnblom, E., Boström, B. & Tranvik, L. (2002) Large differences in the fraction of active bacteria in plankton, sediments, and biofilm. *Microbial Ecology*, **43**, 232–241.
58. Bayles K.W. (2007) The biological role of death and lysis in biofilm development. *Nature Reviews Microbiology*, **5**, 721–726.
59. Domènech, R., Gaudes, A., López-Doval, J., Salvadó, H. & Muñoz, I. (2006) Effects of short-term nutrient addition on microfauna density in a Mediterranean stream. *Hydrobiologia*, **568**, 207–215.

60. Schmid-Araya, J.M. (1994) The temporal and spatial distribution of benthic microfauna in sediments of a gravel streambed. *Limnology and Oceanography*, **39**, 1813–1821.

61. Bott, T.L. & Kaplan, L.A. (1989) Densities of benthic protozoa and nematodes in a Piedmont stream. *Journal of the North American Benthological Society*, **8**, 187–196.

62. Gulis, V., Ferreira, V., & Graça, M.A.S. (2006) Stimulation of leaf litter decomposition and associated fungi and invertebrate by moderate eutrophication: implications for stream assessment. *Freshwater Biology*, **51**, 1655–1669.

63. Pascoal, C. & Cassio, F. (2004) Contribution of fungi and bacteria to leaf litter decomposition in a polluted river. *Applied and Environmental Microbiology*, **70**, 5266–5273.

64. Gessner, M.O. & Newell, S.Y. (2002) Biomass, growth rate, and production of filamentous fungi in plant litter. In: *Manual of Environmental Microbiology* (eds C.J. Hurst, R.L. Crawford, G.R. Knudsen, M.J. McInerney & L.D. Stetzenbach), pp. 390–408. American Society for Microbiology Press, Washington, DC.

65. Artigas, J., Romaní, A.M., Gaudes, A., Muñoz, I. & Sabater, S. (2008) Benthic structure and metabolism in a Mediterranean stream: from biological communities to the whole stream ecosystem function. *Freshwater Biology*, doi:10.1111/j.1365-2427.2008.02140.x.

66. Kemp, P.F. (1990) The fate of benthic bacterial production. *Review of Aquatic Sciences*, **2**, 109–124.

67. Bott, T.L. & Kaplan, L.A. (1990) Potential for protozoa grazing of bacteria in streambed sediments. *Journal of the North American Benthological Society*, **9**, 336–345.

68. Gücker, B. & Fischer, H. (2003) Flagellate and ciliate distribution in sediments of a lowland river: relationships with environmental gradients and bacteria. *Aquatic Microbial Ecology*, **31**, 67–76.

69. Joubert, L.-M., Wolfaardt, G.M. & Botha, A. (2006) Microbial exopolymers link predator and prey in a model yeast biofilm system. *Microbial Ecology*, **52**, 187–197.

70. Matz, C. & Kjelleberg, S. (2005) Off the hook-how bacteria survive protozoan grazing. *Trends in Microbiology*, **13**, 302–307.

71. Hart, D.D. (1992) Community organization in streams: the importance of species interactions, physical factors, and chance. *Oecologia*, **91**, 220–228.

72. Wellnitz, T.A., Rader, R.B. & Ward, J.V. (1996) Light and a grazing mayfly shape periphyton in a Rocky Mountain stream. *Journal of the North American Benthological Society*, **15**, 496–507.

73. Hillebrand, H. & Kahlert, M. (2001) Effect of grazing and nutrient supply on periphyton biomass and nutrient stoichiometry in habitats of different productivity. *Limnology and Oceanography*, **46**, 1881–1898.

74. Mille-Lindblom, C. & Tranvik, L.J. (2003) Antagonism between bacteria and fungi on decomposing aquatic plant litter. *Microbial Ecology*, **45**, 173–182.

75. Romaní, A.M., Fischer, H., Mille-Lindblom, C. & Tranvik, L.J. (2006) Interactions of bacteria and fungi on decomposing litter: differential extracellular enzyme activities. *Ecology*, **87**, 2559–2569.

76. Kuserk, F.T., Kaplan, L.A. & Bott, T.L. (1984) *In situ* measures of dissolved organic carbon flux in a rural stream. *Canadian Journal of Fisheries and Aquatic Sciences*, **41**, 964–973.

77. Meyer, J.L. (1988) Benthic bacterial biomass and production in a blackwater river. *Verhandlungen der Internationalen Vereinigung für Theoretische und Angewandte Limnologie*, **23**, 1832–1838.

78. Volk, C.J., Volk, C.B. & Kaplan, L.A. (1997) Chemical composition of biodegradable dissolved organic matter in streamwater. *Limnology and Oceanography*, **42**, 39–44.

79. Norrman, B., Zweifel, U.L., Hopkinson, C.S. & Fry, B. (1995) Production and utilization of dissolved organic carbon during an experimental diatom bloom. *Limnology and Oceanography*, **40**, 898–907.

80. Amon, R.M.W. & Benner, R. (1996) Bacterial utilization of different size classes of dissolved organic matter. *Limnology and Oceanography*, **41**, 41–51.

81. Docherty, K.M., Young, K.C., Maurice, P.A. & Bridgham, S.D. (2007) Dissolved organic matter concentration and quality influences upon structure and function of freshwater microbial communities. *Microbial Ecology*, **52**, 378–388.

82. Sun, L., Perdue, E.M., Meyer, J.L. & Weis, J. (1997) Use of elemental composition to predict bioavailability of dissolved organic matter in a Georgia river. *Limnology and Ocenography*, **42**, 714–721.

83. Kaplan, L.A. & Newbold, J.D. (2002) The role of monomers in stream ecosystem metabolism. In: *Aquatic Ecosystems, Interactivity of Dissolved Organic Matter* (eds S.E. Findlay & R.L. Sinsabaugh), pp. 97–119. Academic Press, San Diego.

84. Wetzel, R.G. (1993) Microcommunities and microgradients: linking nutrient regeneration, microbial mutualism, and high sustained aquatic primary production. *Netherlands Journal of Aquatic Ecology*, **27**, 3–9.

85. Haack, T.K. & McFeters, G.A. (1982) Nutritional relationships among microorganisms in an epilithic biofilm community. *Microbial Ecology*, **8**, 115–126.

86. Nakano, S. (1996) Bacterial response to extracellular dissolved organic carbon released from healthy and senescent *Fragilaria crotonensis* (Bacillariophyceae) in experimental systems. *Hydrobiologia*, **339**, 47–55.

87. Romaní, A.M. & Sabater, S. (1999) Effect of primary producers on the heterotrophic metabolism of a stream biofilm. *Freshwater Biology*, **41**, 729–736.

88. Espeland, E.M., Francoeur, S.N. & Wetzel, R.G. (2001) Influence of algal photosynthesis on biofilm bacterial production and associated glucosidase and xylosidase activities. *Microbial Ecology*, **42**, 524–530.

89. Sobczak, W.V. (1996) Epilithic bacterial responses to variations in algal biomass and labile dissolved organic carbon during biofilm colonization. *Journal of the North American Benthological Society*, **15**, 143–154.

90. Zhang, X. & Bishop, P.L. (2003) Biodegradability of biofilm extracellular polymeric substances. *Chemosphere*, **50**, 63–69.

91. Chamier, A.C. (1985) Cell-wall-degrading enzymes of aquatic hyphomycetes: a review. *Botanical Journal of the Linnean Society*, **91**, 67–81.

92. Vrba, J., Callieri, C., Bittl, T., *et al.* (2004) Are bacteria the major producers of extracellular glycolytic enzymes in aquatic environments? *International Review of Hydrobiology*, **89**, 102–117.

93. Romaní, A.M. & Sabater, S. (2000) Variability of heterotrophic activity in Mediterranean stream biofilms: A multivariate analysis of physical-chemical and biological factors. *Aquatic Sciences*, **62**, 205–215.

94. Wood, A.M., Rai, H., Garnier, J., *et al.* (1992) Practical approaches to algal excretion. *Marine Microbial Food Webs*, **6**, 21–38.

95. Zlotnik, I. & Dubinsky, Z. (1989) The effect of light and temperature on DOC excretion by phytoplankton. *Limnology and Oceanography*, **34**, 831–839.

96. Søndergaard, M. & Schierup, H.H. (1982) Release of extracellular organic carbon during a diatom bloom in Lake Mossø: molecular weight fractionation. *Freshwater Biology*, **12**, 313–320.

97. Sundh, I. (1992) Biochemical composition of dissolved organic carbon derived from phytoplankton and used by heterotrophic bacteria. *Applied and Environmental Microbiology*, **58**, 2938–2947.

98. Francoeur, S.N. & Wetzel, R.G. (2003) Regulation of periphytic leucine-aminopeptidase activity. *Aquatic Microbial Ecology*, **31**, 249–258.

99. Sekar, R., Nair, K.V.K., Rao, V.N.R. & Venugopalan, V.P. (2002) Nutrient dynamics and successional changes in a lentic freshwater biofilm. *Freshwater Biology*, **47**, 1893–1907.

100. Freeman, C. & Lock, M.A. (1995) The biofilm polysaccharide matrix: A buffer against changing organic substrate supply? *Limnology and Oceanography*, **40**, 273–278.

101. Bowden, W.B., Peterson, B.J., Finlay, J.C. & Tucker, J. (1992) Epilithic chlorophyll a, photosynthesis, and respiration in control and fertilized reaches of a tundra stream. *Hydrobiologia*, **240**, 121–131.

102. Mulholland, P.J., Marzolf, E.R., Hendricks, S.P., Wilkerson, R.V. & Baybayan, A.K. (1995) Longitudinal patterns of nutrient cycling and periphyton characteristics in streams: a test of upstream–downstream linkage. *Journal of the North American Benthological Society*, **14**, 357–370.

103. Horner, R.R. & Welch, E.B. (1981) Stream periphyton development in relation to current velocity and nutrients. *Canadian Journal of Fisheries and Aquatic Sciences*, **38**, 449–457.
104. Frost, P.C., Stelzer, R.S., Lamberti, G.A. & Elser, J.J. (2002) Ecological stoichiometry of trophic interactions in the benthos: understanding the role of C:N:P ratios in lentic and lotic habitats. *Journal of the North American Benthological Society*, **21**, 515–528.
105. Sterner R.W. & Elser, J.J. (2002) *Ecological Stoichiometry: The Biology of Elements from Molecules to the Biosphere*, Princeton University Press, Princeton.
106. Jenkinson, H.F. & Lappin-Scott, H.M. (2001) Biofilms adhere to stay. *Trends in Microbiology*, **9**, 9–10.
107. Stoodley, P., Wilson, S., Hall-Stoodley, L., Boyle, J.D., Lappin-Scott, H.M. & Costerton, J.W. (2001) Growth and detachment of cell clusters from mature mixed-species biofilms. *Applied and Environmental Microbiology*, **67**, 5608–5613.
108. Ács, E. & Kiss, K.T. (1993) Colonization processes of diatoms on artificial substrates in the River Danube near Budapest (Hungary). *Hydrobiologia*, **269/270**, 307–315.
109. Blenkinsopp, S.A. & Lock, M.A. (1992) Impact of storm-flow on electron transport system activity in river biofilms. *Freshwater Biology*, **27**, 397–404.
110. Liu, D., Lau, Y.L., Chau, Y.K. & Pacepavicius, G.J. (1993) Characterization of biofilm development on artificial substratum in natural water. *Water Research*, **27**, 361–367.
111. Neu, T.R., Swerhone, G.D.W., Bockelmann, U. & Lawrence, J.R. (2005) Effect of CNP on composition and structure of lotic biofilms as detected with lectin-specific glycoconjugates. *Aquatic Microbial Ecology*, **38**, 283–294.
112. Romaní, A.M., Giorgi, A., Acuña, V. & Sabater, S. (2004) The influence of substratum type and nutrient supply on biofilm organic matter utilization in streams. *Limnology and Oceanography*, **49**, 1713–1721.
113. Stock, M.S. & Ward, A.K. (1989) Establishment of a bedrock epilithic community in a small stream: microbial (algal and bacterial) metabolism and physical structure. *Canadian Journal of Fisheries and Aquatic Sciences*, **46**, 1874–1883.
114. Sabater, S. & Romaní, A.M. (1996) Metabolic changes associated with biofilm formation in an undisturbed Mediterranean stream. *Hydrobiologia*, **335**, 107–113.
115. Peterson, C.G., Hoagland, K.D. & Stevenson, R.J. (1990) Timing of wave disturbance and the resistance and recovery of a freshwater epilithic microalgal community. *Journal of the North American Benthological Society*, **9**, 54–67.
116. Sabater, S., Gregory, S.V. & Sedell, J.R. (1998) Community dynamics and metabolism of benthic algae colonizing wood and rock substrata in a forest stream. *Journal of Phycology*, **34**, 561–567.
117. Giorgi, A. & Ferreyra, G. (2000) Phytobenthos colonization in a lowland stream in Argentina. *Journal of Freshwater Ecology*, **15**, 39–46.

Chapter 11
Biofilms in Medicine

David A. Spratt, Derren Ready and Jonathan Pratten

The aim of this chapter is to give a brief overview from 'head to toe' of the numerous clinical infections now thought to be associated with the biofilm mode of growth. Highlighted are several aspects of human pathogenesis which are related to biofilm development. These biofilm infections are both associated with host surfaces directly or as a consequence of materials being placed into the body which themselves become colonised. The importance of understanding the biological basis of biofilm-associated infections may allow the effective control of these resistant phenotypes.

11.1　Introduction

The cost, both beneficial and harmful, of the association between microbes and surfaces has long been recognised. As far back as the fourteenth century, Guy de Chauliac, a French surgeon, recorded the relationship between foreign bodies and delayed wound healing [1], while just over a century ago the symbiosis between *Rhizobium* and the roots of leguminous plants was first recorded [2]. Intact host defence systems usually eliminate transient bacterial contamination or colonisation, unless the number of organisms exceeds threshold levels, host defences are impaired, tissue surfaces are traumatised or a foreign body is present [3]. The presence of biofilms in the airways of patients with cystic fibrosis helped to explain the chronic lung infections in this population. There has since been a growing realisation that similar biofilms are a factor in almost every aspect of health care. Indeed, biofilms account for over 65% of human bacterial infections [4] and are a major contributing factor to the difficulty of treating infections.

Although a wide range of bacteria and fungi have been shown to be responsible for human infections, most infections are actually caused by a small number of species. Staphylococci (predominantly coagulase-negative strains) are considered to be the most frequent cause, being responsible for more than half of all infections [5], while streptococci and a collection of gram-negative organisms account (in approximately equal proportions) for most of the remaining infections. Of the gram-negative organisms *Escherichia coli*, *Klebsiella pneumoniae*, *Proteus mirabilis* and *Pseudomonas aeruginosa* are the most commonly isolated [6]. These organisms may originate from the skin of patients or health care workers, tap water to which entry points are exposed or other sources in the environment.

11.2 Infection of the head and neck

11.2.1 Otitis media

Middle-ear infections frequently occur in the paediatric population, and interestingly a high proportion are found to be culture-negative by the use of traditional microbiology methods, whereas data from PCR-based assays consistently suggested a higher incidence of bacterial involvement in otitis media (OM). The presence of attached bacteria to the mucosa of the middle ear, forming a biofilm, rather than the presence of planktonic microorganisms in middle-ear effusions, has been suggested to explain these observations. Scanning electron microscopy was used to investigate the presence of *Haemophilus influenzae* biofilms in an animal model and established that biofilm structures were evident in all specimens from 1 day post-infection and were present for 21 days. Confocal laser scanning microscopy (CLSM) indicated that bacteria within the biofilms were viable [7]. An RT-PCR-based assay system was used by Rayner *et al.* [8] to detect the presence of bacterial mRNA in culture-negative middle ear effusions, demonstrating the presence of viable, metabolically active, intact bacteria in a proportion of culture-negative samples. Mucosal biofilms were visualised by CLSM on 92% of 50 middle-ear mucosa biopsy specimens obtained from 26 children undergoing tympanostomy tube placement for treatment of OM with effusion and recurrent OM suggesting that chronic middle-ear infections are biofilm-related disorders [9].

11.2.2 Microbial keratitis

Infection of the cornea leading to corneal ulceration is referred to as microbial keratitis and is the most serious complication of contact lens wear. Without appropriate and timely treatment, this infection may lead to corneal scarring and perforation leading to reduced vision or blindness. Major factors predisposing to corneal infection include eyes with existing ocular surface disease, corneal trauma and surgery, post-herpetic corneal disease, bullous keratopathy, corneal anaesthesia, corneal exposure and the dry eye syndrome [10, 11]. Historically, these conditions were the most significant risk factors for microbial keratitis; however, contact lens wear is now a major contributory factor in the development of this disease, with contact lens use being a predisposing factor in 56–65% of patients with bacterial keratitis [12]. The annual risk of developing a contact lens-induced keratitis in individuals wearing daily wear soft lenses is estimated at 2.2–4.1 cases per 10 000 and 13.3–20.9 per 10 000 for extended wear contact lens users [13].

In contrast to the predominately gram-positive bacteria isolated from trauma-associated infections, the microorganisms associated with keratitis are most often aerobic or facultatively aerobic gram-negative bacilli or species of the free-living amoeba *Acanthamoeba* [14]. The reasons for the propensity of infections with gram-negative microorganisms may be in part due to the ability of these bacteria to attach to the contact lens surface [15]. *P. aeruginosa* shows significant adherence to contact lenses in vitro, whereas *Staphylococcus aureus* fails to demonstrate any attachment beyond that expected from non-specific adherence, with attachment concentration on areas with large focal deposits.

The events leading to microbial keratitis appear to start with contamination of the contact lens storage case with bacteria and or *Acanthamoeba* spp. from the household mains water supply. Rates of contamination of contact lens cases vary between 24 and 81% [13] and can be found in cases treated with all commercially available disinfection solutions. One study reported contamination of contact lens cases in 29, 50 and 75% of patients using heat, hydrogen peroxide and chemical disinfectants, respectively, suggesting that currently employed contact lens disinfection solutions show poor activity against microbial biofilms found in contact lens storage cases; indeed, these products should be tested against contact lens case biofilms to determine their efficacy. Microorganisms present in the storage case are able to adhere to the contact lens surface during the process of care and storage and may then be transferred to the eye from the lens [15]. Contact lens wear has been shown to lead to corneal hypoxia which interferes with the ability of corneal epithelium to rapidly repair damage to the cell layer [16]. *P. aeruginosa* poorly colonises healthy corneal epithelium; however, adherence is markedly increased when the epithelium was damaged and it is likely that attachment to damaged corneal epithelium is necessary for the development of microbial keratitis. Contact lens use contributes to microbial keratitis by inducing corneal hypoxia and corneal damage and providing a surface, either the storage case or lens, on which bacteria can persist in biofilms, resist disinfection and then come into contact with a susceptible eye.

11.2.3 *Chronic rhinosinusitis*

Chronic rhinosinusitis (CRS) is characterised by inflammation of the mucosa of the nose and paranasal sinuses and is a disease that is poorly controlled with antibiotic therapy, often requiring mechanical debridement to achieve significant clinical improvement [17]. Previous workers have shown that *P. aeruginosa*, *S. aureus* and coagulase-negative staphylococci isolated from patients with CRS are able to form biofilms in an in vitro system [18]. Ferguson and Stolz [19] used transmission electron microscopy to visualise bacterial communities surrounded by a glycocalyx of inert cellular membrane materials in two patients with bacterial CRS. Culture of the material grew *P. aeruginosa* from both patients. Fluorescent in situ hybridisation (FISH) techniques have demonstrated biofilm formation in situ by *Streptococcus pneumoniae*, *S. aureus* and *H. influenza* in patients with CRS [20]. However, they also identified biofilms on healthy control samples, suggesting that colonisation in the absence of infection may also occur. The role that biofilms play in the pathogenic process of CRS still remains to be determined. However, the presence of biofilms on the mucosa of patients with CRS offers an explanation for the failure of antimicrobial therapy currently associated with this condition and highlights the importance of anti-biofilm therapy in the future treatment for patients with chronic rhinosinusitis.

11.2.4 *Adenoid and tonsil tissues*

The presence of bacterial biofilms in adenoid and tonsular tissues is less well studied. The majority of the mucosal surfaces of adenoid tissues removed from children with CRS were shown to be covered with biofilms; however, adenoid tissue from children with sleep apnea had little biofilm coverage, suggesting that these biofilms in the nasopharynx of children with CRS may act as a reservoir for bacteria, perhaps explaining the observed clinical improvement associated with adenoidectomy in these patients [21]. However, further studies are required to support these findings.

Biofilms have been shown to be present within the tissue and crypts of inflamed tonsils [22]. Scanning electron microscopy analysis of tonsil tissues suggested the presence of microbial biofilms by showing bacterial cells in microcolonies. CLSM demonstrated the presence of bacterial cells and a glycocalyx matrix on tonsil samples, providing visual evidence for the presence of biofilms on the tonsils of patients diagnosed with tonsillitis [23]. The presence of biofilms on tonsil tissues may explain the recurrence and chronicity of a number of forms of tonsillitis possibly due to the resistance of biofilm-related infections to antibiotic therapy.

11.2.5   Caries

Dental caries (Plate IV F) can be defined as the localised demineralisation of the tooth tissue by various acids produced by bacterial fermentation of dietary carbohydrates. It has been estimated that 90% of all dentate adults in the UK have at least one restored tooth as a result of caries with a mean frequency of seven per person [24]. Coronal carries can occur on all surfaces of the crown where the supragingival plaque biofilm is allowed to develop and mature. A major question has been which specific bacteria, if any, are involved in the progression of disease. The 'ecological plaque hypothesis' suggests that a shift in the microbiota is brought about by increases in the amount and frequency of dietary fermentable carbohydrates. These substrates are fermented by the bacteria in the supragingival plaque biofilm leading to the production of acid end products like lactic acid. This acid production serves to lower the local pH and favour a shift in the microbial population to acid-tolerant bacteria, e.g. mutans streptococci. Root surface caries, as the name implies, occurs on root cementum or dentine and is also caused by a microbial biofilm. The disease is secondary to gingival recession since in a healthy mouth cementum and dentine are not exposed to the microbiota and therefore unavailable for colonisation. The microbiological nature of the associated plaque biofilm is different from that associated with crown caries even though it is technically still supragingival plaque. The lesion is actively carious and has been shown to have a definite progression since changes in its clinical appearance are observed over time. The progression of the lesion leads to a change in appearance and is categorised as 'leathery'. The microbiology of this biofilm has been the subject of numerous investigations over the years; however, only recently have the problems associated with sampling of the infected underlying dentine, and not the overlying biofilm, been identified and addressed [25, 26]. In their study, Beighton and Lynch [25] showed that the bacterial composition of the carious dentine biofilm associated with 'soft' lesions consisted of significantly more lactobacilli and gram-positive pleomorphic rods but conversely, significantly fewer streptococci compared to the overlying plaque biofilm. Bacteria and bacterial products can also gain access to the pulp chamber, in the majority of cases, as a consequence of caries. Due to significant demineralisation of the enamel, cementum or dentine the pulp can be directly exposed to the biofilm associated with the lesion (Figure 11.1). Root canal infections are invariably polymicrobial in nature; however, mono-infections (such as *Enterococcus* spp.) do occur.

11.2.6   Periodontal diseases

Periodontitis refers to a group of more advanced and related diseases within the broad heading of periodontal disease. It can be defined as 'an apical extension of gingival inflammation to involve the tissues supporting the tooth (periodontal ligament and bone)'. The destruction of the fibre attachment results in a periodontal pocket. This wide spectrum of diseases has been re-classified [27] and at least 48 specific periodontitis categories are now recognised. By far the most common is chronic periodontitis and is the major cause of tooth loss in the adult population. The disease is mediated by the microbiota forming the plaque biofilm on the tooth

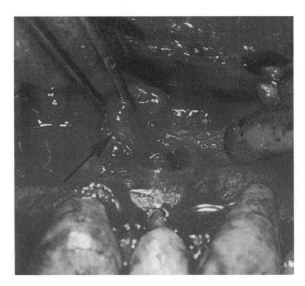

Figure 11.1 Root canal infection, surgical removal of biofilm (arrow) from apex of tooth.

surface. Additionally, and as a consequence of the immune response elicited by the bacteria, further destruction may occur due to the host inflammatory response. The biofilm present in the gingival crevice and later in the periodontal pocket is extremely diverse with up to 100 culturable species from a single pocket [28]. The World Workshop on Clinical Periodontology [29] has designated three species as etiologic agents of periodontitis in a susceptible host, namely *Aggregatibacter actinomycetemcomitans*, *Porphyromonas gingivalis* and *Tannerella forsythensis*. The findings from the majority of other microbiology studies are based on data derived from the culturable microbiota. However, it has been estimated that only 50% of the oral microbiota is culturable [30]. More recently, molecular techniques have been used to detect and identify the unculturable portion of this highly diverse biofilm [31].

The aetiology of periodontal disease is further complicated by a range of predisposing factors. The susceptibility of the host has been shown to be important. A specific genotype of the polymorphic interleukin-1 (a pro-inflammatory cytokine and key regulator of the host responses to microbial infection and a major modulator of extracellular matrix catabolism and bone resorption) gene cluster is associated with severity of periodontitis [32]. Recent data has also suggested that polymorphisms in the host genome may increase patient susceptibility to aggressive periodontitis due to an increase in the presence of specific periodontal pathogens in patients with Fc gamma receptor and IL-6–174 polymorphisms [33]. There is also evidence to suggest that smoking is a significant risk factor for periodontal disease-associated tooth loss [34]. Additionally, diabetes mellitus is a major risk factor for chronic periodontitis and the more severe and rapidly progressing forms [35].

11.3 Respiratory tract

11.3.1 Pneumonia

Pneumonia can be acquired in the community or hospital setting. One previous study using quantitative RT-PCR in a tissue infection model showed increased gene expression of neuraminidases, metalloproteinases, oxidative stress and competence genes in pneumococci. Induction of the competence system by the quorum-sensing peptide (CSP) induced biofilm formation in vitro and also increased virulence in pneumonia in vivo. When pneumococci in the different physiological states were used directly for challenge, sessile cells grown in a biofilm were more effective in inducing meningitis and pneumonia [36].

Pneumonia is a common nosocomial infection and is a leading cause of death due to hospital-acquired infections. Ventilator-associated pneumonia (VAP) occurs in patients receiving mechanical ventilation for periods of more than 48 hours. The incidence of VAP has been reported as 23% in patients receiving mechanical ventilation, with a reported increase of three to tenfold risk for pneumonia in patients receiving mechanical ventilation. The endotracheal tube provides a direct route for bacteria to enter the lower respiratory tract. Bacteria present in the upper airway and oral secretions are able to pool above the cuff of the endotracheal tube and allowing the formation of a biofilm. The biofilm may contain large numbers of bacteria which are able to disseminate into the lungs by ventilator-induced breaths. Dislodging of the biofilm may occur by the instillation of saline into the endotracheal tube, suctioning, coughing or repositioning of the endotracheal tube [37].

11.3.2   Cystic fibrosis

Cystic fibrosis (CF) is the most common lethal genetic disease in the Caucasian population and is considered an infectious disease because of the basic pathophysiology. Chronic lower airway infections cause a progressive pathologic deterioration of lung tissue, a decline in pulmonary function and respiratory failure. Few bacterial species have been implicated as principal CF pathogens. Molecular evidence suggests the presence of a diverse mosaic of bacteria in CF lungs, and infections can be defined as polymicrobial. However, infections are commonly associated with gram-negative bacteria, particularly *P. aeruginosa* and *Burkholderia cepacia*. The capacity to grow in biofilms and the selection of mutants with a mucoid phenotype are major adaptations that allow the persistence of *P. aeruginosa* in the airways making this organism the predominant cause of death in these patients [38]. The extracellular matrix associated with *P. aeruginosa* is composed mainly of the polysaccharide – alginate. The alginate is a key virulence factor of the organism and functions as an adhesion, impeding and is also thought to contribute to the reduced susceptibility of biofilms of the organism [39].

11.4   Gastrointestinal

11.4.1   Peptic ulcer disease

A peptic ulcer is a sore or hole in the lining of the stomach or duodenum. There are two main causes of this disease: (i) use of non-steroidal anti-inflammatory drugs or (ii) infection with a biofilm of *Helicobacter pylori*. *H. pylori* is a gram-negative bacterium that persistently colonises more than half of the human population. While infection almost always results in chronic, active gastritis, patients are often free from obvious clinical symptoms [40]. Up to 95% of duodenal ulcers and 85% of gastric ulcers occur in the presence of *H. pylori* [41]. The area of tissue covered by biofilm is also very high and up to 97% human gastric mucosa is covered with biofilm in urease-positive subjects compared to 1.6% coverage in urease-negative subjects

[42]. Strains of *H. pylori* differ widely in their pathogenic potential. Increased pathogenicity has been linked to the presence of a protein called cytotoxin-associated gene A (CagA) which is present in between 50 and 70% of strains [43]. CagA also acts as a marker for a 40 kb pathogenicity island which codes for up to 31 proteins (depending on strain). Eighteen of these proteins are involved in forming a type IV secretion apparatus which serves to penetrate gastric epithelial cells and delivers CagA, peptidoglycan and other bacterial factors into the host cells. Once inside the cell, the CagA is phosphorylated and interacts with a range of host signalling molecules. About half of *H. pylori* strains also secrete a vacuolating cytotoxin (VacA) which is highly immunogenic [44]. This protein can act to form membrane channels, disrupt endosomal and lysosomal activities, affect integrin receptor-induced cell signalling, interfere with cytoskeleton-dependent cell function, induce apoptosis and modulate the immune system [45, 46]. *H. pylori* infection invariably elicits a strong immune response; however, this seldom clears the infection. Indeed, it is thought that a lot of the pathology arising from the infection is as a result of the immune response. The pathogenicity and immune responses are complex and not entirely understood and certainly beyond the scope of this chapter; for a review see Kusters *et al.* [47].

11.5   Genitourinary

11.5.1   Urinary tract infections

In the urinary tract, bacterial biofilms develop on both living surfaces and artificial implants, producing chronic and often intractable infections. There are up to 175 million urinary tract infections (UTIs) worldwide per annum [48]. Uropathogenic *E. coli* (UPEC) are the most common isolates and may account for 80% of all acute community acquired UTIs [49]. Other bacteria isolated include *K. pneumoniae, P. mirabilis, Enterococcus faecalis, P. aeruginosa, Serratia marcescens* and group B streptococci [49].

Recurring infection is also common and until recently this was thought to be re-infection from vaginal or faecal microbiota [50]. However, it is now known that UPEC can enter, divide and form intracellular bacterial communities (biofilms) in bladder cells [51, 52]. This establishes a reservoir of UPEC inside bladder cells and could conceivably act as a source of further UTIs. UPEC has assembled a number of adherence organelles which serve to bind to different cell types in the urinary tract, e.g. P pili displaying PapG are required for adhesion to human kidney cells [53] and type 1 pili are essential for adhesion to bladder epithelium [54].

UPEC produce a wide range of virulence determinants responsible for the pathology observed. Examples include specific autotransporters termed Sat which are toxic to urinary tract cells [55] and cause extreme vacuolation of host cells [56] and others termed Pic and Tsh which exhibit serine protease activity [57]. Toxins are also produced and include the pore forming alpha-haemolysins [58] and cytotoxic necrotising factor 1 which has been shown to kill human bladder cells and inhibit phagocytosis [59]. Additionally, several iron acquisition systems have been described [60–62]. It is clear that UPEC are exquisitely evolved to form biofilms and cause disease in the urinary tract.

11.5.2   Vaginosis

Bacterial vaginosis is a vaginal infection associated with a malodorous vaginal discharge. In addition to causing irritation, it can cause significant gynaecological and obstetric morbidity and has been associated with pelvic inflammatory disease [63], infections following gynaeco-logical surgery [64] and pre-term birth [65].

The condition is caused by a polymicrobial biofilm which is characterised by a reduction in the proportion of commensal *Lactobacillus* spp. [66], especially hydrogen peroxide pro-ducing taxa [67,68] and a concomitant increase in the levels of commensal vaginal anaerobes including *Gardnerella vaginalis*, *Prevotella* spp., anaerobic gram-positive cocci, *Mobiluncus* spp., *Mycoplasma hominis*, and *Atopobium vaginalis* [69,70]. While the microbiology of bac-terial vaginosis is complex and poorly understood, *G. vaginalis* is isolated from up to 95% of infections and the biofilms are often dominated by this organism. Indeed, the presence of a characteristically antibiotic-resistant biofilm may play a role in colonisation and relapsing of infection [71]. While *G. vaginalis* may be important in the aetiology of the disease, artificial mono-infections of *G. vaginalis* do not usually lead to disease [72,73].

Which environmental factors initiate the bacterial population shift are poorly understood, although the increase in pH is known to be important. Vaginal pH is usually kept at about 4.5 by lactic acid produced by lactobacilli [74,75]; however, this is raised at the onset of menstruation and this may be a factor in disease initiation.

The anaerobic microbiota associated with this disease are able to produce a number of key enzymes (e.g. decarboxylases) and these are thought to degrade proteins and convert the amino acids to amines. This again serves to raise the pH of the vagina to between 5.0 and 5.5 and is accompanied by a characteristic fishy odour. The commensal lactobacilli also produce a range of bacteriocins which are thought to inhibit the growth of the other (disease associated) taxa. The increase in pH may also detrimentally affect the activity of these bacteriocins [76,77] and thereby give further growth advantages to the anaerobic taxa.

11.6   Surgical and nosocomial infections

11.6.1   Medical implants

Over the past five decades, hip or knee replacement surgery has become one of the most frequent prosthetic surgeries in restoring function to disabled arthritic individuals. Currently, more than a million joint replacements are performed worldwide each year; the infection rates for such implants are shown in Table 11.1. Second to aseptic loosening, infection is the most frequent complication leading to long periods of hospitalisation, morbidity, severe functional impairment and sometimes increased mortality. Most infections are caused by staphylococci which gain access to the device during insertion.

The treatment of infection following total joint arthroplasty involves the combination of surgery and antimicrobial therapy. Surgical methods include debridement and prosthesis re-tention, re-implantation with a one or two stage exchange arthroplasty, arthrodesis (knee) and excision arthroplasty (shoulder, hip). Antimicrobial therapy is also combined with surgery. The prolonged course of antibiotics aimed at the causative bacteria can be administered sys-temically or locally. Local delivery can be achieved by implanting antibiotic bead chains or antibiotic loaded bone cements. Novel and challenging therapeutic approaches have been at-tempted, particularly in hip prosthetic infections, based on altering the host immune system, the type and route of infection, the surgical procedure, the bacteria cultured and the antibiotics employed [82,83].

Table 11.1 Annual frequency of implantable devices usage in USA and rate of infection [78–81].

Implant/medical device	Estimated number used annually in USA	Rate of infection (%)
Bladder catheters	30 000 000	10–30
Central venous catheters	5 000 000	3–10
Fracture fixation devices	2 000 000	5–10
Dental implants	1 000 000	5–10
Prosthetic hip/knee joint	600 000	1–3
Vascular grafts	450 000	1
Cardiac pacemakers	300 000	1–7
Mammary implants	130 000	1–2
Prosthetic heart valves	85 000	1–2
Penile implants	15 000	1–3
Heart-assisted devices	700	25–50

11.6.2   Indwelling devices

One of the most frequently used biomaterial devices is the central venous catheter (CVC). This is used in situations as varied as the administration of blood products, gastroenterology (to deliver fluids and nutrients) and oncology (to administer cytotoxic drugs). Infection can occur at any time during the use of the catheter, which can last for several months. Microorganisms, most of which are staphylococci, *Klebsiella* spp. and *Candida albicans* [78], are able to enter and form biofilms on both the inside and outside of the catheter.

Urinary catheters are tubular latex or silicone devices, which when inserted may readily acquire biofilms on the inner or outer surfaces. The organisms commonly contaminating these devices and developing biofilms are *Staphylococcus epidermidis*, *E. faecalis*, *E. coli*, *P. mirabilis*, *P. aeruginosa*, *K. pneumoniae* and other gram-negative organisms. The longer the urinary catheter remains in place, the greater the tendency of these organisms to develop biofilms and result in urinary tract infections. For example, 10–50% of patients undergoing short-term urinary catheterisation (7 days) become infected, whereas virtually all patients undergoing long-term catheterisation (more than 28 days) become infected. Bacterial adhesion to catheter materials is dependent on the hydrophobicity of both the organisms and the surfaces [84]. Catheters display both hydrophobic and hydrophilic regions and thus allow colonisation of the widest variety of organisms. Divalent cations (calcium and magnesium) and increases in urinary pH and ionic strength all result in an increase in bacterial attachment. Certain component organisms of these biofilms produce urease, which hydrolyses the urea in the patient's urine to ammonium hydroxide. The elevated pH results in precipitation of minerals such as struvite and hydroxyapatite. These mineral-containing biofilms form encrustations that may completely block the inner lumen of the catheter.

11.7   Skin and soft tissues

11.7.1   Infective endocarditis

Infective endocarditis can be caused by any organism, bacterial, viral or fungal which can colonise the heart. The majority of infective endocarditis is caused by bacterial biofilms.

Infective endocarditis is potentially fatal and there are between 11 and 50 cases of infective endocarditis per million of the population per annum [85, 86] with a mortality rate between 6 and 40% [87, 88].

The infection usually involves damaged heart valves, e.g. congenital heart defects such as rheumatic heart disease, a ventricular septal defect or a prosthetic valve replacement [89]. These surfaces become colonised by biofilm which damages the valve tissue thereby altering the blood flow conditions and allowing platelet aggregation [90, 91].

Biofilms of streptococci are most common in native valve endocarditis, while prosthetic valve endocarditis tend to be infected with staphylococci [87]. Staphylococcal infections are more associated with fatality than infections caused by other taxa [87]. It is thought that most infective endocarditis is caused by bacteraemia and traditionally it was thought that many cases arose from dental treatment [92]; however, more recently patient oral health and chronic oral infection have been shown to be potentially more important in the initial of bacteriaemia leading to infective endocarditis.

The altered blood flow may allow formation of sterile vegetations composed of platelets these can trap bacteria circulating as bacteraemia and a thrombus can form. The crucial steps in this process are the binding of bacteria to the platelets either directly via bacterial surface proteins or indirectly via plasma bridging molecules. The activation of the platelets is a necessary step in thrombus formation. Bacteria can induce this directly or indirectly, e.g. via secreted bacterial products; however, the process is complex and poorly understood (for a review, see Fitzgerald *et al.* [93]). If untreated or unresponsive to treatment, the thrombus can grow and inhibit valve function and eventually lead to congestive heart failure.

11.7.2   *Wound infections*

There is increasing evidence that bacteria within chronic wounds live within biofilm communities [94]. Chronic wounds arise due to a breakdown in the processes governing normal wound healing and are often caused by one or many underlying conditions the patient may have which ultimately predisposes them to the condition [95]. Around 70% of all chronic wounds originate from one of three underlying causes, these being pressure sores, diabetic foot ulcers and venous ulcers. Although all three conditions have varying aetiology, the end result in the worst scenario is the formation of a chronic wound state which, despite the clinician's best efforts, fails to respond to treatment. A chronic wound environment, laden with characteristic features such as the build-up of wound exudate and necrotic tissue, creates a suitable environment for colonisation and continued proliferation of microorganisms [96]. Once attached, the proliferation and induction of virulence factors enable the microorganisms to overcome the hosts immune defence system and establish an infection state [97].

11.7.3   *Diabetic foot ulcers*

Foot infections are a major complication of diabetes mellitus and contribute to the development of gangrene and lower extremity amputation. Recent evidence indicates that people with diabetes are at greater risk for infection because of underlying neuropathy, peripheral vascular disease and impaired responses to infecting organisms. As we have already discussed throughout this chapter, biofilms are highly resistant to traditional therapy and this resistance has attracted interest as a potential reason why chronic wounds do not heal. This may be especially important for diabetic foot ulcers, which are often characterised by their refractory nature, their predisposition to have associated underlying infection and their improvement with debridement. Diabetic foot ulcers are long-term complications of type 2 diabetes or diabetes mellitus (T2DM). It has been predicted by the World Health Organization that T2DM will soon become a serious health problem worldwide. It is estimated that T2DM will become an epidemic with an estimated 150–220 million of diabetics in 2010 and 300 million in 2025 [98]. Diabetic foot ulcers are associated with high morbidity, they have unpleasant smell and are very painful. Diabetic patients are often unable to attend work, as standard treatment such as debridement needs to be performed on a regular basis. As a consequence, patients suffer psychological trauma and complain that they can no longer interact socially. If diabetic foot ulcers are not treated in time, there is a high risk of amputation and this may lead to re-duced quality of life. *Finegoldia magna* is the most common gram-positive anaerobic coccus (GPAC) that is isolated from diabetic foot ulcers [99] but these ulcers are often associated with polymicrobial infections with several microbes, e.g. *S. aureus* and *P. aeruginosa* also being implicated in infection. However, *F. magna* is a bacterium that is frequently overlooked in clinical samples as it is a slow-growing anaerobe and would often not be isolated if the samples were incorrectly processed.

11.8   Conclusions

- Until recently the significance of biofilms in the clinical setting, especially with regard to their role in medical-related infections, has been underestimated.
- In the last 5 years (up to 2008), the number of publications related to the subject of medical biofilms has risen almost threefold.
- It has been found that several aspects of human pathogenesis within a clinical context are directly related to biofilm development.
- Various types of surfaces in clinical settings are prone to biofilm development and an increased risk of disease may be a direct consequence of their formation.

References

1. Voorhees, A.B. (1985) The development of arterial prostheses. A personal view. *Archives of Surgery*, **120**, 289–295.
2. Beijerinck, M.W. (1888) Die Bacterien der Papilionaceenknöllchen. *Botanische Zeitung*, **46**, 725–804.
3. Schierholz, J.M. & Beuth, J. (2001) Implant infections: a haven for opportunistic bacteria. *Journal of Hospital Infection*, **49**, 87–93.
4. Potera, C. (1999) Microbiology-forging a link between biofilms and disease. *Science*, **283**, 1837–1838.
5. Götz, F. (2002) Staphylococcus and biofilms. *Molecular Microbiology*, **43**, 1367–1378.
6. Donlan, R.M. (2001) Biofilms and device-associated infections. *Emerging Infectious Diseases*, **7**, 277–281.
7. Ehrlich, G.D., Veeh, R., Wang, X., *et al.* (2002) Mucosal biofilm formation on middle-ear mucosa in the chinchilla model of otitis media. *JAMA*, **287**, 1710–1715.
8. Rayner, M.G., Zhang, Y., Gorry, M.C., Chen, Y., Post, J.C. & Ehrlich, G.D. (1998) Evidence of bacterial metabolic activity in culture-negative otitis media with effusion. *JAMA*, **279** (4), 296–299.

9. Hall-Stoodley, L., Hu, F.Z., Gieseke, A., *et al.* (2006) Direct detection of bacterial biofilms on the middle-ear mucosa of children with chronic otitis media. *JAMA*, **296** (2), 202–211.
10. Dart, J.K.G. (1987) Bacterial keratitis in contact lens users. *British Medical Journal*, **295**, 959–960.
11. Dart, J.K.G. (1988) Predisposing factor in microbial keratitis: the significance of contact lens wear. *British Journal of Ophthalmology*, **72**, 926–930.
12. Erie, J.C., Nevitt, M.P., Hodge, D.O. & Ballard, D.J. (1993) Incidence of ulcerative keratitis in a defined population from 1950 through 1988. *Archives of Ophthalmology*, **111** (12), 1665–1671.
13. Liesegang, T.J. (1997) Contact lens-related microbial keratitis: part II: pathophysiology. *Cornea*, **16** (3), 265–273.
14. Clark, B.J., Harkins, L.S., Munrow, F.A. & Devonshire, P. (1994) Microbial contamination of cases used for storing contact lenses. *Journal of Infection*, **28**, 293–304.
15. Aswad, M.I., John, T., Barza, M., Kenyon, K. & Baum, J. (1990) Bacterial adherence to extended wear soft contact lenses. *Opthalmology*, **97**, 296–302.
16. Ren, D.H., Petroll, W.M., Jester, J.V., Ho-Fan, J. & Cavanagh, H.D. (1999) Short-term hypoxia down regulates epithelial cell desquamation *in vivo*, but does not increase *Pseudomonas aeruginosa* adherence to exfoliated human corneal epithelial cells. *CLAO Journal*, **25** (2), 73–99.
17. Palmer, J. (2006) Bacterial biofilms in chronic rhinosinusitis. *The Annals of Otology, Rhinology & Laryngology*, **196**, 35–39.
18. Bendouah, Z., Barbeau, J., Hamad, W.A. & Desrosiers, M. (2006) Use of an *in vitro* assay for determination of biofilm-forming capacity of bacteria in chronic rhinosinusitis. *American Journal of Rhinolology*, **20**, 434–438.
19. Ferguson, B.J. & Stolz, D.B. (2005) Demonstration of biofilm in human bacterial chronic rhinosinusitis. *American Journal of Rhinolology*, **19**, 452–457.
20. Sanderson, A.R., Leid, J.G. & Hunsaker, D. (2006) Bacterial biofilms on the sinus mucosa of human subjects with chronic rhinosinusitis. *Laryngoscope*, **116**, 1121–1126.
21. Zuliani, G., Carron, M., Gurrola, J., *et al.* (2006) Identification of adenoid biofilms in chronic rhinosinusitis. *International Journal of Pediatric Otorhinolaryngology*, **70** (9), 1613–1617.
22. Chole, R.A. & Faddis, B.T. (2003) Anatomical evidence of microbial biofilms in tonsillar tissues: a possible mechanism to explain chronicity. *Archives of Otolaryngology – Head & Neck Surgery*, **129** (6), 634–636.
23. Kania, R.E., Lamers, G.E., Vonk, M.J., *et al.* (2007) Demonstration of bacterial cells and glycocalyx in biofilms on human tonsils. *Archives of Otolaryngology – Head & Neck Surgery*, **33** (2), 115–121.
24. Pine, C.M., Pitts, N.B., Steele, J.G., Nunn, J.N. & Treasure, E. (2001) Dental restorations in adults in the UK in 1998 and implications for the future. *British Dental Journal*, **190**, 4–8.
25. Beighton, D. & Lynch, E. (1995) Comparison of selected microflora of plaque and underlying carious dentine associated with primary root caries lesions. *Caries Research*, **29** (2), 154–158.
26. Schupbach, P., Osterwalder, V. & Guggenheim, B. (1996) Human root caries: microbiota of a limited number of root caries lesions. *Caries Research*, **30**, 52–64.
27. Armitage, G.C. (1999) Development of a classification system for periodontal diseases and conditions. *Annals of Periodontology*, **4**, 1–6.
28. Haffajee, A.D. & Socransky, S.S. (1994) Microbial etiological agents of destructive periodontal diseases. *Periodontology 2000*, **5**, 78–111.
29. Jeffcoat, M.K., McGuire, M. & Newman, M.G. (1997) Evidence-based periodontal treatment. Highlights from the 1996 World Workshop in Periodontics. *Journal of American Dental Association*, **128**, 713–724.
30. Tanner, A., Maiden, M.F., Paster, B.J. & Dewhirst, F.E. (1994) The impact of 16S ribosomal RNA-based phylogeny on the taxonomy of oral bacteria. *Periodontology 2000*, **5**, 26–51.
31. Spratt, D.A., Weightman, A.J. & Wade, W.G. (1999) Diversity of oral asaccharolytic Eubacterium species in periodontitis – identification of novel phylotypes representing uncultivated taxa. *Oral Microbiology & Immunology*, **14**, 56–59.
32. Kornman, K.S., Crane, A., Wang, H.Y., *et al.* (1997) The interleukin-1 genotype as a severity factor in adult periodontal disease. *Journal of Clinical Periodontology*, **24**, 72–77.

33. Nibali, L., Ready, D.R., Parkar, M., *et al.* (2007) Gene polymorphisms and the prevalence of key periodontal pathogens. *Journal of Dental Research*, **86**, 416–420.

34. Holm, G. (1994) Smoking as an additional risk for tooth loss. *Journal of Periodontology*, **65**, 996–1001.

35. Oliver, R.C. & Tervonen, T. (1994) Diabetes – a risk factor for periodontitis in adults? *Journal of Periodontology*, **65**, 530–538.

36. Oggioni, M.R., Trappetti, C., Kadioglu, A., *et al.* (2006) Switch from planktonic to sessile life: a major event in pneumococcal pathogenesis. *Molecular Microbiology*, **61** (5), 1196–1210.

37. Augustyn, B. (2007) Ventilator-associated pneumonia risk factors and prevention. *Critical Care Nurse*, **27** (4), 32–39.

38. Gomez, M.I. & Prince, A. (2007) Opportunistic infections in lung disease: pseudomonas infections in cystic fibrosis. *Current Opinions in Pharmacology*, **7** (3), 244–251.

39. Nichols, W.W., Dorrington, S.M., Slack, M.P. & Walmsley, H.L. (1988) Inhibition of tobramycin diffusion by binding to alginate. *Antimicrobial Agents & Chemotherapy*, **32** (4), 518–523.

40. Blaser, M.J. & Atherton, J.C. (2004) *Helicobacter pylori* persistence: biology and disease. *Journal of Clinical Investigation*, **113**, 321–333.

41. Kuipers, E.J., Thijs, J.C. & Festen, H.P. (1995) The prevalence of *Helicobacter pylori* in peptic ulcer disease. *Alimentary Pharmacology & Therapeutics*, **9**, 59–69.

42. Coticchia, J.M., Sugawa, C., Tran, V.R., Gurrola, J., Kowalski, E. & Carron, M.A. (2006) Presence and density of *Helicobacter pylori* biofilms in human gastric mucosa in patients with peptic ulcer disease. *Journal of Gastrointestinal Surgery*, **10** (6), 883–889.

43. Ching, C.K., Wong, B.C., Kwok, E., Ong, L., Covacci, A. & Lam, S.K. (1996) Prevalence of CagA-bearing *Helicobacter pylori* strains detected by the anti-CagA assay in patients with peptic ulcer disease and in controls. *American Journal of Gastroenterology*, **91**, 949–953.

44. Cover, T.L. & Blaser, M.J. (1992) Purification and characterization of the vacuolating toxin from *Helicobacter pylori*. *Journal of Biological Chemistry*, **267**, 10570–10575.

45. Cover, T.L. & Blanke, S.R. (2005) *Helicobacter pylori* VacA, a paradigm for toxin multifunctionality. *Nature Reviews: Microbiology*, **3**, 320–332.

46. Hennig, E.E., Godlewski, M.M., Butruk, E. & Ostrowski, J. (2005) *Helicobacter pylori* VacA cytotoxin interacts with fibronectin and alters HeLa cell adhesion and cytoskeletal organization *in vitro*. *FEMS Immunology and Medical Microbiology*, **44**, 143–150.

47. Kusters, J.G., van Vliet, A.H. & Kuipers, E.J. (2006) Pathogenesis of *Helicobacter pylori* infection. *Clinical Microbiology Reviews*, **19**, 449–490.

48. Russo, T.A. & Johnson, J.R. (2003) Medical and economic impact of extraintestinal infections due to *Escherichia coli*: focus on an increasingly important endemic problem. *Microbes and Infection*, **5**, 449–456.

49. Ronald, A. (2002) The etiology of urinary tract infection: traditional and emerging pathogens. *American Journal of Medicine*, **113**, 14–19.

50. Stapleton, A. & Stamm, W.E. (1997) Prevention of urinary tract infection. *Infectious Disease Clinics of North America*, **11**, 719–733.

51. Anderson, G.G., Palermo, J.J., Schilling, J.D., Roth, R., Heuser, J. & Hultgren, S.J. (2003) Intracellular bacterial biofilm-like pods in urinary tract infections. *Science*, **301**, 105–107.

52. Justice, S.S., Hung, C., Theriot, J.A., *et al.* (2004) Differentiation and developmental pathways of uropathogenic *Escherichia coli* in urinary tract pathogenesis. *Proceedings of the National Academy of Sciences of the United States of America*, **101**, 1333–1338.

53. Dodson, K.W., Pinkner, J.S., Rose, T., Magnusson, G., Hultgren, S.J. & Waksman, G. (2001) Structural basis of the interaction of the pyelonephritic *E. coli* adhesin to its human kidney receptor. *Cell*, **105**, 733–743.

54. Hung, C.S., Bouckaert, J., Hung, D., *et al.* (2002) Structural basis of tropism of *Escherichia coli* to the bladder during urinary tract infection. *Molecular Microbiology*, **44**, 903–915.

55. Guyer, D.M., Henderson, I.R., Nataro, J.P. & Mobley, H.L. (2000) Identification of sat, an auto-transporter toxin produced by uropathogenic *Escherichia coli*. *Molecular Microbiology*, **38**, 53–66.

56. Guyer, D.M., Radulovic, S., Jones, F.E. & Mobley, H.L. (2002) Sat, the secreted autotransporter toxin of uropathogenic *Escherichia coli*, is a vacuolating cytotoxin for bladder and kidney epithelial cells. *Infection & Immunity*, **70**, 4539–4546.

57. Heimer, S.R., Rasko, D.A., Lockatell, C.V., Johnson, D.E. & Mobley, H.L. (2004) Autotransporter genes pic and tsh are associated with *Escherichia coli* strains that cause acute pyelonephritis and are expressed during urinary tract infection. *Infection & Immunity*, **72**, 593–597.

58. Stanley, P., Koronakis, V. & Hughes, C. (1998) Acylation of *Escherichia coli* hemolysin: a unique protein lipidation mechanism underlying toxin function. *Microbiology & Molecular Biology Reviews*, **62**, 309–333.

59. Mills, M., Meysick, K.C. & O'Brien, A.D. (2000) Cytotoxic necrotizing factor type 1 of uropathogenic *Escherichia coli* kills cultured human uroepithelial 5637 cells by an apoptotic mechanism. *Infection & Immunity*, **68**, 5869–5880.

60. Johnson, J.R. (2003) Microbial virulence determinants and the pathogenesis of urinary tract infection. *Infectious Disease Clinics of North America*, **17**, 261–278.

61. Russo, T.A., Carlino, U.B., Mong, A. & Jodush, S.T. (1999) Identification of genes in an extraintestinal isolate of *Escherichia coli* with increased expression after exposure to human urine. *Infection & Immunity*, **67** (10), 5306–5314.

62. Sorsa, L.J., Dufke, S., Heesemann, J. & Schubert, S. (2003) Characterization of an iroBCDEN gene cluster on a transmissible plasmid of uropathogenic *Escherichia coli*: evidence for horizontal transfer of a chromosomal virulence factor. *Infection & Immunity*, **71**, 3285–3293.

63. Ness, R.B., Kip, K.E., Hillier, S.L., *et al.* (2005) A cluster analysis of bacterial vaginosis-associated microflora and pelvic inflammatory disease. *American Journal of Epidemiology*, **162**, 585–590.

64. Larsson, P.G., Bergstrom, M., Forsum, U., Jacobsson, B., Strand, A. & Wolner-Hanssen, P. (2005) Bacterial vaginosis. Transmission, role in genital tract infection and pregnancy outcome: an enigma. *APMIS*, **113**, 233–245.

65. Hillier, S.L., Nugent, R.P., Eschenbach, D.A., *et al.* (1995) Association between bacterial vaginosis and preterm delivery of a lowbirth-weight infant. The Vaginal Infections and Prematurity Study Group. *New England Journal of Medicine*, **333**, 1737–1742.

66. Alvarez-Olmos, M.I., Barousse, M.M., Rajan, L., *et al.* (2004) Vaginal lactobacilli in adolescents: presence and relationship to local and systemic immunity, and to bacterial vaginosis. *Sexually Transmitted Diseases*, **31**, 393–400.

67. Hillier, S.L., Krohn, M.A., Rabe, L.K., Klebanoff, S.J. & Eschenbach, D.A. (1993) The normal vaginal flora, H_2O_2-producing lactobacilli, and bacterial vaginosis in pregnant women. *Clinical Infectious Diseases*, **16**, 273–281.

68. Eschenbach, D.A., Davick, P.R., Williams, B.L. *et al.* (1989) Prevalence of hydrogen peroxide-producing *Lactobacillus* species in normal women and women with bacterial aginosis. *Journal of Clinical Microbiology*, **27**, 251–256.

69. Catlin, B.W. (1992) *Gardnerella vaginalis*: characteristics, clinical considerations, and controversies. *Clinical Microbiology Reviews*, **5**, 213–237.

70. Sobel, J.D. (2000) Bacterial vaginosis. *Annual Review of Medicine*, **51**, 349–356.

71. Patterson, J.L., Girerd, P.H., Karjane, N.W. & Jefferson, K.K. (2007) Effect of biofilm phenotype on resistance of *Gardnerella vaginalis* to hydrogen peroxide and lactic acid. *American Journal of Obstetrics and Gynecology*, **197** (2), 1–10.

72. Gardner, H.L. & Dukes, C.D. (1955) *Haemophilus vaginalis*: a newly defined specific infection previously classified non-specific vaginitis. *American Journal of Obstetrics and Gynecology*, **69**, 962–976.

73. Criswell, B.S., Ladwig, C.L., Gardner, H.L. & Dukes, C.D. (1969) *Haemophilus vaginalis*: vaginitis by inoculation from culture. *Obstetrics and Gynecology*, **33**, 195–199.

74. Arici, M., Bilgin, B., Sagdic, O. & Ozdemir, C. (2004) Some characteristics of Lactobacillus isolates from infant faeces. *Food Microbiology*, **21**, 19–24.

75. Aroutcheva, A., Gariti, D., Simon, M., *et al.* (2001) Defense factors of vaginal lactobacilli. *American Journal of Obstetrics and Gynecology*, **185**, 375–379.

76. Young, S.E. (1987) Aetiology and epidemiology of infective endocarditis in England and Wales. *Journal of Antimicrobial Agents and Chemotherapy*, **20**, 7–15.

77. Skarin, A. & Sylwan, J. (1986) Vaginal lactobacilli inhibiting growth of *Gardnerella vaginalis*, *Mobiluncus* and other bacterial species cultured from vaginal content of women with bacterial vaginosis. *Acta Pathologica, Microbiologica, et Immunologica Scandinavica*, **94**, 399–403.

78. Dembele, T., Obdrzalek, V. & Votava, M. (1998) Inhibition of bacterial pathogens by lactobacilli. *Zentralblatt für Bakteriologie*, **288**, 395–401.

79. Bayston, R. (1999) Biofilms in medicine and disease: an overview. In: *Biofilms* (eds J. Wimpenny, P. Gilbert, J. Walker, M. Brading & R. Bayston), pp. 1–6. Bioline, Cardiff.

80. Darouiche, R.O. (2003) Antimicrobial approaches for preventing infections associated with surgical implants. *Clinical Infectious Diseases*, **36**, 1284–1289.

81. Furno, F. & Bayston, R. (2004) Antimicrobial/antibiotic (infective resistance) materials. In: *Encyclopedia of Biomaterials and Biomedical Engineering* (eds G.L. Bowlin & G. Wnek), pp. 34–42. Marcel Dekker, New York.

82. Burrows, L.L. & Khoury, A.E. (2004) Infection of medical devices. In: *Encyclopedia of Biomaterials and Biomedical Engineering* (eds G.L. Bowlin & G. Wnek), pp. 839–848. Marcel Dekker, New York.

83. Berbari, E.F., Hanssen, A.D., Duffy, M.C., *et al.* (1998) Risk factors for prosthetic joint infection: case control study. *Clinical Infectious Diseases*, **27**, 1247–1254.

84. Bernard, L., Hoffmeyer, P., Assal, M., Vaudaux, P., Schrenzel, J. & Lew, D. (2004) Trends in the treatment of orthopaedic prosthetic infections. *Journal of Antimicrobial Agents and Chemotherapy*, **53**, 127–129.

85. Brisset, L., Vernet-Garnier, V., Carquin, J., Burde, A., Flament, J.B. & Choisy, C. (1996) *In vivo* and *in vitro* analysis of the ability of urinary catheters to microbial colonization. *Pathologie-Biologie*, **44**, 397–404.

86. Pallasch, T.J. & Slots, J. (1996) Antibiotic prophylaxis and the medically compromised patient. *Periodontology 2000*, **10**, 107–138.

87. Van Der Meer, J.T., Thompson, J., Valkenburg, H.A. & Michel, M.F. (1992) Epidemiology of bacterial endocarditis in The Netherlands. I. Patient characteristics. *Archives of Internal Medicine*, **152**, 1863–1868.

88. Franklin, C.D. (1992) The aetiology, epidemiology, pathogenesis and changing pattern of infective endocarditis, with a note on prophylaxis. *British Dental Journal*, **172**, 369–373.

89. Smith, A.J. & Adams, D. (1993) The dental status and attitudes of patients at risk from infective endocarditis. *British Dental Journal*, **174**, 59–64.

90. Glauser, M.P. & Francioli, P. (1987) Relevance of animal models to the prophylaxis of infective endocarditis. *Journal of Antimicrobial Agents and Chemotherapy*, **20**, 87–93.

91. Fluckiger, U.P., Moreillon, J., Blaser, M., Bickle, M.P. Glauser, P. & Francioli, P. (1994) Simulation of amoxicillin pharmacokinetics in humans for the prevention of streptococcal endocarditis in rats. *Antimicrobial Agents and Chemotherapy*, **38**, 2846–2849.

92. Wahl, M.J. (1994) Myths of dental-induced endocarditis. *Archives of Internal Medicine*, **154**, 137–144.

93. Fitzgerald, J.R., Foster, T.J. & Cox, D. (2006) The interaction of bacterial pathogens with platelets. *Nature Reviews. Microbiology*, **4**, 445–457.

94. Edwards, R. & Harding, K.G. (2004) Bacteria and wound healing. *Current Opinions in Infectious Diseases*, **17** (2), 91–96.

95. Stadelmann, W.K., Digenis, A.G. & Tobin, G.R. (1998) Impediments to wound healing. *The American Journal of Surgical Pathology*, **176**, 39–47.

96. Mertz, P.M. & Ovington, L.G. (1993) Wound healing microbiology. *Clinical Dermatology*, **11**, 739–747.
97. Bowler, P.G., Duerden, B.I. & Armstrong, D.G. (2001) Wound microbiology and associated approaches to wound management. *Clinical Microbiology Reviews*, **14**, 244–269.
98. Zimmet, P., Alberti, K.G.M.M. & Shaw, J. (2001) Global and societal implications of the diabetic epidemic. *Nature*, **414**, 782–787.
99. Wren, M.W.D (1996). Anaerobic cocci of clinical importance. *British Journal of Biomedical Science*, **53**, 294–301.

Chapter 12
Fouling on Artificial Substrata

Antonio Terlizzi and Marco Faimali

Biofouling on artificial substrata results from several processes, whose rate and extent are influenced by the intertwining of numerous physical, chemical and biological factors in the immediate proximity of the surface. The importance of substratum features in influencing species settlement is considered here. An emphasis is given on how biological systems (micro- and macrofouling) can interplay with the nature of substratum in regulating patterns of biofouling development. The environmental issues related to the deployment of man-made structures in coastal waters are also discussed, and some guidelines on the use of artificial substrata in the management strategies for controlling fouling in industry are provided.

12.1 The influence of the nature of artificial substrata on fouling assemblages

When exposed to water, artificial substrata are quickly conditioned by adsorbing organic compounds. The impact of conditioning films on microbial attachment and growth has been widely debated. Some evidence suggests that conditioning films have a role in the degree and pattern of oral biofilm development (see Chapter 12; [1]). However, other studies have demonstrated that, even in the presence of adsorbed nutrients, the nature of the substratum is more important to bacterial adhesion than the conditioning film and that there is little or no correlation between protein adsorption on surfaces and bacterial adhesion [2, 3].

Substratum properties, in terms of roughness, wettability, surface tension and polarisation, may regulate patterns of cell accumulation and cell distribution during the early stages of biofilm development. Gerchakov *et al.* [4] demonstrated that in sea water the first bacterial attachment is faster on glass and 304 stainless steel surfaces than on copper-zinc brass and copper-nickel alloy. Pasmore *et al.* [5] studied the effects of polymer surface roughness, hydrophobicity and surface charge on the attachment of *Pseudomonas aeruginosa* in the early stages of biofilm development and concluded that cells were most readily removed from the smoothest, most hydrophilic, neutral surfaces, but that these properties did not influence the amounts of developed biofilm. Bacteria are known to attach more rapidly to hydrophobic, non-polar surfaces such as Teflon and other plastics than to hydrophilic materials such as glass or metals [6].

Although reduced adhesion to rough surfaces has been reported, high surface roughness is generally considered to increase the extent of bacterial accumulation. Adhesion occurs at surface irregularities: the absence of micro cracks and crevices can significantly decrease microbial biomass. Nickels *et al.* [7] reported differences in biomass and structure of biofilm assemblages colonising silica grains of the same size and water pore space and differing in microtopography only. Characklis *et al.* [8] noted that when higher surfaces are available, microbial colonisation increases with surface roughness. Medilanski *et al.* [9] evaluated the influence of surface topography on colonisation of stainless steel surface finishes at roughness values between 0.03 and 0.89 μm and reported a higher rate of microbial colonisation on the roughest surfaces.

The increased surface area provided at the microorganism–material interface might facilitate film attachment by providing more contact points [10]. Little *et al.* [11] demonstrated that porous welds increase sites for colonisation in respect to smooth pipe surfaces. Surface roughness is also crucial in cell retention on surfaces [12]. If surface irregularities are much larger than the microorganisms, passive retention is minimal. During an experiment evaluating the impact of a shift of applied potential between −0.5 and +0.5 V on the adsorption of two bacterial strains to platinum, titanium, stainless steel, copper, aluminium alloy and carbon steel, Armon *et al.* [13] found that bacterial adsorption gradually decreases towards either the positive or negative direction of polarisation.

The physical nature of surfaces – in terms of roughness, thermal capacity, colour, charge, elemental and organic composition – has been also shown to affect the settlement of a variety of macro- and microscopic marine organisms during biofouling community development [14–19].

Various surface features, namely texture [20–22], composition [23–27], colour [28, 29], structure [30], surface chemistry [31–33], surface tension [34], mechanical properties [35], low surface energy [36], interfacial alkalinity [37] and surface wettability [38] influence the species composition and relative abundance of macrofouling (see Chapter 3). Most of the above studies concur that a given effect of a particular material can be identified under laboratory conditions using a single model species. However, in situ experiments show that substratum nature influences the settlement of micro- and macrofoulers only in early stages of colonisation, being masked and complicated soon after the first exposure, and becoming modulated by complex interactions of environmental variables including biological, chemical and physical cues, light, food availability and the presence of conspecific adults. Furthermore, because the effect of substratum varies with the period of submersion, it is difficult to compare studies on fouling assemblages using different natural and artificial substrata and for varying lengths of time [25]. Glasby [17] also suggested caution in generalising about effects of surface composition and orientation because they may interact with each other and/or with other factors, being quite different for different taxa and between sites.

12.2   Environmental, physical–chemical and biological interactions during fouling colonisation of artificial substrata

The establishment of biofouling is a competitive race for a substratum by various organisms, determined not only by the relative concentration of potential biofoulers in the water column near the surface [39], but also by several complex environmental, physical–chemical and biological interactions during substratum colonisation.

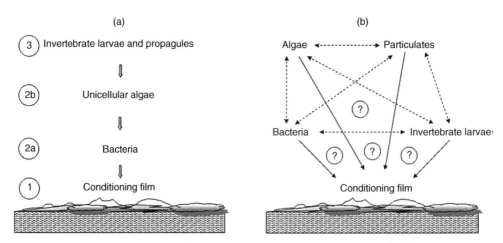

Figure 12.1 Diagrams of two models for biofouling succession. (a) The classical view of successional biofouling on a substratum. Each fouling stage is a step that precedes the following stage and causality between levels is implied. (b) Surface biofouling based on the probability that a particular biofouling organism will encounter the substratum. After formation of the conditioning film, further colonisation will depend on which type of organism encounters the surface. In the absence of a substratum, molecules and organisms may attach to each other and participate in the formation of marine snow. (Adapted from Reference [39].)

Two main models describe the establishment of biofouling. The classical view (Figure 12.1(a)) refers to fouling as a succession of three stages [40, 41]: molecular biofouling adsorption, biofilm formation and macrofouling development. The successional model, implying causality from stage to stage, may only be found under particular circumstances [42]. A widely accepted model of fouling processes is dynamic, or probabilistic (Figure 12.1(b)) [39, 41]. All fouling stages run continuously, leading to dynamic and complex interactions between water and substratum, water and specific biofouling organisms and interspecifically among biofouling organisms, which, again, may interact with physical forces such as water flow or gravitation (see also Chapters 4 and 7).

Prediction of the environmental, physical–chemical and biological interaction during microfouling colonisation processes in the field is almost impossible. These consist of attached bacteria, yeasts, unicellular algae, fungi and protozoa, all enmeshed in a matrix of extracellular polymers [43]. The extracellular matrix is generally composed of water and microbial macromolecules [44] and provides a complex array of microenvironments, such as open spaces or channels [45] (see also Chapters 9–11). Microorganisms within biofilms can maintain radically different microenvironments at biofilm–surface interfaces from the surrounding ones in terms of pH, dissolved oxygen and other organic and inorganic species. In some cases, these interfacial conditions could not be maintained in the bulk medium at room temperature near atmospheric pressure, showing that microorganisms within biofilms produce reactions that are not predicted by the chemistry of the bulk medium [46]. Biofouling, therefore, cannot be avoided by eliminating the initial stages of colonisation, but changes in the initial biofilm may influence the quality and/or composition of its communities. For sessile organisms, larval settlement determines future growth and survival [47]. Settlement responses of invertebrate larvae to natural chemical cues vary from induced settlement and metamorphosis to settlement inhibition [48]. Studies on the biotic factors influencing larval settlement on substrata

(a)

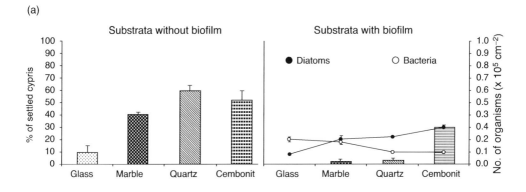

(b)

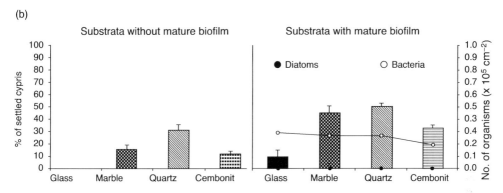

Figure 12.2 Effect of biofilm developed in laboratory (a) and field (b) on settlement of *Balanus amphitrite* larvae on different kinds of natural (marble, quartz) and artificial (glass, cembonit) substrata.

intensively concentrated on biofilm presence [49–57]. The (either negative or positive) effect of biofilms on larval settlement could depend on substratum features. However, few authors have stressed the interaction between substratum and biofilm in affecting larval settlement [58]. The cypris of *Balanus amphitrite* (Crustacea, Cirripedia) is an excellent model to study the mechanisms underlying chemical and physical cue detection during settlement [23,59–61] including the role of biofilms [54,62–64] (see Chapter 3).

Recently, experimental studies on the interplay between the nature of the substratum and biofilm formation in regulating *B. amphitrite* larval settlement suggested that settlement selectivity of cyprids for the different substrata is affected mainly by experimental conditions [19]. A laboratory-developed biofilm, dominated by diatoms, determines an inhibition of settlement, whereas a field-developed biofilm, dominated by bacteria, stimulates settlement (Figure 12.2). A further experiment, testing settlement at sea, showed that after 1 month of immersion settlement is influenced not only by the presence of biofilm, but also by competition with other macroorganisms (Figure 12.3). Serpulids settle first, showing a preference for rougher substrata (marble, quartz and cembonit), while barnacles settle later on smoother substrata (glass) with higher bacterial density and lower serpulid settlement. Under natural condition, thus, barnacle settlement is affected by interspecific competition that modifies direct larva–substratum interactions, and every mutual interaction can play a key role. The effect of substratum is

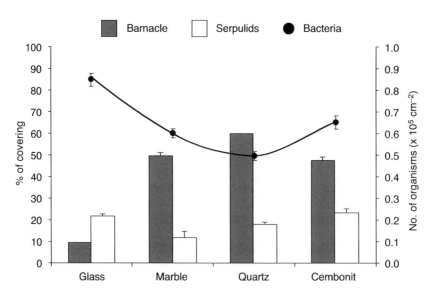

Figure 12.3 Settlement of *Balanus amphitrite* and *Hydroides elegans* on different kinds of natural (marble, quartz) and artificial (glass, cembonit) substrata after 1 month of immersion in the field.

modulated by the presence of the biofilm under both laboratory and natural conditions (Figure 12.4).

Differences in natural versus laboratory conditions can explain some contrasting results obtained in laboratory and field studies [59, 64]. Signal molecules from conspecifics, or substances (also toxic) present in the aqueous system, could be absorbed by the extracellular polymeric substances produced by biofilm, interfering with settlement [65]. This could modify their chemical–physical properties of substrata, consequently changing the outcome of their interaction with larvae.

Besides substratum features, physical factors influencing biofouling development include water flow, light, gravity, temperature, salinity, pressure and spatial orientation (see Chapters 2 and 3). Water flow may affect dispersal and settlement, exerting hydrodynamic forces on settling propagules, as a settlement cue inducing active behaviour of motile propagules, and as a mediating factor affecting various other settlement cues [66]. Larval supply is more important than substratum microhabitat in determining the development of fouling assemblages [67, 68]. Investigations on the effects of light on marine organisms often focused on plants, simply because light intensity directly influences photosynthesis. Early discussions about light influences on subtidal organisms were based on comparisons of assemblages along depth gradients [69, 70]. However, many factors other than light intensity can change with depth [71, 72]. Glasby [73] remarked that numerous urban structures, such as marinas, have been added to the bays and estuaries around coastal cities, increasing natural shading of surrounding marine habitats. Moreover, marinas provide many shaded hard substrata in the form of pilings, pontoons and boats. Assemblages of epibiota on pilings at marinas differ markedly and consistently from those on nearby rocky substrata [74]. Different degrees of shading could explain differences in the cover of many organisms growing on pier pilings and on adjacent rocky reefs. Glasby and Connell [75] demonstrated the importance of substratum orientation in the development of epibiotic assemblages on vertical versus horizontal surfaces.

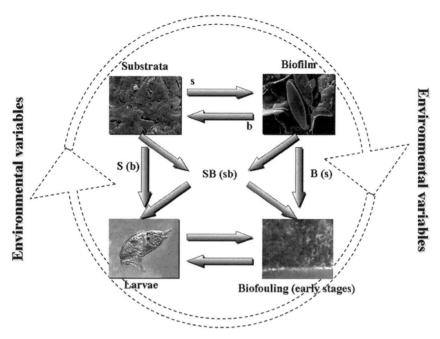

Figure 12.4 A model of interactions between larvae, biofilm and substratum during larval settlement. Environmental variables have the potential to modify the biofilm–substratum interaction, affecting the presence and growth of biofilm (b) which, in turn, limits the larval perception to substratum cues, altering the direct effect of substratum on settlement (S). At the same time, the substratum effect on biofilm specific composition (S) can differ depending on environmental variables and, consequently, modify the direct effect of biofilm on larval settlement (B). In the natural environment, every submerged substratum is rapidly (hours) covered by biofilm, making undistinguishable the role of biofilm and substratum in regulating larval settlement. The interaction with larval settlement should be therefore interpreted as the overlapping of both factors (SB).

The season of first submersion and the length of submersion affect the colonisation of artificial substrata because of seasonal fluctuations in larval abundances and biological interactions between earlier and later settlers. The effects of substratum, its submersion season and the submersion period on the development of subtidal biofouling assemblages on four types of artificial surfaces, including steel, stainless steel, cathodically protected steel and concrete, have been recently examined for a period of 18 months in a subtropical harbour by Lin and Shao [26]. Results indicate that both submersion season and submersion period are more important than substratum type in the development of subtidal biofouling assemblages.

The spatial scale influences the structure of assemblages recruiting to artificial substrata [76]. Variation in recruitment patterns of biofoulers over both small and large spatial scales is common in biofouling assemblages, calling for consideration in all experiments with artificial substrata. Small-scale differences in the development of biofouling assemblages on experimental plates are often found in field experiments on efficacy of antifouling coatings [77] (Figure 12.5). Therefore, only carefully designed experiments and adequate replication can ensure reliable results (see Chapter 22).

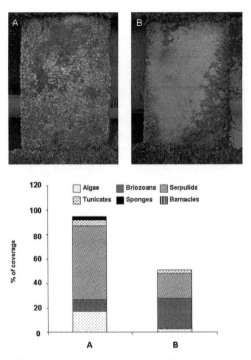

Figure 12.5 Small-scale differences in the development of biofouling assemblages after 236 days of immersion (Genoa harbour, Italy) on two replicates of experimental panels (painted with experimental AF paint) placed in the same frame at 80 cm of distance during efficacy test (raft test).

12.3   Man-made structures as extra habitat for biofouling organisms

Concrete dikes, artificial reefs, aquaculture cages, undersea storage tanks, offshore platforms, pontoons, retaining walls, pipelines, industrial cooling water systems and many artificial structures are continuously added to coastal waters worldwide. Their surfaces produce 'extra-habitat', quite distinct from natural rocky reefs [78]. In some cases, man-made structures, in form of artificial reefs, can be adopted as a useful tool in the conservation and management of coastal ecosystems [79]. In the Mediterranean Sea, for instance, *Posidonia oceanica* seagrass beds are severely damaged by illegal fish trawling. Hence, submerged artificial reefs have been adopted as an effective physical deterrent to trawling activities. The structures do not just control illegal fishing activities but have the potential to enhance fish biomass and diversity [80]. Large limestone boulders, along with sponges and coral transplanting, have been also successfully adopted in coral reef restoration policies [81].

Biofouling assemblages on artificial substrata have been studied for many decades, not only to prevent biofouling [82, 83], but also as empirical models to study community succession and its underlying mechanisms [25, 84–92]. Artificial substrata have been widely used to test hypotheses stemming from ecological theory. Such studies typically involve suspending settlement panels from floating rafts or pontoons to test the effects of particular factors on the development of biofouling assemblages [93–97].

Care should be paid when using artificial substrata to infer on naturally occurring assemblages. We have seen, in fact, that assemblages on artificial surfaces often differ from those on nearby natural substrata [24, 74, 76, 98, 99]. Being an artificially selected subset of adjacent, natural communities, biofouling assemblages may not be representative of natural situations [100, 101].

Local communities, in their turn, might be influenced by allochthonous species coming from artificial substrata. Ocean-going vessels, for instance, are artificial islands hosting species that dwell in harbours and estuaries around the world, transporting them across different biogeographic regions [102, 103]. The overall surface covered by these artificial islands is enormous. The US bioregions, for instance, are interested each year by the arrivals of 438 million square metre of wetted hull surface area, which is comparable to the total surface of the islands of the Seychelles archipelago. Vessels represent settlement substratum for biofouling species, provide protected recesses that both sessile and mobile fauna can occupy, and their ballast tanks can entrain everything from plankton to fish [104]. Invasions can occur when these transported organisms come in contact with structures in a new port or release larvae into its waters. Under proper conditions, these invaders may establish in the new port and spread to nearby areas. Such introductions represent a threat to biodiversity, alter local community composition, influence the performance of ecosystems and can cause significant economic costs. Vectors contributing to the introduction of alien species are poorly understood and there have been few quantitative accounts of the role that ship traffic plays in moving species from foreign ports [105]. Biofouling invasions are covered in more detail in Chapter 24.

The influence of artificial substrata in modifying the patterns of competition between autochthonous and allochthonous species has been recently highlighted. Tyrrell and Byers [106] predicted that exotic species on average should be more abundant on artificial substrata than on natural ones. The experimental test of potential systematic trends in exotic versus native species richness and abundance in biofouling communities as a function of settlement substratum showed that exotic species increased dramatically, whereas native species declined. Artificial surfaces may therefore provide a novel context for competitive interactions, giving additional chances to exotic species in an environment for which they otherwise might not be as well adapted compared to native species. Additions of artificial substrata to nearshore environments may disproportionately favour exotic species by increasing local sources of exotic propagules to colonise all types of substrata [107].

In order not to further enhance the spread of exotic species, coastal managers should consider limiting the amount of submerged artificial substrata and avoiding destruction of natural hard substrata in coastal and estuarine habitats [108].

12.4   Artificial substrata for biofouling monitoring in cooling water systems

Biofouling can cause serious operational and maintenance problems to the cooling water system of power plants (see Chapter 19). A wide range of organisms colonise cooling water plants. Microorganisms with the lowest grade of organisation, especially bacteria, are the first to colonise heat exchangers and other surfaces. The resulting layer of slime reduces heat exchange efficiency thereby significantly increasing energy costs. Corrosion is also increased within the anaerobic conditions under slime layers. Macrofouling usually colonises intake structures, cooling water intake tunnels, culverts, condenser tube plates and occasionally discharge tunnels (Plate IV A–D). A wide range of macrofouling organisms is found, but the main culprits are mussels, barnacles and serpulids (hard biofouling).

Problems caused by hard biofoulers are of multiple natures: encrusting growths may be so thick to reduce water flow, while pump head losses are increased. The detachment of mussel shells lodged in condenser tubes leads to long-term erosion-corrosion, hence to salt contamination of inflowing water which, in turn, causes corrosion of boiler tubes and turbine blades. Shell filters prevent mussel shells from reaching condenser tubes. However, the actual impact on plant efficiency is not yet fully understood. Changes in the type of treatment, changes in environmental factors, or accidental plant stoppages often allow for the settlement of fouling even in a very short time.

In some cases, the presence of macrofouling in the intake and/or dumping structures represent the main issue, while in others, the biofilm growth on the condenser tubes (microfouling) is more relevant. Biofouling in the plants considerably increases running costs resulting from a serious decline in plant efficiency. Injection of biocides (chemical control) is the most used among the available biofouling control technologies of industrial water treatment.

Consolidated antifouling strategies for water treatment in industrial plants are often ineffective against the adaptability of a benthic population to changed environmental features. The lack of adequate knowledge on the life cycles and population dynamics of macrofoulers, at a specific site, can cause antifouling treatment failure. Monitoring the presence of biofouling organisms, both before and after treatment, is crucial for a successful antifouling programme. With a biological macrofouling monitoring plan, online adjustments are possible to the chemical treatment strategy followed by the plant management, by changing the type of treatment, antifouling products and treatment schedule.

Faimali *et al.* [109] proposed a 'basic model' for proper water treatment management of plants with different technical, geographical and environmental characteristics and subject to different regulations. The procedure described is just an example of how the use of artificial substrata can be of primary importance in defining environmentally acceptable antifouling technologies. The model involves weekly and monthly collection of benthic and planktonic samples from several sites located within and outside the industrial plants to characterise the time of settlement and to plan the optimisation of antifouling treatment with specific biocides. Different monitoring plans can be built to meet local requirements. This approach has both economic and environmental benefits, allowing the drastic reduction of the environmental impact of water treatment on coastal marine ecosystems.

Selection of artificial substrata to use for a quantitative and qualitative characterisation of settlers is crucial in the monitoring program. The choice of unsuitable substrata often fails to adequately represent the time of settlement and can make biocide treatment useless. Substrata made with different materials and of different types depending on the life cycle of the target organism to check can be used. For barnacles and serpulids, panels immersed with a special support structure are generally used. For mussels, collectors for larvae are employed. They consist of support frames for ropes and nets in mixed fibre (synthetic and natural), which are very similar to those employed in mussel farming. Substrata must be kept in natural sea water before exposure to allow a sufficient microfouling growth to stimulate larval metamorphosis and settlement.

As a general guideline, the multi-substratum approach is suggested (Plate IV E) to obtain a reliable portrait of the potential settlers and their changes in abundance over time. Results obtained in an Italian petrochemical plant (ERG Raffinerie Mediterranee, Priolo Gargallo, Siracusa, Italy) following this monitoring strategy for the optimisation of the treatment with a specific product for adult settled molluscs used for shock treatment (MT-200$^{\circledR}$, NALCO) revealed the importance of the substratum/larva interaction. In monthly samplings, ropes and net proved to be the substrata better recording bivalve larval settlement (target organism). Conversely, for 3-monthly, 6-monthly and annual sampling, hard substrata such as cembonit and plastic were more efficient in recording settlement. The progressive decay of ropes and net, in fact, prevented the quantification of settlement over longer periods. The use of different materials, therefore, allows optimising sampling in relation to the time of immersion of sampling surfaces.

12.5   Conclusions

- The environmental issues posed by the deployment of artificial substrata in natural systems call for a better understanding of the ecological causes underlying patterns of biofouling development on artificial substrata. This is a fascinating field of research, requiring unravelling the complex interactions regulating biofilm development, settlement of macrofoulers and their recruitment within the assemblages.
- Key to the effort needed for the understanding of these interactions is the integration of different expertises.
- Under laboratory conditions, elucidating how the interplay between artificial substrata and biofilm can regulate larval settlement cannot ignore contributions by physicist, microbiologists and ecologists.
- Results from laboratory experiments need to be integrated with fieldwork.
- Under natural conditions, the substratum–biofilm role in shaping macrofouling assemblages is easily overshadowed by a complex suite of biotic and abiotic factors.
- Well-designed inspections provided by in situ experiments are thus a logical requirement to quantify the consistency of the outcomes obtained in the laboratory.

References

1. Bradshaw, D.J., Marsh, P.D., Watson, G.K. & Allison, C. (1997) Oral anaerobes cannot survive oxygen stress without interacting with facultative/aerobic species as a microbial community. *Letters in Applied Microbiology*, **25**, 385–387.
2. Poleunis, C., Compere, C. & Bertrand, P. (2002) Time-of-flight secondary ion mass spectrometry: characterisation of stainless steel surfaces immersed in natural seawater. *Journal of Microbiological Methods*, **48**, 195–205.
3. Ostuni, E., Chapman, R.G., Liang, M.N., *et al.* (2001) Self-assembled monolayers that resist the adsorption of proteins and the adhesion of bacterial and mammalian cells. *Langmuir*, **17**, 6336–6343.
4. Gerchakov, S.M., Roth, F.J., Sallman, B., Udey, L.R. & Marszalek, D.S. (1977) Observations on microfouling applicable to OTEC systems. In: *Proceedings of the Ocean Thermal Energy Conversion (OTEC) Biofouling and Corrosion Symposium* (ed. H. Gray), pp. 63–75. US Department of Energy, Division of Ocean Energy Systems, Seattle, WA.

5. Pasmore, M., Todd, P., Pfiefer, B., Rhodes, M. & Bowman, C.N. (2002) Effect of polymer surface properties on the reversibility of attachment of *Pseudomonas aeruginosa* in the early stages of biofilm development. *Biofouling*, **18**, 65–71.

6. Bendinger, B., Rijnaarts, H.H.M., Altendorf, K. & Zehnder, A.J.B. (1993) Physicochemical cell surface and adhesive properties of coryneform bacteria related to the presence and chain length of mycolic acids. *Applied Biochemistry and Biotechnology*, **59**, 3973–3977.

7. Nickels, J.S., Bobbie, R.J., Martz, R.F., Smith, G.A., White, D.C. & Richards, N.L. (1981) Effects of silicate grain shape, structure, and location on the biomass and community structure of colonizing marine microbiota. *Applied and Environmental Microbiology*, **41**, 1262–1268.

8. Characklis, W.G., McFeters, G.A. & Marshall, K.C. (1990) Physiological ecology in biofilm systems. In: *Biofilms* (eds W.G. Characklis & K.C. Marshall), pp. 341–394. John Wiley & Sons, New York.

9. Medilanski, E., Kaufmann, K., Wick, L.Y., Wanner, O. & Harms, H. (2002) Influence of the surface topography of stainless steel on bacterial adhesion. *Biofouling*, **18**, 193–203.

10. Korber, D.R., Chai, A., Woolfaardt, G.M., Ingham, S.C. & Cadwell, D.E. (1997) Substratum topography influences susceptibility of *Salmonella enteritidis* biofilms to sodium phosphate. *Applied and Environmental Microbiology*, **63**, 3352–3358.

11. Little, B.J., Wagner, P.A. & Jacobus, O.J. (1988) The impact of sulfate-reducing bacteria on welded copper-nickel seawater piping systems. *Materials Performance*, **27**, 57–61.

12. Verran, J. & Boyd, R.D. (2001) The relationship between substratum surface roughness and microbiological and organic soiling: a review. *Biofouling*, **17**, 59–71.

13. Armon, R., Starosvetsky, J., Dancygier, M. & Starosvetsky, D. (2001) Adsorption of *Flavobacterium breve* and *Pseudomonas fluorescens* P17 on different metals: electrochemical polarization effect. *Biofouling*, **17**, 289–301.

14. Meadows, P.S. & Campbell, J.I. (1972) Habitat selection by aquatic invertebrates. *Advances in Marine Biology*, **10**, 271–382.

15. Crisp, D.J. (1984) Overview of research on marine invertebrate larvae, 1940–1980. In: *Marine Biodeterioration: An Interdisciplinary Study* (eds J.D. Costlow & R.C. Tipper), pp. 103–126. US Naval Institute Press, Maryland.

16. Fletcher, R.L. & Callow, M.E. (1992) The settlement, attachment and establishment of marine algal spores. *British Phycological Journal*, **27**, 303–329.

17. Glasby, T.M. (2000) Surface composition and orientation interact to affect subtidal epibiota. *Journal of Experimental Marine Biology and Ecology*, **248**, 177–190.

18. Terlizzi, A., Conte, E., Zupo, V. & Mazzella, L. (2000) Biological succession on silicone fouling-release surface: long term exposure tests in the harbour of Ischia, Italy. *Biofouling*, **15**, 327–342.

19. Faimali, M., Garaventa, F., Terlizzi, A., Chiantore, M. & Cattaneo-Vietti, R. (2004) The interplay of substratum nature and biofilm formation in regulating *Balanus amphitrite* Darwin, 1854 larval settlement. *Journal of Experimental Marine Biology and Ecology*, **306**, 37–50.

20. Crisp, D.J. & Ryland, J.S. (1960) Influence of filming and of surface texture on the settlement of marine organisms. *Nature*, **185**, 119.

21. Berntsson, K.M., Jonsson, P.R., Lejhalla, M. & Gatenholmb, P. (2000) Analysis of behavioural rejection of micro-textured surfaces and implications for recruitment by the barnacle *Balanus improvisus*. *Journal of Experimental Marine Biology and Ecology*, **251**, 59–83.

22. Zhang, H., Lamb, R. & Lewis, J. (2005) Engineering nanoscale roughness on hydrophobic surface-preliminary assessment of fouling behaviour. *Science and Technology of Advanced Materials*, **6**, 236–239.

23. Rittschof, D., Branscomb, E.S. & Costlow, J.D. (1984) Settlement and behavior in relation to flow and surface in larval barnacles, *Balanus amphitrite* (Darwin). *Journal of Experimental Marine Biology and Ecology*, **82**, 131–146.

24. McGuinness, K.A. (1989) Effect of some natural and artificial substrata on sessile marine organisms at Galeta Reef, Panama. *Marine Ecology Progress Series*, **52**, 201–208.

25. Anderson, M.J. & Underwood, A.J. (1994) Effects of substratum on the recruitment and development of an intertidal estuarine fouling assemblage. *Journal of Experimental Marine Biology and Ecology*, **184**, 217–236.

26. Lin, H.J. & Shao, K.T. (2002) The development of subtidal fouling assemblages on artificial structures in Keelung Harbor, Northern Taiwan. *Zoological Studies*, **41**, 170–182.

27. Brown, C.J. (2005) Epifaunal colonization of the Loch Linnhe artificial reef: influence of substratum on epifaunal assemblage structure. *Biofouling*, **21**, 73–85.

28. Pomerat, C.M. & Reiner, E.R. (1942) The influence of surface angle and of light on the attachment of barnacles and other sedentary organisms. *Biological Bulletin*, **82**, 14–25.

29. James, R.J. & Underwood, A.J. (1994) Influence of colour of substratum on recruitment of spirorbid tubeworms to different types of intertidal boulders. *Journal of Experimental Marine Biology and Ecology*, **181**, 105–115.

30. Perkol-Finkel, S., Shashar, N. & Benayahu, Y. (2006) Can artificial reefs mimic natural reef communities? The roles of structural features and age. *Marine Environmental Research*, **61**, 121–135.

31. Knight-Jones, E.W. & Crisp, D.J. (1953) Gregariousness in barnacles in relation to the fouling of ships and to anti-fouling research. *Nature*, **171**, 1109–1110.

32. Holm, E.R., Cannon, G., Roberts, D., Schmidt, A.R., Sutherland, J.P. & Rittschof, D. (1997) The influence of initial surface chemistry on the development of the fouling community at Beaufort, North Carolina. *Journal of Experimental Marine Biology and Ecology*, **215**, 189–203.

33. Harder, T., Lau, C.K.S., Dobretsov, S., Fang, Tsz. K. & Qian, P.Y. (2003) A distinctive epibiotic bacterial community on the soft coral *Dendronephthya* sp. and antibacterial activity of coral tissue extracts suggest chemical mechanism against bacterial epibiosis. *FEMS Microbiology, Ecology*, **43**, 337–347.

34. Becker, K., Siriratanachai, S. & Hormchong, T. (1997) Influence of initial substratum surface tension on marine micro- and macro-fouling in the Gulf of Thailand. *Helgoländer Meeresuntersuchungen*, **51**, 445–461.

35. Gray, N.L., Banta, W.C. & Loeb, G.I. (2002) Aquatic biofouling larvae respond to differences in the mechanical properties of the surface on which they settle. *Biofouling*, **18**, 269–273.

36. Callow, M.E. & Fletcher, R.L. (1994) The influence of low surface energy materials on bio-adhesion – a review. *International Biodeterioration and Biodegradation*, **34**, 333–348.

37. Eashwar, M., Subramanian, G., Chandrasekaran, P., Manickam, S.T., Maruthamuthu, S., Balakrishnan, K. (1995) The interrelation of cathodic protection and marine macrofouling. *Biofouling*, **8**, 303–312.

38. Dahlström, M., Jonsson, H., Jonsson, P.R. & Elwing, H. (2004) Surface wettability as a determinant in the settlement of the barnacle *Balanus improvisus* (Darwin). *Journal of Experimental Marine Biology and Ecology*, **305**, 223–232.

39. Clare, A.S., Rittschof, D., Gerhart, D.J. & Maki, J.S. (1992) Molecular approaches to non-toxic antifouling. *Invertebrate Reproduction and Development*, **22**, 67–76.

40. Davis, A.R., Targett, N.M, McConnell, O.J. & Young, C.M. (1989) Epibiosis of marine algae and benthic invertebrates: natural products chemistry and other mechanisms inhibiting settlement and overgrowth. In: *Bioorganic Marine Chemistry*, Vol. **3** (ed. P.J. Scheuer), pp. 85–114. Springer, Berlin.

41. Maki, J.S. & Mitchell, R. (2002) Biofouling in the marine environment. In: *Encyclopedia of Environmental Microbiology* (ed. G. Bitton), pp. 610–619. John Wiley & Sons, New York.

42. Bers, A.V. (2006) *Antifouling Protection at Different Scales – Multiple Defence in Mytilus Edulis and the Global Performance of Mytilid Microtopographie*. PhD thesis, University of Kiel, Germany.

43. Lam, C., Harder, T. & Qian, P.Y. (2005) Growth conditions of benthic diatoms affect quality and quantity of extracellular polymeric larval settlement cues. *Marine Ecology Progress Series*, **294**, 109–116.

44. Allison, D. (2003) The biofilm matrix. *Biofouling*, **19**, 139–150.

45. Davey, M.E. & O'Toole, G. (2000) Microbial Biofilms: from ecology to molecular genetics. *Microbiology and Molecular Biology Reviews*, **64**, 847–867.

46. Little, B.J. & Lee, J.S. (2007) *Microbiologically Influenced Corrosion*. John Wiley & Sons, New York.

47. Hills, J.M. & Thomason, J.C. (1998) The effect of scales of surface roughness on the settlement of barnacle (*Semibalanus balanoides*) cyprids. *Biofouling*, **12**, 57–69.

48. Hadfield, M.G. & Paul, V.J. (2001) Natural chemical cues for settlement and metamorphosis of marine invertebrate larvae. In: *Marine Chemical Ecology* (eds J. McClintock & B. Baker), pp. 431–461. CRC Press, Boca Raton.

49. Maki, J.S., Rittschof, D., Schmidt, A.R., Snyder, A.G. & Mitchell, R. (1989) Factors controlling attachment of bryozoan larvae: a comparison of bacterial films and unfilmed surfaces. *The Biological Bulletin*, **177**, 295–302.

50. Hadfield, M.G., Unabia, C., Smith, C.M. & Micheal, T.M. (1994) Settlement of the ubiquitous fouler *Hydroides elegans*. In: *Recent Development in Biofouling Control* (eds M.F. Thompson, R. Sarojini & R. Nagabhushanam), pp. 65–74. A. A. Balkema, Rotterdam.

51. Keough, M.J. & Raimondi, P.T. (1995) Responses of settling invertebrate larvae to bioorganic films: effects of different types of films. *Journal of Experimental Marine Biology and Ecology*, **185**, 235–253.

52. Mitchell, R. & Maki, J.S. (1998) Microbial surface films and their influence on larval settlement and metamorphosis in the marine environment. In: *Marine Biodeterioration: Advanced Techniques Applicable to the Indian Ocean* (eds M.F. Thompson, R. Sarojini & R. Nagabhushanam), pp. 489–497. Oxford & IBH Publishing Company, New Delhi.

53. Wieczorek, S.K. & Todd, C.D. (1998) Inhibition and facilitation of settlement of epifaunal marine invertebrate larvae by microbial biofilm cues. *Biofouling*, **12**, 81–118.

54. Wieczorek, S.K., Clare, A.S. & Todd, C.D. (1995) Inhibitory and facilitatory effects of microbial films on settlement of *Balanus amphitrite* larvae. *Marine Ecology Progress Series*, **119**, 221–228.

55. Unabia, C.R.C. & Hadfield, M.G. (1999) Role of bacteria in larval settlement and metamorphosis of the polychaete *Hydroides elegans*. *Marine Biology*, **133**, 55–64.

56. Qian, P.Y., Thiyagarajan, V., Lau, S.C.K. & Cheung, S.C.K. (2003) Relationship between bacterial community profile in biofilm and attachment of the acorn barnacle *Balanus amphitrite*. *Aquatic Microbial Ecology*, **33**, 225–237.

57. Dobretsov, S.V., Dahms, H.U. & Qian, P.Y. (2006) Inhibition of biofouling by marine microorganisms and their metabolites. *Biofouling*, **22**, 43–54.

58. O'Connor, N.J. & Richardson, D.J. (1994) Comparative attachment of barnacle cyprids (*Balanus amphitrite* Darwin, 1854; *B. improvisus* Darwin, 1854; and *B. eburneus* Gould, 1841) to polystyrene and glass substrata. *Journal of Experimental Marine Biology and Ecology*, **183**, 213–225.

59. Wethey, D.S. (1986) Ranking of settlement cues by barnacle larvae: influence of surface contour. *Bulletin of Marine Science*, **39**, 393–400.

60. Raimondi, P.T. (1988) Rock-type affects settlement, recruitment, and zonation of the barnacle *Chthamalus anisopoma* Pilsbury. *Journal of Experimental Marine Biology and Ecology*, **123**, 253–267.

61. Holmes, S.P., Sturgess, C.J. & Davies, M.S. (1997) The effect of rock-type on the settlement of *Balanus balanoides* (L.) cyprids. *Biofouling*, **11**, 137–147.

62. Maki, J.S., Rittschof, D., Costlow, J.D. & Mitchell, R. (1988) Inhibition of attachment of larval barnacles, *Balanus amphitrite*, by bacterial surface film. *Marine Biology*, **97**, 199–206.

63. Maki, J.S., Rittschof, D., Samuelsson, *et al.* (1990) Effect of marine bacteria and their exopolymers on the attachment of cypris larvae. *Bulletin of Marine Science*, **46**, 499–511.

64. Thompson, R.C., Norton, T.A. & Hawkins, S.J. (1998) The influence of epilithic microbial films on the settlement of *Semibalanus balanoides* cyprids – a comparison between laboratory and field experiments. *Hydrobiologia*, **375/376**, 203–216.

65. Decho, A.W. (1990) Microbial exopolymer secretions in ocean environment: their role(s) in food webs and marine processes. *Oceanography and Marine Biology: An Annual Review*, **28**, 73–153.

66. Abelson, A. & Denny, M. (1997) Settlement of marine organisms in flow. *Annual Review of Ecology and Systematics*, **28**, 317–339.

67. Minchinton, T.E. & Scheibling, R.E. (1991) The influence of larval supply and settlement on the population structure of barnacles. *Ecology*, **72**, 1867–1879.
68. Bertness, M.D., Gaines, S.D., Stephens, E.G. & Yund, P.O. (1992) Components of recruitment in populations of the acorn barnacle Semibalanus balanoides (Linnaeus). *Journal of Experimental Marine Biology and Ecology*, **156**, 199–215.
69. Levring, T. (1966) Submarine light and algal shore zonation. In: *Light as an Ecological Factor* (eds R. Bainbridge, G.C. Evans & O. Rackham), pp. 305–318. Blackwell, Oxford.
70. Hiscock, K. & Mitchell, R., 1980. The description and classification of sublittoral epibenthic ecosystems. In: *The Shore Environment. Vol. 2: Ecosystems* (eds J.H. Price, D.E.G. Irvine & W.F. Farnham), pp. 323–370. Academic Press, London.
71. Kain, J.M., Drew, E.A. & Jupp, B.P. (1975) Light and the ecology of Laminaria hyperborea II. In: *Light as an Ecological Factor* (eds G.C. Evans, R. Bainbridge & O. Rackham), pp. 63–92. Blackwell, Oxford.
72. Stachowitsch, M., Kikinger, R., Herler, J., Zolda, P. & Geutebrück, E. (2002) Offshore oil platforms and fouling communities in the southern Arabian Gulf (Abu Dhabi). *Marine Pollution Bulletin*, **44**, 853–860.
73. Glasby, T.M. (1999) Interactive effects of shading and proximity to the seafloor on the development of subtidal epibiotic assemblages. *Marine Ecology Progress Series*, **190**, 113–124.
74. Glasby, T.M. (1999) Differences between subtidal epibiota on pier pilings and rocky reefs at marinas in Sydney, Australia. *Estuarine, Coastal and Shelf Science*, **48**, 281–290.
75. Glasby, T.M. & Connell, S.D. (2001) Orientation and position of substrata have large effects on epibiotic assemblages. *Marine Ecology Progress Series*, **214**, 127–135.
76. Rule, M.J. & Smith, S.D.A. (2005) Spatial variation in the recruitment of benthic assemblages to artificial substrata. *Marine Ecology Progress Series*, **290**, 67–78.
77. Wahl, M. (2001) Small scale variability of benthic assemblages: biogenic neighbourhood effects. *Journal of Experimental Marine Biology and Ecology*, **258**, 101–114
78. Glasby, T.M. & Connell, S.D. (1999) Urban structures as marine habitats. *Ambio*, **28**, 595–598.
79. Pickering, H., Whitmarsh, D. & Jensen, A. (1998) Artificial reefs as a tool to aid rehabilitation of coastal ecosystems: investigating the potential. *Marine Pollution Bulletin*, **37**, 505–514.
80. Jensen, A.C., Collins, K.J. & Lockwood, A.P.M. (2000) *Artificial Reefs in European Seas*. Springer, Heidelberg.
81. Jaap, W.C. (2000) Coral reef restoration. *Ecological Engineering*, **15**, 345–364.
82. Holmstrom, C. & Kjelleberg, S. (1994) The effect of external biological factors on settlement of marine invertebrate and new antifouling technology. *Biofouling*, **8**, 147–160.
83. Abarzua, S. & Jakubowski, S. (1995) Biotechnological investigation for the prevention of bio-fouling. I. Biological and biochemical principles for the prevention of biofouling. *Marine Ecology Progress Series*, **123**, 301–312.
84. Schoener, A. (1974) Colonization curves for planar marine islands. *Ecology*, **55**, 818–827.
85. Greene, C.H., Schoener, A. & Corets, E. (1983) Succession on marine hard substrata: the adaptive significance of solitary and colonial strategies in temperate fouling communities. *Marine Ecology Progress Series*, **13**, 121–129.
86. Oshurkov, V.V. (1992) Succession and climax in some fouling communities. *Biofouling*, **6**, 1–12.
87. Butler, A.J. & Connolly, R.M. (1996) Development and long term dynamics of a fouling assemblage of sessile marine invertebrates. *Biofouling*, **9**, 187–209.
88. Bos, R., Van Der Mei, H.C. & Busscher, H.J. (1999) Physico-chemistry of initial microbial adhesive interactions – its mechanisms and methods for study. *FEMS Microbiology Reviews*, **23**, 179–230.
89. Raghukumar, S., Anil, A.C., Khandeparker, L. & Patil, J.S. (2000) Thraustochytrid protists as a component of marine microbial films. *Marine Biology*, **136**, 603–609.
90. Frederick, J.A., Jacobs, D. & Jones, W.R. (2000) Biofilms and biodiversity: an interactive exploration of aquatic microbial biotechnology and ecology. *Journal of Industrial Microbiology and Biotechnology*, **24**, 334–338.

91. Baine, M. (2001) Artificial reefs: a review of their design, application, management and performance. *Ocean and Coastal Management*, **44**, 241–259.

92. Rice, A.R., Hamilton, M.A. & Camper, A.K. (2003) Movement, replication, and emigration rates of individual bacteria in a biofilm. *Microbial Ecology*, **45**, 163–172.

93. Sutherland, J.P. & Karlson, R.H. (1977) Development and stability of the fouling community at Beaufort, North Carolina. *Ecological Monographs*, **47**, 425–446.

94. Dean, T.A. & Hurd, L.E. (1980) Development in an estuarine fouling community: the influence of early colonists on later arrivals. *Oecologia*, **46**, 295–301.

95. Schmidt, G.H. & Warner, G.F. (1984) Effects of caging on the development of a sessile epifaunal community. *Marine Ecology Progress Series*, **15**, 251–263.

96. Osman, R.W. & Whitlatch, R.B. (1995) The influence of resident adults on recruitment: a comparison to settlement. *Journal of Experimental Marine Biology and Ecology*, **190**, 169–198.

97. Brown, K.M & Swearingen, D.C. (1998) Effects of seasonality, length of immersion, locality and predation on an intertidal fouling assemblage in the Northern Gulf of Mexico. *Journal of Experimental Marine Biology and Ecology*, **225**, 107–121.

98. Connell, S.D. & Glasby, T.M. (1999) Do urban structures influence local abundance and diversity of subtidal epibiota? A case study from Sydney Harbour, Australia. *Marine Environmental Research*, **47**, 373–387.

99. Connell, S.D. (2000) Floating pontoons create novel habitats for subtidal epibiota. *Journal of Experimental Marine Biology and Ecology*, **247**, 183–194.

100. Smith, S.D.A. & Rule, M.J. (2002) Artificial substrata in a shallow sublittoral habitat: do they adequately represent natural habitats or the local species pool? *Journal of Experimental Marine Biology and Ecology*, **277**, 25–41.

101. Glasby, T.M. (1998) Estimating spatial variability in developing assemblages of epibiota on subtidal hard substrata. *Marine and Freshwater Research*, **49**, 429–437.

102. Bax, N., Williamson, A., Aguero, M., Gonzalez, E. & Geeves, W. (2003) Marine invasive alien species: a threat to global biodiversity. *Marine Policy*, **27**, 313–323.

103. Occhipinti-Ambrogi, A. & Savini, D. (2003) Biological invasions as a component of global change in stressed marine ecosystems. *Marine Pollution Bulletin*, **46**, 542–551.

104. Wonham, M.J., Carlton, J.T., Ruiz, G.M. & Smith, L.D. (2000) Fish and ships: relating dispersal frequency to success in biological invasion. *Marine Biology*, **136**, 1111–1121.

105. Boero, F. (2002) Ship-driven biological invasions in the Mediterranean Sea. *Alien Marine Organisms Introduced by Ships in the Mediterranean and Black Seas*. CIESM Workshop Monographs, **20**, 87–92.

106. Tyrrell, M.C. & Byers, J.E. (2007) Do artificial substrata favour nonindigenous fouling species over native species? *Journal of Experimental Marine Biology and Ecology*, **342**, 54–60.

107. Bulleri, F. & Airoldi, L. (2005) Artificial marine structures facilitate the spread of a non-indigenous green alga, Codium fragile ssp. tomentosoides, in the north Adriatic Sea. *Journal of Applied Ecology*, **42**, 1063–1072.

108. Airoldi, L., Abbiati, M., Beck, M.W., *et al.* (2005) An ecological perspective on the deployment and design of low-crested and other hard coastal defence structures. *Coastal Engineering*, **52**, 1073–1087.

109. Faimali, M., Garaventa, F. & Geraci, S. (2001) Macrofouling monitoring in industrial plants (Part two). *European Coatings*, **77**, 59–64.

Chapter 13
Paint and Coatings Technology for the Control of Marine Fouling

Alistair A. Finnie and David N. Williams

With a primary focus on fouling control coatings that are designed for use in the commercial shipping and pleasure craft markets, this chapter provides an overview of the most common technologies that are currently available to protect immersed structures from biofouling. These technologies are classified according to whether they utilise a biocidal or non-toxic approach to control fouling. General mechanistic descriptions of the mode of operation are provided for each, along with a consideration of the performance attributes of the main compositional sub-classes and the relative cost to users. Finally, the chapter reviews some emerging potential alternative approaches, and considers the main influences that are driving the development of new and improved technologies to control marine fouling.

13.1 Introduction

In the context of commercial ships, boats, recreational yachts and other pleasure craft, fouling can be defined as unwanted biological growth on the exposed underwater surfaces of these vessels. As a direct consequence of this biological growth, the flow of water along the immersed surface is perturbed and the hydrodynamic drag is increased: the more severe the fouling growth, the greater the increase in the associated hydrodynamic drag [1]. For the vessel operator, this increased drag leads to a reduction in manoeuvrability and loss of speed or, alternatively, an increase in the power required to maintain speed. For all powered vessels, loss of speed and increased power requirements mean greater fuel consumption which in turn mean higher operating costs and greater environmental emissions of greenhouse gasses.

The negative consequences of fouling and the benefits of a clean hull have been recognised for millennia: as reported by Plutarch in the first century AD [2], a vessel 'glides lightly, and as long as it is clean, easily cuts the waves; but when it is thoroughly soaked, when weeds, ooze, and filth stick upon its sides, the stroke of the ship is more obtuse and weak; and the water, coming upon this clammy matter, doth not so easily part from it; and this is the reason why they usually scrape the sides of their ships.' It is now well-recognised that the consequences of macrofouling, particularly hard-shell calcareous animal fouling, are more detrimental to the ship operator than algal and bacterial microfouling slimes and much research effort has been expended to more precisely characterise the effect of fouling through laboratory, field and full ship trials (e.g. [3–5]). As noted by Schultz [6], the results of these studies consistently demonstrate that fouling causes a significant increase in the frictional drag of the hull, and the size of this increase depends on the biology (i.e. the organisms present and coverage), as well as the hull-form and its operating speed. As an example, a US Navy frigate sailing at 30 knots (15.4 m s^{-1}) would be expected to require a 55% increase in the shaft horsepower to overcome the effect of heavy calcareous fouling, or else suffer a 10.7% loss of speed for a fixed power output [6].

In Plutarch's time, the available hull treatments – waxes, tars, pitches [1] and lead sheathing [7] – were primarily intended to waterproof the hull rather than to prevent fouling, hence the need for frequent scraping or careening. Nowadays, the underwater surfaces of most commercial vessels and pleasure craft are protected by paints that have been specifically designed to negate the accumulation of biological growth. Historical overviews of the evolution of modern fouling control paints have been presented by a number of authors [1, 8–10], so will not be considered in detail here. These modern paints fall under two distinct mechanistic classes: biocidal antifouling paints, which function through the use of toxic agents to prevent the settlement and growth of fouling species; and non-biocidal foul release coatings (FRCs), which make use of favourable surface energetics to minimise adhesion of organisms and facilitate easy removal by water flow.

13.2   Biocidal antifouling paints

By definition [11], biocidal antifouling paints contain one or more active ingredients (biocides) which are used to control the growth and settlement of fouling organisms (microbes and higher forms of plant or animal species) on vessels, aquaculture equipment and other structures in the marine environment. In principle, the biocidal moiety can be tethered to the surface in order to prevent its release into the surrounding waters and avoid any associated negative environmental impact. However, to-date this approach has been largely unsuccessful and the biocidal antifouling paints that are in current commercial use almost exclusively function by releasing biocides at a more or less controlled rate into the boundary layer of the immersed surface, from where they are dispersed into the bulk sea water.

More than 4000 species of fouling organism have been recorded [12], including hard- and soft-shelled animals, algae and bacteria [1]. Each class of organism, and to some extent each individual species, has its own unique biology and susceptibility to biocides. In other words, some biocides are more potent against certain fouling types than others. The efficacy of copper as an antifouling biocide has been recognised since ancient times and it continues to be in widespread use today. Copper is active against a wide range of fouling organisms but it is most active against sessile invertebrates such as barnacle and tubeworm species. Consequently, the amount of copper released by a paint film (known as the leaching rate or the release rate) must be much higher for the paint to be effective against algal fouling as well as hard fouling. It has been reported [13] that a copper release rate of 10 µg cm^{-2} day^{-1} is required to be effective against barnacles, but a release rate of 20 µg cm^{-2} day^{-1} is required to be additionally effective against diatom, although such values should be taken as indicative rather than absolute limiting values [14], particularly as the determined release rate is very dependent on the method of measurement [15]. Nevertheless, acceptable fouling control can therefore be achieved through the use of a copper-based biocide alone provided that relatively high copper release rates are maintained for the intended lifetime of the paint. However, it is common

$$\tfrac{1}{2}Cu_2O + H^+ + 2Cl^- \Leftrightarrow CuCl_2^- + \tfrac{1}{2}H_2O$$

$$CuCl_3^{2-} + Cl^- \Leftrightarrow CuCl_2^- + Cl^- \Leftrightarrow CuCl^0 + Cl^- \Leftrightarrow Cu^+ + Cl^-$$

Figure 13.1 The dissolution of cuprous oxide in sea water [16, 17].

practice to use copper in combination with one or more co-biocides which are selected for their complementary biocidal activities [12].

In biocidal antifouling paints, copper is most usually used in pigmentous form as cuprous oxide (Cu_2O). As shown in Figure 13.1, when cuprous oxide comes into contact with sea water, it generates soluble hydrated Cu(I) chloride complexes which are then rapidly oxidised to liberate cupric ions, the main biocidal species [16].

Cuprous oxide is usually supplied as a red or purple powder, where the most substantive difference between these forms is the primary particle size. This means that it is difficult to formulate light or brightly coloured paints using cuprous oxide as the principal biocide. In such cases, cuprous thiocyanate can be substituted where cosmetic considerations are particularly important, for example in the yacht or pleasure craft markets. In the commercial shipping sector, cosmetics are generally less important and so reds, browns and other darker colours have traditionally been favoured as these colours are more easily formulated for cuprous oxide-based paints. However, cuprous thiocyanate is also often used in preference to cuprous oxide for products that are designed specifically for use on aluminium-hulled vessels as it reduces the risk of galvanic corrosion that can occur if cuprous oxide comes into contact with exposed aluminium [12].

The co-biocide is generally selected in order to boost the performance of the paint by providing enhanced activity against macroalgal fouling and/or algal and bacterial biofilms (slime). These co-biocides are usually small organic or organometallic compounds with low water solubilities and many were originally developed principally as agricultural pesticides or as fungicides [18]. The choice of co-biocide depends in part on geographical location as biocidal antifouling paints are increasingly regulated on a national or international level. For example, in the European Union, biocidal antifouling paints are now regulated under the Biocidal Products Directive (BPD) [11] (see Chapters 21 and 26) and, as shown in Table 13.1, only a limited list of biocides are currently permitted [19]. In contrast, the Japanese market is less highly regulated and a wider range of biocides are used [21] including all but one of the biocides (tolylfluanid) that are permitted in Europe. In further contrast, many countries which have economically significant ship building or ship repair industries, for example in Asia and the Middle East, currently have no specific biocide regulation and any biocide can be legally used irrespective of its perceived environmental impact.

The move towards increasing regulation of the industry has been driven in a large part by recognition of the environmental effects that resulted from the widespread use of tributyltin (TBT) in previous decades (see subsequent discussion in Section 13.2.1.3; [22]). None of those biocides that remain in common use are perceived to give rise to environmental effects that are comparable to those of TBT. While there is a slight trend towards the development of paints that are free of copper-based biocides, this is perhaps as much driven by the approximately fivefold increase in copper metal prices between 2003 and 2008 [23], as it is by environmental concerns. Currently, copper-free biocidal antifouling paints remain niche products which are

Table 13.1 Antifouling biocides in current use in the European Union and Japan [18, 20].

Chemical name	CAS Number	In use in Japan?	In use in EU?
Cuprous oxide/dicopper oxide	1317-39-1	Yes	Yes
Cuprous thiocyanate/copper thiocyanate	1111-67-7	Yes	Yes
Copper	7440-50-8	No	Yes
Zinc ethylenebis (dithiocarbamate)/zineb	12122-67-7	Yes	Yes
Zinc dimethyl dithiocarbamate/ziram	137-30-4	Yes	No
Zinc 2-pyridinethiol-1-oxide/pyrithione zinc	13463-41-7	Yes	Yes
Copper, bis(1-hydroxy-2(1H)-pyridinethionato-O, S)-/pyrithione copper	14915-37-8	Yes	Yes
Naphthenic acids, copper salts	1338-02-9	Yes	No
3-(3,4-Dichlorophenyl)-1,1-dimethyl urea/diuron	330-54-1	Yes	No
Pyridine-triphenylborane	971-66-4	Yes	No
2,3,5,6-Tetrachloro-4-(methylsulphonyl) pyridine	13108-52-6	Yes	No
N-(2,4,6-Trichlorophenyl) maleimide	13167-25-4	Yes	No
Dichloro-N-[(dimethylamino)sulphonyl] fluoro-N-(p-tolyl)methanesulphenamide/tolylfluanid	731-27-1	No	Yes
N,N-dimethyl-N'-phenyl-N'-(dichlorofluoromethylthio) sulfamide/ dichlofluanid	1085-98-9	Yes	Yes
2-Methylthio-4-tert-butylamino-6-cyclopropylamino-s-triazine	28159-98-0	Yes	Yes
4,5-Dichloro-2-n-octyl-4-isothiazolin-3-one	64359-81-5	Yes	Yes
2,4,5,6-Tetrachloroisophthalonitrile	1897-45-6	Yes	No
Tetramethylthiuram disulphide	137-26-8	Yes	No

used to meet a particular demand or to comply with local regulatory requirements. It can be reasonably expected that the widespread use of copper-based biocides will continue for the foreseeable future.

13.2.1 Biocidal antifouling paint types

13.2.1.1 Contact leaching paints

Perhaps the oldest and simplest route to formulating an antifouling paint is simply to combine a mechanically robust and durable film forming material (binder) with as much pigmentous biocide as possible. Such paints rely upon the biocide loading in the paint film being above the critical pigment volume content (CPVC) [24] so that the biocide pigment surface is not fully wetted by the binder and each pigment particle is in direct contact with one or more other pigment particles. Typically, cuprous oxide is combined with an inert polymer such as a vinyl or acrylic copolymer. This means that when the paint film is exposed to sea water, the biocide particles at the film surface dissolve from the film–sea water interface, creating pores in the film surface which fill with sea water and, in turn, expose further biocide particles. As this process progresses deeper and deeper into the paint film, the diffusion path length for the biocide from the surface of the exposed pigment to the film surface rapidly increases and so, as determined by Fick's law of diffusion, the flux of biocide (i.e. the biocide release rate) falls exponentially with time [9]. As a result, the useful lifetime of such paints is usually limited by the build-up of this skeletal leached layer to 12 months or so. Nowadays, such paints are largely restricted to the lower performance end of the market and have limited economic importance. Therefore they will not be discussed in more detail here.

13.2.1.2   Soluble matrix and controlled depletion polymer paints

The principal problem of contact leaching paints is that the uncontrolled increase in leached layer size leads inexorably to a decline in the biocide release rate. Once established, the leached layer is essentially locked in place as the binder polymers that are used are essentially inert and insoluble in sea water. In soluble matrix paints, a portion of the insoluble binder polymer is replaced by a sea water soluble binder component. Consequently, dissolution of the soluble binder component promotes the formation of a more porous and mechanically weak leached layer, which can then be partially removed by the flow of water along the hull when the vessel is active. For this reason, such paints are sometimes known as ablative or erodible paints because the dissolution of the binder component means that the paint film becomes thinner with time [17]. The more modern versions of such paints are also often referred to as controlled depletion polymer (CDP) paints in order to differentiate them from self-polishing copolymer (SPC) paints in the marketplace.

The most commonly used soluble binder component is rosin or a rosin derivative [17]. Rosin is a naturally occurring material that is harvested industrially from various species of pine tree [25]. It largely consists of a mixture of C-20 unsaturated diterpene monocarboxylic acids, such as abietic, pallustric and pimaric acids and their isomers, which are marginally soluble at the pH of sea water (pH 8.0–8.2 is typical). Rosin can be used directly or, alternatively, it can be used in the form of various metal soaps, or in chemically modified forms such as hydrogenated, disproportionated or polymerised derivatives. Rosin and its derivatives tend to be hard, brittle and friable material with relatively low film strength. Such paints therefore rely upon the conventional insoluble and inert film-forming polymers (such as vinyl or acrylic resins) that are also present in the formulation in order to achieve the requisite mechanical strength. High levels of rosin will increase the erosion rate and enhance the antifouling performance, but the impaired film integrity may lead to cracking and detachment [26]. On the other hand, low levels of rosin will provide mechanically robust films but, because there is less soluble material in the binder, then the erosion rate is reduced and, because a stronger leached layer is produced, the lifetime and performance of the paint is compromised. In practice, there is always a balance to be achieved and there is much formulating skill in achieving the right balance of properties for any particular set of performance targets.

The use of a soluble film-forming component, such as rosin, allows a degree of control to be exercised over the rate of formation of the leached layer for such paints. However, even though the leached layer is initially weak and easily removed, paints of this type are still subject to leached layer build-up [17]. The underlying physical and chemical processes are not fully understood, but it is thought that contributing factors to leached layer build-up include oxidative degradation of the rosin (which will lower its solubility), and the precipitation of basic copper carbonate and related insoluble copper complexes within the leached layer itself (which will reduce the porosity of the leached layer) [17]. Because of this, the useful lifetime of CDP paints is generally limited to around 36 months, although this can be extended by regular underwater scrubbing to remove the leached layer and rejuvenate the paint film.

An interesting attempt to mathematically model the leaching and erosion behaviour of rosin-based paints has recently been made by Yebra *et al.* [27], building on their earlier success in modelling TBT-SPC paint behaviour [28, 29]. The authors were unable to develop a universal model to predict the sea water behaviour of such rosin-based paints from *a priori* mechanistic assumptions; they conclude, however, that further development of the mathematical modelling approach in conjunction with an experimental design approach and simplified test procedures may be a useful tool for optimising future formulation development and paint performance.

13.2.1.3   Self-polishing copolymer antifouling paints

The development of TBT-based SPC paints in the 1970s introduced a revolutionary step-change to both the performance and the lifetime of antifouling paints [30]. The main binder component in these paints was a copolymer which carried a number of pendant tributyltin carboxylate ester functional groups. By virtue of the ability of the tributyltin carboxylate ester groups to undergo hydrolysis and ion-exchange reactions in sea water, the hydrophobic and highly sea water insoluble TBT copolymer becomes solubilised as the corresponding polycarboxylic acid or polycarboxylate soap. This reaction is confined to the leached layer [28], and as each molecular layer of TBT copolymer reacts and dissolves, the layer below can then react and dissolve and so on. Dissolution of the polymer is substantially controlled by the rate of reaction of the TBT copolymer, rather than by the physical action of water on the film surface, and so it is essentially a spontaneous process leading to continuous surface renewal. As there are no significant quantities of insoluble binder components in the paint film, there is nothing to stabilise the leached layer, which remains thin and allows the biocide release rate to remain virtually constant for the lifetime of the paint, when the final molecular layers of polymer have reacted and dissolved [31]. Moreover, this process results in a film surface that becomes hydrodynamically smoother with time, leading to significant additional economic benefits for the shipowner.

The overall rate of film thickness reduction for the paint – conventionally termed the polishing rate – can be simply controlled by varying the level of tributyltin carboxylate ester functionality in the copolymer [31]: for any given biocide content, higher TBT content equates to faster polishing rates and higher biocide release rates, meaning that it was a relatively simple matter to tailor paint performance attributes to suit a full range of vessel operating characteristics. The useful lifetime of the paint is determined solely by the film thickness of paint that is applied and so, for the first time, guaranteed performance in excess of 3 years could be routinely achieved and a 5-year scheme became the norm for many vessel types.

From the performance standpoint, TBT-SPC paints were widely regarded as the benchmark standard for fouling control technology [32]. However, in addition to release of the afore-mentioned polycarboxylates and conventional biocides, the reaction of the TBT copolymer also liberated TBT species. Such TBT species are highly effective biocides for marine organisms in their own right (see Chapters 16 and 17) and this no doubt aided the impressive antifouling performance of these paints. In fact, these paints evolved from the common use of tributyltin oxide (TBTO) and other organotin compounds as antifouling biocides in the

Table 13.2 Major tin and tin-free SPC polymer types.

SPC type (polymer-CO_2-X)	Functional group
TBT copolymers	$X = Sn(C_4H_9)_3$
Metal acrylate copolymers	$X = M\text{-}O_2R$
	$M = Cu, Zn$
Silyl acrylate copolymers	$X = SiR_3$

1950s and 1960s. Nevertheless, the introduction of TBT-SPC paints and their subsequent near-universal adoption as the high performance antifouling technology of choice meant that unprecedented quantities of TBT were released into the marine environment. The environmental consequences of this are well documented and included imposex in marine gastropod molluscs and infertility and severe shell deformation in oysters [33]. In response to this, TBT-SPC antifoulings were voluntarily withdrawn from the market, firstly by International Paint and subsequently by many other manufacturers in 2002/2003 in anticipation of action by the International Maritime Organization (IMO). Even though the IMO's Antifouling Systems Convention only entered into legal force on 17 September 2008, in effect it led to a worldwide ban on the application of TBT- and other organotin-based paints from 1 January 2003, and a ban on the exposed presence of such paints on ship hulls from 1 January 2008 [34] (see Chapters 21 and 26).

Well ahead of this IMO action, several countries imposed their own limits on the application of TBT-based antifoulings, and Japan imposed a ban as early as 1990 [33]. The effect of this was that paint companies, in particularly Japanese paint companies, were stimulated to develop TBT-free copolymer-based technologies that combined the positive performance attributes of TBT-SPCs with an improved environmental profile.

In chemical terms, TBT-copolymers can be thought of as a protected water-soluble carboxylic acid-functional polymer whereby the deprotection step – hydrolysis and/or ion-exchange – facilitates the dissolution of the parent polycarboxylate at a predetermined rate and acts to mediate the controlled release of the pigmentous biocides from the underlying paint film [28]. Much effort was expended in investigating chemistries that would potentially mimic this process, and this led to the development and commercial introduction of the two main high-performance polymer-based tin-free SPC technologies that are in current use, namely metal acrylate copolymers and silyl acrylate copolymers.

Metal acrylate copolymer SPCs, specifically copper and zinc acrylate copolymer SPCs, were first developed by Nippon Paint Company (NPC) [35] and were subsequently widely introduced by NPC and International Paint in advance of the IMO's action to regulate the use of TBT-SPCs. As illustrated in Table 13.2, the parent polycarboxylate copolymer is generally protected as a copper or zinc carboxylate complex which can undergo hydrolysis and ion exchange in sea water to liberate the soluble products [36]. Related metal acrylate technologies were later developed and commercialised by Kansai Paint and Chugoku Marine Paints (CMP) among others [17]. The performance of certain current copper acrylate SPC products has been clearly shown to be at least equivalent to the TBT-SPC products that they replaced (see Chapter 14), although this should not necessarily be assumed to be the case for all metal acrylate products.

For silyl acrylate copolymer SPCs, the parent polycarboxylate copolymer is protected as a hydrolysable triorganosilyl ester. Commercially, such systems were initially developed by

CMP and by Nippon Oil and Fats (now part of NKM Coatings) and are also sold by Jotun Paint and others. Unlike metal acrylates, most commercial silyl acrylate-based SPC antifouling paints routinely contain an amount of rosin or a rosin derivative [37,38], albeit at a lower level than a CDP, and so the purist may be tempted to classify metal acrylate SPCs as the more classical SPC technology. In reality though, the term SPC has evolved from its original use as a quasi-technical term to differentiate high-performance reactive polymer-based technologies from cheaper, lower performing rosin-based products. Nowadays, this distinction has become somewhat blurred by its increasing use as a marketing tool by some companies for any product that has a claimed lifetime of more than 3 years, irrespective of the underlying technology. A number of products are available that are described by suppliers as self-polishing that are based on rosins or rosin derivatives (or sometimes synthetic rosin alternatives) and would arguably be more traditionally classed as CDP products.

13.2.1.4   Other biocidal antifouling paint technologies

As mentioned above, relatively small levels of rosin are often routinely added to silyl acrylate SPC copolymer paints [37, 38]. Some manufacturers have taken this a stage further and combined larger amounts of rosin with TBT-free SPC copolymers or other polymers that are designed to gradually hydrolyse or solubilise in sea water, such as metal acrylate copolymers [9, 39] or non-aqueous dispersions of carboxylate-functional acrylic copolymers [27]. In essence, such products – often called hybrid paints – can be considered to be an amalgam of CDP and SPC technologies and as such, the cost and performance expectations generally fall between those of the two parent technologies. Consequently, the lifetime of the product generally lies between 3 and 5 years, depending on the product and its use profile.

13.2.2   Future developments of biocidal antifouling paints

Paint manufacturers will, of course, continue to refine and optimise their existing products in the short term and seek to gain sustained competitive advantage over their competitors by developing improved technologies in the medium and long term. An additional common driver for longer term future development will be to further minimise the environmental impact associated with the release of biocides. It can be reasonably expected that the trend for increasing legislation in this area (see Chapter 26) will eventually lead to the removal from the market of some the biocides that are in current use, either because of their perceived environmental impact, or because biocide manufacturers are unable to bear the substantial expenditure associated with the rigorous and extensive testing that is needed to demonstrate the absence of adverse environmental effects and obtain product registrations. As an illustration of the magnitude of such expenditure, it has been estimated that it took 10 years and cost around $10 million for Rohm and Haas to achieve regulatory approval for 4,5-Dichloro-2-*n*-octyl-4-isothiazolin-3-one in the USA alone [40].

There has been significant research in the area of natural product biocides for a number of years and a large number of materials from sources as diverse as marine algae, sponges, trees and toads have been investigated [10] (see also Chapter 8). While many of these compounds indeed show significant apparent antifouling activity, there is often an implicit assumption that natural product biocides are, by definition, 'environmentally friendly'. While this is likely to be the case for some, it is equally likely not to be the case for others and, in both environmental and legislative terms, these materials are in principle no different from synthetic biocides. As such they require the same level of scrutiny with regard to human toxicity and environmental fate.

While the prospects for the widespread commercial exploitation of natural product antifoulants are probably limited by the costs associated with product registrations, there remains considerable academic interest in this area. The chemistries of the vast range of potential natural product antifouling biocides that have been studied are described in detail elsewhere [41], but perhaps the most widely studied group is the halogenated furanones [42]. These materials, produced by the marine algae *Delisea pulchra*, have been found to inhibit the formation of biofilms on surfaces, but have yet to be registered as biocides for antifouling end-use. Zosteric acid, which is a sulphoxy-phenolic acid derived from the marine algae *Zostera marina*, has also received much attention as a potential natural product antifoulant [43]. Indeed, zosteric acid is claimed to be an effective biocide in conventional antifouling paints and in silicones [44, 45], but the compound has high water solubility and its release from conventional antifouling paints is difficult to control.

Another class of natural compounds that have been considered as environmentally friendly alternatives to metal-based antifouling agents are enzymes. The use of enzymes as antifoulants has been investigated continuously for over 20 years and an excellent review by Olsen *et al.* [46], which assesses the state of the art of enzymatic antifouling technology, has recently been published. Enzymes, for example proteases [47], can act in a direct way on fouling organisms by a biocidal effect or by affecting an organism's ability to adhere, or in an indirect way by interacting with the coating or the surroundings to produce a biocide that inhibits fouling. There remain many technical issues associated with the use of enzymes in marine antifouling products including retention of enzyme activity and achieving broad-spectrum efficacy. However, the biggest hurdle is that of legislation as, at least within the European Union, they are likely to be considered biocides [46] and will therefore require registration similar to that for conventional biocides.

13.3   Non-biocidal coatings

Concerns over the environmental impact of antifouling biocides have led to an increasing interest in the development of a biocide-free fouling control solution. In particular, the concept of developing a surface which reduces the adhesion strength of fouling organisms and therefore self-cleans under the weight of fouling or by water flow has been the focus of much attention [48]. Despite the fact that commercial products based on this concept have been on the market since 1995 [49], it is fair to say that the mechanism for release of fouling is not well understood.

13.3.1   *Foul release surfaces*

In order to create a surface that has minimal adhesion to the adhesives produced by marine organisms, the common adhesive principles need to be considered [50]. On this basis there is an apparent requirement to minimise:

(1)   Mechanical interlocking, by producing a surface that is smooth and non-porous
(2)   Wetting of the surface, by selection of surface functional groups
(3)   Chemical bonding, by producing a surface free of reactive functional groups

(4) Electrostatic interactions, by producing a low polarity surface with or without ionic groups

(5) Diffusion of molecular chains from the marine adhesive into the surface, by producing a surface with closely packed functional groups

The consideration of polymer surface properties as a means to reduce wetting was considered in the context of bioadhesion by Baier [51]. Baier proposed that a minimum in bioadhesion occurs in a critical surface tension range between 20 and 30 mN m^{-1}. However, subsequent studies have shown that surface properties alone do not dictate the performance of foul release surfaces and that fracture mechanics play an important role in the breaking of the surface marine adhesive joint: Chaudhury and co-workers [52] explained the observation that adhesive strengths of viscoelastic adhesives could not be predicted from the surface free energy of the substrate by invoking a model of adhesion based on slippage at the interface; in studies by Brady [53], relative adhesion data was shown to correlate with the square root of the product of surface energy and elastic modulus; and Kohl and Singer [54] on barnacle release mechanisms showed that release forces decrease with increasing film thickness in accordance with the model developed by Kendall for release of a rigid cylinder from a substrate.

13.3.2   Foul release surface design

Many generic polymer types have been tested for foul release properties and the relative performance of materials has been ranked according to barnacle adhesion strength. Only two groups of materials appear to fulfil the requirements for foul release, namely silicones and fluoropolymers [55]. Silicone FRCs have already become a commercial success story [48], but it is not until recently that fluoropolymer-based coatings have become viable on a performance and cost basis [56].

13.3.2.1   Fluoropolymers

Polymers containing fluorinated groups are an obvious choice for FRCs due to the potential for low surface energy and general chemical stability. A wide range of fluoropolymer-based materials have been investigated for use as FRCs. An overview of the types of fluorine-containing fouling control coatings is provided in Table 13.3. The first type of fluoropolymer-based coatings extensively studied were fluorinated polyurethane coatings based in fluorinated polyols [57]. Trials of these materials were undertaken on two US Navy patrol vessels in 1986 [70]. These materials showed limited success with hand scrubbing required to remove shell fouling. Fluorinated polyurethanes, like many of the fluoropolymer-containing systems that have been tested are of relatively high modulus and therefore do not encourage adhesive bond fracture through peeling.

More recently, research has been focused on the preparation of fluoropolymer-containing amphiphilic network coatings which are designed to provide compositional and topographical heterogeneity with the aim to reduce adhesive interaction with complex marine adhesives. Gudipati *et al.* [62] have shown that cross-linked hyperbranched fluoropolymer and poly(ethylene) glycol amphiphilic networks provide improved release of algal spores and fluoropolymer systems based on a similar principle are now commercially available (see the discussion on International Paint's Intersleek 900 below).

Table 13.3 Fluorine-containing foul release coating technology summary.

Chemistry type	Description	Year	Remarks	Reference
Fluorinated polyurethanes	Polyols with trifluoromethyl moieties crosslinked with aliphatic isocyanates	1976	Settlement on US Navy vessels, but relatively easy to remove	[57]
	Polyols with fluoroalkyl moieties with aliphatic isocyanates	1998	No data provided	[58]
	Range of fluorinated polyols crosslinked with isocyanates including an isophorone diisocyanate encapped perfluoropolyether	2003	Settlement, but some foul release behaviour observed	[59]
Fluorinated block copolymers	Poly[(2-isopropenyl-2-(oxazoline)] and perfluoroalkyl surfactants	1999	Settlement, but some foul release behaviour observed	[60]
	Poly(styrene-b-isoprene) polymers with semi-fluorinated and polyethylene glycol side-chains	2003	Reduced level of *Ulva* zoospore settlement	[61]
	Hyperbranched fluoropolymer with polyethylene glycol	2005	Reduced level of *Ulva* zoospore settlement and sporeling release demonstrated	[62]
Fluorinated silicones	Tridecafluoro-1,1,2,2-(tetrahydrooctyl)triethoxysilane crosslinked PDMS	1999	No data provided	[63]
	Poly(methylpropenoxy-fluoroalkylsiloxane)s	2000	Reduction in bacterial settlement	[64]
	Polymethylhydridosiloxane with perfluoroalkyl groups	1998	Settlement, but some foul release behaviour observed	[65]
	Silicone containing fluoropolymer	2007	Low 'pseudo barnacle' adhesion recorded	[66]
Perfluoropolyethers	Perfluoropolyether (PFPE) with siloxy functionality	2000	Reduced level of fouling on immersion; low barnacle adhesion demonstrated	[67]
	PFPE macromonomer containing terpolymer crosslinked with an isocyanate	2006	Reduced level of *Ulva* zoospore settlement and sporeling release demonstrated	[68]
Fluoroacrylates	Poly(1H, 1H, 2H, 2H -perfluorodecanoyl diitaconate)	1999	Reduced level of fouling observed	[69]

13.3.2.2   Silicones

Silicones (more accurately called polysiloxanes) are polymers comprising an inorganic silicon–oxygen backbone and organic side groups. However, in the context of FRCs, silicone generally refers to cross-linked polydimethylsiloxane (PDMS). The physicochemical properties of cross-linked PDMS make it an ideal candidate for FRCs. The Si—O bond in the polymer backbone is strong but has a low energy barrier to rotation, which results in a high polymer chain flexibility and elastomeric properties when lightly cross-linked. The non-polar methyl organic side groups provide 'non-stick' properties [71].

The first patents in the area of silicone FRCs date as far back as 1955 [72], but initial attempts to use these materials were hampered by application issues, poor adhesion to substrates and cost [10]. Groundbreaking work on room temperature vulcanised (RTV) silicone elastomers was carried out by Milne in the 1970s [73] and subsequent work found that the addition of non-reactive siloxane fluids, such as phenylmethyl siloxanes, to silicone elastomers improved the performance of silicone FRCs [74]. Contrary to popular speculation, fluid additives do not act to plasticise PDMS coatings and do not leach from coatings to any significant extent, rather they weaken the adhesive joint between the coating and fouling organisms, and studies have shown that substantial levels remain in the film after extended periods due to their very low water solubility [75].

Much work has focused on the optimisation of silicone FRCs. Many studies on the addition of fluids were carried out in the 1990s through US Office of Naval Research (ONR) funding [76] (and more recently high-throughput experimentation has been used to optimise the foul release performance of silanol-terminated PDMS and PDMS-polyurethane-based coatings [77]). Nevertheless, since the 1980s, the major advances in the use of silicone FRCs have not been associated with improved foul release performance, instead methods for improving application and adhesion of FRCs to substrates [78, 79] have led to the increasing commercial importance of such systems.

13.3.3   The development of commercial foul release products

Despite the early discovery that silicones were effective FRCs, it took until 1990s for their commercial benefit to be realised. Silicones were perceived to have inadequate mechanical properties, and adhesion and contamination problems were initially difficult to overcome. In addition, the availability of low-cost SPC technology meant that until the proposed ban of TBT-based antifoulings silicone technology was not economically viable.

The first full vessel application of a silicone FRC was carried out on the 'Tropic Lure' in 1993 by International Paint. In the same year, the first large-scale application to a Navy vessel, the submarine *HMAS Collins*, took place [49]. Intersleek 425 was then commercially launched by International Paint for high-speed vessels (>30 knots) and following the success of early applications, where only microfouling was evident after 5 years of service [80], Intersleek 700 was launched in 1999 for high-activity scheduled ships travelling at 15–30 knots. Similar products followed Intersleek into the market during the 1990s including SigmaGlide (Sigma Coatings) and BioClean SPG (Chugoku Marine Paints). Table 13.4 summarises the main silicone fouling release products on the market today. A review of the performance of a range of these first-generation silicone FRCs was undertaken by Watermann [81].

Table 13.4 Summary of current commercially available foul release coating technologies.

Product type	Market	Company	Product name	Target market
Silicone foul release	Marine	Ameron	ABC Release	All vessel types
		International Paint	Intersleek 425	High-speed vessels, >30 kts
			Intersleek 700	High-activity vessels, >15 kts
		Hempel	Hempasil 77100	Activity > 50% and >25 kts
			Hempasil 77500	Activity > 75% and >15 kts
		Jotun	SeaLion	High-speed, high-activity vessels
		Chugoku Marine Paint	Bioclean	Ocean going vessels, >15 kts
			Bioclean S	Coastal vessels
			Bioclean C	High-speed aluminium craft, >30 kts
		Kansai Paint	Biox	Fast vessels
		KCC	Lo-Frick A/F100	Fast vessels
		SigmaKalon	SigmaGlide 890	Fast vessels
		Transocean	Ultima System	Fast vessels
	Pleasure craft	Oceanmax	Propspeed	Propellers
		International Paint	Veridian	Hulls and propellers
		Chugoku	Bioclean S	Hulls
			Seajet Pellerclean	Propellers
Fluoropolymer foul release	Marine	International Paint	Intersleek 900	All vessel types, >10 kts
Hybrid foul release	Pleasure craft	Microphase Coatings Inc.	Phasecoat UFR	Propellers and outdrives

Following the removal of TBT-SPC antifouling paints in the wake of the IMO action (see Chapter 21), silicone FRCs became more commercially attractive in terms of both cost (in comparison to tin-free SPC paints) and application. As technology for over-coating old TBT antifouling paints became available and with further restrictions on the use of biocides, marine paint manufacturers have intensified their research into FRCs that will be effective at lower vessel speeds. A successful outcome of this research has been the introduction in 2007 by International Paint of Intersleek 900, a fluoropolymer-based product which is specified for vessels with speeds of 10 knots and above [82].

A highly significant additional benefit of FRCs over most biocidal antifouling technologies is that they can bring demonstrable economic benefits to the shipowner through the reduction in hydrodynamic drag, arising in large part from the very smooth surface properties of such paint films [83]. Fuel savings of up to 10% have been claimed by some manufacturers [84]. However, opportunities remain for the development of products with even better foul release properties, particularly to microfouling, with the ultimate aim being to produce foul release performance under static conditions suitable for the pleasure craft market and delivering even greater fuel savings to the commercial sector. Such products may also have particular benefits in the aquaculture and oil-production/offshore sectors, where the use of biocidal products and/or regular mechanical cleaning still predominate [85, 86] (see Chapters 18 and 19).

13.3.4 Alternative technologies

As well as the development of fouling control technologies based on conventional coating components an enormous number of coating and non-coating alternative technologies have been explored [87]. None of these alternative fouling control technologies have achieved widespread commercial success. However, with an increasing awareness of perceived environmental issues associated with biocidal antifouling products, there continues to be much research activity into alternative technologies. In the context of this review, alternative technologies are broadly divided into two areas:

- Non-fouling surfaces – non-biocidal alternatives to foul release
- Non-coating solutions

13.3.4.1 Non-fouling surfaces

There have been many studies aimed at developing a truly non-fouling surface based on a range of concepts from varying surface wettability to producing a stimulus responsive 'smart' surface. The concept of amphiphilic surfaces has already been touched upon, but coatings having an extremely hydrophilic or hydrophobic nature have also been considered in the context of fouling control [88]. Highly hydrophilic surfaces such as hydrogels have been considered as media for release of bioactive materials [89], but also as fouling control surfaces in their own right [90]. However, despite studies showing inhibition of settlement [91], coatings based on hydrogels have not been exploited for marine end-use due to their relatively poor mechanical properties and the practicality of application. More recently, the other extreme of wettability has been considered as a route to non-fouling surfaces [92]. The concept of 'super-hydrophobic' coating derived from hydrophobic, physically rough surface is inspired by nature and in particular the lotus leaf [93]. Numerous scientific papers are dedicated to the understanding behind wettability and surface roughness [92, 94], and it has been shown that there is a critical value of the fraction of a surface that is in contact with a liquid droplet below which 'air pockets' will exist beneath that droplet. This so-called Cassie–Baxter regime is related to the roughness and surface energy of the surface [95]. While there are studies which show that microtopographical surfaces do reduce macrofouling [96], even if a practical 'super-hydrophobic' surface was available, it is still unclear whether it would provide a sustained non-fouling effect in the marine environment.

Stimulus responsive or 'smart' surfaces have also been studied for fouling resistance. In particular, poly(N-isopropyl acrylamide) (PNIPAAm) has been investigated as a temperature responsive material as it exhibits a lower critical solution temperature (LCST) behaviour [97]. Polymers that exhibit a LCST in aqueous solutions 'switch' from being hydrophilic and soluble at low temperature to hydrophobic and insoluble at a specific critical temperature (T_c). This concept has been shown to have some validity by Ista *et al.* [98] who found that the marine microorganisms were removed from a surface grafted with PNIPAAm at the T_c phase transition. Light-stimulated wettability changes have also been considered in the context of non-fouling stimulus responsive surfaces [99]. However, it seems unlikely that such systems will be of use on the bottom of large marine vessels.

13.3.4.2   Non-coating solutions

Over the last 50 years, a number of non-coating solutions to the problem of marine fouling have been considered. These alternatives include the use of electricity [100], magnetism [101], sound [102], heat [87] and even radiation [103]. Electrical techniques, probably the most successful non-coating alternative, work either by directly inhibiting organism settlement or by the generation of biocide from electrolysis of sea water [104]. However, the relative complexity of such systems has so far limited their application to heat exchangers and sea water inlets [105].

The use of mechanical cleaning devices for pleasure craft and ship hulls, sometimes known as hull-husbandry, is not so much a preventative measure as a short-term cure for fouling. However, with increasing environmental pressure on the use of biocides, the use of underwater cleaning in combination with specialised non-biocidal fouling control systems can become a viable option, and there are a number of companies who offer cleaning solutions for ship hulls coated with silicones and glass-flake epoxies [106]. The use of boat scrubbing machines in the pleasure craft market has also been considered and there are an increasing number of available commercial systems [107]. In studies on the use of boat scrubbers, it is clear that an automated boat cleaner using brushes or high-pressure water is a potential solution, but its acceptance would rely upon the willingness of the customer to adopt scrubbing over the available coating options [108].

13.4   Paint selection – economic considerations

Perhaps understandably, paint manufacturers are unwilling to openly share information on the pricing of their products, but it is probably fair to assume that, in general, longer lasting and higher performing paints have a higher selling price than shorter lifetime, lower performing products. According to Eliasson [20], all of the major paint types are more expensive than the previously available TBT-SPC antifouling paints and on a per litre basis FRCs are more expensive than current SPC technologies, which are more expensive in turn than hybrid antifoulings and CDPs. However the cost of the paint is only one of the contributory factors to the overall cost of painting, particularly in the commercial sector. As detailed by Eliasson [20], other pertinent costs include the off-hire for the vessel, hull cleaning, surface preparation, paint application and dry-dock hire. As shown in Table 13.5, when these additional costs are taken into account along with the expected lifetime of the paint, then the estimated overall cost of a FRC is actually less than for SPC antifouling products on a cost per square m per year basis, even though the per litre cost is substantially higher. Eliasson has assumed a 10-year lifetime for the FRC, whereas, in practice, a lifetime of 5 years is more normally specified for such products. Nevertheless, when the aforementioned potential fuel economy benefits are

Table 13.5 Estimated typical paint and painting costs for biocidal antifouling and foul release coatings [97].

Technology	Assumed paint lifetime (years)	Relative paint cost, (TBT-SPC = 1)	Estimated overall application cost	
			$ m^{-2}	$ m^{-2} year^{-1}
CDP	3	1.5	50	15.2
Hybrid	5	1.75	50	10
Tin-free SPC	5	2–3	75	15
Foul release	10	4–6	116	11.6

also taken into account, FRCs can be expected to be seen as an increasingly attractive option to many ship owners.

13.5 Future perspectives

Numerous technologies have been tested throughout history in an attempt to develop the ultimate solution to the marine biofouling problem. The introduction of TBT-SPC technology in the late 1970s led to a step-change in performance, which resulted in a belief that the problem had been solved and raised customer expectations on performance of subsequent fouling control coating technologies. The well-documented impacts of tributyl tin compounds on non-target organisms (see Chapters 16 and 17) resulted in the IMO action to ban new applications by 2003 [34] (see Chapters 21 and 26) and led to the introduction of equally efficient tin-free SPC antifouling paints (see Chapter 14). Large rises in fuel prices have led to increasing value of fouling control coatings to the customer and a need for even higher performance products. However, there is also increasing regulatory pressure on any system that relies on the release of biocides. Indeed, the BPD has already reduced the number of available biocides for use within Europe and in the USA further restrictions on biocide use are possible through Reregistration Eligibility Decisions (REDs) and Uniform National Discharge Standards (UNDS).

In addition to restrictions on the use of biocides, the forthcoming Registration, Evaluation and Authorisation of Chemicals (REACH) legislation and solvent emissions regulations, for example the Solvent Emission Directive (SED) in Europe, further constrain the chemist in developing new coating technologies. With this in mind, it is likely that future developments will lie in more efficient use of registered biocides in antifouling paints and in the area of non-biocidal fouling control coating solutions.

One of the more challenging areas for new fouling control technologies is the pleasure craft sector, where vessels spend the majority of their time under static conditions. In this market sector, there is an immediate need for new environmentally benign fouling control technologies as local legislation has often particularly focused on sensitive environments such as the Baltic, or areas where pleasure craft use is particularly heavy, such as Shelter Island in San Diego Bay [109]. It is therefore likely that in the short term some pleasure craft owners in certain locations may be forced to accept the need for biocide-free coating technologies in conjunction with regular and frequent boat cleaning.

In the marine sector, the use of FRCs can be expected to increase [110], especially with the development of new and more effective systems for slower vessels. In the longer term, there will also be continued focus on the development of improved foul release surfaces for even lower vessel speeds, with the ultimate goal being coatings that spontaneously release at any speed or none. The development of such coatings would of course also have obvious benefits in the fields of aquaculture, offshore and other static industrial applications where fouling control is required.

In addition to the search for new fouling release technology, the search for a non-fouling surface is bound to continue, with one area of increasing interest being the control of surface physicochemical structure at the micro- and nanoscale. Indeed, much of the technologies currently under investigation through the US Office of Naval Research are aimed at surface engineering, and the EU Framework 6 Project *Advanced Nano Structured Surfaces for the Control of Biofouling* (AMBIO) [111] specifically targets this area. While fouling control technology has undoubtedly come a long way since Plutarch's time, it undoubtedly also still has a long way to go.

13.6   Conclusions

- The detrimental consequences of fouling on ship and boat hulls have been recognised for millennia and include loss of speed and increased power requirements to maintain speed. Paint and coating technology is routinely used to control fouling and such products can be generally classified as either biocidal paints, which function through the use of toxic agents to prevent the settlement and growth of fouling species and non-biocidal FRCs, which make use of favourable surface energetics to minimise adhesion of organisms and facilitate easy removal by water flow.
- The most commercially successful biocidal antifouling paints can be mechanistically sub-classified as contact leaching, soluble matrix/CDP, SPC or hybrid antifouling paints.

References

1. WHOI (1952) *Marine Fouling and its Prevention*. Woods Hole Oceanographic Institute, United States Naval Institute, Annapolis.
2. Plutarch (1870) *Plutarch's Morals* (translated from the Greek by several hands corrected and revised by William W. Goodwin with an introduction by Ralph Waldo Emerson). Little, Brown & Co, Boston.
3. Holm, E.R., Schultz, M.P., Haslbeck, E.G., Talbott, W.J. & Field, A.J. (2004) Evaluation of hydrodynamic drag on experimental fouling-release surfaces, using rotating discs. *Biofouling*, **20**, 219–226.
4. Lethwaite, J.C., Molland, A.F. & Thomas, K.W. (1985) An investigation into the variation of ship skin frictional resistance with fouling. *Transactions of the Royal Institute of Naval Architects*, **127**, 269–284.
5. Schultz, M.P. (2004) Frictional resistance of antifouling coating systems. *Journal of Fluids Engineering*, **126**, 1039–1047.
6. Schultz, M.P. (2007) Effects of coating roughness and biofouling on ship resistance and powering. *Biofouling*, **23**, 331–341.

7. Steffy, J.R. (1985) The Kyrenia ship: an interim report on its hull construction. *American Journal of Archaeology*, **89**, 71–101.

8. Bertram, J. (2000) Past, present and prospects of antifouling. In: *32nd WEGEMT School on Marine Coatings*, pp. 87–97. WEGEMT, New York.

9. Anderson, C. (2004) Coatings, antifoulings. In: *Kirk-Othmer Encyclopedia of Chemical Technology*, Vol. **7**, 5th edn (ed. R.E. Kirk-Othmer), pp. 150–167. Wiley-Interscience, New Jersey.

10. Almeida, E., Diamantino, T.C. & de Sousa, O. (2007) Marine paints: the particular case of antifouling paints. *Progress in Organic Coatings*, **59**, 2–20.

11. European Union (1998) European Directive 98/8/EC of the European Parliament and of the Council of 16 February 1998 concerning the placing of biocidal products on the market. *Official Journal of the European Union*, L123/1.

12. Lewis, J.A. (1998) Marine biofouling and its prevention on underwater surfaces. *Materials Forum*, **22**, 41–61.

13. Banfield, T.A. (1980) Marine finishes: part 2. *Journal of the Oil and Colour Chemists Association*, **69**, 93–100.

14. de Wolf, P. & Londen, A.M. (1966) Anti-fouling compositions. *Nature*, **209**, 272–274.

15. Finnie, A.A. (2006) Improved estimates of environmental copper release rates from antifouling products. *Biofouling*, **22**, 279–291.

16. Sharma, V.K. & Millero, F.J. (1988) Oxidation of copper (I) in seawater. *Environmental Science and Technology*, **22**, 768–771.

17. Yebra, D.M., Kiil, S. & Dam-Johansen, K. (2004) Antifouling technology – past, present and future steps towards efficient and environmentally friendly antifouling coatings. *Progress in Organic Coatings*, **50**, 75–104.

18. Callow, M.E. (1999) The status and future of biocides in marine biofouling prevention. *Recent Advances in Marine Biofouling Prevention*, **3**, 109–126.

19. European Commission (2007) Commission regulation (EC) No. 1451/2007 of 4 December 2007 on the second phase of the 10-year work programme referred to in Article 16(2) of Directive 98/8/EC of the European Parliament and of the Council concerning the placing of biocidal products on the market. *Official Journal of the European Union*, L325/3.

20. Eliasson, J. (2003) Economics of coatings/corrosion protection of ships – selecting the correct type of anticorrosion protection for underwater applications on new buildings. In: *Lloyd's List Event Conference: Prevention and Management of Marine Corrosion*. 2–3 April 2003. Lloyds, London.

21. JPMA (2008) http://www.toryo.or.jp/eng/index.html, Japan Paint Manufacturers Association, Japan. Accessed 5 July 2009.

22. Thomas, K.V., Blake, S.J. & Waldock, M.J. (2000) Antifouling paint booster biocide contamination in UK marine sediments. *Marine Pollution Bulletin*, **40**, 739–745.

23. LME (2008) http://www.lme.co.uk/copper_graphs.asp, London Metal Exchange, London. Accessed 5 July 2009.

24. Caprari, J.J., Slutzky, O., Pessi, P.L. & Rascio, V. (1986) A study of the leaching of cuprous oxide from vinyl antifouling paints. *Progress in Organic Coatings*, **13**, 431–444.

25. Conner, A.H. (1989) Chemistry of other components in naval stores. In: *Naval Stores* (eds D.F. Zinkel & J. Russel), pp. 441–475. Pulp Chemicals Association, New York.

26. Anderson, C.D. (1993) *Self-Polishing Antifoulings: A Scientific Perspective*. Ship Repair & Conversion, London.

27. Yebra, D.M., Kiil, S., Dam-Johansen, K. & Weinell, C.E. (2006) Mathematical modelling of tin-free chemically-active antifouling paint behaviour. *AIChE Journal*, **52**, 1926–1940.

28. Kiil, S., Weinell, C.E., Pederson, M.S. & Dam-Johansen, K. (2001) Analysis of self-polishing antifouling paints using rotary experiments and mathematical modeling. *Industrial and Engineering Chemistry Research*, **40**, 3906–3920.

29. Kiil, S., Weinell, C.E., Pederson, M.S. & Dam-Johansen, K. (2002) Mathematical modelling of a self-polishing antifouling paint exposed to seawater: a parameter study. *Transactions of the Institution of Chemical Engineers, Part A*, **80**, 45–52.

30. Milne, A. (1991) Ablation and after: the law and the profits. In: *Polymers in a Marine Environment: The Institute of Marine Engineers Third International Conference*. Marine Management Ltd, London.

31. Anderson, C. (1993) Self-polishing antifoulings: a scientific perspective. In: *Ship Repair and Conversion 93, BML Business Meetings*. BML Business Meetings, Rickmansworth.

32. Anderson, C.D. & Hunter, J.E. (2001) TBT-free antifouling coating technologies and performance – a technical review. In: *Pollution Prevention from Ships and Shipyards Symposium: Oceanology International 2001*, pp. 81–86. ATRP Corporation, Miami.

33. Champ, M.A. (2000) A review of organotin regulatory strategies, pending actions, related costs and benefits. *The Science of the Total Environment*, **258**, 21–71.

34. IMO (2001) *International Convention on the Control of Harmful Anti-Fouling Systems on Ships*. International Maritime Organisation, London.

35. Ohsugi, H.Y., Matsuda, M., Eguchi, Y. & Ishikura, S. (1989) Antifouling behaviour of copper-containing polymer. In: *Thirteenth International Conference on Organic Coatings Science and Technology 1989*, pp. 185–191. Technomic Publishing Co., Lancaster, PA.

36. Shilton, C. (1997) Mechanism of action of tin-free antifouling paints: Intersmooth 360 Ecoloflex. *Pitture e Vernici*, **73** (9), 10–18.

37. Anon (2004) *Antifouling Product and Technology Guide*. International Marine Coatings, Felling, UK.

38. Silverman, G.S. & Aubart, M. (2006) Polymer binders represent a major breakthrough for self-polishing marine antifoulant paints. *Paint & Coatings Industry*, **2006**, 40–46.

39. Fox, J. & Finnie, A.A. (2001) Antifouling Paint. Patent EP1144518-A1, Akzo Nobel NV.

40. Rittschof, D. (2000) Natural product antifoulants: one perspective on the challenges related to coatings development. *Biofouling*, **15**, 119–127.

41. Clare, A.S. (1993) Towards nontoxic antifouling. *Journal of Marine Biotechnology*, **6**, 3–6.

42. De Nys, R., Steinberg, P.D., Willemsen, P., Dworjanyn, S.A., Gabelish, C.L. & King, R.J. (1995) Broad spectrum effects of secondary metabolites from the red alga *Delisea pulchra* in antifouling assays. *Biofouling*, **8**, 259–271.

43. Todd, J.S., Zimmerman, R.C., Crews, P. & Alberte, R. (1993) The antifouling activity of natural and synthetic phenolic acid sulphate esters. *Phytochemistry*, **34**, 401–444.

44. Alberte, R.S. & Zimmerman, R.C. (1999) Antifouling Agents. Patent WO0016623-A1, Phycogen Inc.

45. Barrios, C.A., Xu, Q., Cutright, T. & Newby, B.Z. (2004) Zosteric acid: an effective antifoulant for reducing bacterial attachment on coatings. *Polymer Preprints*, **45**, 227–228.

46. Olsen, S.M., Pedersen, L.T., Laursen, M.H., Kiil, S. & Dam-Johansen, K. (2007) Enzyme-based antifouling coatings: a review. *Biofouling*, **23**, 369–383.

47. Bonaventura, C., Bonaventura, J. & Hooper, I.R. (1991) Antifouling Methods using Enzyme Coatings. Patent US5998200, Duke University.

48. Candries, M. & Anderson, C.D. (2003) Estimating the impact of new-generation antifoulings on ship performance: the presence of slime. *Journal of Marine Engineering and Technology (Part A)*, **2003** (2), 13–22.

49. Anon (1998) *Intersleek 700: First Biocide-Free Fouling Control System for Scheduled Shipping*. Propeller Direct, International Paint Ltd, Felling, UK.

50. Fourche, G. (1995) An overview of the basic aspects of polymer adhesion 1: fundamentals. *Polymer Engineering and Science*, **35**, 957–967.

51. Baier, R.E. (1973) Influence of the initial surface condition of materials on bioadhesion. In: *Proceedings of the 3rd International Congress on Marine Corrosion and Fouling 1973* (eds R.F. Acker, B.F. Brown, J.R. DePalma & W.P. Iverson), pp. 633–639. Northwestern University Press, Evanston.

52. Chaudhury, M.K., Newby, B.Z. & Brown, H.R. (1995) Macroscopic evidence of the effect of interfacial slippage on adhesion. *Science*, **269**, 1407–1409.

53. Brady, R.F. (2000) Clean hulls without poisons: devising and testing nontoxic marine coatings. *Journal of Coatings Technology*, **72**, 45–56.

54. Kohl, J.G. & Singer, I.L. (1999) Pull-off behaviour of epoxy bonded to silicone duplex coatings. *Progress in Organic Coatings*, **36**, 15–20.

55. Swain, G. (1999) Redefining antifouling coating. *Paint Coat Europe*, July, 18–25.

56. Anon (2007) *INTERSLEEK^® 900 Fluoropolymer Foul Release Coating*. International Paint Ltd., London.

57. Field, D.E. (1976) Fluorinated polyepoxy and polyurethane coatings. *Journal of Coatings Technology*, **48**, 43–47.

58. Brady, R.F. & Aronson, C.L. (2003) Elastomeric fluorinated polyurethane coatings for nontoxic fouling control. *Biofouling*, **19**, 59–62.

59. Brady, R.F. & Bonafede, S.J. (1998) Compositional effects on the fouling resistance of fluoroethane coatings. *Surface Coatings International*, **4**, 181–185.

60. Brady, R.F., Bonafede, S.J. & Schmidt, D.L. (1999) Self-assembled water-borne fluropolymer coatings for marine fouling resistance. *Surface Coatings International*, **12**, 582–585.

61. Youngblood, J., Andruzzi, L., Ober, C.K., *et al.* (2003) Coatings based on side-chain ether-linked poly(ethylene glycol) and fluorocarbon polymers for the control of marine biofouling. *Biofouling*, **19**, 91–98.

62. Gudipati, C.S., Finlay, J.A., Callow, J.A., Callow, M.E. & Wooley, K.L. (2005) The antifouling and foul-release performance of hyperbranched fluoropolymer (HBFP) – poly(ethylene glycol) (PEG) composite coatings evaluated by adsorption of biomacromolecules and the green fouling alga *Ulva*. *Langmuir*, **21**, 3044–3053.

63. Johnston, E., Bullock, S.U.J., Gatenholm, P. & Wynne, K.J. (1999) Networks from dihydroxypoly(dimethylsiloxane) and (tridecafluoro-1,1,2,2-tetrahydrooctyl)-triethoxysilane: surface microstructures and surface characterisation. *Macromolecules*, **32**, 8173–8182.

64. Tsibouklis, J., Peters, V., Smith, J.R., Nevell, T.G. & Thorpe, A.A. (2000) Poly(methylpropenoxyfluoroalkylsiloxane)s: a class of fluoropolymers capable of inhibiting bacterial adhesion onto surfaces. *Journal of Fluoropolymer Chemistry*, **104**, 37–45.

65. Mera, A., Fox, R.B., Johnston, E., Bullock, S. & Wynne, K. (2001) Toward minimally adhesive surfaces utilizing siloxanes. In: *Conference: Silicones in Coatings II*. Paper 22. Paint Research Association, Hampton, UK.

66. Dahling, M., Lien, E.M., Orsini, L.M., Galli, G. & Chiellini, E. (2007) Fouling Release Composition. Patent WO 2007/102741 A1, Jotun AS.

67. Williams, D. & Lines, R. (2001) Antifouling Coating Composition comprising a fluorinated Resin. Patent A1287056B1, International Coatings Ltd, Ausimont SPA.

68. Yarbrough, J.C., Rolland, J.P., DeSimone, J.M., Callow, M.E., Finlay, J.A. & Callow, J.A. (2006) Contact angle analysis, surface dynamics, and biofouling characteristics of cross-linkable, random perfluoropolyether-based Graft terpolymers. *Macromolecules*, **39**, 2521–2528.

69. Pullin, R.A., Nevell, T.G. & Tsibouklis, J. (1999) Surface energy characteristics and marine antifouling performance of poly(1H, 1H, 2H, 2H perfluorodecanoyldiitaconate) film structures. *Material Letters*, **39**, 142–148.

70. Brady, R.F. & Griffith, J.R. (1987) Non-toxic alternatives to antifouling paints. *Journal of Coating Technology*, **59**, 113–119.

71. Vincent, H.L. & Bausch, G.G. (1997) Silicon fouling release coatings. *Naval Research Reviews*, **4**, 39–45.

72. Robbart, E. (1955) Ship's Hull coated with Anti-Fouling Silicone Resin and Method of Coating. Patent US2986474A.

73. Milne, A. & Callow, M. (1985) Non-biocidal antifouling processes. In: *Polymers in the Marine Environment* (ed. R. Smith), pp. 229–233. The Institute of Marine Engineer, London.

74. Milne, A. (1975) *Coated Marine Surfaces*. Patent GB1470465A, International Paint Ltd.

75. Burnell, T., Carpenter, J., Truby, K., Serth-Guzzo, J., Stein, J. & Wiebe, D. (2000) Advances in non-toxic silicone biofouling release coatings. *ASC Symposium*, **729**, 180–193.

76. Truby, K., Darkangelo Wood, C., Stein, J., *et al.* (2000) Evaluation of the performance enhancement of silicone biofouling-release coatings by oil incorporation. *Biofouling*, **15**, 141–150.

77. Webster, D.C., Chisholm, B.J. & Stafslien, S.J. (2007) Mini review: combinatorial approaches for the design of novel coating systems. *Biofouling*, **23**, 179–192.
78. Groenlund Scholten, M., Martin, A., Weinrich Thorlaksen, P., Oxfeldt Andresen, A. & Nielsen, J. (2003) *A Tie-Coat Composition Comprising at Least Two Types of Functional Polysiloxane Compounds and a Method for Using the Same for Establishing a Coating on a Substrate*. Patent EP1670866B1, Hempel A/S.
79. Hamilton, T., Green, G.E. & Williams, D.N. (1998) *Fouling Inhibition*. Patent EP 1042413 B1, Akzo Nobel Coatings International BV.
80. Anon (1999) *Intersleek 700: First Biocide-Free Fouling System for Scheduled Shipping*. International Paint Ltd, Felling, UK.
81. Watermann, B., Berger, H.D., Sönnichsen, H. & Willemsen, P. (1997) Performance and effectiveness of non-stick coatings in seawater. *Biofouling*, **11**, 101–118.
82. Anon (2007) *Intersleek 900: Fluoropolymer Foul Release Coating for all Vessel Types*. International Paint Ltd, Felling, UK.
83. Candries, M., Altar, M. & Anderson, C.D. (2000) Considering the use of alternative antifoulings: the advantages of foul-release systems. In: *Conference Proceedings ENSUS 2000*, pp. 88–95. Departments of Marine Technology and Marine Sciences, Newcastle University, UK.
84. Westergaard, C.H. (2007) *Comparison of fouling Control Coating Performance to Ship Propulsion Efficiency*. Hempel A/S.
85. Walker, I. (1998) Non-toxic fouling control systems. *Pitture e Vernici Europe*, **74**, 17–22.
86. Willemsen, P. (2005) Biofouling in European aquaculture: is there an easy solution? *European Aquaculture Society Special Publication*, **35**, 82–87.
87. Swain, G.W. (1998) Biofouling control – a critical component of drag reduction. In: *Proceedings of the International Symposium on Seawater Drag Reduction, 22–24 July* (ed. J.C.S. Meng), pp. 155–161. Newport, USA.
88. Genzer, J. & Efimenko, K. (2006) Recent developments in superhydrophobic surfaces and their relevance to marine fouling: a review. *Biofouling*, **22**, 339–360.
89. Gatenholm, P., Holmstrom, C., Maki, C. & Kjelleberg, S. (1995) Toward biological antifouling surface coatings: marine bacteria immobilized in hydrogel inhibit barnacle larvae. *Biofouling*, **8**, 293–301.
90. Cowling, M.J., Hodgkiess, T., Parr, A.C.S., Smith, M.J. & Marrs, S.J. (2000) An alternative approach to antifouling based on analogues of natural processes. *The Science of the Total Environment*, **258**, 129–137.
91. Rasmussen, K., Willemsen, P.R. & Ostgaard, K. (2002) Barnacle settlement on hydrogels. *Biofouling*, **18**, 177–191.
92. Marmur, A. (2006) Super-hydrophobicity fundamentals: implications to biofouling prevention. *Biofouling*, **22** (2), 1–9.
93. Neinhuis, C. & Barthlott, W. (1997) Characterization and distribution of water-repellent, self-cleaning plant surfaces. *Annals of Botany*, **79**, 667–677.
94. Marmur, A. (2006) Soft contact: measurement and interpretation of contact angles. *Soft Matter*, **2**, 12–17.
95. Callies, M. & Quéré, D. (2005) On water repellency. *Soft Matter*, **1**, 55–61.
96. Scardino, A.J., Harvey, E. & De Nys, R. (2006) Testing attachment point theory: diatom attachment on microtextured polyimide biomimics. *Biofouling*, **22**, 55–60.
97. Cunliffe, D., De las Heras Alarcon, C., Peters, V., Smith, J.R. & Alexander, C. (2003) Thermoresponsive surface grafted poly(*N*-Isopropylacrylamide) copolymers: effect of phase transitions on protein and bacterial attachment. *Langmuir*, **19**, 2888–2899.
98. Ista, L.K., Pérez-Luna, V.H. & Lopèz, G.P. (1999) Surface-grafted, environmentally sensitive polymers for biofilm release. *Applied and Environmental Microbiology*, **65**, 1603–1609.
99. Nobuhiko, K., Tsuneo, A., Mizuno, T. & Norikazu, N. (1997) Prevention of Fouling and Beltlike Unit for Preventing Fouling for Ship or Immersion Structure. Patent JP11092315 A, Ishikawajima Harima Heavy Ind.

100. Corp, B. (1981) *Multilayered Submersible Structure with Fouling Inhibiting Characteristic.* Patent US7025013B1, Brunswick Corp.
101. Mitsuru, K.K. (1981) *Antifouling agent for Aquatic Life.* Patent JP63057503, Koryu Kogyo KK.
102. Aksel'band, A.M. (1960) Ultrasonic protection of ships from fouling. *Transactions of the Oceanographic Commission, The Academy of Sciences, USSR,* **13**, 7–9.
103. Morley, C.O., Clarke, H.J., Bowe, H.J.M. & Arnold, M.H.M. (1958) The use of radioactivity against marine fouling. *Journal of the Oil and Colour Chemists' Association,* **41**, 445–452.
104. Omae, I. (2003) General aspects of tin-free antifouling paints. *Chemical Reviews,* **103**, 3431–3448.
105. Nakamatsu, S., Harada, H., Shinomiya, Y. & Omizu, T. (1978) *An Electrolytic Cell for Electrolysis of Sea Water.* Patent US4173525, Chlorine Eng Corp Ltd.
106. Anon (1997) Could underwater cleaning replace antifoulants? *Marine Engineers' Review,* February, 28–29.
107. Weber, M. (1997) *Device for External Cleaning of Ships Hulls.* Patent WO9741026.
108. Anon (2006) Contactless cleaning of underwater hulls. Marina Port Zélande B.V.
109. Anon (2007) *Antifoulings: The Legislative Position Key Points Summary.* International Paint Ltd, Felling, UK.
110. Lewis, J.A. (2003) TBT antifouling paints are now banned! What are the alternatives and what of the future? *Surface Coatings Australia,* **40**, 12–15.
111. AMBIO (2008) *Advanced nanostructured Surfaces for the Control of Biofouling, FP6 European Project, Contract NMP4-CT-2005–011827.* Available from http://www.ambio.bham.ac.uk. Accessed 5 July 2009.

Chapter 14
Fouling on Shipping: Data-Mining the World's Largest Antifouling Archive

Jeremy C. Thomason

The aim of this chapter is to describe how modern marine antifouling coatings perform relative to each other. As there is almost no pertinent literature to review, this is achieved through the presentation of an analysis of the world's largest data archive of biofouling on the operational global marine fleet. Hence, this chapter mostly describes results new to the field and emphasis is placed on how hull factors (speed, area and size) and time as in-service period and year of application affect the efficacy of different coating technologies. A major objective is to determine if the ecotoxicological concerns over the use of organotin-based biocidal coatings has been traded for a greater impact on the world's climate through greater atmospheric emissions from ships with poorly performing coatings: have we gone from the frying pan to the fire?

14.1 Introduction

The historical perspective on marine biofouling and how modern technologies function is presented in Chapters 13 and the impacts of the organotin-based biocidal coatings are considered in great detail in Chapter 16 and 17, therefore the purpose of this short introduction is to set the scene for the subsequent analysis.

As chemistry advanced in the early twentieth century, so did the nature of the coatings, and the simple traditional antifouling coatings were replaced with tributyl tin self-polishing coatings (TBT-SPC) where hydrolysis in sea water controls the release of the biocide and maintains a smooth surface. The combination of a very powerful biocide with a self-polishing smooth surface made these products economically very beneficial: they offered direct fuel savings for up to 5 years, increased dry dock intervals with a knock-on benefit of reduced downtime for a very expensive hull. After their introduction to the market in 1975, their use reached a peak in 1991 when ~77% (calculated as total deadweight tons [DWT] of vessels coated) of marine coatings applied were TBT-SPCs. Following worldwide measurable coastal concentrations of TBT and demonstrated impacts on non-target organisms, such as oysters and keystone gastropods, with effects detectable at 1 ng L^{-1}, e.g. [1–4], it was not surprising that concerns were raised about their continued use. From 1982 onwards, national legislation for many countries restricted the use of TBT-SPCs to marine vessels longer than 25 m, followed by a proposed International Maritime Organization (IMO) restriction on reapplication by 2003 and total removal, or sealing in, of TBT-SPCs by 2008 (see Chapter 21). Despite vigorous support from many quarters for these legal instruments, there has been some concern that without high-performance replacement coatings for TBT-SPCs a direct toxic impact on marine organisms would simply be traded for an indirect climate change impact through additional CO_2 emissions [5, 6]: the classic out of the frying pan and into the fire scenario. As TBT-SPCs have been removed from the marketplace, there has been considerable research and development by the dominant antifouling companies such as International Paint, PPG, Jotun, Hempels, Chugoku, Nippon Oil and Fat, Korean Chemical Company and Nippon Paint Marine Coatings to provide marine antifouling coatings with similar economic benefits without any associated collateral environmental impacts. There has been some scepticism that this is not achievable [7, 8] and there is widespread opinion in the end-user community (ship owners, agents and yards) that the 'best coatings' have been banned and anything that is subsequently available is not as good. Dealing (either supporting or countering) such anecdotal opinion is difficult without robust data. The two most recent commercially available innovations for the global marine fleet have been foul release (FR) and metal acrylate self-polishing coatings (MA-SPC). The former work by removing biofouling under shear as the vessel steams and the latter work in a similar fashion to the TBT-SPCs they were designed to replace, namely biocide delivery and a smooth self-polishing surface (further details can be found in Chapter 13). There are other commercially available niche products of different technologies, but these are targeted at leisure craft and hence not covered in this analysis.

To assess both the performance of traditional coatings, TBT-SPCs, and their replacements is challenging as data are proprietary and not readily available to academia. The peer-reviewed literature records progress at the cutting edge of the coatings development field, with papers on novel antifouling chemicals and potentially useful strategies (e.g. [9–31]), antifouling surface characteristics (e.g. [23, 32–36]), release rates of biocides and mechanisms for their estimation (e.g. [37–39]) and aspects of their performance in the laboratory and field (e.g. [18, 20, 21, 25, 31, 40–43]) being described in approximately 120 papers in the last 20 years. During the same period (1988–2008), there have also been 520 world patents granted for antifouling coatings, of which 266 are solely for marine coatings. Yet despite this wealth of data it is almost impossible to obtain reliable and representative data on biofouling on commercially available antifouling coatings on operating vessels. There are a few notable examples of such studies, though these have been largely concerned with how ship hulls act as vectors for invasive species (see Chapter 24), rather than focus on how the coatings are performing. Only three studies have been global in scope and none in the last 100 years has surveyed more than 250 vessels [44]. This is likely to be due to the lack of access to commercial vessels, perhaps due to few collaborations between scientists and vessel owners, difficulty of timing access to dry docks (with often short notice this makes hull inspections expensive without the availability of local technical services), recent antiterrorism legislation hindering access to dockyards in many countries, and the fact that brand names of commercial products may not reflect changes in coating technology because of commercial secrecy. The latter is a fact of life in the commercial world as all manufacturers seek to protect their developments (cf. Chapter 27). Thus, amongst the wealth of published material on antifouling coatings, there has been no analysis of the relative performance of marine antifouling coatings: until now.

14.2   Digging the data-mine

Since 1974 an archive of technical reports has been collated from vessel assessments undertaken around the world by a global team of highly trained technical inspectors who survey vessels as they are dry-docked for initial application or renewal of their antifouling and anticorrosion coatings. A standardised form is filled out which records the type of vessel (46 different types), vessel size (DWT), operating speed (knots), date when coatings were applied and when they were assessed, from which in-service interval (months) is calculated, whether the vessel has since voyaged in ice or been cleaned in the water prior to docking, the type of coating used, which part of the vessel has been coated (sides, flats, i.e. the bottom of the hull, or the boot-top, i.e. the waterline), and data on animal, weed and slime biofouling. Currently this archive (as at June 2007) is the largest available data set in the world on marine biofouling on operating vessels, containing over 242 000 records, of which 152 746 are usable after data cleaning (see below). This is a sample of approximately 2.3% of the global schedule and deep-sea fleet as assessed by DWT inspected in 2005 [45].

To ensure maximum validity of the analysis a rigorous cross-checked cleaning of the data set was undertaken to give the final 152 746 samples. This involved removing records for all those vessels that had voyaged in ice or that had been scrubbed in the water prior to docking, records for coating only applied above the waterline (boot-top), records for coating primers, coatings of unknown technology, vessels where the in-service period was recorded as zero months (new build records), vessels with no size specified, static vessels such as oil storage tanks, vessels with spurious speed records of greater than 50 knots, and a few vessels with first generation (pre-1992) MA-SPCs. These latter coatings were of an identifiably different technology from the recent commercially available MA-SPCs, and were only applied for a very short period and only on very few vessels which made it technically very difficult to incorporate them into the analysis.

The database records biofouling data as the extent (e) and severity (s) of both animal (A) and weed (W) biofouling. From the biofouling assessments an index of coating performance (C_p) was derived:

$$C_p = A_s \times A_e + W_s \times W_e$$

where C_p ranges between 0 and 100. However as a response variable this index of coating performance is difficult to deal with as the data are strongly U-shaped and skewed, non-transformable to normality, and thus preclude the use of the usual parametric statistics. To deal with this problem, C_p was made binary where a score of <10 is taken to be a satisfactory (1) performance and >10 a coating failure (0). These benchmarks are industrially relevant.

Such a binary response variable, coating satisfaction (S_c), enabled the use of logistic regression modelling which due to its great flexibility can include both continuous and categorical predictor variables. This was undertaken in SPSS v15, as this program provides a powerful programming language (syntax) and sophisticated memory management that permits the parsing and manipulation of very large data sets that include a considerable amount of alphanumeric (string) data.

Further manipulation of the data was undertaken to aggregate the number of vessel types to just four to prevent over-specification of the model:

- Naval vessels such as destroyers, aircraft carriers and frigates. These are peculiar in comparison to the rest of the global fleet as they operate at high speed but spend most of their time at anchor
- Coastal vessels which was taken to be all vessels $<10 000$ DWT, plus those such as barges, auxiliaries, tugs and trawlers

- Schedule vessels which includes container ships, ferries, gas carriers and liners
- Deep sea vessels which are tankers, bulkers, ore carriers and general cargo carriers

These categories map onto current naval architectural and market views of the marine fleet and thus this approach should permit future comparisons to be readily made.

An initial multi-factorial model was specified which modelled the probability of S_c occurring (i.e. $C_p < 10$) as a function of the following predictors:

- vessel size (DWT),
- speed (knots),
- in-service duration (months),
- area of vessel coated (sides or flats),
- vessel category (naval, coastal, schedule or deep sea),
- coating applied,
- year of application (1973–2006).

The initial model was over-specified with all seven-way interactions between predictors estimated. This took nearly 24 hours to run on a 3.06 GHz Intel processor running at 100% capacity with 2GB RAM which was enough reason to reduce the model as well as being required by good modelling practice. Diagnostic statistics were used to reduce the over-specification of the model to give a parsimonious fit with the fewest predictors and interactions but with the maximum goodness of fit. To this end, examination of the log-likelihood and Wald statistics (see [46]) was used to step-wise reduce a series of intermediate models the combination of predictors to a very simple final model:

$$P(S_c) = 1 + e^z$$

Where P = probability, S_c = coating satisfaction, and z = size + speed + in-service duration + area + vessel category + coating + year of application.

Predicted probabilities estimated by the model were stored and used to explore the relationships within the data. The advantage of this approach is that questions about coating performance can be addressed according to each of the model parameters whilst the other parameters are held constant. Simple aggregation of the data cannot do this and leaves the analyst with the niggle that any simple result has a considerable number of what-ifs attached. Furthermore, for all predictors an odds ratio as well as the predicted probability is calculated. The odds ratio gives the likelihood of S_c occurring given a change of 1 unit in a given continuous predictor (i.e. for speed this would be per knot) or for a categorical predictor the chance of S_c occurring compared to the occurrence of the reference level. For further details on logistic regression, see [46,47].

14.3   The first nuggets

It would be futile to present results in terms of the many different coatings that have been sold to ship owners since 1974, and it is this much more amenable to aid the understanding of relative performance to aggregate the coatings into their respective technology types, namely:

- Traditional coatings: These work simply by uncontrolled dissolution of a biocide from a coating.

- TBT-SPC: TBT acrylate biocide self-polishing co-polymers where hydrolysis controls biocide release and polishing.
- MA-SPC: Copper, zinc and silyl acrylate biocide self-polishing co-polymers where hydrolysis also controls biocide release and polishing.
- CDP: Controlled depletion polymers, also described as eroding, ablative, polishing, self-polishing or hydration coatings in which soluble resins allow the coating to disintegrate slowly. This is physical ablation, not chemical.
- Hybrids: Blends of CDP and SPC technologies.
- Foul release: Non-stick technology based on silicone chemistries that allow biofouling to be removed with hydrodynamic shear.

Much more detail on these different coating technologies and how they work is given in Chapter 13.

14.3.1   Ship-scale variation in performance

The performance of the five different types of coating varied by area they were applied to on the hull (Plate V A). It is quite clear that the biofouling challenge of the sides of a vessel, where light obviously has a strong influence, is significantly greater than the flats of the hull which are mostly shaded. Thus, on the flats, FR, MA-SPC and TBT-SPC coatings all have a mean predicted $S_c > 80\%$, yet on the sides of vessels only FR coatings have a mean predicted $S_c > 80\%$, with MA-SPCs at approximately 79% and TBT-SPCs at approximately 78%. This exemplifies the success of the modern FR coatings in dealing with the algal biofouling challenge without recourse to any biocide. The largest differences in performance between flats and sides are for CDPs and hybrids, both of which show $>11\%$ reduction in mean predicted S_c due to the effect of light. Other recent work on small samples of vessels have highlighted differences between general hull biofouling and small pockets of biofouling around rudders, rope guards, propeller posts, sea chests and bulbous bows (see [48] and Chapter 24 and references therein). These pockets of biofouling may be due to the structural complexity offering a hydrodynamic refuge or because of the difficulty in applying coatings in these areas.

The major point to be made from these results is that although the biofouling challenge is cosmopolitan and vessels trade globally, there are still very large differences in coating performance at the ship scale. The practical consequence is that this may result in different coatings being used on different parts of the same vessel: this is a simple outcome of economics and it also requires that ship scale research and development testing needs to account for differences in potential performance on the flats and sides.

Not only does the part of the vessel coated have an impact on performance, but also the size of the vessel. Overall, there is a very small (1.00000035), but significant ($p = 0.004$), odds ratio showing that as vessel tonnage increases there is a small increase in the chance of S_c. Looking at this relationship by coating type (Plate V B–D) shows a degree of variation in the linear relationship between log(DWT) and mean predicted S_c. For all coatings the relationships are statistically very weak (the r^2 values range from only 0.04 to 0.17), but because of the large sample sizes the regression slopes are all significantly different from 0 (all $p \leq 0.004$). The biggest effect of vessel size is for traditional coatings and the two SPC technologies (determined simply as the magnitude of the slope of log(DWT) versus mean predicted S_c). All these three are significantly different ($p < 0.05$) from each other. The weakest effect of size was seen for hybrids, FRs and CDPs, with no difference between FRs and hybrids, and FRs and CDPs but between CDPs and hybrids (Plate V B–D). This relationship may be related to the operating profiles of vessels as they get larger, with the biggest, and most costly to operate, vessels having shorter turnaround times, thus reducing the potential of biofouling at anchor. Analysis of operating profiles is needed to address this question. However given the grouping of the different technologies, this analysis may be detecting the effect of Reynolds numbers (i.e. the effect of hull length × speed).

14.3.2   Time-dependent performance

With increasing in-service duration (months) the probability of S_c declines: the odds ratio is 0.979 ($p < 0.001$). This is not surprising given the way the different technologies work, though what is surprising is that this is a fairly similar ratio for all coatings (Plate V B–D), though there are small but significant differences between them ($p < 0.05$), with CDPs, FRs, traditional and TBT-SPCs all having statistically similar slopes, different from both MA-SPCs which have the smallest change in mean predicted S_c, and hybrids with the fastest with increasing in-service interval.

A possibly complicating factor in this analysis is that different thicknesses of coatings are often used by coatings manufacturers to give a desired service life. Thus implicit within this analysis is that the vessels analysed showed $C_p > 10$ due to coating performance and not total coating depletion. There is a flag variable for each record in the database that indicates coating depletion and all records with a positive flag were removed prior to analysis. Thus the linear relationship between mean predicted S_c and in-service interval is likely to be statistically robust.

14.3.3   Speed-dependent performance

The odds ratio overall for speed (knots) was 1.063, though the effect of speed was also coating-dependent ($p < 0.05$), with the slopes between speed and mean predicted S_c for hybrids and MA-SPCs having similar relationships (Plate V B–D), FRs having the shallowest slope, traditional coatings and CDPs the steepest slope, with TBT-SPCs somewhere between hybrid and traditional coatings. These relationships are somewhat surprising. Both CDPs and FRs work by the effect of hydrodynamic shear, removing polymer from CDPs and biofouling from FRs, but CDPs are the most speed-dependent and FRs the least. As the performance of FRs appears to be the least speed-dependent, this is good evidence that they could work on a wider range of vessels than they are currently applied to. It is also interesting to note the differences between the two SPC systems, indicative perhaps of a difference in mode of action (see Chapter 13). The performance of all the coatings converge at higher speeds and thus on very fast vessels selection of coating type should probably be driven by environmental concerns: thus, FRs should be the coating of choice.

14.3.4   Performance history

Careful readers of the above will have the impression that there are two broad groups of antifouling coatings separated by their performance, i.e. the two SPCs and FR coatings form a group of high performers and traditional, CDPs and hybrids form a group of lower performers. To fully appreciate the difference in performance between these different coating technologies, examination of mean predicted S_c by coating, by year of application, for each vessel category, namely naval, coastal, deep sea and schedule, is a thorough approach.

First it should be noted that deductions from the patterns found near the beginning (<1976) and end of the model (>2004) should be treated with caution due to small sample sizes in these locations and overall poorer fit towards the outer perimeters of the model.

The results show that the implementation of FR technology has generally been very successful (Plate V E–H) with FRs showing consistent improvement over time for all vessel categories except naval, though the sample size for naval vessels is very small ($n = 24$). Similarly the introduction of MA-SPCs since the restrictions on the use of TBT-SPCs has been implemented (see Chapter 21) has also been shown high-quality performance, so much so that MA-SPCs are much better performers than the later generation TBT-SPCs. Indeed, both FRs and MA-SPCs are now similar in performance to TBT-SPCs at their prime in the early 1980s. Thus there is no evidence to support the conclusion that the maritime industry has jumped from the frying pan (TBT-SPCs with high performance but concomitant high environmental toxicity) to the fire (low-performing coatings which impacts global climate change but lower environmental toxicity) [5, 6]. Conversely it appears that the coatings industry has risen to the challenge and has provided high performance coatings with lower or non-existent toxicity (see Chapters 16, 17 and 27). As expected from their initial concept and design, hybrid coatings and CDPs show intermediate performance for all vessel types except naval vessels where low sample sizes, no vessels with hybrid coatings and a great variation in performance due to operating conditions (see above) make the patterns very difficult to interpret.

The performance patterns for deep sea and schedule vessels is broadly similar, though schedule vessels show overall higher performances for each coating type. This is a reflection of operating profiles of these vessels, with schedule vessels working on average at 5 knots faster than deep-sea vessels. Thus from what is described above for the effect of speed all coatings should work better on schedule vessels: which is what we see in Plate V E–H. Given recent price hikes in crude oil (approximately $90 per barrel in January 2008) several large owners have reduced the operating speed of their fleets by 4–6 knots. This has a considerable fuel saving for the fleet but given the relationship between ship speed, operating profiles and coating technology then it may have a significant impact on coating performance.

In the mid 1970s to early 1980s there was a consistent difference of approximately 30% between traditional coatings and TBT-SPCs (Plate V E–H). From the early 1980s all types of vessel show a steady but gradual decline in performance of TBT-SPCs, except naval vessels where the decline is more dramatic but happens later on in the early 1990s. The reasons for this may have multiple causes, including changes in vessel operating profiles or adaptation of marine biofouling organisms to TBT [49]. At the moment it is impossible to tell which, if any, is correct until a more in-depth analysis of a sub-sample of the database is carried out. However, what can be concluded at this moment is that the coatings industry had by the late 1980s an urgent need to develop better performing coatings whether TBT-SPCs were withdrawn from the market or not.

14.4   Conclusions

- This is the first analysis of biofouling and coating performance on the global maritime fleet.

- The introduction of TBT-SPCs in the late 1970s had a huge positive impact on the performance of antifouling coatings, and thus became the de facto standard by which all other coating technologies were judged.
- There was a steady decline in the performance of TBT-SPCs from the mid-1980s leading to a need for replacement technologies, whether or not TBT-SPCs were removed from the market.
- MA-SPC and FR technologies were introduced to replace TBT-SPCs and their current performance is comparable to that of early-generation TBT-SPCs.
- It can therefore be concluded that the withdrawal of TBT-SPCs from the market has not resulted in a negative impact on global climate change via increased CO_2 emissions from vessels due to lower performance of the newer coating technologies: we have left the frying pan and have avoided the fire at the same time.

Acknowledgements

The analyses presented in this chapter would have been impossible without the foresight of the scientists who originally set it up in the 1970s, and the many hundreds of technical service agents who have undertaken nearly 200 000 surveys over the last 30 years. This work was supported by an Industrial Fellowship from the Royal Society.

References

1. Gibbs, P.E., Bryan, G.W. & Pascoe, P.L. (1991) TBT-induced imposex in the Dogwhelk, *Nucella lapillus* – geographical uniformity of the response and effects. *Marine Environmental Research*, **32**, 79–87.
2. Langston, W.J., Bryan, G.W., Burt, G.R. & Gibbs, P.E. (1990) Assessing the impact of tin and TBT in estuaries and coastal regions. *Functional Ecology*, **4**, 433–443.
3. Bryan, G.W., Gibbs, P.E., Burt, G.R. & Hummerstone, L.G. (1987) The effects of TBT accumulation on adult dogwhelks, *Nucella lapillus* – long-term field and laboratory experiments. *Journal of the Marine Biological Association of the United Kingdom*, **67**, 525–544.
4. Bryan, G.W., Gibbs, P.E., Hummerstone, L.G. & Burt, G.R. (1986) The decline of the gastropod *Nucella lapillus* around Southwest England – evidence for the effect of tributyltin from antifouling paints. *Journal of the Marine Biological Association of the United Kingdom*, **66**, 611–640.
5. Evans, S.M., Birchenough, A.C. & Brancato, M.S. (2000) The TBT ban: out of the frying pan into the fire? *Marine Pollution Bulletin*, **40**, 204–211.
6. Evans, S.M. (1999) TBT or not TBT?: that is the question. *Biofouling*, **14**, 117–129.
7. Karlsson, J. & Eklund, B. (2004) New biocide-free anti-fouling paints are toxic. *Marine Pollution Bulletin*, **49**, 456–464.
8. Voulvoulis, N., Scrimshaw, M.D. & Lester, J.N. (2002) Comparative environmental assessment of biocides used in antifouling paints. *Chemosphere*, **47**, 789–795.
9. Yee, L.H., Holmstrom, C., Fuary, E.T., Lewin, N.C., Kjelleberg, S. & Steinberg, P.D. (2007) Inhibition of fouling by marine bacteria immobilised in kappa-carrageenan beads. *Biofouling*, **23**, 287 294.
10. Fay, F., Linossier, I., Langlois, V. & Vallee-Rehel, K. (2007) Biodegradable poly(ester-anhydride) for new antifouling coating. *Biomacromolecules*, **8**, 1751–1758.

11. Wang, S.J., Fan, X.D., Si, Q.F., *et al.* (2006) Preparation and characterization of a hyper-branched polyethoxysiloxane based anti-fouling coating. *Journal of Applied Polymer Science*, **102**, 5818–5824.

12. Statz, A., Finlay, J., Dalsin, J., Callow, M., Callow, J.A. & Messersmith, P.B. (2006) Algal antifouling and fouling-release properties of metal surfaces coated with a polymer inspired by marine mussels. *Biofouling*, **22**, 391–399.

13. Tang, Y., Finlay, J.A., Kowalke, G.L., *et al.* (2005) Hybrid xerogel films as novel coatings for antifouling and fouling release. *Biofouling*, **21**, 59–71.

14. Loschau, M. & Kratke, R. (2005) Efficacy and toxicity of self-polishing biocide-free antifouling paints. *Environmental Pollution*, **138**, 260–267.

15. Hellio, C., Tsoukatou, M., Marechal, J.P., *et al.* (2005) Inhibitory effects of Mediterranean sponge extracts and metabolites on larval settlement of the barnacle *Balanus amphitrite*. *Marine Biotechnology*, **7**, 297–305.

16. Gudipati, C.S., Finlay, J.A., Callow, J.A., Callow, M.E. & Wooley, K.L. (2005) The antifouling and fouling-release performance of hyperbranched fluoropolymer (HBFP)-poly(ethylene glycol) (PEG) composite coatings evaluated by adsorption of biomacromolecules and the green fouling alga *Ulva*. *Langmuir*, **21**, 3044–3053.

17. Fay, F., Linossier, I., Langlois, V., Haras, D. & Vallee-Rehel, K. (2005) Study of bioactive surfaces for antifouling marine coating. *Abstracts of Papers of the American Chemical Society*, **230**, U4336–U4337.

18. Faimali, M., Garaventa, F., Mancini, I., *et al.* (2005) Antisettlement activity of synthetic analogues of polymeric 3-alkylpyridinium salts isolated from the sponge *Reniera sarai*. *Biofouling*, **21**, 49–57.

19. Dong, B.Y., Manolache, S., Somers, E.B., Wong, A.C.L. & Denes, F.S. (2005) Generation of antifouling layers on stainless steel surfaces by plasma-enhanced crosslinking of polyethylene glycol. *Journal of Applied Polymer Science*, **97**, 485–497.

20. Sjogren, M., Dahlstrom, M., Goransson, U., Jonsson, P.R. & Bohlin, L. (2004) Recruitment in the field of *Balanus improvisus* and *Mytilus edulis* in response to the antifouling cyclopeptides barettin and 8,9-dihydrobarettin from the marine sponge *Geodia barretti*. *Biofouling*, **20**, 291–297.

21. Pettitt, M.E., Henry, S.L., Callow, M.E., Callow, J.A. & Clare, A.S. (2004) Activity of commercial enzymes on settlement and adhesion of cypris larvae of the barnacle *Balanus amphitrite*, spores of the green alga *Ulva linza*, and the diatom *Navicula perminuta*. *Biofouling*, **20**, 299–311.

22. Diers, J.A., Pennaka, H.K., Peng, J.N., Bowling, J.J., Duke, S.O. & Hamann, M.T. (2004) Structural activity relationship studies of zebra mussel antifouling and antimicrobial agents from verongid sponges. *Journal of Natural Products*, **67**, 2117–2120.

23. Scardino, A., De Nys, R., Ison, O., O'Connor, W. & Steinberg, P. (2003) Microtopography and antifouling properties of the shell surface of the bivalve molluscs *Mytilus galloprovincialis* and *Pinctada imbricata*. *Biofouling*, **19**, 221–230.

24. Rittschof, D., Lai, C.H., Kok, L.M. & Teo, S.L.M. (2003) Pharmaceuticals as antifoulants: concept and principles. *Biofouling*, **19**, 207–212.

25. Nogata, Y., Yoshimura, E., Shinshima, K., Kitano, Y. & Sakaguchi, I. (2003) Antifouling substances against larvae of the barnacle *Balanus amphitrite* from the marine sponge, *Acanthella cavernosa*. *Biofouling*, **19**, 193–196.

26. Kitano, Y., Yokoyama, A., Nogata, Y., *et al.* (2003) Synthesis and anti-barnacle activities of novel 3-isocyanotheonellin analogues. *Biofouling*, **19**, 187–192.

27. Faimali, M., Sepcic, K., Turk, T. & Geraci, S. (2003) Non-toxic antifouling activity of polymeric 3-alkylpyridinium salts from the Mediterranean sponge *Reniera sarai* (Pulitzer-Finali). *Biofouling*, **19**, 47–56.

28. Burgess, J.G., Boyd, K.G., Armstrong, E., *et al.* (2003) The development of a marine natural product-based antifouling paint. *Biofouling*, **19**, 197–205.

29. Brady, R.F. & Aronson, C.L. (2003) Elastomeric fluorinated polyurethane coatings for nontoxic fouling control. *Biofouling*, **19**, 59–62.

30. Schoenfeld, R.C., Conova, S., Rittschof, D. & Ganem, B. (2002) Cytotoxic, antifouling bromo-tyramines: a synthetic study on simple marine natural products and their analogues. *Bioorganic & Medicinal Chemistry Letters*, **12**, 823–825.
31. Berglin, M., Larsson, A., Jonsson, P.R. & Gatenholm, P. (2001) The adhesion of the barnacle, *Balanus improvisu*s, to poly(dimethylsiloxane) fouling-release coatings and poly(methyl methacrylate) panels: the effect of barnacle size on strength and failure mode. *Journal of Adhesion Science and Technology*, **15**, 1485–1502.
32. Yu, J. (2003) Biodegradation-based polymer surface erosion and surface renewal for foul-release at low ship speeds. *Biofouling*, **19**, 83–90.
33. Gay, C. (2003) Some fundamentals of adhesion in synthetic adhesives. *Biofouling*, **19**, 53–57.
34. Akhremitchev, B.B., Bemis, J.E., Al-Maawali, S., Sun, Y.J., Stebounova, L. & Walker, G.C. (2003) Application of scanning force and near field microscopies to the characterization of minimally adhesive polymer surfaces. *Biofouling*, **19**, 99–104.
35. Berntsson, K.M., Andreasson, H., Jonsson, P.R., *et al.* (2000) Reduction of barnacle recruitment on micro-textured surfaces: analysis of effective topographic characteristics and evaluation of skin friction. *Biofouling*, **16**, 245–261.
36. Becker, K., Hormchong, T. & Wahl, M. (2000) Relevance of crustacean carapace wettability for fouling. *Hydrobiologia*, **426**, 193–201.
37. Handa, P., Fant, C. & Nyden, M. (2006) Antifouling agent release from marine coatings-ion pair formation/dissolution for controlled release. *Progress in Organic Coatings*, **57**, 376–382.
38. Al-Juhni, A.A. & Newby, B.M.Z. (2006) Incorporation of benzoic acid and sodium benzoate into silicone coatings and subsequent leaching of the compound from the incorporated coatings. *Progress in Organic Coatings*, **56**, 135–145.
39. Yebra, D.M., Kiil, S., Dam-Johansen, K. & Weinell, C. (2005) Reaction rate estimation of controlled-release antifouling paint binders: rosin-based systems. *Progress in Organic Coatings*, **53**, 256–275.
40. Dobretsov, S.V. & Qian, P.Y. (2003) Pharmacological induction of larval settlement and metamorphosis in the blue mussel *Mytilus edulis* L. *Biofouling*, **19**, 57–63.
41. Afsar, A., De Nys, R. & Steinberg, P. (2003) The effects of foul-release coatings on the settlement and behaviour of cyprid larvae of the barnacle *Balanus amphitrite amphitrite* Darwin. *Biofouling*, **19**, 105–110.
42. Thomason, J.C., Hills, J.M. & Thomason, P.O. (2002) Field-based behavioural bioassays for testing the efficacy of antifouling coatings. *Biofouling*, **18**, 285–292.
43. Wood, C.D., Truby, K., Stein, J., *et al.* (2000) Temporal and spatial variations in macrofouling of silicone fouling-release coatings. *Biofouling*, **16**, 311–322.
44. Mineur, F., Johnson, M.P., Maggs, C.A. & Stenenga, H. (2007) Hull fouling on commercial ships as a vector of macroalgal introduction. *Marine Biology*, **151**, 1299–1307.
45. Anon. (2006) *World Fleet Statistics*. Lloyds Register Fairplay, London.
46. Hosmer, D.W. & Lemeshow, S. (2000) *Applied Logistic Regression*. John Wiley & Sons, New York.
47. Field, A. (2000) *Discovering Statistics using SPSS for Windows*, 2nd edn. SAGE Publications, Los Angeles.
48. Otani, M., Oumi, T., Uwai, S., *et al.* (2007) Occurrence and diversity of barnacles on international ships visiting Osaka Bay, Japan, and the risk of their introduction. *Biofouling*, **23**, 277–286.
49. Vogt, C., Nowak, C., Diogo, J.B., Oetken, M., Schwenk, K. & Oehlinann, J. (2007) Multi-generation studies with *Chironomus riparius* – effects of low tributyltin concentrations on life history parameters and genetic diversity. *Chemosphere*, **67**, 2192–2200.

Chapter 15
Consequences of Fouling on Shipping

Robert Edyvean

The aim of this chapter is to give a brief history of ship biofouling and then to describe the effects of biofouling on the external hull (in terms of corrosion and drag) and the internal operating systems (in terms of corrosion and other effects).

15.1 Introduction

The effects of biofouling on ships have been known and combated by mariners and shipwrights for millennia. Two main problems were encountered; one, now rarely affecting commercial shipping, was the effect of wood boring molluscs and associated wood rotting fungi which could rapidly destroy wooden hulls below the waterline. The second problem, which is still a considerable factor in commercial shipping, is the effect of biofouling on speed and manoeuvrability. In addition, there are effects of corrosion and blockage of sea water systems and effects on fuel and cargo.

For as long as man has used the sea, he has been seeking ways of combating these problems. Coatings have long been popular – traditionally based on tar or natural resins and now sophisticated antifouling systems. Claddings have a more recent history, though there are reports of closely spaced copper nails being used in Roman times, copper cladding came into widespread use in the eighteenth century as both a barrier to the effects of biofouling on wooden hulls and as an antifoulant preventing the settlement of many biofouling species (the term 'copper-bottomed' – defining a sound investment – derived from the use of such cladding to protect trading ships). Apart from a brief problem when copper cladding was first used on 'iron-hulled' ships, copper, in one form or other, has been the main antifouling agent to this day. The problem encountered with the first iron-hulled ships was that the copper cladding was not only ineffective in preventing biofouling (as it became cathodic and thus protected in a galvanic couple with the iron and was no longer able to release toxic copper ions), but also caused corrosion of the iron hull (as iron became the anode in the couple – a phenomenon investigated and explained by Sir Humphry Davy).

David Houghton, then of H.M. Naval Base, Portsmouth, UK, was a pioneer of the modern understanding of the effects of biofouling on ships and, with his co-author S.A. Gage, published an important summary of the, then current, understanding of the subject in 1979 [1]. In their paper, Houghton and Gage divide the problems into 'extra-hull' and 'intra-hull' and essentially cover all the main areas that are still a problem to this day. Another summary, this time of the early years of US Navy research into biofouling, was published in 1992 [2]. While there have been considerable advances in materials selection and protective and antifouling coatings in the last 30 years, the potential effects of biofouling on ships have remained the same.

15.2   Biofilms and their effect on the external hull of ships

Marine biofilms can develop to their fullest extent on the external hull below the water line. Thereafter in this chapter (only) biofilm is used in its broader sense and is taken to include bacteria, fungi and other microorganisms, including the settling stages of the larger marine biofouling organisms such as barnacles, tunicates, hydroids, mussels and seaweeds [3]. These large biofouling organisms can grow to maturity if the ship is in port, stationary or slow moving for any length of time. If the ship is moving at more than 2 knots, then the animal species will be reduced. The ultimate (mature or 'climax') community of this external biofilm will be similar to that found on the adjacent shoreline or dock and, while the rate and amount of biofouling varies in different parts of the world and with season, distance from shore and water depth [4] (see Chapters 4 and 12). Distance from shore does not necessarily prevent the development of the mature community but may alter its composition (tankers stopping to carry out cleaning operations at sea can suffer considerable biofouling development). One more recent problem is an increase in the amount and diversity of biofouling that can settle and develop while a ship is docked due to docks and harbour areas now being less polluted and enabling more species to survive. Biofouling organisms are very tenacious and can survive considerable distances and changes of environment. Not only does this mean that ships hulls need to be well maintained with good antifouling and cleaning regimes, but has also resulted in ships being the unwitting transmission vectors moving species around the world to the detriment of the indigenous marine flora and fauna. The use of antifouling coatings or claddings and the movement and changes in water conditions experienced as the ship travels from place to place means that a ships hull rarely becomes as extensively biofouled as a stationary offshore structure such as an oil platform.

While dead weight of biofouling organisms and the biofilm obscuring the hull are rarely the problems on ships as they are on offshore platforms [5, 6], the cost of biofouling and its effects are still considerable for shipping. These costs are due to corrosion, now rare on the external hull, and hydrodynamic loading or drag. As ships are sailing with the water and are designed to travel through it, they tend not to suffer the same damaging effects of increased hydrodynamic loading that have been found for static structures such as offshore platforms where biofouling, by increasing the surface roughness and diameter of these structures causes an increase in the effects of wave and current loading and drag forces [7,8]. Heaf [9] calculated that a layer of biofouling 150 mm thick would increase loading on an offshore platform by 42.5% and this would reduce predicted life to failure by 54%.

However, on the hulls of ships even a small biofouling development (a slime layer of microorganisms) can noticeably affect performance due to surface roughness. Such a biofilm slime can reach 2 mm in thickness with some 30×10^6 bacteria g^{-1} and more than 0.25×10^6 diatoms cm^{-2} [10, 11]. The increasing roughness causes the ship to have more difficulty in passing through the water (it is less slippery). This results in either a greater power requirement and hence fuel consumption, or a slower speed for the same power. In either case, it costs money and increases environmental damage.

In her review of the use of copper alloy sheathing for ships hulls, Powell [12] reports that a conventionally painted ships hull has a roughness of 125 μm and that this roughness increases by 50–70 μm every year [13]. Powell also reports that for every 10 μm increase in surface roughness there is an initial increase in power requirement (the amount of power required to maintain the expected cruising speed) of 1% up to 230 m roughness followed by an increase of 0.5% for each 10 μm thereafter. While it could be argued that initial bacterial settlement could actually reduce roughness, it is now understood that even microbial biofilms are very uneven and patchy and hence are likely to increase roughness after only a short development time. If larger biofouling organisms become established then considerable loss of efficiency will be experienced. Software to calculate the effects (as cost) of hull roughness has been developed [14] and is available through the major antifouling coatings supply companies.

Powell [12] also reports results of trials comparing copper alloy clad ships with conventionally finished hulls (with antifouling coatings). The copper alloys are not only antifouling but also smoother, resulting in fuel cost savings of 24.8 and 27% in two examples. In 1979, Houghton and Gage [1] estimated that 19% of the UK Royal Navy's fuel bill was used in overcoming the effects of roughness and biofouling, while in 1992, Alberte *et al.* [2] estimated that the costs for drag-related fuel increases in the US Navy were running at $75 to 100 million per year and that dry docking costs of removing biofouling were £360 million in 1981. Thus, even with advances in antifouling coatings since then, the cost of ship hull biofouling is still enormous. Additionally, we are now aware that such costs go beyond the financial cost of the extra fuel and should take into account the environmental effects of the carbon released.

The cost of dry docking and cleaning the biofouling from the hull of a large cargo vessel is huge, as is the cost of applying an antifouling system. There is thus a considerable industry supplying such coatings and significant research efforts are continuing to improve the effectiveness and life of antifouling systems.

Recent studies have produced new calculations of the effects of biofouling and surface roughness on power consumption both in terms of overall effects, for example penalties of up to 86% for heavy calcareous biofouling [15], and in terms of the effects of microsurface topography [16]. In addition, whereas previously the chemical effects of antifouling surfaces have been extensively studied, more recently the effects of the physical surface topography of coatings have been investigated (e.g. Townsin's [17] investigation of the latest low surface energy coatings). It has now been shown that microtopographies can be engineered which reduce spore settlement of biofouling algae by up to 58% with multi-featured microtopographies (2 μm pillars and 10 μm pyramids) [18] and by 77% by the replication of the microtopography of sharkskin [18, 19]. These studies open up a new era in our ability to reduce the effects, particularly the drag effects, of biofouling. We now have the ability to produce coatings and protection systems which either physically or chemically prevent corrosion of ships hulls and a growing ability to prevent the drag effects of biofouling using low- or non-toxic systems.

15.3   The effects of biofilms on the internal operating systems of ships

'Both fungi and bacteria are important with regard to problems occurring on board ships. They will attack singly, or in combinations, a whole range of carbon-based materials and can damage others by products produced during their metabolism. Conditions within a ship are usually fairly ideal for microorganism development, whether they be within closed systems, e.g. fuel tanks, or the general environs of the vessel. Microorganism development will proceed in the presence of small traces of water, e.g. condensation, and nutrients'. So wrote Houghton and Gage in 1979 [1] and their words are just as relevant today. Bacterial and fungal biofilms can cause extensive, serious and costly damage to ships systems and cargo to the extent of total failures of the ships propulsion systems.

The adsorption of organic molecules, the settlement of bacteria and the development of the biofilm to include other groups of microorganisms together with inorganic debris will occur in most water dominated systems and, to the extent of biofouling by bacteria and fungi, in systems where water is only a minor component. Thus, there are many systems on board ship which are prone to the effects of biofilms. These include water systems (desalination plant, ballast tanks, for fire prevention, heat exchangers and other uses) and systems which may not be immediately associated with biofilms and their effects, such as hydraulics and fuels.

Sea water is a well-known corrosive environment and any biological activity can enhance its aggressiveness. The presence of marine macrofouling can cause damage to protective coatings and the underlying metal and provide conditions in which microorganisms capable of enhancing corrosion can thrive [20,21]. Localised corrosion is thus a major problem caused (or rather enabled and enhanced) by microbial biofilms. Often the more hostile the environment to the microorganisms, the more damage can be caused. This is due to a greater non-uniformity in the biofilm and to the specialised nature of the microorganisms, their metabolism and the environment they create to protect themselves. Thus, for example, on a copper alloy which is potentially toxic, any biofilm that does establish is likely to be highly non-uniform and specialised and can lead to severe localised corrosion. The biofilm can create differential aeration, pH or other electrochemical cells on both a macroscopic and microscopic scale, produce corrosion-promoting metabolites such as acids and sulphur compounds and influence the rate of movement of metal and other ions.

The presence of a biofilm will thus modify the interface and can affect corrosion by the following mechanisms [22]:

- The influence of microbial metabolism
- The influence of microbial products
- The influence of extracellular polymeric substances
- Influencing the electrochemistry at the metal surface
- Chelation of metal ions
- Destabilisation of corrosion inhibitors and anti-corrosion systems

In addition, there are interactions between the biofilm and the surface. The surface can have a considerable effect on the biofilm and vice versa. For example, more extracellular material is produced by the diatom *Amphora coffeaeformis* when attaching to low surface energy silicon elastomers than when attaching to high surface energy glass yet fewer cells attach to the elastomer [23].

As the biofilm system, cells and extracellular polymeric substances, is largely water (see Chapters 9 and 10), the whole can act as a modifying electrolyte [24] and the bacteria of the biofilm can control the passage of ions in both directions, both metal ions leaving the surface and electrolyte and other ions reaching it. The biofilm is a regulator to the exchange of ions between the metal surface and the aqueous environment and many reactions between the products of bacterial metabolism and the metal will take place within the biofilm thickness. The metal/solution interface can be changed to such a degree that corrosion rates can be accelerated by factors of 1000–100 000 [25].

Thus, on the surface of a copper heat exchanger tube, a localised, tightly adherent consortia of bacteria may develop which protect themselves from exposure to too many copper ions by utilising extracellular polysaccharides and have metabolisms which actively encourage the dissolution of the alloy beneath them by pH, use of metal ions in their metabolism or simply enabling the establishment of a strong differential oxygen corrosion cell.

Concentration cells are particularly important where pH can range from 1 to 10.6 [26] and oxygen from saturated to none at all. Such changes, occurring at the biofilm/metal interface cannot be detected in the bulk liquid phase [27, 28]. Whatever the mechanism, the outcome of microbiologically influenced corrosion is nearly always vigorous localised corrosion, which can be catastrophic in pressurised water or other systems.

15.4   The effects of biofilms in water systems

Both fresh and sea water are used extensively on board ships, for cooling, in bilges, fire fighting systems, washing and drinking. Thus, sea water may be drawn in as required from wherever the ship happens to be, both sea and fresh water will be stored in a variety of conditions from relatively clean in fire fighting and desalination equipment and for human consumption to very dirty in bilges and as displacement in tanks and fuel bunkers. While clean fresh sea water systems can suffer severely with the development of macrobiofilms (i.e. biofilms similar, or worse than those on the outer hulls as often no antifouling can be present) resulting in blockages and corrosion, it is often those areas least expected to be able to support a biofilm which cause the most problems such as in hydraulic oils and fuel systems.

Water-cooled heat exchangers on board ships have many uses including air conditioning and engine cooling. Biofouling, and its subsequent effects on corrosion of the waterside of shell and tube heat exchangers, is still a major problem, particularly in cases where untreated cooling water is used (and especially if this water is taken from polluted sources in estuaries or docks). The water quality and subsequent bacterial growth, coupled with design details that do not fully take corrosion into account, can lead to microbiologically induced corrosion events which, if aggravated by localised stresses, can lead to failures within a few weeks of service.

An example of such a failure is shown in the case of a three-pass heat exchanger system on board a ship in long-term harbour anchorage. The heat exchangers consisted of a steel shell and 90/10 cupro/nickel heat exchanger tubes [29]. These tubes had an internal surface area enhanced by rifling. The coolant water was untreated brackish water from the harbour. Failure was due to corrosion perforation brought about by a patchy biofilm formed due to inadequate filtration and treatment of the cooling water and low flow velocities. The water may have been allowed to stand in the tubes for long periods of time which would exacerbate the problem by allowing a wider variety of bacteria, such as anaerobic sulphate-reducing bacteria (SRBs), to find niches.

The SRBs are one of the most deleterious groups of bacteria involved in microbiologically influenced corrosion [28]. Not only can they influence corrosion locally by their metabolic activity, but they release sulphide into the environment which can cause sulphide cracking and enhance corrosion-fatigue as well as enhancing surface corrosion, degrading lubricating oils and moving surfaces. Sulphide-polluted water is highly deleterious to copper alloys (particularly copper/nickel alloys) and pitting can become severe if levels reach 0.01 ppm [30]. Localised levels of sulphide can easily reach 30 ppm in the presence of SRBs [31].

It should also be noted that many corrosion/stress failures influenced by biofilms have occurred at sites of last minute design changes or less than satisfactory repairs (producing weak spots in the design which are the first to succumb to the effects). Such on-site repairs and, indeed, certain manufacturing processes impart considerable residual stresses to the alloy being used and can allow microbiologically influenced corrosion attack to develop into stress corrosion cracking or, with the addition of cyclic stresses (which can range from low-frequency wave action to high-frequency stresses caused by pumps and motors), to corrosion-fatigue [31, 32]. Cracking can lead from a potentially salvageable situation of a small leak to a catastrophic failure of a pipe along its whole length.

15.5   Biofilm effects in fuel and hydraulic systems

On first consideration, the fact that bacteria and fungi can cause severe problems in hydraulic and fuel oils is not immediately obvious. However, these microorganisms can be well adapted and as long as there is a source of nutrients (the oil) and access to water (even if only by condensation), certain bacterial and fungi will grow, often vigorously.

Houghton and Gage [1] describe how the Royal Navy encountered problems with aviation fuel stores in the mid- to late-1950s where a thick microbial mat would form at the fuel/water interface if aviation fuel was stored in tanks where water was present and allowed to separate out. While they describe that, by using biocidal additives and good housekeeping, keeping, the fuel dry was the solution, such problems with aviation fuel storage keep reoccurring to this day.

More pertinent to shipping is the effects of similar growths of fungi and bacteria in fuel oil, which was particularly noted due to the finer specifications and greater filtration required by turbine powered ships in the 1970s [1]. Again this is a problem that, while now well known, continues to occur and exact a high cost on shipping. Keeping fuel dry is not always an option in ships, especially where the fuel has to be displaced by sea water in the bunkers or transportation tanks and even when the fuel can be kept largely free of water it has to be well below 1% to restrict microbial growth [33]. Any water present, even if only condensation caused by changes in temperature, will enable some microorganisms to gain a hold and this initial microbial growth will produce more water [34]. In addition, as sea water is usually present, this provides some essential mineral salts for microbial growth that might otherwise be missing [35].

Gaylarde *et al.* [34] list the following effects of microbial growth in fuels. It:

- increases the water content,
- causes microbially induced corrosion of the storage tanks and pipework,
- causes blockage of filters and pipelines, increase wear in pumps and incorrect readings from fuel probes,
- causes the formation of sludges,
- causes emulsification of the fuel,
- breakdowns hydrocarbons and increased sulphur content,

- penetrates protective coatings in tanks and pipelines,
- shortens engine life,
- causes health problems.

Corrosion can be both localised and enhanced by the biofilm and cause more general effects, for example by SRBs producing iron sulphides and hydrogen sulphide. Hydrogen sulphide will sour the fuel allowing sulphide stress corrosion cracking of stressed and vibrating parts, while iron sulphides and other general corrosion products will add to the potential for blockage of pipes, filters, injectors, etc. The blockage of filters and the need to replace disposable filters more frequently is often the first indication of a problem in a piped system. While in some cases this is just a minor irritation resulting in a need to clean or replace the filter, it can be the precursor to failure of, for example, a fire-fighting system or an engine.

Further background to the effect of microorganisms on fuels and fuel systems can be found in several papers [36–40].

15.6   Other areas of biofilm effects

There is an area which could be classed as either external or internal to the ship and this is the effects of biofouling on a ship's superstructure and fittings. While tending to only cause minor damage, there can occasionally be potentially serious consequences if, for example, materials in lifeboats/rafts are damaged by bacterial or fungal growth. While such effects are less with modern materials than previously with natural materials, the biodeterioration of plastics and other man-made materials is well known, if uncommon. Equally, sensitive measuring equipment, including optical glass and various sensor materials that may be in use on ships, can become damaged by biofilm, particularly fungal growth.

15.7   Conclusions

- Biofilms can cause damage to both the internal and external systems in ships.
- Biofilms can have a severe detrimental impact on ship hull surface drag.
- Biofilms can enhance corrosion in ship systems.
- Good quality coatings and electrochemical systems can prevent, or at least retard, corrosion.
- While there are very effective toxic antifouling systems, these have become restricted. However, recent advances in understanding low surface energy coatings and coating microtopography may herald a new era of effective low-toxicity antifouling systems and result in considerable improvements in ship power efficiency.

References

1. Houghton, D.R. & Gage, S.A. (1979) Biology in ships. *Transactions I. Marine Ecology*, **91**, 1–7.
2. Alberte, R.S., Snyder, S., Zahuranec, B.J. & Whetstone, M. (1992) Biofouling research needs for the United States Navy: programme history and goals. *Biofouling*, **6**, 91–96.
3. Callow, M.E. & Edyvean, R.G.J. (1990) Algal fouling and corrosion. In: *Introduction to Applied Phycology* (ed. I. Akatsuka), pp. 367–387. SPB Academic Publishing, The Hague.

4. Richmond, M.D. & Seed, R. (1991) A review of marine macrofouling communities with special reference to animal fouling. *Biofouling*, **3** (2), 151–168.
5. Haderlie, E.C. (1984) A brief overview of the effects of macrofouling. In: *Marine Biodeterioration: An interdisciplinary Study* (eds J.D. Costlow & R.C. Tipper), pp.163–166. US Naval Institute, Annapolis.
6. Edyvean, R.G.J. (1987) Biodeterioration problems of North Sea oil and gas production – a review. *International Biodeterioration*, **23**, 199–232.
7. Kingsbury, R.W.S.M. (1981) Marine fouling of North Sea installations. In: *Marine Fouling of Offshore Platforms*, Vol. **1**. Society for Underwater Technology, London.
8. Wolfram, J. & Theophanatos, A. (1985) The effects of marine fouling on the fluid loading of cylinders: some experimental results. In: *Proceedings 17th Offshore Technology Conference*, pp. 517–526. Houston, Texas.
9. Heaf, N.J. (1981) The effect of marine growth on the performance of fixed offshore platforms in the North Sea. In: *Marine Fouling of Offshore Platforms*, Vol **II**. Society for Underwater Technology, London.
10. Edyvean, R.G.J. & Moss, B.L. (1986) Microalgal communities on protected steel substrata in seawater. *Estuarine, Coastal and Shelf Science*, **22**, 509–527.
11. Hendey, N.I. (1951) Littoral diatoms of Chichester Harbour with special reference to fouling. *Journal of the Royal Microbiological Society*, **71**, 1–86.
12. Powell, C.A. (1994) Copper-nickel sheathing and its use for ships hulls and offshore structures. *International Biodeterioration and Biodegradation*, **34**, 321–332.
13. Neilson, R., Palumbo, E. & Sedat, R. (1983) *Comparison of Three Methods of Coating Ships Surfaces to Control Underwater Roughness*. Centre for Maritime studies, Webb Institute of Naval Architecture, USA.
14. O'Leary, C. & Anderson, C.D. (2003) A new hull roughness penalty calculator, International Marine Coatings. Available from http://www.international-marine.com. Accessed 6 July 2009.
15. Schultz, M.P. (2007) Effects of coating roughness and biofouling on ship resistance and powering. *Biofouling*, **23** (5), 331–341.
16. Weinell, C.E., Olsen, K.N., Christoffersen, M.W. & Kiil, S. (2003) Experimental study of drag resistance using a laboratory scale rotary set-up. *Biofouling*, **19** (Suppl. 1), 45–51.
17. Townsin, R.L. (2003) The ship hull fouling penalty. *Biofouling*, **19** (Suppl. 1), 9–15.
18. Schumacher, J.F., Carman, M.L., Estes, T.G., *et al.* (2007) Engineered antifouling microtopographies – effect of feature size, geometry, and roughness on settlement of zoospores of the green alga *Ulva*. *Biofouling*, **23**, 55–62.
19. Carman, M.L., Estes, T.G., Feinberg, A.W., *et al.* (2006) Engineered antifouling microtopographies – correlating wettability with cell attachment. *Biofouling*, **22**, 11–21.
20. Edyvean, R.G.J., Terry, L.A. & Picken, G.B. (1985) Marine fouling and its effects on offshore structures in the North Sea – a review. *International Biodeterioration*, **21**, 277–284.
21. Terry, L.A. & Picken, G.B. (1986) Algal fouling in the North Sea. In: *Algal Biofouling* (eds L.V. Evans & K.D. Hoagland), pp. 211–230. Elsevier, Amsterdam.
22. Characklis, W.G. (1988) Fouling biofilms and corrosion. In: *Microbial Corrosion* (eds C.A.C. Sequeria & K. Tiller), pp. 95. Elsevier, Amsterdam.
23. Callow, M.E. & Evans, L.V. (1974) Studies on the fouling alga *Enteromorpha*. III. Cytochemistry and autoradiography of adhesive production. *Protoplasma*, **80**, 15–27.
24. Edyvean, R.G.J. & Videla, H.A. (1994) Biological corrosion. In: *Recent Advances in Biodeterioration and Biodegradation*, Vol. **II**, Chapter VI (eds K.L. Garg, N. Garg & K.G. Mukerji), pp. 81–116. Naya Prokash, Calcutta.
25. Costello, J.A. (1969) The corrosion of metals by microorganisms. A literature survey. *International Biodeterioration*, **5**, 101–118.
26. Terry, L.A. & Edyvean, R.G.J. (1981) Microalgae and corrosion. *Botanica Marina*, **24**, 177–183.
27. Case, L.C. (1977) *Water Problems in Oil Production – An Operator's Manual*, 2nd edn. PPC books, Tulsa.

28. Edyvean, R.G.J. (1990) The effects of microbiologically generated hydrogen sulphide in marine corrosion. *Marine Technology Society Journal*, **24**, 5–9.
29. Eames, I.W., Edyvean, R.G.J. & Brook, R. (1992) Corrosion of refrigerator condensers cooled with estuarine water. *Building Services Engineering Research and Technology*, **13** (2), 101–105.
30. Schweitzer, P.A. (1983) *Corrosion and Protection Handbook*. Marcel Dekker, New York.
31. Thomas, C.J., Edyvean, R.G.J., Brook, R. & Austen, I.M. (1988) The effects of microbially-produced hydrogen sulphide on the corrosion fatigue of offshore structural steels. *Corrosion Science*, **27**, 1197–1204.
32. Thomas, C.J., Brook, R. & Edyvean, R.G.J. (1988) Biologically enhanced corrosion-fatigue. *Biofouling*, **1**, 65–77.
33. Hill, E.C. & Hill, G.C. (1993) Microbiological problems in distillate fuels. *Transactions of the Institute Marine Engineering*, **104**, 119–130.
34. Gaylarde, C.C., Bento, F.M. & Kelley, J. (1999) Microbial contamination of stored hydrocarbon fuels and its control. *Reviews in Microbiology*, **30**, 1–10.
35. Bento, F.M. & Gaylarde, C.C. (1998) Effect of additives on fuel stability – a microbiological study. In: *LABS 3 – Third Latin American Biodegradation and Biodeterioration Symposium* (eds C.C. Gaylarde, T.C. Barbosa & N.H. Gabilan), paper no. 10. The British Phycological Society, UK.
36. Smith, R.N. (1987) Fuel testing. In: *Microbiology of Fuels* (ed. R.N. Smith), pp. 49–54. Institute of Petroleum, London.
37. Hill, E.C. (1987) Fuels. In: *Microbial Problems in the Offshore Oil Industry* (eds E.C. Hill, J. Shennan & R. Watkinson), pp. 219–230. John Wiley & Sons, New York.
38. Smith, R.N. (1991) Developments in fuel microbiology. In: *Biodeterioration and Biodegradation*, 8th edn (ed. H.R. Rossmoore), pp. 112–124. Elsevier, London.
39. Haggatt, R.D. & Morchat, R.M. (1992) Microbiological contamination: biocide treatment in naval distillate fuel. *International Biodeterioration and Biodegradation*, **29**, 87–99.
40. Neihof, R.A. & May, M. (1983) Microbial and particulate contamination in fuel tanks on naval ships. *International Biodeterioration Bulletin*, **19**, 59–68.

Chapter 16
Consequences of Antifouling Coatings – The Chemist's Perspective

Dickon Howell and Brigitte Behrends

The aim of this chapter is to give an overview about the currently available antifouling (AF) biocides, including a general overview about the constituents of AF coatings and their historical evolution. Although it is claimed that tributyltin (TBT) no longer poses a problem after the 2008 International Maritime Organization (IMO) ban, a section about TBT has been included because of its persistence in the marine environment and its potential occurrence on ships AF systems even after 2008. Emphasis is on copper including information about its environmental fate, physicochemical behaviour and bioavailability. Copper is regarded as an evolving problem in marinas, harbours, shipyards and major shipping lanes, as it is not degradable. The chapter further covers the most widely accepted organic biocides and their fate in the environment. The possible pathways of AF biocides entering the marine environment are also described as related to shipyard processes and ships in service.

16.1   Introduction

In the marine environment, all surfaces are affected by the attachment of biofouling organisms, such as bacteria, algae and invertebrates, including barnacles and mussels (see Chapters 3, 7, 12 and 15). These biofouling organisms also contribute to the corrosion of submerged surfaces. The speed of ships travelling through the open sea is significantly reduced by biofouling (see Chapter 15), thereby increasing fuel consumption and atmospheric emissions of ships. Over 90% of world trade is carried by the international shipping industry [1]. The United Nations Conference on Trade and Development (UNCTAD) estimates that the operating costs of merchant ships alone contributes about US$380 billion in freight rates within the global economy, equivalent to about 5% of total world trade [2]. In 2004, the industry shipped around 6.8 thousand million tonnes over a distance of about 4.1 million miles, giving roughly 27.6 thousand billion tonne–miles of total trade [3]. To put this into perspective, a 1 mm thick layer of algal slime could increase hull friction by 80%, fuel consumption by 17% and cause a 15% loss in ship speed [4]. It is estimated that on average fuel consumption increases by 6% for every 100 μm increase in the average hull roughness caused by biofouling organisms [5]. Reduced fuel costs and the less frequent need to dry-dock and repaint, due to the use of AF coatings, were estimated to be worth US$5.7 billion per annum to the shipping industry in the mid-1990s [6].

16.2   History of AF technologies

The ancient Carthaginians and Phoenicians used pitch, and possibly copper sheathing, on the bottoms of ships to prevent biofouling, and later coated hulls with sulphur and arsenical compounds, whereas the Greeks and Romans both introduced lead sheathing. Copper cladding was then 're-invented' in the seventeenth and eighteenth centuries, but it became redundant after the introduction of steel vessels at the end of the nineteenth century, because of problems caused by galvanic corrosion. A new technology was needed to protect ships' hulls from biofouling, but it took until the mid-twentieth century for the development of AF coatings to provide it. The most successful of these coatings were the self-polishing copolymer (SPC) coatings containing tributyltin (TBT), which not only had highly effective AF properties, but also were self-polishing and smoothing, meaning that they became more hydrodynamically efficient with increased length of service. Although TBT was invented *ca* 150 years ago, it was only used as an AF biocide in the early 1970s in combination with various copper compounds. Triphenyltin (TPhT) was also used as an active ingredient in AF paints. Nowadays, TPhT is mainly used in agriculture as fungicide, but it can still be found in deeper sediments of repair shipyards (B. Behrends, personal communications) [7].

The challenge for AF technologies is to be toxic towards a broad range of marine organisms at the point of release (ship hulls), but non-toxic to non-target organisms, combined with a high degradation rate in the marine environment and a low bioaccumulation rate. Organotin compounds like TBT did not fulfil these requirements. TBT was found to be both, toxic to a range of non-target organisms and to be far more persistent in the environment than originally thought [8–17]. 'TBT was perhaps the most toxic substance ever deliberately introduced to the marine environment by mankind' [18].

The IMO Convention on the Control of Harmful AF Systems (see Chapter 21) has placed a ban on the application of TBT in AF coatings after 2008 [19], that has stimulated coating manufacturers to devise new ways of producing the same performance with non-TBT coatings. Research on alternative coatings has followed two major lines of enquiry: (1) the development of AF coatings that use the same binder delivery mechanism as the banned TBT coatings but contain different biocidal products and (2) the development of non-biocidal alternatives.

After the TBT ban, advanced effective alternative chemical biocides have been developed. The original SPC coating contained acrylic or methacrylic copolymers that are easily hydrolysed in sea water. Biofouling organisms attached on the surface of a paint film are eliminated together with the copolymer film, which is decomposed by the hydrolysis of the copolymers. This hydrolysis/erosion process is supposed to result in a smooth surface; hence these copolymers are called self-polishing and when blended with biocides confer an ability to control biocide leaching rates [1]. If the carboxylic group of the polymer bonds to a biocidal pendant group, the copolymer will show biocidal activity by the hydrolysis of the pendant group in sea water.

16.3   Constituents of biocidal coatings

The constituents of AF coatings normally include (1) polymer binders or a combination of components that constitute the binder, (2) pigments and extenders or fillers, (3) carrier components (volatile liquids in liquid paints, air in powder coatings), (4) a number of functional additives and (5) biocides [20]. Of these, binders, pigments and biocides are the most important.

Table 16.1 Inorganic pigments used in marine coatings [21].

Compound	Colour	Comments
Lead chromate	Used for green, yellow or red paints	Highly toxic yet difficult to replace as no other pigments have similar light fastness and brightness. Can cause nervous system damage, anaemia, memory loss, achy joints and concentration difficulties
Chromates	Used for green, yellow or orange paints	Causes skin rashes, ulcers of the upper respiratory tract and nose irritation as well as lung cancer
Cadmium	Used for green, yellow, orange or red paints	Cadmium is extremely toxic to many organisms. As it binds to sulphur more strongly than zinc, it can replace zinc in some enzymes, rendering them useless
Cuprous oxide	Used for red, brown or orange paints	Cuprous oxide is a highly efficient biocide and is used in most antifouling coatings. Causes irritation to respiratory tract, which may result in ulceration and perforation. Causes irritation to skin

Note: Highly toxic inorganic pigments include barium chromate (barium and soluble chromates), chrome yellow (lead and soluble chromates), zinc yellow (soluble chromates), Naples yellow (lead antiminate) and flake white (lead).

Binders and resins are the glue that holds the coatings together. Chosen based on the physical and chemical properties of the desired finished film, the binder often classifies the paint (alkyd, oil or plasticiser). In general, metal coatings are dominated by alkyds, although water-based acrylics, epoxies, polyurethanes and polyesters are also used for certain types of coatings. AF coatings generally either have hydrolysing binders that hydrolyse upon immersion and self-polish, or have hydrating binders that do not self-polish to the same degree.

Pigments add colour to a coating (e.g. the deep red colour seen below the plimsoll line of most container ships arising from the high Cu pigment concentration in the AF paints), but they also perform a variety of other functions such as anticorrosion, film permeability, surface adhesion, weather resistance and controlling the level of gloss. Both inorganic and organic pigments are used in marine coatings, and all have different uses, colours and toxicity as can be seen in Table 16.1.

Many organic pigments are now replacing metal-based pigments, most of which are derived from products of the petrochemical industry such as benzene, toluene, the xylenes, naphthalene, anthracene and phenols. They are intensely coloured, particulate organic solids that are physically and chemically unreactive with, and soluble in the carrier substrate. As they are chemically bound in the paint matrix and do not leach out of the paint unintentionally, their in-service environmental impact is minimal, unless they erode with the polishing binder.

16.4   Biocides in AF paints

Copper compounds such as cuprous oxide (Cu_2O), copper thiocyanate (CuSCN) or metallic copper have already been used in combination with TBT as the principal biocides in AF coatings. There are, however, several algal species (e.g. *Enteromorpha* spp., *Ectocarpus* spp.,

Achnanthes spp.) that show marked physiological tolerance to copper [21], so additional booster biocides are included in many new AF coatings for the replacement of TBT. The ideal booster biocide must have broad-spectrum toxic activity to target organisms, low water solubility for performance longevity, a history of safe usage, and must be non-persistent when released into the aquatic environment [22].

These boosters include both copper and zinc pyrithione (ZPT) (antifungal agents used in anti-dandruff shampoos and outdoor paints), diuron (a substituted urea herbicide), Chlorothalonil^{TM} (tetrachloroisophthalonitrile), dichlofluanid (a phenylsulfamide acaricide), Irgarol 1051^{TM} (a triazine), Sea Nine 211^{TM} (dichloro-isothiazolone), zineb and maneb (both dithiocarbonate fungicides containing zinc and manganese, respectively). Many of them, including zineb, maneb, Chlorothalonil^{TM} and dichlofluanid have a history of use as agricultural pesticides. Eight of these biocides (Chlorothalonil^{TM}, dichlofluanid, Irgarol 1051^{TM}, Sea Nine 211, 2,3,5,6-tetrachloro-4-methylsulphonyl pyridine, 2-hiocyanomethylthiobenzothiazole, ZPT, zineb) are approved by the UK Health and Safety Executive for use in amateur and professional AF products marketed in the UK [23] and only coatings containing dichlofluanid, ZPT or zineb as the active ingredients can be applied on vessels less than 25 m in length.

In a survey carried out in the UK of boat yards and chandlers [24], the rank order in terms of number of vessels treated with a particular biocide was cuprous oxide > diuron > copper thiocyanate > Irgarol 1051^{TM} and ZPT > dichlofluanid. Quantities of AF biocides distributed in the UK in a 1-year period ranged from 59 kg (minimum estimate for Irgarol 1051^{TM}) to 331 769 kg (maximum estimate for Cu). This study also attempted to calculate a leaching rate for these biocides assuming that all biocide was leached from the paint during a 9-month period, the assumed lifetime of the paint, and estimated that the tonnages of biocide leaching into the environment (based on estimated boat numbers in UK waters) ranged from 0.98 tonnes (Irgarol 1051^{TM}) to 425.7 tonnes (cuprous oxide). Unfortunately, the methodology behind these figures is flawed as it is not expected that all biocide will be leached from the paint, and most AF coatings are designed to last for 3–5 years and will not necessarily spend all of this time in UK waters.

16.4.1   *Organotin*

Tin has a larger number of its organometallic derivatives in commercial use than any other element. The first systematic studies of organotins were done by Sir Edward Frankland (1825–1899), who synthesised diethyltin diiodide in 1853 and tetraethyltin in 1859 and today more than 800 compounds are known [11]. Use of tin expanded in the 1940s with the advent of the plastics industry when it was found that the addition of certain organotin derivatives to PVC polymers could prevent discolouration and embrittlement caused by heat and light. Nowadays, the use of mono/dialkyltin derivatives as heat and light stabiliser additives in PVC remains the major use of organotins (about 70%), although it is also used in fungicides, miticides, molluscicides, nematocides, ovicides, rodent repellents, wood preservatives and AF paints [11].

Organotin compounds comprise a group of organometallic moieties characterised by an Sn atom covalently bound to one or more organic substances. The Sn$^-$C bonds are stable in the presence of water, atmospheric O_2 and heat (up to 200°C), although UV radiation, strong acids and electrophilic agents readily cleave the bond, meaning that the principal abiotic removal mechanisms of TBT are volatilisation, photolysis (typical Sn$^-$C bond cleavage energies are about 210 kJ mol^{-1} compared with the energy of blue light at 400 nm which is approximately

300 kJ Einstein^{-1} [15, 25]), and adsorption followed by sedimentation. In his review, de Mora [15] states that volatilisation is not an important loss mechanism as the vapour pressure of tributyltin oxide is quite low, with estimates ranging from 6.4×10^{-7} to 1.2×10^{-4} mm Hg at 20°C. This may not always be the case, however, as scavenging by bubbles in the water column and subsequent loading of the surface microlayer followed by ejection to the atmosphere has been shown to be three orders of magnitude more important than previously thought [25]. Biotic pathways include biosorption, uptake and biodegradation.

The fate of organic compounds in the aquatic environment is closely linked to their partitioning between aqueous media and sediment. If partitioning occurs to water or to sediment, the biocide is subsequently available for uptake into the primary trophic levels and the rest of the food chain, although the likelihood of uptake from sediment is lower than from water, with detrivores primarily affected. Organic pollutants can be sorbed onto particulate matter or exist in solution. Adsorption onto particulate matter and flocculation, in which newly formed particulate matter traps the species, are often termed as sorption and are responsible for reducing the concentration and toxicity of biocides in the water column and accumulation in the sediments. Soluble pollutants are more mobile than particulate-associated pollutants and are consequently more likely to undergo bioaccumulation rather than being trapped in the sedimentary phase [26]. The degree of sorption to sediments is generally influenced by the surface area available for contaminant binding, the presence of other organic and inorganic compounds competing for those adsorbing sites and the structure and charge of the biocides in the water. Silts and clays contain charged minerals known as montmorillonite, which impart a negative charge to the particles, thereby attracting cationic species. In contrast, sands are composed of neutrally charged minerals and have a lower specific surface area and thus lower binding capacity.

TBT is the most persistent of all organic biocides [21]. Crustaceans and fish accumulate much lower amounts than bivalves due to efficient enzymatic mechanisms that degrade TBT in the body. In natural waters, TBT has a short residence time, with a half-life in the range of several days to several weeks, whereas degradation studies of TBT associated with sediments suggest half-lives at least an order of magnitude longer than those found in water [11]. As TBT exhibits a tendency to accumulate in sediments, degradation processes in this phase are more likely to control the overall persistence of TBT in the environment.

Degradation trends suggest that TBT either debutylates to dibutyltin (DBT) and monobutyltin (MBT) in aerobic sediments or degrades to DBT, which subsequently desorbs to the overlying water column. In anaerobic sediment, the half-life of TBT is not discernible and appears to be in the order of tens of years [13]. TBT bioaccumulates with an accumulation factor as high as 10 000 and is chronically and acutely toxic [27]. Past studies have focused on coastal organisms such as molluscs [16, 17, 28–32], although there is evidence that TBT is affecting organisms higher up the food chain in the open ocean such as bluefin tuna [33] and killer whales [34]. In a review in the late 1990s of contamination of French coastal waters by organotin compounds, it was noted [14] that the improvement in organotin pollution seen during the 1980s had ceased and the situation had stagnated at all levels which were still far too high and the development of non-toxic AF paints was necessary.

16.4.2 Copper

Since the Royal Navy first clad their ships with copper sheathing in 1761, copper has been used for AF. Copper compounds, such as cuprous oxide (Cu_2O), copper thiocyanate (CuSCN)

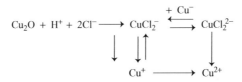

Figure 16.1 Dissolution of copper in sea water.

or metallic copper are now used as principal AF biocides. Transition metals such as copper and zinc are essential to life because of the catalytic and structural roles they play in proteins and other biomolecules [35]. At the same time, high levels of copper can be detrimental to life, thus providing a means for controlling unwanted organisms [36]. To avoid metal induced toxicity, most organisms use a redundant combination of metal regulated import inhibition, sequestration and enhanced export mechanisms. Combinations of these mechanisms are used to form detoxification pathways controlled through metal-binding proteins (metallothioneins) at transcriptional, translational or enzymatic levels [35]. Sufficiently high levels of these metals, or interference from other biocides, can cause a breakdown in these detoxification processes, resulting in mortality.

Although some copper is found naturally in the environment in its elemental form, most is commercially produced from sulphide or oxide minerals [36]. Copper occurs in one of four oxidation states: Cu^0, Cu^+, Cu^{2+} and Cu^{3+}, although trivalent Cu is very rare. Cu^{2+} (the cupric ion) is the most important oxidation state of Cu and the one most commonly found in water [36]. The Cu^+ ion is converted to Cu^0 and Cu^{2+} by a disproportionate reaction: $2Cu^+ = Cu^{2+} + Cu^0$. Cu^0 in water is then converted to the Cu^{2+} ion by an oxidation reaction [7]. Dissolution of Cu_2O in sea water is proportional to both the H^+ and Cl^- activity and results in a sequence of reactions as shown in Figure 16.1. The Cl^- and H^+ ions react with the Cu_2O particles at the pigment front leading to the production of chloro-copper complexes. These are then oxidised to Cu^{2+} in the leach layer and released as ionic copper or labile copper complexes. The first reaction shown in Figure 16.1 is irreversible and influenced by kinetics, whereas the reaction from $CuCl_2^-$ to $CuCl_3^{2-}$ is reversible and instantaneous and can be considered in equilibrium at all times [37]. The Cu(II) ion is believed to be the most bioavailable and toxic species [38].

Cu^+ is found primarily in anaerobic conditions and is readily oxidised to Cu^{2+} which is chemically labile, reacting with a variety of inorganic and organic ligands as well as with many types of particles. Complexes are formed with bases (ammonium, carbonate, chloride, hydroxide, nitrate and sulphate), and labile Cu species include ions, ion pairs, readily dissociable inorganic and organic complexes and easily exchangeable Cu sorbed on colloidal inorganic or organic matter. Organic metal-complexing ligands are produced as by-products of metabolism as well as of the breakdown of biological material [39]. The typical ratio of labile (free copper ions and inorganically bound copper) to total dissolved copper is between 10 and 30% [40]. If only total copper is determined, the toxicity will be overestimated by a factor of 4 on average.

Although the redox process is undeniably important [40], the pH level is probably the single chemical parameter influencing metal speciation most significantly (Table 16.2). With respect to the inorganic dissolved fraction of the metals, the H^+ ion will compete with the metals for ligands such as OH^-, Cl^-, CO_3^{2-}, HCO_3^-, HS^-, S^{2-}, sulphates and phosphates [41]. It has been shown by the authors that the percent fraction of metals adsorbed on particles rises steeply from almost zero to nearly 100% within a narrow and element-specific pH range. At low pH,

Table 16.2 Postulated physicochemical forms of copper in sea water under natural conditions [43].

	Name of species	Example
1.	Hydrated ion	$Cu^{2+}{}_{aq}$
2.	Dissolved inorganic complexes	$CuCl^+$, $CuSO_4$, $Cu(OH)^+$
3.	Dissolved organic complexes	
	Labile	Cu-glycine
	Inert	Cu-(humic substance)
4.	Colloidal complexes	
	Organic	Cu-polysaccharides
	Organic/inorganic	Cu bound to organic matter adsorbed onto hydrated metal oxides
5.	Particulate Cu	Cu adsorbed by fine clays retained on a 0.45 μm filter

most of the metals are dissolved since H^+ replaces metals sorbed on particles, whereas at high pH, the metals occur predominantly as colloidal or particulate. Furthermore, the presence of organic compounds influences metal speciation by moving the pH to metal adsorption curve to the left (lower pH) by the order of one pH unit. Higher than 20% sorbed fractions of Cu are generally reported only in cases where pH is higher than 6.5 or TOC is higher than 6 mg L^{-1} [41]. This is an important factor when discussing freshwater systems, as both pH and TOC can change dramatically over the course of a riverine system. Marine systems, however, are buffered by the carbonate alkalinity capacity at a pH of 8.0–8.4 and have high TOC meaning that metals would be less inclined to exist in the aqueous phase than they would in freshwater systems.

In considering the interactions of trace metals with aquatic biota, one can identify three levels of concern: (1) metal speciation in the external environment, (2) metal interactions with the biological membrane separating the organism from its environment and (3) metal partitioning with the organism and the attendant biological effect [42]. As Borgmann [43] notes, biological impacts such as alteration of in situ communities and demonstration of toxicity in environmental samples often occur at sites with elevated metal concentrations, but this does not prove that metals are actually responsible for these effects. Metal-induced biological effects cannot usually be inferred from measured environmental concentrations because metal bioavailability can vary dramatically from site to site. These differences can lead to differences in metal bioaccumulation, which in turn lead to differences in metal-induced effects.

The bioavailability of a substance in the marine environment depends on chemical and physical factors acting outside organisms and biological factors acting within or on the surface of organisms. The former affect most biota in the same way [43], and can include complexation of metal ions by inorganic and organic complexing agents, adsorption to particulate matter, precipitation and binding within soluble matrices. The biological factors are less well understood and there are several models that have been developed to investigate these effects.

16.4.2.1   The free-ion activity model

The free-ion activity model (FIAM) [42, 44] states that the biological response elicited by a dissolved metal is usually a function of the free-metal ion concentration, M^{z+} (H_2O), which is determined not only by the total dissolved metal concentration, but also by the concentration and nature of the ligands present in solution. If it is assumed that the cell surface is in equilibrium with various metal species in the bulk solution, and that this equilibrium precedes the expression of the biological response, it follows that the identity of the metal form(s) reacting with the cell surface is of no biological significance – no single species in solution can be considered more (or less) available than any other. In a system at equilibrium, the free-metal ion activity reflects the chemical reactivity of the metal. It is this reactivity that determines the extent of the metal's reactions with surface cellular sites, and hence its bioavailability. The surface equilibrium assumption also implies that other (cationic) species, notably the hardness cations, Ca^{2+} and Mg^{2+} and the hydrogen ion H^+, may compete with the metal of interest, M^{z+}, for binding at the surface complexation site. Such antagonistic interactions would tend to reduce the equilibrium concentration of the surface complex, and thus diminish the biological response [42].

16.4.2.2   *The gill surface interaction model and the biotic ligand model*

The gill surface interaction model (GSIM) [45] forms the basis of the more recently developed biotic ligand model (BLM) [42]. The BLM, an adaptation of the GSIM and the FIAM, is a mechanistic approach that greatly improves the ability to generate site-specific ambient water quality criteria for metals in the natural environment relative to conventional relationships based only on hardness. Using an equilibrium geochemical modelling framework, the BLM incorporates the competition of the free metal ion with other naturally occurring cations (e.g. Ca^{2+}, Na^+, Mg^{2+}, H^+), together with complexation by abiotic ligands (e.g. DOM [dissolved organic matter], chloride, carbonates, sulphide) for binding with the biotic ligand, the site of toxic action on the organism [46]. The critical metal concentration on the biotic ligand, which is associated with, for example, 50% mortality (or any other endpoint), is assumed to be independent of the water quality characteristics, meaning that the BLM approach can be used to predict the toxicity of a metal over a range of water quality characteristics [47].

In consequence, metal toxicity decreases with increasing water hardness and therefore increasing competition by the Ca^{2+} ions on the free metal ion concentration M^{z+} (H_2O) (Figure 16.2). One might assume that in waters that are very hard, or indeed are very saline with a large number of free cationic species, the lethal concentration of metals that one can put into that ecosystem would be higher than in freshwater. This, however, is not always the case, due to the joined concepts of bioaccumulation and subsequent bioconcentration.

Discovered by Cornec [49], the concepts of bioaccumulation and bioconcentration were generalised to all marine organisms by Vinogradov [50] who, in effect, definitively established that the majority of species could accumulate not only those organic or inorganic elements needed for their survival, but also those that had no known physiological role. All aquatic organisms take up trace metals, both essential and non-essential and accumulate these metals to different degrees, showing different patterns of bioaccumulation. This accumulated metal can then be divided into two fractions: metabolically available metal and detoxified metal. Metabolically available metal can play essential roles in metabolism but is also potentially toxic, there being a threshold maximum concentration of this metal fraction, beyond which toxic effects come into play. Detoxified metal is biologically inert and so long as the combined rate of excretion and detoxification of metabolically available metal has exceeded the rate of total metal uptake, it would be possible for the detoxified metal concentration in the body, and therefore the total metal concentration, to increase to any range of values [51].

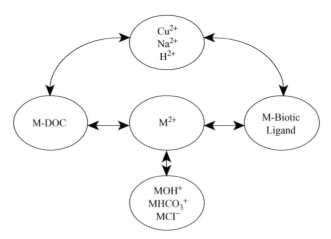

Figure 16.2 Conceptual diagram of the biotic ligand model (BLM) showing inorganic and organic complexation in the water and interaction of metals and cations on the biotic ligand. DOC, dissolved organic carbon. (After Allen *et al.* [48].)

If an organism has a net accumulation strategy, it could build up concentrations of toxins that are not toxic to the organism itself, but could be potentially lethal to any predator, including humans. This is a familiar concept to all ecosystems whereby the top predators can be poisoned by a seemingly non-lethal concentration of xenobiotic entering the environment and being bioconcentrated by lower organisms. This bioconcentration is especially noticeable in aquatic ecosystems, as aquatic organisms take up metals not only from food but also from the surrounding aquatic medium or hydrosphere.

16.4.2.3   Occurrence of copper in the environment

The hydrosphere comprises marine, estuarine, river and other freshwater environments, both surface and underground, and copper is present in each within the dissolved, suspended, particulate or sedimentary phases (Table 16.3) [52]. Only a fraction of the copper in the dissolved phase exists as free or hydrated Cu^{2+}, with the balance being complexed with organic or inorganic ligands, or associated with a colloidal phase that includes microparticles and macromolecules too small to settle under gravitational force [36].

Within marinas, harbours and estuaries, which are the focus of many predicted environmental concentrationmodels, re-suspending bottom sediment and its influence on the partitioning of substances in overlying water is of great importance. Bottom sediments are generally anoxic and contain greater concentrations of sulphides, which have a strong affinity for trace metals. This means that as the suspended sediment concentration increases so will sulphide levels and the proportion of metal being associated with the sulphide phase [53].

In their paper summarising a plenary lecture given at the SECOTOX 2001 World Conference in Cracow, Poland, Janssen *et al.* [47] stated that one of the main shortcomings of present environmental risk assessment procedures is the inadequacy, or total lack, of incorporating metal bioavailability and that current water quality standards and risk assessment procedures of metals are predominately based on total or dissolved metal concentrations, despite evidence to the contrary. The result of this is that environmental copper may exceed environmental quality standard levels if it is measured using the total dissolved form, even though the bioavailable

Table 16.3 Summary of copper concentrations in the hydrosphere as reported in the Copper Sourcebook 1998 [54] (covering years 1993–1996).

Environmental media		Concentration	Units
Hydrosphere			
Coastal	Dissolved	0.06–4.3	ppb
	Total	0.5–13.8	ppb
	Suspended solids	0.6–37 000	ppm
Estuarine	Dissolved	0.02–4.7	ppb
	Total	1.2–71.6	ppb
	Suspended solids	0.38–72	ppm
Oceanic	Dissolved	nd–10	ppb
	Total	0.04–10	ppb
	Suspended solids	0.01–2.8	ppm
Lake	Dissolved	0.1–15.6	ppb
	Total	0.1–15.6	ppb
River	Dissolved	0.18–3000	ppb
	Total	0.5–5800	ppb
Groundwater	Dissolved	0.003–70	ppb
	Total	1–1160	ppb
Drinking water	Total	0.3–1352	ppb
Hydrosphere – sediments			
Coastal	Particulate	0.03–3789	ppm
	Interstitial water	25.5–32.7	ppb
Estuarine	Particulate	0.3–2895	ppm
	Interstitial water	0.3–100	ppb
Ocean	Particulate	3.1–648	ppm
	Interstitial water	22–45	ppb
Lake	Particulate	0.4–796	ppm
	Interstitial water	45.6–52	ppb
River	Particulate	5.3–4570	ppm

copper is actually below the approved limits. Although because of this copper is seen by some as being more environmentally friendly than TBT, there is nevertheless concern about levels of this metal in the marine environment, and regulatory bodies are already turning their attention to contamination in ports, harbours and coastal zones. For instance, the use of copper has been banned as an AF biocide on pleasure craft in Baltic Sea and West Coast areas of Sweden, and release rate limits have been imposed in Canada, Denmark and are about to be enforced in the USA following studies by the US Navy [48] investigating copper sources and loading in US Navy harbours. AF coatings were seen to constitute a significant portion of the total copper loading into Navy harbours, although civilian craft were seen to have release rates almost twice those of military vessels ($8 \ \mu g \ cm^{-2} \ day^{-1}$ vs. $3.9 \ \mu g \ cm^{-2} \ day^{-1}$). The highest copper load was at San Diego Bay, which had $23\,000 \ kg \ year^{-1}$.

16.4.3 Organic biocides

16.4.3.1 ECONEA^{TM}

ECONEA^{TM} (1H-Pyrrole-3-carbonitrile, 4-bromo-2-(4-chlorophenyl)-5-trifluoromethyl) from Janssen Preservation & Material Protection is, like most organic biocides, based on an insecticide called chlorfenapyr [54], which has an additional N-ethoxymethyl group at the pyrrole ring. This compound might be a replacement for copper, as it is active against a wide range of biofouling organisms including barnacles. This non-metal AF agent hydrolyses rapidly when in solution to form very low toxicity breakdown products that are biodegradable, according to the manufacturer.

16.4.3.2   Irgarol 1051™

Irgarol 1051™ (N-Cyclopropyl-N'-(1,1-dimethylethyl)-6-(methylthio)-1,3,5-triazine-2,4-diamine) belongs to the s-triazine group of compounds which act as photosystem II (PSII) inhibitors, with the inhibition of photosynthetic electron transport in chloroplasts as their biochemical mode of action [55]. Light promotes the damaging capacity of PSII inhibitors, because radicals and singlet oxygen are formed when chloroplasts with reduced electron transport capacity are irradiated. The normal production of these reactive species can be kept under control by protective systems, but when formed in excess they give rise to the destructive effects observed during and after herbicide exposure [56–58]. Therefore, algae will suffer from a double assault: a decrease in electron transport and an increased production of radicals. As it is the most hydrophobic of the family of triazines, it is expected that this compound would be detected at higher concentrations in sediment samples than in the aqueous phase. Its water solubility of 7 mg L^{-1} implies a higher affinity to particulate matter than the corresponding atrazine, which exhibits higher water solubility (33 mg L^{-1}) [23].

It has been shown [53] that the partitioning of Irgarol 1051™ to suspended solids increases with increasing sediment concentration and that at very low-suspended solids concentration (1 g L^{-1}) very little is adsorbed to the particulate phase. From this it can be concluded that in the natural environment where suspended sediment concentrations are likely to be lower than 1 g L^{-1} the biocides will be predominantly in the dissolved form. This has also been shown by fugacity modelling that predicted that Irgarol 1051™ would predominantly reside in the aqueous phase (>98%) [26]. Due to this preference for the aqueous phase, dissolved concentrations do increase after dredging in marinas [59] due to remobilisation into the water column. Studies [27,60,61] have shown that Irgarol 1051™ rapidly photodegrades (more than 80% degradation in sea water over 15 weeks) to yield a stable metabolite 2-methylthio-4-tert-butylamino-6-amino-s-triazine (M1) which has been shown to be from 2.5 to 10 times less toxic than Irgarol 1051™ itself [27].

16.4.3.3   Diuron

In a study looking at marine macroalgae, diuron (N'-(3,4-dichlorophenyl)-N,N-dimethylurea) was found to be the least toxic but most persistent biocide tested [62]. Diuron has been shown [21] to exhibit a removal range of 4–10% over particulate matter concentration from 10 to 10 000 mg L^{-1} that was not influenced by pH. It has been shown to be persistent in sea water and to degrade much quicker under anaerobic conditions, with an anaerobic half-life of 14 days in marine sediments at 15°C. When associated with paint particles in sediments, however, it showed no signs of degradation under anaerobic conditions [63].

16.4.3.4   Sea Nine®

Sea Nine (4,5-dichloro-2-n-octyl-3(2H)-isothiazolone) was found to be the most toxic out of all the booster biocides on the market today [27] (EC50 > 0.003 mg L^{-1}). In view of this extreme toxicity, it would not appear that Sea Nine offers any clear advantage over TBT as a replacement biocide. However, Sea Nine® has been shown to completely degrade within 24 hours [63] and biological degradation is considered to be over 200 times faster than hydrolysis and photolysis [64]. The rapid degradation reduces the concentration significantly below toxic levels and its metabolites are ring opened structures with a toxicity 4–5 orders of magnitude less than Sea Nine® [23]. This short half-life is much lower than TBT and most other booster biocides, thus reducing its bioaccumulation potential and suitability for use in AF paint. It also has a much lower chronic or reproductive toxicity to marine mammals. Recent results, however [65], show that nominal concentrations of 32 and 100 nM are able to exert a long-term effect on natural phytoplankton communities.

16.4.3.5   Zinc/copper pyrithione

ZPT is well known for its bactericidal and fungicidal effects in anti-dandruff shampoo and other cosmetic products. ZPT and the related substance, copper pyrithione (CPT), are marketed as environmentally neutral, non-persistent AF biocides because they photolyse and rapidly degrade into less toxic compounds. In water, ZPT and CPT degrade through photolysis and biodegradation with a half-life of around 30 minutes [66]. ZPT is also known to undergo transchelation with copper and possibly manganese replacing the zinc ion, and due to this rapid degradation in the water column, dissolved zinc is seen as being the greatest risk [53]. It has been suggested that in waters where UV degradation is poor, it may accumulate in sediments as copper or manganese complexes prior to biodegradation or hydrolysis [12, 66]. It has been shown [66] that for all sediments the adsorption of zinc onto the sediment is rapid and a steady state is reached in 10–20 days. The degree of adsorption is largely related to the proportion of the 63 μm fraction of the sediment present. The 18-day LC50 values of herbicides such as Irgarol and Diuron are about 1000 times higher than those of pyrithiones due to the rapid degradation time of ZPT and CPT [67].

16.4.3.6   Dichlofluanid

Dichlofluanid (N-dichlorofluoromethylthio-N',N'-dimethyl-N-phenylsulfamide) had the greatest tendency towards solid phase partitioning when compared to Irgarol, Chlorothalonil and Diuron over a range of particulate matter concentration from 10 to 10 000 mg dm^{-3}, with a removal range of 8–52% [26]. This sorption was enhanced at high pH [7] and increased suspended matter, conditions such as one might find in an estuary or marina. Dichlofluanid was also found to be strongly bound to particulate matter with a desorption percentage of less than 1%. Despite this, it has been shown to completely degrade within 24 hours [63].

16.5   Distribution of biocides in the environment

In order to protect the environment against any possible impact from AF coatings, it is important to ascertain different sources of AF biocides, what the possible fluxes could be and what their fate will be in order to establish a management strategy for their input into the aquatic environment. There are four main ways in which AF biocides can enter the environment:

(1) during the application of AF coatings, (2) in-service leaching from coating, (3) during paint removal and (4) during paint disposal [68]. In the case of small particles of paints entering the aquatic environment during hosing or from rainwater run-off, each particle can become a new site for biocide release, and in certain locations may be an important source of booster biocide contamination.

Over the past few years, hydroblasting has become the preferred method to remove AF coatings from a ship's hull due to the human health risks from breathing sandblasted material. This hydroblasting can break up removed paint into paint chips and then into 10 μm particles which can be widely distributed in waterways [69], and act as a point source as the biocides continue to leach from them. Contaminated washdown waters from a large ship can exceed 375 000 L from hydroblasting and can contain up to 6 million ppt TBT [70]. Paint particles that are bound in the bottom sediment can get resuspended due to dredging, storm effects or bioturbation and thus cause contamination many years after they originally entered the marine environment. Without national regulations on discharge requirements, this practice will continue and significantly increase after 2008 as a result of the IMO treaty that requires all TBT coatings to be removed or sealed [69].

Biocides can also enter the marine environment through ship grounding. In a study conducted around the Great Barrier Reef World Heritage Centre, Australia, although TBT was undetectable at all outer reef and mid-shelf mooring sites, concentrations detected at the Heath Reef grounding site were grossly elevated as a consequence of AF paint being abraded from the ships hull at the time of grounding and during subsequent refloating [71].

Leaching from the sides of hulls continues to be the major input, however, and is the most common source used in risk analysis models. If the source data for these models is incorrect then the predicted environmental concentration levels they give will be similarly skewed. In their (MCA) study of the environmental fate of booster biocides, Voulvoulis et al. [21] found that extracting the data from the literature in order to compare the biocides was a difficult task, and even when those data exist in most cases they were not comparable, as a different methodology was used in each case. The same problem applies to all environmental models used presently, if there is a wide range in the input data then there is going to be a concomitant wide range in the output data. This means that not only can regulatory authorities set limits at whichever end of the range they want, but also that a lack of trust in the regulatory limits will arise.

16.6   Conclusions

- The use of AF technologies dates back to the Carthaginians and Phoenicians who used copper sheathing, sulphur and arsenic containing coatings, and lead sheathing on their vessels. Copper compounds are still the most widely used biocides in self-polishing AF coatings today.
- There is an acknowledged issue with the use of copper in AF coatings in that it accumulates in the sediments of marinas, shipyards and harbours and does not degrade. The only way to remove copper from these sediments therefore is dredging.
- Since the mid-twentieth century, the most successful AF technology has been the SPC coating containing TBT, which not only had highly effective AF properties, but also had

harmful effects on non-target organisms. With raising environmental awareness, this no longer became acceptable and TBT has been banned by IMO.

- TBT has been replaced by numerous organic biocides, which act together with copper compounds. Some of the new organic biocides have rapid degradation in sea water.
- Alternative coatings (silicone, ceramic, hard and nano-coatings) all have shortcomings such as speed of application or low mechanical strength, although silicone coatings are currently being applied to commercial vessels.

References

1. IMO (2006) *Global Ballast Water Management Programme: The International Response.* Available from http://globallast.imo.org/index.asp?page=internat_response.htm& menu=true. Accessed 6 July 2009.
2. United Nations Conference on Trade and Development (2004) *Review of Maritime Transport.* UNCTAD, United Nations, New York and Geneva.
3. United Nations Conference on Trade and Development (2005) *Review of Maritime Transport.* UNCTAD, United Nations, New York and Geneva.
4. Evans, S.M., Birchenough, A.C. & Brancato, M.S. (2000) The TBT ban: out of the frying pan into the fire? *Marine Pollution Bulletin,* **40** (3), 204–211.
5. Townsin, R.L. (1987) Development in the calculation of rough underwater surface power penalties. In: *Celena 25th Anniversary Symposium.* Genoa.
6. Rouhi, A.M. (1998) The squeeze on tributyltins. *Chemical & Engineering News,* **76** (14), 41–42.
7. Omae, I. (2003) General aspects of tin-free antifouling paints. *Chemical Reviews,* **103**, 3431–3448.
8. Tolosa, I., Readman, J.W., Blaevoet, A., Ghilini, S., Bartocci, J. & Horvat, M. (1996) Contamination of Mediterranean (Cote d'Azur) coastal waters by organotins and Irgarol 1051 used in antifouling paints. *Marine Pollution Bulletin,* **32** (4), 335–341.
9. Batley, G.E., Scammell, M.S. & Brockbank, C.I. (1992) The impact of the banning of tributyltin-based antifouling paints on the Sydney rock oyster, *Saccostrea commercialis. Science of The Total Environment,* **122** (3), 301–314.
10. Minchin, D., Oehlmann, J., Duggan, C.B., Stroben, E. & Keatinge, M. (1995) Marine TBT antifouling contamination in Ireland, following legislation in 1987. *Marine Pollution Bulletin,* **30** (10), 633–639.
11. Hoch, M. (2001) Organotin compounds in the environment – an overview. *Applied Geochemistry,* **16** (7–8), 719–743.
12. Thomas, K.V., Blake, S.J. & Waldock, M.J. (2000) Antifouling paint booster biocide contamination in UK marine sediments. *Marine Pollution Bulletin,* **40** (9), 739–745.
13. Dowson, P.H., Bubb, J.M. & Lester, J.N. (1996) Persistence and degradation pathways of tributyltin in freshwater and estuarine sediments. *Estuarine, Coastal and Shelf Science,* **42** (5), 551–562.
14. Michel, P. & Averty, B. (1999) Contamination of French coastal waters by Organotin compounds: 1997 Update. *Marine Pollution Bulletin,* **38** (4), 268–275.
15. de Mora, S.J. (1996) The tributyltin debate: ocean transportation versus seafood harvesting. In: *Tributyltin: Case Study of an Environmental Contaminant* (ed. S.J. de Mora), pp. 1–20. Cambridge University Press, Cambridge.
16. Alzieu, C. (2000) Environmental impact of TBT: the French experience. *Science of The Total Environment,* **258** (1–2), 99–102.
17. Alzieu, C., Sanjuan, J., Michel, P., Borel, M. & Dreno, J.P. (1989) Monitoring and assessment of butyltins in Atlantic coastal waters. *Marine Pollution Bulletin,* **20** (1), 22–26.
18. Goldberg, E. (1986) TBT an environmental dilemma. *Environment,* **28** (8), 42–44.
19. IMO (2002) *Anti-fouling Systems.* International Maritime Organisation, London.

20. CEPE (1999) Utilisation of more environmentally friendly antifouling coatings. Report No.: 96/559/3040/DEB/E2. CEPE, Brussels.
21. Voulvoulis, N., Scrimshaw, M.D. & Lester, J.N. (2002) Comparative environmental assessment of biocides used in antifouling paints. *Chemosphere*, **47** (7), 789–795.
22. Turley, P. & Waldron, C. (1999) Smooth sailing: the role of co-biocides in marine antifouling paint. *Paint and Coatings Industry*, July 1999, 47.
23. Konstantinou, I.K. & Albanis, T.A. (2004) Worldwide occurrence and effects of antifouling paint booster biocides in the aquatic environment: a review. *Environment International*, **30** (2), 235–248.
24. Boxall, A.B.A., Comber, S.D., Conrad, A.U., Howcroft, J. & Zaman, N. (2000) Inputs, monitoring and fate modelling of antifouling biocides in UK estuaries. *Marine Pollution Bulletin*, **40** (11), 898–905.
25. Saint-Louis, R. & Pelletier, E. (2004) Sea-to-air flux of contaminants via bubbles bursting. An experimental approach for tributyltin. *Marine Chemistry*, **84**, 211–244.
26. Voulvoulis, N., Scrimshaw, M.D. & Lester, J.N. (2002) Partitioning of selected antifouling biocides in the aquatic environment. *Marine Environmental Research*, **53** (1), 1–16.
27. Fernandez-Alba, A.R., Hernando, M.D., Piedra, L. & Chisti, Y. (2002) Toxicity evaluation of single and mixed antifouling biocides measured with acute toxicity bioassays. *Analytica Chimica Acta*, **456** (2), 303–312.
28. Pellizzato, F., Centanni, E., Marin, M.G., Moschino, V. & Pavoni, B. (2004) Concentrations of organotin compounds and imposex in the gastropod *Hexaplex trunculus* from the Lagoon of Venice. *Science of the Total Environment*, **332** (1–3), 89–100.
29. Strand, J. & Asmund, G. (2003) Tributyltin accumulation and effects in marine molluscs from West Greenland. *Environmental Pollution*, **123** (1), 31–37.
30. Evans, S.M., Kerrigan, E. & Palmer, N. (2000) Causes of imposex in the dogwhelk *Nucella lapillus* (L.) and its use as a biological indicator of tributyltin contamination. *Marine Pollution Bulletin*, **40** (3), 212–219.
31. Ellis, D.V. & Agan, P.L. (1990) Widespread neogastropod imposex: a biological indicator of global TBT contamination? *Marine Pollution Bulletin*, **21** (5), 248–253.
32. Strand, J., Glahder, C.M. & Asmund, G. (2006) Imposex occurrence in marine whelks at a military facility in the high Arctic. *Environmental Pollution*, **142** (1), 98.
33. Ueno, D., Inoue, S., Takahashi, S., *et al.* (2004) Global pollution monitoring of butyltin compounds using skipjack tuna as a bioindicator. *Environmental Pollution*, **127** (1), 1.
34. Kajiwara, N., Kunisue, T., Kamikawa, S., Ochi, Y., Yano, S. & Tanabe, S. (2006) Organohalogen and organotin compounds in killer whales mass-stranded in the Shiretoko Peninsula, Hokkaido, Japan. *Marine Pollution Bulletin*, **52** (9), 1066–1079.
35. Dameron, C.T. & Harrison, M.D. (1998) Mechanisms for protection against copper toxicity. *American Journal of Clinical Nutrition*, **67**, 1091–1097.
36. Georgopoulos, P.G., Roy, A., Yonone-Lioy, M.J., Opiekun, R.E. & Lioy, P.J. (2001) Environmental copper: its dynamics and human exposure issues. *Journal of Toxicology and Environmental Health Part B*, **4**, 341–394.
37. Yebra, D.M., Kiil, S. & Dam-Johansen, K. (2006) Mathematical modelling of tin-free chemically-active antifouling paint behaviour. *AIChE Journal*, **52** (5), 1926–1940.
38. Zirino, A. & Seligman, P.F. (2002) Copper chemistry, toxicity and bioavailability and its relationship to regulation in the marine environment. Office of Naval Research second workshop report. Office of Naval Research, San Diego.
39. Jones, B. & Bolam, T. (2007) Copper speciation survey from UK marinas, harbours and estuaries. *Marine Pollution Bulletin*, **54** (8), 1127–1138.
40. Zwolsman, J.J.G. & van Eck, G.T.M. (1990) The behaviour of dissolved Cd, Cu, and Zn in the Scheldt estuary. In: *Estuarine Water Quality Management: Monitoring, Modelling, and Research* (ed. W. Michaelis), pp. 413–420. Springer, Berlin.
41. Gunderson, P. & Steinnes, E. (2003) Influence of pH and TOC concentration on Cu, Zn, Cd, and Al speciation in rivers. *Water Research*, **37**, 307–318.

42. Campbell, P.G.C. (1995) Interactions between trace metals and aquatic organisms: a critique of the free-ion activity model. In: *Metal Speciation and Bioavailability in Aquatic Systems* (eds A. Tessier & D.R. Turner), p. 679. John Wiley & Sons, Chichester.

43. Borgmann, U. (2000) Methods for assessing the toxicological significance of metals in aquatic ecosystems: bio-accumulation–toxicity relationships, water concentrations and sediment spiking approaches. *Aquatic Ecosystem Health and Management*, **3** (3), 277–289.

44. Sunda, W.G. & Guillard, R.L.L. (1976) The relationship between cupric ion activity and the toxicity of copper to phytoplankton. *Journal of Marine Research*, **34**, 511–529.

45. Pagenkopf, G.K. (1983) Gill surface interaction model for trace-metal toxicity to fishes: role of complexation, pH and water hardness. *Environmental Science & Technology*, **17**, 342–347.

46. Niyogi, S. & Wood, C.M. (2004) Biotic Ligand Model, a flexible tool for developing site-specific water quality guidelines for metals. *Environmental Science & Technology*, **38** (23), 6177–6192.

47. Janssen, C.R., Heijerick, D.G., De Schamphelaere, K.A.C. & Allen, H.E. (2003) Environmental risk assessment of metals: tools for incorporating bioavailability. *Environment International*, **28** (8), 793–800.

48. Allen, H.E., Ditoro, D., Paquin, P. & Santore, S. (2002) Water quality criteria/Cu concentrations/Cu activity. In: *Copper Chemistry, Toxicity and Biovailability and Its Relationship to Regulation in the Marine Environment*, Office of Naval Research second workshop report (eds A. Zirino & P.F. Seligman), p. 79. Office of Naval Research, San Diego.

49. Cornec, E. (1919) Spectrographic studies of the ash of marine plants. *Academie Science Paris*, **168**, 513–514.

50. Vinogradov, A.P. (1953) *The Elementary Chemical Composition of Marine Organisms*. Sears Foundation for Marine Science, Yale University, New Haven, CT.

51. Rainbow, P.S. (1999) Bioaccumulation of trace metals: biological significance. *Oceanis*, **25** (4), 547–561.

52. Harrison, B.J. & Lewis, A.G. (1998) *Copper Information Sourcebook – 1998: The World's Scientific Literature on Copper in the Environment and Health*. International Copper Association, Vancouver.

53. Comber, S.D.W., Franklin, G., Gardner, M.J., Watts, C.D., Boxall, A.B.A. & Howcroft, J. (2002) Partitioning of marine antifoulants in the marine environment. *Science of The Total Environment*, **286** (1–3), 61–71.

54. Tomlin, C.D.S. (2003) *The Pesticide Manual*. British Crop Protection Council, Alton, UK.

55. Gatidou, G., Kotrikla, A., Thomaidis, N.S. & Lekkas, T.D. (2004) Determination of two antifouling booster biocides and their degradation products in marine sediments by high performance liquid chromatography-diode array detection. *Analytica Chimica Acta*, **505** (1), 153–159.

56. Chesworth, J.C., Donkin, M.E. & Brown, M.T. (2004) The interactive effects of the antifouling herbicides Irgarol 1051 and Diuron on the seagrass *Zostera marina* (L.). *Aquatic Toxicology*, **66** (3), 293.

57. Dahl, B. & Blanck, H. (1996) Toxic effects of the antifouling agent irgarol 1051 on periphyton communities in coastal water microcosms. *Marine Pollution Bulletin*, **32** (4), 342–350.

58. Jones, R. (2005) The ecotoxicological effects of Photosystem II herbicides on corals. *Marine Pollution Bulletin*, **51** (5–7), 495.

59. Bowman, J.C., Readman, J.W. & Zhou, J.L. (2003) Seasonal variability in the concentrations of Irgarol 1051 in Brighton Marina, UK; including the impact of dredging. *Marine Pollution Bulletin*, **46** (4), 444–451.

60. Liu, D., Maguire, R.J., Lau, Y.L., Pacepavicius, G.J., Okamura, H. & Aoyama, I. (1997) Transformation of the new antifouling compound Irgarol 1051 by *Phanerochaete chrysosporium*. *Water Research*, **31** (9), 2363–2369.

61. Okamura, H., Watanabe, T., Aoyama, I. & Hasobe, M. (2002) Toxicity evaluation of new antifouling compounds using suspension-cultured fish cells. *Chemosphere*, **46** (7), 945–951.

62. Myers, J.H., Gunthorpe, L., Allinson, G. & Duda, S. (2006) Effects of antifouling biocides to the germination and growth of the marine macroalga, *Hormosira banksii* (Turner) Desicaine. *Marine Pollution Bulletin*, **52** (9), 1048–1055.

63. Thomas, K.V., McHugh, M., Hilton, M. & Waldock, M. (2003) Increased persistence of antifouling paint biocides when associated with paint particles. *Environmental Pollution*, **123** (1), 153–161.

64. Jacobson, A.H. & Willingham, G.L. (2000) Sea-nine antifoulant: an environmentally acceptable alternative to organotin antifoulants. *Science of The Total Environment*, **258** (1–2), 103.

65. Larsen, D.K., Wagner, I., Gustavson, K., Forbes, V.E. & Lund, T. (2003) Long-term effect of Sea-Nine on natural coastal phytoplankton communities assessed by pollution induced community tolerance. *Aquatic Toxicology*, **62** (1), 35–44.

66. Turley, P., Fenn, R.J., Ritter, R.C. & Callow, M.E. (2005) Pyrithiones as antifoulants: environmental fate and loss of toxicity. *Biofouling*, **21** (1), 31–40.

67. Okamura, H. (2002) Photodegradation of the antifouling compounds Irgarol 1051 and Diuron released from a commercial antifouling paint. *Chemosphere*, **48** (1), 43–50.

68. Boxall, A.B.A., Conrad, A.U. & Reed, S. (1997) Environmental problems from antifouling agents. Report no. P2F(97)03, Environment Agency, Bristol.

69. Champ, M.A. (2003) Economic and environmental impacts on ports and harbors from the convention to ban harmful marine anti-fouling systems. *Marine Pollution Bulletin*, **46** (8), 935–940.

70. Fox, T., Beacham, T., Schafran, G. & Champ, M. (1999) Advanced technologies for removing TBT from ship washdown and drydock runoff wastewaters. In: *Oceans '99, MTS/IEEE Conference Proceedings: Riding the Crest into the 21st Century* (ed. M. Society), pp. 17–26. Marine Technology Society, Washington, DC.

71. Haynes, D. & Loong, D. (2002) Antifoulant (butyltin and copper) concentrations in sediments from the Great Barrier Reef World Heritage Area, Australia. *Environmental Pollution*, **120** (2), 391–396.

Chapter 17
Consequences of Antifouling Systems – An Environmental Perspective

Cato C. ten Hallers-Tjabbes and Simon Walmsley

This chapter aims to highlight the complex ecological context affecting the impact of antifouling systems (AFS), in particular of those that are based on biocide action and to discuss the ecological consequences of the AFSs which have been explained from a chemical perspective in the previous chapter. We demonstrate how the consequences of antifouling are largely intertwined with the biological and abiotic context given by the marine and freshwater environments where the AFS are operating. We also highlight a crucial development in alternative antifouling, the biocide-free alternatives.

17.1   Introduction

A clear demarcation exists between the antifouling that is now globally recognised and regulated as harmful, tributyltin (TBT) has been banned along with other organotins as AFSs by the AFS Convention [1], and the alternatives that largely became developed once a ban on TBT was on the horizon. Amongst the latter, there is an increasing trend to develop and use biocide-free AFSs, which are consistently increasing their market share. This accommodates the challenge for alternative AFS to be non-harmful, i.e. biocide-free. Apart from shipping, aquaculture also used several different organtins for antifouling. Once the International Maritime Organization agreed to develop a mandatory instrument to ban all organotins as antifouling [1], the TBT vendors announced that they would cease bringing organotins for AFSs to market upon adoption of the mandatory instrument. TBT has been officially banned on the date of entry into force of the AFS Convention on 18 September 2008 (Chapter 21).

Environmental impact knowledge about AFS indicates a severe ecotoxicity of TBT, with little remaining uncertainty, while other biocide-containing AFS are found to be less ecotoxic, though the remaining uncertainty is higher than for TBT [2]. Concentrating on the chemistry of AFS for predicting environmental impact does not reflect the influence of the many aquatic environmental processes, i.e. biotic processes, such as bioconcentration/magnification, food web mechanisms, reproduction strategies and indirect or interactive effects, and abiotic processes, such as hydrographical conditions, interaction with organic matter, salinity, pH, temperature and aeration, that are crucial for fate and behaviour of AFS in natural systems. The impact of

TBT is not restricted to aquatic biota only and the impact on humans has been documented. In the following paragraphs, we give a general explanation about the impact of AFS as we see it and refer to specific issues when and where appropriate.

The chemical pathways of AFSs as described in the previous chapter are complex. Where the previous chapter mentions biocide-based AFS only, we want to stress that monitoring the development of AFS should give ample attention to a future window in which effective biocide-free formulations play a major, if not a total, role.

17.2   Antifouling systems

We will highlight TBT, an organotin and the main alternative AFS for shipping, including the minor contribution of organotin triphenyltin (TPT). AFS containing organotins were banned from ships for all EU ports as of 01 January 2008 and world-wide upon Entry into Force of the IMO AFS Convention, as of 18 September 2008. AFS fall into two categories: those that contain biocides to fend off biofouling and those that do not. In the biocide-based category, AFSs are developed with chemically synthesised compounds or based on natural antifouling compounds (see Chapters 25). Such natural AFSs are based on substances excreted by settler organisms to keep their direct environment free from other settling species; although naturally produced, the substances often are far from harmless and can be potent toxins (see Chapter 27). While alternative biocide-containing AFSs have been amply developed in lieu of TBT, an increasing market share is supplied by biocide-free, non-stick coatings. Such coatings, although at first only suitable for fast-moving, high-turnover vessels (>25 knots), now are applicable to vessels that sail at lower speeds; at present, the market share of such AFSs is about 15% and is expected to increase (see Chapters 13 and 14).

As early as the late 1980s, TBT was called 'the most toxic compound ever brought into the marine environment' [3]. TBT is by its properties an outlier in the realm of organochemical contaminants and hence differs from other AFS as well. Depending on the pH of the water, the molecule is polar or non-polar. The latter form penetrates more readily through biological membranes than the former. As sea water is well buffered at a pH around 8 (slightly alkaline), TBT is five times more toxic (membrane penetrable) in marine than in fresh waters, which tend to be slightly acidic. Apart from the pH dependency, salinity and the presence and character of sediment particles influence the environmental behaviour and fate of TBT. TBT in marine waters tends to adhere to particles and biota, by which it has an easy route to be deposited in the sediment. The behaviour of TBT in both the abiotic and biotic environment cannot be predicted from parameters such as quality structure activity relationships (QSARs) in the way behaviour and fate of other organic contaminants (such as polyaromatic hydrocarbons (PAHs) and polychlorinated biphenyls (PCBs)) can be. In the abiotic compartments of the ecosystem, TBT may last in the sediment for decades [4, 5], in particular if the sediment is anoxic, while in the water degradation depends very much on temperature. While in tropical waters, TBT degradation can be a matter of weeks and in colder waters TBT can last much longer. TBT can volatilise into the air [6] and has been found to precipitate with rain [7].

17.3   Tributyltin – behaviour and fate in natural systems

Likewise, the behaviour of TBT and the metabolising, detoxifying and excretion pathway in biota is not predictable. The distribution pattern of TBT and its metabolites DBT (dibutyltin) and MBT (monobutyltin) in whelks (*Buccinum undatum*) and its prey mussels (*Mytilus edulis*) from the same area is totally different [8]. Moreover, the body concentration of TBT in the different organs of *B. undatum* differs over the season, with the highest concentration of TBT in the reproductive organs at the onset of the breeding season.

While most animals are able to metabolise TBT and so partly detoxify it, this detoxification comes at a cost, and the development of imposex is only one of them. Among the many neuroendocrine and other effects of TBT [9, 10], the increase in levels of testosterone, the male sexual hormone in females [11], is of particular concern, as this phenomenon is not restricted to gastropods and may occur throughout the animal kingdom, including in humans (Oehlmann, personal communications); endocrine disruption effects of TBT have also been found in human tissue [12, 13].

While in several coastal intertidal areas the levels of TBT have dropped since the measures to ban TBT from small ships in certain areas (such as Europe and the USA), no recovery has been found in offshore areas, where merchant ships continued to leach TBT from antifouling paints. Ten years after the partial ban on TBT from ships less than 25 m, estuaries in Eastern UK contained low concentrations of butyltins in sediments, but high concentrations of BTs in mussels [14]. The epifauna had improved between 1987 and 1997 [15–17]. In Swedish estuaries sediment, TBT and TPT levels were higher in the upper 10 cm, posing a risk of remobilisation of TBT bound in sediment [18]. In offshore seas, gastropods continue to show considerable levels of imposex and butyltin burdens in the vicinity of shipping routes [19–22]. TBT burdens in the offshore sea remain high [23, 24]. In coastal small-shipping areas, illegal use of TBT has been demonstrated [25, 26]. TBT contamination of biota has been found widely throughout the animal kingdom in the world seas as well as in algae, and with sometimes high concentrations in top predators, such as marine mammals, where TBT and metabolites were found both in blubber and in the liver. The latter indicates that TBT is likely to be involved in toxic reaction chains.

17.4 Environmental risks associated with AFS and alternative AFS

AFSs based on biocide action, including those based on TBT, use booster biocides in their formulation (see Chapter 16); many of which have been found to be persistent in anaerobic conditions and anoxic sediments.

Far less data exist on the risk posed by alternative AFS. TBT remains the most obvious compound present at levels that pose a risk, although other AFSs, such as irgarol and diuron, may pose a risk in present concentrations in the water [27]; seagrass beds and other plant communities appear to be particularly vulnerable. The current ecotoxicological risk profiles of five antifouling biocides, copper, TBT, Irgarol 1051, Sea-Nine 211 and pyrithionate, showed that the spatio-temporal impact of TBT was higher than that of other AFSs. The impact profile of TBT is also much better defined with less remaining uncertainty. The potential risk of other AFSs suffers from a much higher uncertainty [28]. Anoxic sediments in general slow down the degradation process of organic biocides. Pyrithione that degrades by exposure to UV may be far less degradable when stored in sediments in the absence of UV penetration.

The bioavailability of copper is more complex than the statement reflects – that levels may exceed EQS but not in true bioavailability. Although this might be true in some cases, there are many other factors that influence fate and behaviour of copper [29] and hence could increase the toxicity; EQS need therefore be set at a high precautionary level.

17.5   AFS ecotoxicity

Many different effects of TBT have been found additional to the sentinel ones (imposex and decline of populations in gastropods and shell-malformation and reduction of growth and reproduction in oysters); the effects cover a wide range of endocrine disruptive, immunotoxic, neurotoxic and reproduction effects. Bioturbation as a means of reworking sediments allows metals and xenobiotics to become bioavailable and so affects and often enhances sediment toxicity.

On the way to biocide-free AFS, biocide-based systems, including those containing Cu, can only be intermediate technologies. Pending such development for evaluating impacts of antifouling and knowing the remaining uncertainties, one should apply Predicted Environmental Concentration/Predicted No Effect Concentration, QSARs and expert opinions with a substantial range of precaution to prevent entering the same path as was walked with TBT and risk of making the same mistakes and to learn the lesson from the history of the use and policies for TBT.

In the further developments, a crucial prerequisite throughout should be that antifouling substances, as they are continuously in contact with the waters around the ship, should neither be endocrine disrupting nor genotoxic, reprotoxic or carcinogenic.

17.6   AFS in the presence of other contaminants

Areas where ships sail and release antifouling substances most often also contain a mixture of contaminants, which can interact with the AFS. Evidence is mounting that single contaminant toxicity assessments underestimate the level of toxic action in the presence of other contaminants due to not-recognised interactive effects (synergistic, antagonistic or additive) in contaminant mixtures; this phenomenon is seldom investigated further [30]. Whenever the joint toxicity in contaminant mixtures is investigated, the interaction is almost always found to add to the toxic action of single contaminants. This applies specifically at low contaminant concentrations, usually present in the marine environment. Additivity alone as a single explanation of combined toxicity underestimates the effects of complex effluents [31].

While sensitivity to contaminants differs between species, the differences between the individual sensitivities of five species to a mixture of contaminants were far less than the differences in sensitivity to the individual contaminants [32]. Sensitivity differences to the antifoulings Irgarol 1051, Kathon 5287, chlorothalonil, diuron, dichlofluanid, 2-thiocyanomethylthio-benzothiazole (TCMTB) and TBT are partly species-specific. Degradation products of Irgarol 1051 were less toxic to crustaceans and microalgae, but more toxic to a bacterium, while degradation products of diuron were less toxic to microalgae than to a bacterium. The toxicity of 33% of the compound mixtures was additive; 21% of the mixtures was less toxic than expected on additive grounds (antagonistic effect), but most of the mixtures (46%) showed a synergistic enhancements of toxicity [33].

Sediment quality (the content of nitrogen and amino acids) influences accumulation and metabolising of organic contaminants. Bioaccumulation of contaminants associated with sediment, and subsequent trophic transfer of contaminants, can be affected by ecosystem events, such as a labile organic matter input after phytoplankton blooms. Sedimentary distribution of contaminants and microbial biodegradation depends on the quality of sediment organic matter (SOM) as much as on the quantity and the macrofauna present [34, 35]. A relatively small decrease in the total volatile solids content of the sediment increased the concentration of contaminants (metals and organic) in interstitial water and increased sediment toxicity [36]. A sediment-bound mixture of Cd, Cu, Ni, Pb and Zn had a significantly greater than additive effect on the survival of an estuarine meiobenthic harpacticoid copepod, with the mixture being 1.4 times more toxic than that expected by simple additivity [37].

TBT in combination with PCB-126 (polychlorbiphenyl-126) acts moderately additive on immunotoxicity in channel catfish [38]. A diet containing low quantity of PCBs and TBT acted additively on spawning of fish, resulting in reduced spawning frequency, number of eggs and spawning success. TBT alone also reduced larval survival, indicating transgenerational toxicity [39]. A mixture of PCB and TBT (1:1) acted synergistically on swimming behaviour of carp, while TBT and PCB acted additive in decreasing growth [40].

Salinity can affect metal toxicity. Cadmium and lead respond to a salinity increase (5–25 PSU) with lower concentrations of the free ion, due to complexation with chloride ions, while the toxicity of nickel, copper and zinc was less or not affected by salinity changes. Increasing salinity decreased toxicity of a metal mixture until a salinity of 15, while at higher salinities (25) the toxicity of the mixture remained constant [41].

17.7   Sensory–behavioural effects

For marine organisms that use chemoreception as predominant mechanism for sensing the world around them by functional stimuli, chemoreception and chemotaxis (bacteria) have been found to be affected by a suite of contaminants [42–44]. Apart from disrupting chemosensory capacity of the affected organism, body odour may be affected as a consequence of endocrine disruption such as the androgenic action of TBT, and hence may impair recognition of conspecifics, as in mating attraction or brood care processes [45]. To date, such effects have little been studied. At sites with high levels of imposex in female snails, Straw and Rittschof [46] found a strongly decreased response by males to female pheromones, while females did not respond at all to male pheromones. Trimethyltin inhibits chemotaxis of amoeba [47]; TMT completely inhibited the chemotactic response of the cells towards all tested chemoattractants, although cell motility was only slightly inhibited. Two Atlantic fish species avoided TBTO from concentrations above 5.5. μg L^{-1} [48].

Zinc sulphate and copper sulphate block olfaction in rainbow trout (and other fish and birds), probably through destroying olfactory neurons [49]; the copper ion is acute toxic to olfactory sense in Atlantic salmon [50]. Rainbow trout were unable to detect their homing water after exposure to copper (22 μg L^{-1}); the effect was partly recovered after 2–10 weeks [51]. Copper (94 μg L^{-1}) reduced the filtration rate of *M. edulis* by half, probably through reducing the ciliary beating via the branchial nerves [52]. Short-term exposure of fish to atrazine (belonging to the same chemical group as irgarol) or diuron at 5 μg L^{-1} can affect different behaviours in fish, directly or through altering the chemical perception of natural substances of eco-ethological importance [53].

17.8 Conclusions

- A clear difference exists between the internationally recognised harmful AFS (TBT and other organotins, Antifouling Convention [1]) and alternative AFSs.
- Alternative AFSs can still be biocide-based, yet an increasing market share is being taken by biocide-free AFSs.
- Behaviour, fate and toxicity of TBT are much better studied, while little uncertainty remains about its potential to act in a harmful way.
- TBT is more toxic than has been found for any of the present alternatives and is known to affect endocrine and reproduction mechanisms throughout the animal kingdom.
- The impact of alternative biocide-based AFSs, being less extensively documented than TBT, leaves considerable remaining uncertainty.
  - Consequences of antifouling are largely intertwined with the biological and abiotic context of the aquatic environments where AFS is used and should take into account food web and reproductive processes, biomagnification and accumulation mechanisms, as well as abiotic processes related to hydrography and sediment–water interactions.
  - Evaluation studies on the impact of alternative AFSs should consider interactive effects (such as additive, synergistic or antagonistic effects resulting from the presence of other chemicals, interactions with sediments and with biota) and ecosystem interactions, indirect effects, food web relationships and sensory–behavioural effects.
- The development of biocide-free AFSs should be strongly encouraged.

References

1. IMO (2001) *International Convention on the Control of harmful Anti-Fouling Systems on Ships.* Adopted 5 October 2001. International Maritime Organisation, London.
2. Ranke, J. (2002) Persistence of antifouling agents in the marine biosphere. *Environmental Science & Technology*, **36**, 1539–1545.
3. Ward, J. (1988) Antifouling paints threaten fisheries resources. *Naga, The ICLARM Quarterly 15*, International Center for Living Aquatic Resources Management, Philippines.
4. MAFF (1993) *Aquatic Environment Monitoring Report*, Vol. **36**, pp. 1–78. Ministry of Agriculture, Fisheries and Food, Dir. Fisheries Research, Lowestoft.
5. Maguire, R.J. (2000) Review of the persistence, bioaccumulation and toxicity of tributyltin in aquatic environments in relation to Canada's toxic substances management policy. *Water Quality Research Journal Canada*, **35**, 633–679.
6. Amouroux, D., Tessier, E. & Donard, O.F.X. (2000) Volatilization of organotin compounds from estuarine and coastal environments. *Environmental Science & Technology*, **34**, 988–995.
7. Ariese, F., Burgers, I., Van Hattum, B., Van Der Horst, B., Swart, K. & Ubbels, G. (1997) *Chemische monitoring Loswal Noord-West; aanvangssituatie.* IVM rapport nr. 97-05, in opdracht van het Rijksinstituut voor Kust en Zee. Vrije Universiteit, Amsterdam.
8. Mensink, B.P., Boon, J.P., Ten Hallers-Tjabbes, C.C., Van Hattum, B.G.M. & Koeman, J.H. (1997) Bioaccumulation of organotin compounds and imposex occurrence in a marine food chain (Eastern Scheldt, The Netherlands). *Environmental Technology*, **18** (12), 1235–1244.

9. Snoeij, N.J., Penninks, A.H. & Seinen, W. (1987) Biological activity of organotin compounds – an overview. *Environmental Research*, **44**, 335–353.

10. Fent, K. (1996) Ecotoxicology of organotin compounds. *Critical Reviews in Toxicology*, **26**, 1–117.

11. Spooner, N., Gibbs, P.E., Bryan, G.W. & Goad, L.J. (1991) The effect of tributyltin upon steroid titres in the female dog-whelk, *Nucella lapillus*, and the development of imposex. *Marine Environmental Research*, **32**, 37–49.

12. Heidrich, D., Steckelbroeck, S., Bidlingmaier, F. & Klingmueller, D. (1999) Effect of tributyltin (TBT) on human aromatase activity. Paper presented at *Congress of the Endocrine Society (USA)*, 1999.

13. Whalen, M.M., Logathan, B.G. & Kannan, K. (1999) Immunotoxicity of environmentally relevant concentrations of butyltins on human natural killer cells *in vitro*. *Environmental Research*, **81**, 108–116.

14. Harino, H., O'Hara, S.C.M., Burt, G.R., Chesman, B.S., Pope, N.D. & Langston, W.J. (2003) Organotin compounds in Mersey and Thames estuaries a decade after UK TBT legislation. *Journal of the Marine Biological Association of the United Kingdom*, **83**, 11–22.

15. Rees, H.L., Waldock, R., Matthiessen, P. & Pendle, M.A. (1999) Surveys of the epibenthos of the Crouch Estuary (UK) in relation to TBT contamination. *Journal of the Marine Biological Association of the United Kingdom*, **79**, 209–223.

16. Rees, H.L., Waldock, R., Matthiessen, P. & Pendle, M.A. (2001) Improvements in the epifauna of the Crouch Estuary (United Kingdom) following a decline in TBT concentrations. *Marine Pollution Bulletin*, **42**, 137–144.

17. Waldock, R., Rees, H.L., Matthiessen, P. & Pendle, M.A. (1999) Surveys of the benthic infauna of the Crouch Estuary (UK) in relation to TBT contamination. *Journal of the Marine Biological Association of the United Kingdom*, **79**, 225–232.

18. Voulvoulis, N., Scrimshaw, M.D. & Lester, J.N. (2002) Partitioning of selected antifouling biocides in the aquatic environment. *Marine Environmental Research*, **53**, 1–16.

19. Ten Hallers-Tjabbes, C.C., Kemp, J.F. & Boon, J.P. (1994) Imposex in whelks (*Buccinum undatum*) from the open North Sea: relation to shipping traffic Intensities. *Marine Pollution Bulletin*, **28**, 311–313.

20. Ten Hallers-Tjabbes, C.C., Everaarts, J.M., Mensink, B.P. & Boon, J.P. (1996) The decline of the North Sea whelk (*Buccinum undatum* L.) between 1970 and 1990: a natural or human induced event? *Marine Ecology Pubblicazioni della Stazione Zoologica di Napoli I*, **17**, 333–343.

21. ten Hallers-Tjabbes, C.C., Wegener, J.-W., Van Hattum, A.G.M., *et al.* (2003) Organotin levels and imposex development in the benthic gastropods *Buccinum undatum, Neptunea antiqua* from the North Sea. Relations to shipping density and hydrography. *Marine Environmental Research*, **55**, 203–233.

22. Swennen, C., Ruttanadakul, N., Ardseungnern, S., Singh, H.R., Mensink, B.P. & Ten Hallers-Tjabbes, C.C. (1997) Imposex in sublittoral gastropods from the Gulf of Thailand and Strait of Malacca in relation to shipping. *Environmental Technology*, **18**, 1245–1254.

23. MEPC 52/15 (2005) *Harmful anti-fouling systems for ships; Highlighting evidence for the need to urgently ratify the AFS Convention*. Submitted by the World Wildlife Fund (WWF), Friends of the Earth International (FOEI), the World Conservation Union (IUCN) and Intertanko, International Marine Organisation, Marine Environmental Protection Committee, London.

24. MEPC 55/INF.4 (2007) *Harmful anti-fouling systems for ships; Evidence of the continuing global impact of organotin highlighting the need to urgently ratify the AFS Convention*, Submitted by the World Wildlife Fund (WWF), Friends of the Earth International (FOEI), the World Conservation Union (IUCN), Intertanko and Bulgaria, International Marine Organisation, Marine Environmental Protection Committee, London.

25. Kettle, B. (2000) Case Study. Current TBT issues in Australia. Presented at *LC//SG23, LC Secretariat*, Townsville, Australia.

26. Ten Hallers-Tjabbes, C.C., Gee, D. & Guedes Vaz, S. (2006) Precaution as an invigorating context for scientific input in policy processes. In: *Interfaces between Science and Society* (eds Â. Pereira, S. Vaz & S. Togneli), pp. 118–135. Greenleaf Publishing, Sheffield.

27. Thomas, K.V., Fileman, T.W., Readman, J.W. & Waldock, M.J. (2001) Antifouling paint booster biocides in the UK coastal environment and potential risks of biological effects. *Marine Pollution Bulletin*, **42**, 677–688.

28. Ranke, J. & Jastorff, B. (2002) Risk comparison of antifouling biocides. Environmental Control and Policy. *Fresenius Environmental Bulletin*, **11**, 769–772.

29. Helland, A. & Bakke, T. (2002) Transport and sedimentation of Cu in a microtidal estuary, SE Norway. *Marine Pollution Bulletin*, **44**, 149–155.

30. Hall, M.J. & Brown, M.T. (2002) Copper and manganese influence the uptake of cadmium in marine macroalgae. *Bulletin of Environmental Contamination and Toxicology*, **68**, 49–55.

31. Ross, K.E. & Bidwell, J.R. (2003) Assessing the application of an additive model to estimate toxicity of a complex effluent. *Journal of Environmental Quality*, **32**, 1677–1683.

32. Pedersen, F. & Petersen, G.I. (1996) Variability of species sensitivity to complex mixtures. *Water Science & Technology*, **33**, 109–119.

33. Fernandez-Alba, A.R., Hernando, M.D., Piedra, L. & Chisti, Y. (2002) Toxicity evaluation of single and mixed antifouling biocides measured with acute toxicity bioassays. *Analytica Chimica Acta*, **456**, 303–312.

34. Granberg, M.E., Hansen, R. & Selck, H. (2005) Relative importance of macrofaunal burrows for the microbial mineralization of pyrene in marine sediments: impact of macrofaunal species and organic matter quality. *Marine Ecology Progress Series*, **288**, 9–74.

35. Selck, H., Granberg, M.E. & Forbes, V.E. (2005) Impact of sediment organic matter quality on the fate and effects of fluoranthene in the infaunal brittle star *Amphiura filiformis*. *Marine Environmental Research*, **59**, 19–45.

36. Swartz, R.C., Kemp, P.F., Schults, D.W. & Lamberson, J.O. (1988) Effects of mixtures of sediment contaminants on the marine infaunal amphipod, *Rhepoxynius abronius*. *Environmental Toxicology and Chemistry*, **7**, 1013–1020.

37. Hagopian-Schlekat, T., Chandler, G.T. & Shaw, T.J. (2001) Acute toxicity of five sediment-associated metals, individually and in a mixture, to the estuarine meiobenthic harpacticoid copepod *Amphiascus tenuiremis*. *Marine Environmental Research*, **51**, 247–264.

38. Regala, R.P., Rice, C.D., Schwedler, T.E. & Dorociak, I.R. (2001) The effects of Tributyltin (TBT) and 3,3,4,4,5-Pentachlorobiphenyl (PCB-126) mixtures on antibody responses and phagocyte oxidative burst activity in Channel Catfish, *Ictalurus punctatus*. *Archives of Environmental Contamination and Toxicology*, **40**, 386–391.

39. Nirmala, K., Oshima, Y., Lee, R., Imada, N., Honjo, T. & Kobayashi, K. (1999) Transgenerational toxicity of tributyltin and its combined effects with polychlorinated biphenyls on reproductive processes in Japanese medaka (*Oryzias latipes*). *Environmental Toxicology and Chemistry*, **18**, 717–721.

40. Schmidt, K., Staaks, G.B.O., Pflugmacher, S. & Steinberg, C.E.W. (2005) Impact of PCB mixture (Aroclor 1254) and TBT and a mixture of both on swimming behavior, body growth and enzymatic biotransformation activities (GST) of young carp (*Cyprinus carpio*). *Aquatic Toxicology*, **71**, 49–59.

41. Verslycke, T., Vangheluwe, M., Heijerick, D., De Schamphelaere, K., Van Sprang, P. & Janssen, C.R. (2003) The toxicity of metal mixtures to the estuarine mysid *Neomysis integer* (Crustacea: Mysidacea) under changing salinity. *Aquatic Toxicology*, **64**, 307–315.

42. Atema, J. (1985) Chemoreception in the sea: adaptations of chemoreceptors and behaviour to aquatic stimulus conditions. In: *Physiological Adaptations of Marine Animals*, Vol. **39** (ed. M.S. Laverack), pp. 387–423. Society of Experimental Biology, Company of Biologists, University of Cambridge.

43. Blaxter, J.H.S. & Ten Hallers-Tjabbes, C.C. (1992) The effects of pollutants on sensory systems and behaviour of aquatic animals. *Netherlands Journal of Aquatic Ecology*, **26**, 43–58.

44. Ten Hallers-Tjabbes, C.C. (1997) Tributyltin and Policies for Antifouling. *Environmental Technology*, **18**, 1265–1268.

45. Ten Hallers-Tjabbes, C.C. (1997) Chemosensory mediated orientation in marine snails and in humans in relation to environmental Conditions. In: *Proceedings Third International Conference*

on Orientation and Navigation. Birds, Humans and Other Animals. Royal Institute of Navigation, London.

46. Straw, J. & Rittschof, D. (2004) Responses of mud snails from low and high imposex sites to sex pheromones. *Marine Pollution Bulletin*, **48**, 1048–1054.

47. Sroka, J., Madeja, Z., Galanty, A., *et al.* (2001) Trimethyltin inhibits the chemotaxis of *Dictyostelium discoideum* amoebae. *European Journal of Protistology*, **37** (3), 313–326.

48. Hall, L.W., Jr., Pinkney, A.E., Zeger, S., Burton, D.T. & Lenkenvich, M.J. (1984) Behavioral responses to two estuarine fish species subjected to bis (tri-n-butyltin) oxide. *Water Resources Bulletin*, **20**, 235–239.

49. Starcevic, S.L. & Zielinski, B.S. (1997) Neuroprotective effects of glutathione on rainbow trout olfactory receptor neurons during exposure to copper sulphate. *Comparative Biochemistry and Physiology C*, **117**, 211–219.

50. Bjerselius, R., Winberg, S., Winberg, Y. & Zeipel, K. (1993) Ca super(2+) protects olfactory receptor function against acute Cu(II) toxicity in Atlantic salmon. *Aquatic Toxicology*, **25**, 125–138.

51. Saucier, D., Astic, L. & Rioux, P. (1991) The effects of early chronic exposure to sublethal copper on the olfactory discrimination of rainbow trout, *Oncorhynchus mykiss*. *Environmental Biology of Fishes*, **30**, 345–351.

52. Howell, R., Grant, A.M. & Maccoy, N.E. (1984) Effect of treatment with reserpine on the change in filtration rate of *Mytilus edulis* subjected to dissolved copper. *Marine Pollution Bulletin*, **15**, 436–439.

53. Saglio, P. & Trijasse, S. (1998) Behavioral responses to atrazine and diuron in goldfish. *Archives of Environmental Contamination and Toxicology*, **35**, 484–491.

Chapter 18
Fouling and Antifouling in Oil and Other Offshore Industries

Henry M. Page, Jenifer E. Dugan and Fred Piltz

The purpose of this chapter is to provide an overview of biofouling and related issues in the offshore petroleum industry, specifically on offshore oil and gas platforms. Our chapter focuses on macroorganisms attached to the platform structure. We review important fouling taxa, abiotic and biotic factors that influence assemblage structure and rate of development, and the role of fouling assemblages in marine food webs. We consider the engineering design, maintenance and cost implications of biofouling for the industry, including environmental issues such as decommissioning and exotic species. We also identify data gaps and opportunities where further research on oil platforms will not only improve understanding of the general ecology of biofouling assemblages, but could also provide information on regional or global change in ocean climate.

18.1   Introduction

Offshore oil and gas platforms include some of the largest man-made structures in the marine environment. Currently, there are more than 8000 fixed offshore oil and gas structures worldwide with another 1500 under construction or planned [1]. The expansion of the offshore oil and gas industry in the 1970s and 1980s increased interest in the impact of fouling organisms on the functional operation of offshore platforms and in the general ecology of these biotic assemblages [2–6].

Biofouling is of concern in the design and operational functioning of offshore platforms. Fouling organisms can increase the mass of the platform structure, affecting its response to wave and earthquake loading [7]. Even if the weight of the fouling assemblage is insignificant relative to the weight of the structure, the accumulation of fouling organisms can increase the diameter and surface roughness of platform support members, which increases drag on the members and thus the hydrodynamic loading of the structure when subjected to waves and currents [3]. In an analysis of the impact of fouling on hydrodynamic loading of a platform in the North Sea, a layer of fouling organisms 15 cm thick, while only causing a 0.15% increase in weight of the structure, was projected to increase hydrodynamic loading by 17.5% ([8]).

In the USA, the Minerals Management Service (MMS), the agency responsible for managing offshore energy resources, regulates and guides the assessment and control of hydrodynamic loading on platforms through regulations [9] that refer to Recommended Practices RP 2A-WSD established by the American Petroleum Institute [10]. Platforms in the North Sea operate under Offshore Installations regulations that require structural integrity and therefore assessment of the fouling assemblage on platforms in that environment [3, 11].

In addition to concerns regarding the impact of fouling organisms on hydrodynamic loading, accumulations of these organisms may also interfere with other aspects of platform operations [8, 11]. The fouling assemblage obscures the platform surface and may need to be removed prior to visual inspection and maintenance of the platform jacket by divers or remotely operated vehicles (ROVs). Some taxa with sharp shells, such as oysters and barnacles, may damage underwater cables and other equipment. Finally, the presence of fouling organisms can promote microbially influenced corrosion of steel by providing microorganisms with appropriate conditions for growth or by mechanically disrupting antifouling surface coatings [3, 11].

18.2   The biofouling assemblage

18.2.1   General characteristics

Offshore oil and gas platforms are typically located in soft benthic habitat, but the structures are colonised by species that require hard substrata for attachment. The species of animals and algae that become established on platforms must arrive originally from the plankton as pelagic larvae or spores from breeding populations located elsewhere (see Chapter 2) or, for those species that possess propagules of limited dispersal ability, through transport to platforms via boat or barge hull fouling or perhaps on floating objects (e.g. floating mats of macroalgae).

Although the specific biological composition of the fouling assemblage varies with biogeographical region, and can vary within a region [6, 12–14], the general physical structure of the assemblage is typically defined by a few dominant space-holding taxa (Figures 18.1 and 18.2; Plates VI E, F). These taxa include those with hard exteriors, such as mussels, *Mytilus* sp., encrusting bivalves (oysters, scallops, chamas), barnacles, encrusting bryozoans and hard corals, as well as those with 'soft' exteriors such as sponges, anemones, foliose bryozoans, soft corals and hydroids [4–6, 18]. At some locations, macroalgae can comprise a significant component of the fouling assemblage near the water surface [5, 6], whereas at other locations macroalgae can be relatively sparse [8, 14].

Coastal oil and gas platforms can be located in relatively close proximity (on a scale of one to tens of kilometres) to natural rock outcrops (reefs). Where this occurs, biotic assemblages on platforms have been found to differ from those of natural reefs [14]. For example, prominent space holders on platforms in the Santa Barbara Channel, California, include mytilid mussels (*Mytilus californianus*, *Mytilus galloprovincialis*), barnacles (e.g. *Megabalanus californicus*, *Balanus trigonus*) and anemones (*Corynactis californica*, *Metridium senile*). In contrast, natural rocky reefs in the region are covered at equivalent depths to a greater degree with filamentous and foliose brown and red macroalgae, coralline algae, giant kelp (*Macrocystis pyrifera*) and encrusting and branching bryozoans; the subtidal mussel zone present on platforms is absent [14, 15]. In the unusual case of a platform being installed on rocky seafloor, the fouling assemblage on the platform also differed from that attached to the seafloor [19].

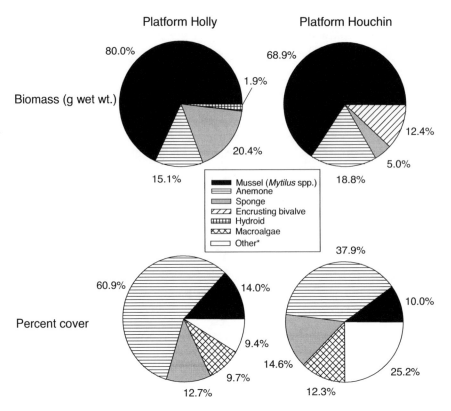

Figure 18.1 Biomass and percent cover composition of attached invertebrates and macroalgae at 12 m depth on two offshore platforms in the Santa Barbara Channel [15, 16]. *Other category includes primarily barnacles, bryozoans and tunicates.

The platform fouling assemblage can reach several centimetres in thickness [7, 14, 18]. The thickest fouling assemblages occur on those platforms that are colonised by mytilid mussels. For example, assemblage thickness on platforms offshore of California often exceeds 15 cm through the layering of mussels (*M. californianus*, *M. galloprovincialis*) [14, 20] with 120 cm reported from one nearshore platform [21]. Similarly, platforms in the North Sea that are colonised by *Mytilus edulis* may have fouling assemblages of 15–20 cm thick [11]; masses of *M. galloprovincialis* on a platform in the Adriatic Sea exceeded 20 cm in thickness [22]. The three-dimensional structure provided by larger, sessile organisms, such as mussels, barnacles and encrusting bivalves, provides habitat for smaller mobile and sessile species that include amphipod and isopod crustaceans, crabs, snails and nestling bivalves [6, 19, 22, 23].

Fouling species can grow faster and attain larger sizes on offshore platforms compared with natural habitats [4, 23–25]. For example, the bay (Mediterranean) and sea mussel, *M. galloprovincialis* and *M. californianus*, attain shell lengths at or near their maximum reported sizes on platforms in the Santa Barbara Channel (so-called platform gigantism [26]). Growth rate was more rapid and to a larger maximum size for the intertidal stalked barnacle (*Pollicipes polymerus*) on a platform in the Santa Barbara Channel compared with natural rocky habitat [24]. Mussels (*M. edulis*) on a platform in the North Sea were also reported to have rapid growth rates, comparable to those of mussels cultivated inshore [12].

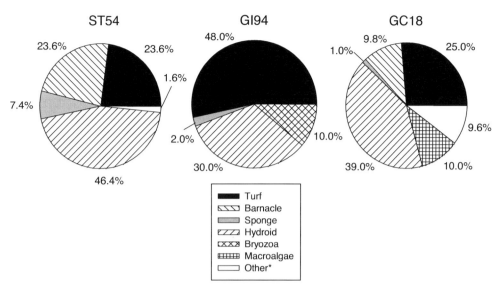

Figure 18.2 Percent cover composition of attached invertebrates and macroalgae at 10 m depth on three offshore platforms in the Gulf of Mexico [17]. Platform identifications are provided above the diagrams. *Other category includes primarily anemones, tunicates and bivalves.

18.2.2   *Vertical zonation*

Changes with depth in the composition, cover, thickness and biomass of attached organisms are a characteristic feature of platform fouling assemblages. Some dominant fouling organisms are found typically within particular depth ranges. Mytilid mussels, found on platforms offshore of California and in the North and Irish Seas, tend to be most abundant in water depths of less than 20 m [12, 14, 23, 27]. Large masses of algae are also found primarily near the surface on some platforms in the Gulf of Mexico [17] and North Sea [5]. In contrast, some anemones (e.g. *M. senile, Metridium farcimen*) may be most abundant at deeper depths [14, 27].

It has been proposed that some platform taxa may occupy a more limited depth range on natural substrata, where they are commonly exposed to predators and competitors not necessarily present at the platforms [6]. This is clearly the case for mytilid mussels in the Santa Barbara Channel, which can be abundant intertidally, but are rarely found subtidally in natural habitats. The thickness and biomass of the platform assemblage generally decreases with depth, but may also be less near the water surface [6, 7, 14, 20]. However, exceptions to this pattern occur that may relate to the water depth in which the platform is located. Fouling biomass increased with depth at shallow platforms (−12 to −14 m) in the Adriatic Sea [22].

Discrete vertical zones based on the dominant taxa present have been assigned to platform assemblages. For example, four to six zones were recognised for assemblages on platforms in the Santa Barbara Channel and Santa Maria Basin, California, with more zones for platforms in deeper water [14]. Mytilid mussels are characteristically found in the shallowest zone; however, the vertical extent and lower boundary of the mussel zone varies among platforms. Encrusting bivalves, sponges and anemones predominate deeper. For one platform, Grace (Santa Barbara Channel), these zones were characterised by *Mytilus* spp. (−1.8 to −13.7 m), barnacles/scallops (−13.7 to −27.4 m), anemones/encrusting species, such as sponges and bivalves (−27.4 to −70.1 m), and encrusting species/sea stars (−70.1 to −96.9 m) [14].

Three to five zones have been recognised for platforms in the North Sea [5, 11, 27]. In one area, zones were defined by mussels in the shallowest depths (to −20 m), anemones in middle depths and more diversity deeper, including hydroids and a soft coral [27]. The five zones recognised for a platform in another area of the North Sea were characterised by macroalgae (to −10 m), hydroids and arborescent bryozoans (to −31 m), calcareous bryozoans (−31 to −51 m), encrusting bryozoans (−31 to −71 m) and aggregate tubeworms and deep-water barnacles (−71 m to mud line) [5].

Variation in platform assemblages with depth reflects patterns of larval recruitment [28], and the response of fouling taxa to depth-related gradients in physical and biological factors (see Chapter 6). Physical factors include light, temperature and wave action, whereas biological factors include species interactions (e.g. competition and predation) [29, 30], and processes that relate to the succession or change in species assemblages over time (e.g. facilitation, inhibition [31], see Chapter 4). Although general patterns of zonation can often be recognised, the species composition, abundance and biomass of assemblage species as a function of depth can vary greatly among platforms even within the same region in response to local physical and biological factors (e.g. [6, 12, 13, 17]. As a result, it is generally not possible to identify specific depth boundaries that will consistently define a particular biotic zone.

18.2.3   *Biogeographic patterns*

The taxonomic composition of the fouling assemblage will depend to a large extent on the species pool available within the biogeographic region of the platform. A major distinction in the composition of fouling assemblages on platforms exists between regions where mussels of the genus *Mytilus* occur and those where mussels are absent. Where *Mytilus* sp. occurs, such as offshore of California [4, 14], the North [5, 12] and Irish [23] Seas, the Bohai Sea, located in northern China [8], and the Adriatic Sea [22], this taxon is a major component of the fouling assemblage on offshore platforms, particularly in water depths of less than 20 m on the structure. Mussels are competitive dominants [30], capable of settling on and overgrowing other taxa (e.g. barnacles, bryozoans, encrusting bivalves) and of establishing multi-layered masses on platform support members and conductor pipes.

In regions where mussels are absent, other hard-shelled taxa, such as barnacles and cementing bivalves, which would be overgrown by mussels in the areas above, are often assemblage dominants [6, 18, 19]. For example, in waters off Louisiana and Texas, barnacles of the genus *Balanus* are dominant fouling organisms on the coastal platforms, whereas cementing bivalves predominate on platforms further offshore [6, 17]. These taxa provide the physical structure of the assemblage, and secondary substratum for the attachment of macroalgae and other invertebrates such as sponges, hydroids and bryozoans. Because mytilid mussels are absent from the Gulf of Mexico, hydrodynamic loading considerations appear to be much less of a concern for platforms here compared with those offshore of California or in the North Sea [7]. In tropical waters, such as the Gulf of Guinea, hard corals may form important components of platform fouling assemblages [18].

The specific composition of the fouling assemblage varies among platforms within geographic regions. For example, several biogeographic provinces (Tamaulipan, Texan, Austroriparian, Caribbean) are represented along the Gulf coasts of Louisiana and Texas. Species assemblages on platforms located within these provinces reflect local oceanographic conditions that include extremes in temperature, salinity, dissolved oxygen concentration and the distribution of turbid water layers [6, 13].

The influence of biogeography in structuring fouling assemblages was also suggested by results from platforms in the Santa Barbara Channel, a transition zone between the Oregonian and Californian biotic provinces [32]. The relative abundance of different invertebrates (as percent cover) on platforms arrayed along that channel varied such that platforms in closest proximity to one another tended to have invertebrate assemblages more similar to each other than to platforms located further away (in terms of percent cover). The along-channel variation in platform assemblages probably reflects the interaction of regional oceanographic gradients of water temperature, phytoplankton abundance and delivery rate of larvae to the platforms within this transition zone with more local physical and biological processes that influence post-settlement growth and survival, such as disturbance from storm swell and competition for space [32].

18.2.4 *Influence of distance from shore and water depth*

The biotic structure of fouling assemblages can vary with distance from shore and water depth of the platform (see also Chapter 6). This phenomenon is well-documented for assemblages on platforms in the north-western Gulf of Mexico [6, 13, 17] where two to three distinctive biotic groupings have been identified. Platforms closest to shore in water depths to $\sim$30 m tend to have fouling assemblages dominated by one to several species of barnacles that are often covered by other species, including hydroids, bryozoans and sponges (the coastal assemblage). For platforms located further offshore, in water depths of from 30 to 60 m, fouling assemblages are dominated by bivalves rather than barnacles (the offshore assemblage). A third category, (the bluewater assemblage) was recognised for platforms in water depths of greater than 60 m [6]. This assemblage was characterised by low biomass compared with coastal and offshore assemblages, and the presence of algae and stalked barnacles near the surface and bivalves at greater depths. However, fouling assemblages of platforms in this last category can be similar to those of the offshore assemblage depending on location in the Gulf [13, 17]. Biotic assemblages on platforms in the Gulf of Mexico are likely influenced by a combination of interrelated environmental factors that vary with distance from shore, including water mass properties (water temperature, salinity, turbidity), primary productivity in the water column, disturbance (from wave action) and proximity to Caribbean water masses [6, 17].

Distance from shore may also interact with local current patterns and the proximity of larval source populations to influence platform-fouling assemblages. Platforms located closest to the coast in the central and northern North Sea may be exposed to greater numbers of larvae from source populations in natural coastal habitats, leading to higher densities of fouling species on these platforms compared with platforms further offshore [11, 12]. Since larvae are transported offshore by currents, those species with a longer larval lifespan, such as *M. edulis*, may be favoured over other species with a shorter larval life, to reach platforms further offshore [12]. Similarly, the maximum thickness of assemblages tends to decrease with distance from shore for platforms in the Santa Barbara Channel [26]. This pattern may be related to inshore–offshore differences in the recruitment and growth rate of mussels, and/or to differences in the rate of dislodgement of mussels from the platforms by storm swell.

18.3 Fouling assemblage development

Following installation, the intertidal and subtidal portions of the platform structure are rapidly colonised by marine organisms. Early colonists are microscopic forms, such as bacteria, diatoms, dinoflagellates and filamentous algae, which form a microbial film [33] (see also Chapters 4, 9 and 12). This film may facilitate the subsequent colonisation of the structure by larger macroalgae and invertebrates [34]. The composition of the fouling assemblage then changes over time. This change in species composition during the colonisation of open space is broadly defined as succession and the concept of succession has been used to predict how an assemblage will change over time and to understand processes responsible for these changes [31]. Typically, early colonists are opportunistic species that have high reproductive rates, short lifespans and broad habitat requirements, but are poor competitors for space. In contrast, later colonists may have more specific habitat requirements, longer lifespans and are better competitors for space [35,36]. The general process of succession in the fouling assemblage has been documented following platform installation in a variety of regional settings. As would be expected, the specific taxa involved, successional sequence and accumulation rates of biomass vary among and within regions [6, 11, 26, 27].

Water depth, season and year have been identified as important factors that can influence the rate of assemblage development. Typically, rates of development are highest in shallow depths and during the spring and summer months [6, 28]. Development of the fouling assemblage on ceramic tiles deployed at three depths (6, 12 and 18 m) at an offshore platform in the Santa Barbara Channel was largely predictable in terms of the composition and sequence of occurrence of dominant taxa over a 2-year period, but rates of recruitment and expansion of the assemblage were most rapid at the shallowest depth (6 m). In addition, development of the assemblage was more rapid during the second year of the study (2000–2001), suggesting that oceanographic conditions influenced invertebrate recruitment and growth [28]. Early successional taxa included colonial tunicates and encrusting bryozoans, whereas barnacles, sponges and mussels (*M. californianus*) dominated later. The rate at which biomass accumulated directly onto a scraped area of the same platform was also measured at the three depths over consecutive 12-month periods (Figure 18.3b). Biomass increase was most rapid at a depth of 6 m, and was also higher the second year of the study (2000–2001) (equivalent to 2.75 kg m^{-2} after 12 months at 6 m; Figure 18.3b) as observed on the tiles.

Succession may lead to the development of a climax community that reaches a steady state, in terms of growth and loss, with its environment [17, 26]. In the case of fouling assemblages on platforms offshore of California, the biomass would consist primarily of the competitively dominant sea mussel, *M. californianus*, in shallow depths and encrusting bivalves and anemones deeper [26]. There are no long-term data available with which to examine whether a steady state is reached, and it is unlikely that such a state is ever achieved in shallow water for California platforms (see also Chapter 4). Mussel clumps increase in size through recruitment and individual growth. Eventually, disturbance from storm swell dislodges large clumps, opening up space that is re-colonised by mussels and other species [37]. The persistence and stability in species composition and abundance of the fouling assemblage in deeper waters is unknown.

A scenario was developed for Gulf of Mexico platforms which viewed the biofouling crust (assemblage) as a system in equilibrium between accretion and shedding [17]. Accretion of the crust was postulated as dependent on the passing ocean water for food and new larval settlement, whereas biotic interactions such as predation, competition and bioerosion all

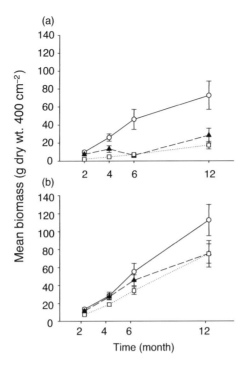

Figure 18.3   Mean ($x \pm$ 1SE, $n = 8$) accumulation of biomass at 6 (○), 12 (▲) and 18 m (□) on the surface of conductor pipes on Platform Houchin in the Santa Barbara Channel (a) from 1999 to 2000 and (b) from 2000 to 2001. Surfaces of pipes were initially scraped to remove attached organisms in July 1999.

contribute to crust loss directly or in concert with wave action. However, long-term repetitive measurements of species composition, densities, biomass and thickness of platform assemblages will be needed to establish whether equilibrium is reached over the long term.

18.4   Ecological effects of the fouling assemblage

18.4.1   Impacts on the benthos

In some locations, the sloughing of fouling organisms from the platform structure to the seafloor can affect the distribution and abundance of benthic organisms. Off the coast of central and southern California where the principal space holding taxa at depths affected by waves are mussels (*M. californianus*, *M. galloprovincialis*), swell and platform maintenance cleaning operations dislodge mussels and associated organisms, which fall to the seafloor as 'faunal litterfall' [4, 20]. The amount of faunal litterfall, measured over time following maintenance cleaning of support members and conductor pipes on one platform, increased with the growth in thickness of the fouling assemblage (Figure 18.4). Mussels account for nearly all (more than 90%) of the dislodged invertebrate biomass [20]. Faunal litterfall alters the physical characteristics of the seafloor by contributing to the formation of a 'shell mound' of hard substrate that can extend 70 m or more in diameter and 7 m or more in maximum height [38]. The phenomenon of shell mound formation has been documented at most oil platforms off the coast of California [38] and may occur in other regions.

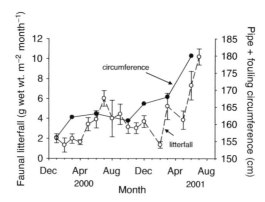

Figure 18.4 Relationship between the monthly mean biomass ($\pm$1SE, $n = 8$) of fouling taxa dislodged by wave action and the thickness of the fouling assemblage on conductor pipes over time at Platform Hogan in the Santa Barbara Channel. Fouling taxa (faunal litterfall) were sampled in traps of 0.0113 m^2 diameter deployed at a depth of 18 m and retrieved monthly. Thickness of the fouling assemblage is reported as circumference of the conductor pipe plus the attached assemblage.

The effects of faunal litterfall on benthic biota are dramatic. There was little overlap in the composition of species that occur on shell mounds compared with the adjacent soft seafloor; shell mound species are more characteristic of hard substratum habitat in studies from California waters [39]. Furthermore, the food subsidy provided by the faunal litterfall is crucial to sustaining the shell mound biotic assemblage. This is evident when mound assemblages under platforms are compared to those in which platforms have been removed [39]. Predatory and omnivorous sea stars (*Pisaster* spp., *Asterina miniata*), in particular, were much less abundant and of a smaller size at shell mound-only sites compared with the mounds under extant platforms.

18.4.2   The fouling assemblage and platform food webs

Conceptual models have been constructed presenting potential pathways of energy flow among the platform algal and invertebrate assemblages, fish assemblages and the surrounding pelagic and benthic environment [6, 26]. However, there are few empirical data on these pathways. Studies in the Gulf of Mexico [6] and the Santa Barbara Channel [15] have found that the matrix of sessile space-holding taxa provide habitat and food for small invertebrates important in the diet of microcarnivorous fishes. At two platforms in the Santa Barbara Channel, amphipod crustaceans comprised the highest proportion of potential prey, from 89 to 98% (by number) of the diet of a resident microcarnivorous fish (*Oxylebius pictus*) [15]. However, this study suggested that food web pathways of platforms may differ from those of natural reefs because the amphipod assemblage contained a high percentage (more than 50% by number) and abundance of two exotic species, *Caprella mutica*, and *Erichthonius brasiliensis*, that were either absent or rare on the natural reefs.

18.5   Novel and exotic species

There are several studies indicating that the fouling assemblage of offshore oil and gas platforms can contain exotic species and it has been proposed that platforms may facilitate the expansion of these species into new areas (see Chapter 24). The introduction and spread of exotic species by way of platforms could occur through two mechanisms. First, exotic species may colonise mobile drilling platforms at the point of origin or during transit to their destination. For example, 12 species of exotic barnacles were identified from the fouling assemblage on the portion of the 'Maui' oil platform that was submerged during its tow from Japan to New Zealand [40]. Six of these species, *Balanus improvisus*, *Balanus albicostatus*, *Balanus reticulatus*, *Megabalanus volcano*, *Megabalanus rosa* and *Tetraclita squamosa japonica*, had not previously been reported from New Zealand.

Second, as islands of hard substratum habitat, fixed platforms may serve as 'stepping stones' that facilitate the expansion of exotic species across expanses of soft seafloor (see also Chapter 24). Species of coral have been found attached to some platforms in the Gulf of Mexico, leading to the conclusion that platforms have facilitated the expansion of coral populations there, including the exotic species, *Tubastraea coccinea* [41]. Ciguatera is a common human disease of tropical coral reef ecosystems acquired by consuming ciguatoxins in fish tissue. It has been suggested that platforms may facilitate the spread of ciguatera in the Gulf of Mexico through the provision of habitat in the upper euphotic zone for the source dinoflagellate, *Gambierdiscus toxicus* [42]. There are few records of this disease in the north-western Gulf of Mexico, a region of soft seafloor that is considered poor habitat for *G. toxicus*. However, this dinoflagellate has recently been found on platforms in the north-western Gulf of Mexico.

Exotic invertebrate species have been reported from offshore oil platforms in the Santa Barbara Channel [43]. Conspicuous exotic species (a bryozoan, *Watersipora subtorquata*, and anemone, *Diadumene* sp.) were found on two of seven platforms surveyed. Inconspicuous exotic species (amphipods, *C. mutica* and *Ericthonius brasiliensis*) were detected on two platforms surveyed for these smaller species. None of these species had been previously reported from platforms. It is expected that additional exotic species will be detected with further surveys and taxonomic analyses of these platform assemblages.

Once established, exotic species can become dominant members of the fouling assemblage perhaps to the detriment of native species and serve as a potential source of propagules to natural reef habitats [43]. In addition, the presence of exotic species in fouling assemblages may influence the degree to which oil and gas platforms provide ecological functions (e.g. biodiversity, food chain support) similar to those of natural reefs. The presence of exotic species on platforms also has consequences for various platform decommissioning options in California and elsewhere, including the removal and transport of platforms for use as artificial reefs, particularly if removals are conducted without regard for the potential transport/dispersal of these species. In some cases, where platforms may provide habitat for rare or unusual species (e.g. coral populations in the Gulf of Mexico), such platforms and their associated fouling assemblages may possess an intrinsic environmental value that could influence decommissioning options [41].

18.6   Economic value of the biofouling assemblage

The biofouling assemblage on offshore platforms can have economic value. In California waters, harvest of mussels from platforms provided a nearly continuous commercial source of this shellfish for over 20 years. Driven by the need to frequently clean platform support members of layers of mussels up to 90 cm thick, Phillips Petroleum, Texaco, and other companies have partnered in the past with one company (Ecomar Inc.) to transform the annual expense of mussel removal into a commercially viable business [44]. The use of mussels cleaned from platforms for human consumption has also been reported for the Adriatic Sea [22].

Some fouling organisms may also contain potentially important natural products. Harvesting of marine organisms with important natural products in sufficient quantities for applied uses could potentially have negative ecological consequences for natural reef ecosystems [45, 46]. For instance, 37 854 l of wet bryozoan, *Bugula neritina*, is needed to extract multigram quantities of bryostatin 1, under study as a potential anti-cancer agent [45]. The bryostatin-producing clade of *B. neritina* is found on oil platforms off the coast of California [16]. Platform species could potentially act as sources of marine natural products, thereby reducing the need to exploit populations on natural reefs [47].

18.7   Management of biofouling and economic impacts

The petroleum industry has used several approaches to manage biofouling on offshore oil and gas platforms. One approach has been to take biofouling into consideration in the design of platform structures. Thus, platforms are constructed of much thicker steel under the assumption that structural members and conductor pipes will be fouled and present a much larger surface area to wave action with than without fouling [48]. A second approach is to plan for the periodic removal of the fouling assemblage, wherein platform members and conductor pipes are cleaned periodically as needed. Manual scraping and high-pressure water blasting are used to dislodge fouling species. In the Santa Barbara Channel, platforms are cleaned irregularly, often annually, depending on the weather [14] to facilitate visual inspection of the platform structure and reduce hydrodynamic loading of the support members. The maintenance removal of fouling organisms is generally restricted to the upper 20 m of structure for platforms in California waters. The cost to manually clean these platforms of accumulated organisms is approximately US$30 000–$100 000 per cleaning cycle [48] (S. Benson, personal communication).

The application of an antifouling covering to the platform surface has been a traditional method of controlling marine fouling, at least for the first few years following platform installation. Antifouling paints and other antifouling materials such as plates or panels containing a biocide (e.g. toxic copper-nickel alloy) are applied to clean members to inhibit the attachment of fouling species. In one example, a protective copper-nickel-coated elastomer antifouling sheath [48, 49] was deployed on the 60 conductor pipes of a platform newly installed in 214 m depth of water offshore of Huntington Beach, southern California, in 1984. It was estimated that 153 mt of structural steel, which would have been required to handle the additional loading from biofouling, was eliminated through the application of the antifouling sheath. After 10 months, no growth had occurred on the protected areas, whereas 2.5–5.0 cm of barnacle and mussel fouling was observed on the unprotected areas [48].

More recently, an apparatus that attaches around structural members and conductor pipes, the Marine Growth Preventor (MGP), produced by Innovative Engineering Ventures (IEV) [50], has been used to inhibit the development of biofouling on platforms. The MGP is a single or multiple-ring apparatus that fits around support members or conductor pipes (Plate VI F). The MGP moves longitudinally and rotates with water motion, providing a continuous rolling action preventing the settlement of fouling organisms. IEV states that the MPG can work at all water depths and all orientations provided that there is suitable water motion to move the apparatus. MGPs were first used in Australia in 1990, followed by Indonesia in 1991, the UK in 1992 and the USA in 2004.

A significant ecological and financial concern arises during the final stages in the life of an offshore platform when the platform is decommissioned. In the majority of cases in the USA, the platform must be removed and the seafloor restored to the original condition per Federal regulations. When a platform is decommissioned, the fouling assemblage must be removed prior to the scrapping or recycling of the steel structure. The biomass of this assemblage can be considerable [51]. For two platforms that were removed from the Santa Barbara Channel in 1988 (Helen, Herman), a total of ~907 mt wet weight of marine organisms were removed from the structures, which had a combined steel weight of approximately ~2721 mt. The thickness of the fouling assemblage on these platforms was ~38 cm, with 30 cm of hard growth. For four other platforms that were removed in 1996 (Hazel, Hilda, Heidi, Hope), total fouling biomass removed was estimated at more than 2449 mt, compared to the steel mass of all structures combined of 9072 mt.

Cost estimates for cleaning the fouling assemblage from platforms being decommissioned and removed range from US$50 000 for the smallest platforms (Category I, one well, shallow water) to over US$100 million (Category V deep water multi-well) [52]. Fouling assemblage removal costs ranging from US$150 000 to $1 500 000 are estimated for deep-water platforms offshore of California [53].

18.8   Conclusions

- Biofouling assemblages of platforms are typically characterised by a few dominant space-holding taxa. This fouling assemblage can reach tens of centimetres in thickness, with the greatest accumulation generally in shallowest depths and the thickest assemblages occurring on platforms colonised by mytilid mussels.
- Changes in species composition, biomass and thickness of fouling assemblages occur with water depth creating biotic zones. However, local physical and biological factors influence patterns of zonation such that the depth boundaries of each zone can be variable within and across platforms even within the same region.
- A major distinction in the fouling assemblage of platforms exists between regions where mussels occur (e.g. California, North Sea, Irish Sea, Adriatic Sea) and regions where mussels are absent (e.g. Gulf of Mexico). A number of physical and biological factors have been postulated to influence platform fouling assemblages within regions, and from inshore to offshore, including light, water temperature, salinity, dissolved oxygen concentration, disturbance from storm swell, larval supply and species interactions, such as competition and predation.
- Following installation, the submerged portion of the platform structure is rapidly colonised by marine taxa. Changes in species composition, abundance and biomass of the fouling assemblage over time are influenced by water depth, season, year and other factors. However, evaluating whether platform assemblages reach a steady state or climax community will require additional long-term data.

- Conceptual models of potential pathways of energy flow among platform fouling assemblages and the surrounding benthic and pelagic environment exist, but empirical data are needed to evaluate the ecological importance of these exchanges.
- More information is also needed on the composition of fouling assemblages, particularly the presence of exotic species, which has implications for the degree to which platforms provide ecological functions comparable to natural habitats, and of species that may have economic value, for example, as sources of natural products.
- Approaches used to accommodate biofouling include designing platforms with thicker steel, manual cleaning, antifouling coverings and MPGs.
- Studies of fouling assemblages on offshore platforms, which typically have lifespans of decades, may provide insights into changes in ocean climate. Short-term, inter-annual changes in oceanographic conditions may influence platform assemblages through effects on invertebrate recruitment and growth. Longer-term (decades) climatic changes may shift the composition of fouling assemblages. Monitoring of fouling assemblages on platforms could permit an evaluation of the potential impact of these longer-term changes on the oceanic environment.

References

1. Anon (2007) *Offshore Engineer*. August, 72.
2. Bascom, W.A., Mearns, J. & Moore, M.B. (1976) A biological survey of oil platforms in the Santa Barbara Channel. *Journal of Petroleum Technology*, **28**, 1280–1284.
3. Goodman, K. & Ralph, R. (1979) Fouling – the marine growth industry. *Offshore Engineer*, September 1979, 113–117.
4. Wolfson, A., Van Blaricom, G., Davis, N. & Lewbel, G.S. (1979) The marine life of an offshore oil platform. *Marine Ecology Progress Series*, **1**, 81–89.
5. Forteath, G.N.R., Picken, G.B., Ralph, R. & Williams, J. (1982) Marine growth studies on the North Sea oil platform Montrose Alpha. *Marine Ecology Progress Series*, **8**, 61–68.
6. Gallaway, B. & Lewbel, G. (1982) *The Ecology of Petroleum Platforms in the Northwestern Gulf of Mexico: A Community Profile*. BLM Open File Report 82-03, US Department of the Interior, Fish and Wildlife Service, Biological Services Program FWS/OBS-82/72.
7. Heideman, J.C. & George, R.Y. (1981) Biological and engineering parameters for macrofouling growth on platforms offshore Louisiana. *Oceans*, **13**, 550–557.
8. Yan, T. & Yan, W.X. (2003) Fouling of offshore structures in China – a Review. *Biofouling*, **19** (Suppl.), 133–138.
9. CFR (2006) *Platforms and Structures*. Minerals Management Service, Outer Continental Shelf, Subpart 1, 250.920. July 1. United States Code of Federal Regulations.
10. American Petroleum Institute (2000) *Recommended Practices for Planning, Designing and Constructing Fixed Offshore Platforms*, RP-2A–WSD, 21st edn. Working Stress Design, Washington, DC.
11. Ralph, R. & Troake, R.P. (1980) Marine growth on North Sea oil and gas platforms. *Proceedings of the 12th Annual Offshore Technology Conference*, Houston, **4**, 49–51.
12. Forteath, G.N.R., Picken, G.B. & Ralph, R. (1983) Interaction and competition for space between fouling organisms on the Beatrice Oil Platforms in the Moray Firth, North Sea. *International Biodeterioration Bulletin*, **19**, 45–52.
13. Dokken, Q.R., Withers, K., Childs, S. & Riggs, T. (2000) *Characterization and Comparison of Platform Reef Communities Off the Texas Coast*. TAMU-CC-0007-CCS, Texas Parks and Wildlife Department. Texas A&M University, Corpus Christi.

14. CSA (2005) *Survey of Invertebrate and Algal Communities on Offshore Oil and Gas Platforms in Southern California.* OCS Study MMS 2005–070, Continental Shelf Associates, US Department of the Interior, Minerals Management Service, Camarillo.

15. Page, H.M., Dugan, J.E., Schroeder, D.M., Nishimoto, M.M., Love, M.S. & Hoesterey, J. (2007) Ecological performance and trophic links: comparison among offshore oil platforms and natural reefs for a selected fish and their prey. *Marine Ecology Progress Series*, **334**, 245–256.

16. Schmitt, R., Page, H.M., Dugan, J.E., *et al.* (2007) *Advancing Marine Biotechnology: Use of OCS Oil Platforms as Sustainable Sources of Marine Natural Products.* Draft Final Study Report, OCS Study, MMS 2006-054.

17. Carney, R.S. (2005) *Characterization of Algal-Invertebrate Mats at Offshore Platforms and the Assessment of Methods for Artificial Substrate Studies.* OCS Study MMS 2005-038, US Department of the Interior, Minerals Management Service, New Orleans.

18. Boukinda, M.L., Schoefs, F., Quiniou-Ramus, V., Birades, M. & Garretta R (2007) Marine growth colonization process in Guinea Gulf: data analysis. *Journal of Offshore Mechanics and Arctic Engineering*, **129**, 97–106.

19. Stachowitsch, M., Kikinge, R., Herler, J., Zolda, P. & Guetebruck, E. (2002) Offshore oil platforms and fouling communities in southern Arabian Gulf (Abu Dhabi). *Marine Pollution Bulletin*, **44**, 853–860.

20. Page, H.M., Dugan, J.E., Dugan, D. & Richards, J. (1999) Effects of an offshore oil platform on the distribution and abundance of commercially important crab species. *Marine Ecology Progress Series*, **185**, 47–57.

21. Simpson, R.A. (1977) *The Biology of Two Offshore Oil Platforms.* IMR Ref 76-13, Institute of Marine Resources, University of California.

22. Relini, G., Tixi, F., Relini, M. & Torchia, G. (1998) The macrofouling on offshore platforms at Ravenna. *International Biodeterioration and Biodegradation*, **41**, 41–55.

23. Southgate, T. & Myers, A.A. (1985) Mussel fouling on the Celtic Sea Kinsale field gas platforms. *Estuarine, Coastal and Shelf Science*, **20**, 651–659.

24. Page, H.M. (1986) Differences in population structure and growth rate of the stalked barnacle, *Pollicipes polymerus* between a rocky headland and an offshore oil platform. *Marine Ecology Progress Series*, **29**, 157–164.

25. Page, H.M. & Hubbard, D.M. (1987) Temporal and spatial patterns of growth in mussels, *Mytilus edulis*, on an offshore platform: relationships to water temperature and food availability. *Journal of Experimental Marine Biology and Ecology*, **111**, 159–179.

26. MBC Applied Environmental Sciences (1987) *Ecology of Oil/Gas Platforms Offshore California.* OCS Study/MMS 86-0094, US Department of the Interior, Minerals Management Service, Los Angeles.

27. Whomersley, P.G.B. & Picken, O.J. (2003) Long-term dynamics of fouling communities found on offshore installations in the North Sea. *Journal of the Marine Biological Association of the United Kingdom*, **83**, 897–901.

28. Bram, J.B., Page, H.M. & Dugan, J.E. (2005) Spatial and temporal variability in early successional patterns of an invertebrate assemblage at an offshore platform. *Journal of Experimental Marine Biology and Ecology*, **317**, 223–237.

29. Connell, J.H. (1961) The influence of interspecific competition and other factors on the distribution of the barnacle *Chthamalus stellatus*. *Ecology*, **42**, 710–723.

30. Paine, R.T. (1974) Intertidal community structure: experimental studies of the relationship between a dominant competitor and its principal predator. *Oecologia*, **15**, 93–120.

31. Connell, J.H. & Slatyer, R.O. (1977) Mechanisms of succession in natural communities and their role in community stability and organization. *American Naturalist*, **111**, 1119–1144.

32. Page, H.M., Dugan, J.E., Culver, C.S. & Maridan, B. (2007) Oceanographic gradients and patterns in invertebrate assemblages on offshore oil platforms. In: *Advancing Marine Biotechnology: Use of OCS Oil Platforms as Sustainable Sources of Marine Natural Products* (eds R. Schmitt, H.M. Page, J.E. Dugan, *et al.*), Draft Final Study Report, OCS Study, MMS 2006-054.

33. Railkin, A.I. (2004) *Marine Biofouling: Colonization Processes and Defenses*. CRC Press, Boca Raton.

34. WHOI (1967) *Marine Fouling and its Prevention*. Woods Hole Oceanographic Institution, United States Naval Institute, Annapolis.

35. Osman, R.W. (1977) The establishment and development of a marine epifaunal community. *Ecological Monographs*, **47**, 37–63.

36. Underwood, A.J. & Anderson, M.J. (1994) Seasonal and temporal aspects of recruitment and succession in an intertidal estuarine fouling assemblage. *Journal of the Marine Biological Association of the United Kingdom*, **74**, 563–584.

37. Harger, J.R.E. & Landenberger, D.E. (1971) The effect of storms as a density dependent mortality factor on populations of sea mussels. *Veliger*, **14**, 195–201.

38. MEC Analytical Systems (2003) *An Assessment and Physical Characterization of Shell Mounds Associated with Outer Continental Shelf Platforms Located in the Santa Barbara Channel and Santa Maria Basin, California*. MMS Contract No. 1435-01-02-CT-85136, US Department of the Interior, Minerals Management Service, Camarillo.

39. Bomkamp, R.E., Page, H.M. & Dugan, J.E. (2004) Role of food subsidies and habitat structure in influencing benthic communities of shell mounds at sites of existing and former offshore oil platforms. *Marine Biology*, **146**, 201–211.

40. Foster, B.A. & Willan, R.C. (1979) Foreign barnacles transported to New Zealand on an oil platform. *New Zealand Journal of Marine and Freshwater Research*, **13**, 143–149.

41. Sammarco, P.W., Atchison, A.D. & Boland, G.S. (2004) Expansion of coral communities within the Northern Gulf of Mexico via offshore oil and gas platforms. *Marine Ecology Progress Series*, **280**, 129–143.

42. Villareal, T.A., Hanson, S., Qualia, S., Jester, E.L.E., Granade, H.R. & Dickey, R.W. (2007) Petroleum production platforms as sites for the expansion of ciguatera in the northwestern Gulf of Mexico. *Harmful Algae*, **6**, 253–259.

43. Page, H.M., Dugan, J.E., Culver, C.S. & Hoesterey, J. (2006) Exotic invertebrate species on offshore oil platforms. *Marine Ecology Progress Series*, **325**, 101–107.

44. Ocean Star Drilling Museum (2007) Available from http://www.oceanstaroec.com/fame/1999/mariculture.htm. Cited 1 November 2007.

45. Schaufelberger, D.E, Koleck, M.P., Beutler, J.A., *et al.* (1991) The large-scale isolation of bryostatin-1 from *Bugula neritina* following current good manufacturing practices. *Journal of Natural Products*, **54**, 1265–1270.

46. Thorpe, J.P., Sole-Cava, A.M. & Watts, P.C. (2000) Exploited marine invertebrates: genetics and fisheries. *Hydrobiologia*, **420**, 165–184.

47. Culver, C.S., Page, H.M. & Dugan, J.E. (2005) Oil, gas platforms, sources for marine natural products? *Global Aquaculture Advocate*, June 2005, 60–61.

48. Engel, R. & Ray, J. (1985) Bio-Shield: an anti-fouling system for offshore platforms that works! *Oceans*, **17**, 62–70.

49. Mark Tool Company (2007) Available from http://www.marktool.com/bioshield.htm. Cited 1 November 2007.

50. Innovative Engineering Ventures (IEV) (2007) Available from http://www.iev-group.com/oilNgasDivision.htm. Cited 1 November 2007.

51. Culwell, A.S. (1997) Removal and disposal of deck and jacket structures. In: *Proceedings: Public Workshop Decommissioning and Removal of Oil and Gas Facilities Offshore California: Recent Experiences and Future Deepwater Challenges, September 23–25* (eds F. Manago & B. Williamson), MMS OCS Study 98-0023. Ventura.

52. National Research Council (1985) *Disposal of Offshore Platforms*. National Academy of Sciences, USA

53. US DOI MMS (2004) *Offshore facility decommissioning costs*. Pacific OCS Region, March 31, 2004 Report, Table 6–1, US Department of the Interior, Minerals Management Service, Pacific Outer Continental Shelf Region, USA.

Chapter 19
Biofouling and Antifouling in Aquaculture

Simone Dürr and Douglas I. Watson

This chapter aims to explain the differences in the problems that face the aquaculture sector in terms of biofouling and antifouling compared with other industries. The crucial difference from other industries to bear in mind is that in aquaculture live produce is reared and harvested for human consumption. Fouling impacts both on structures as well as the produce itself. We present the major fouling groups and explain common antifouling strategies. Costs of fouling and antifouling strategies are indicated.

19.1 The cost of fouling in aquaculture

Aquaculture is an increasingly more important industry because of declining global fisheries. In 2004 world aquaculture production totalled 59.4 million tonnes with an industry value of US$70.3 billion worldwide [1]. Of this quantity the Western European Region contributed only 3.5% in production, but was valued at US$5.4 billion (equating to 7.68% of global value). Naturally, biofouling affects all aquaculture structures. Additionally, stock species, such as shellfish and salmon, can be severely impacted both directly and indirectly by fouling organisms.

Exact actual costs created by the biofouling problem in the aquaculture industry are not known because they have not been accurately assessed, and so only estimates are available. Difficulties exist for operators to determine which losses are related to biofouling and which to other factors. Estimates are that approximately 5–10% in industry value is spent on dealing with biofouling-related problems every year [2]. For Europe this is estimated to be up to €260 million per year. Economically, biofouling can decrease the product value by up to 90%. Costs to the finfish sector relate primarily to the control of biofouling growth on nets and other structures, while in shellfish aquaculture a major cost factor is biofouling on the stock species themselves by both epibionts and endobionts with fouling on structures such as trays being an additional problem. Fouling generally results in increased labour costs through increased requirements for cleaning and often reductions in the value of the product. Labour costs alone can increase by up to 20% through biofouling on structures such as nets or on stock organisms. In total, biofouling may result in add-on annual farm production costs of up to €120 000 per farm [2]. For Norway, Olafsen [3] estimated that fouling costs the salmon industry between €0.02 and €0.09 kg^{-1} of salmon produced. This may have resulted in a total cost of antifouling strategies for the Norwegian Salmon industry of up to €45 630 000 in 2003 (total Norwegian salmon production in 2003: 507 000 tonnes [4]). The cost to the Scottish mussel rope industry is estimated to be between €380 000 and €640 000 per annum [5]. Biofouling in Australian pearl oyster aquaculture is estimated to represent 30% of total operational costs [6]. Some researchers, for example Enright *et al.* [7], have produced estimates related to the cost of biofouling within the oyster trade. Their research indicated that 20% of the final market price relates to issues surrounding biofouling. The CRAB project [8] estimated losses of 30% within the European oyster market related to cleaning costs and mussel stock losses of 20% during the mussel cleaning process (both figures relate to total operating costs). Claereboudt *et al.* [9] assumed the costs to be slightly higher, at 30% of total operating costs, for the scallop sector. Other than generating increased operational and cleaning costs, shell quality also plays a part. Fouled stock organisms can be significantly downgraded in price for purely aesthetic reasons. Fouling on the blue mussel *Mytilus edulis* leads to a yield loss of 10–20% (J. Maguire, personal communications). Shell quality is particularly important for bivalves sold on the half shell such as oysters. Any reduction in the aesthetics of the shells can result in a lower market value for the product. Additionally, such biofouling can alter shell shape and result in thinner and, therefore, weaker shells which has production and processing implications. Ross *et al.* [10] showed there to be instances where biofouling can facilitate growth (e.g. scallop culture) through, for example, reduced water flow and the promotion of organic matter availability. Possible production savings have been estimated to be as much as €130 000 per annum for a farm producing 3 million scallops [11].

19.2   The problems with fouling in aquaculture

19.2.1   What is affected?

Biofouling affects equipment, infrastructure and even stock in the aquaculture industry. Equipment and infrastructure are primarily boats and barges, buoys and ropes, nets and cages in finfish operations, plus trays, pearl nets, ropes and bags in shellfish aquaculture. Besides the number and variety of affected materials, the very special problem that aquaculture has with biofouling is that it can occur on shellfish stock (e.g. oysters, mussels, scallops, prawns). Crustaceans and bivalves are live organisms that enter the human food chain, therefore any antifouling treatment may have an impact on human health (see below).

19.2.2   Specific issues within aquaculture

19.2.2.1   Structures, materials and stock

Biofouling in aquaculture differs from other industries in various ways. Biofouling in aquaculture is very much structure, material, site and stock-specific. Fouling pressure on infrastructure varies with equipment deployed and situations may occur where adjacent farm sites stocking the same species but utilising different equipment will be affected by an alternative fouling assemblage (D. Watson and S. Dürr, personal observations). Characteristics and types of affected materials are extremely diverse, ranging from hard plastics such as PVC, soft plastics like nylon and various metals such as steel which may have flat, smooth or very heterogeneous surfaces. Plastics such as nylon may also be woven or knotted. Materials can be flexible as in nettings or hard as in cage structures. Affected surfaces can be mobile such as boats and netting or stationary such as anchors and cage structures. This high variability in types of material and their surface characteristics influences the accumulation of fouling by providing a high diversity in habitats to satisfy the different settlement requirements of fouling species (see Chapter 3) and leads to different fouling assemblages (see Chapter 4) even at the same aquaculture farm. Recent studies by the authors (D. Watson and S. Dürr, unpublished data) have shown, for example, that hard foulers prefer to recruit to harder substrata, such as culture trays for scallops, whereas algae and other soft foulers (ascidians, hydroids, sponges) recruit preferentially to ropes and nets. Hence antifouling strategies need to be adapted to the high diversity of materials and fouling organisms present at any given site.

19.2.2.2   *Environmental factors and larval availability*

Fouling can differ within and between sites and geographical regions. Structures at the same site can be exposed to the same environmental factors (e.g. constant depth if floating) or to variable factors (e.g. variable depth if anchored to the seafloor due to tidal movement). Shade and depth may affect community composition [12].

The extent and the type of fouling vary greatly with timing and duration of immersion. Stock and equipment permanently immersed in the marine environment will generally be more fouled than those in the intertidal zone. Biofouling impacts in the shellfish sector also vary in nature and intensity between intertidal and subtidal culturing techniques and also between suspended shellfish and infaunal organisms such as clams. Infaunal species are affected by few or no biofouling organisms, since they are not generally exposed to settling larvae.

Fouling communities in aquaculture can often appear as monocultures of certain species such as the ascidian *Ciona intestinalis* or the blue mussel *M. edulis* [13] (S. Dürr and D. Watson, unpublished observations). This is often more apparent with introduced fouling species, probably due to the high disturbance regime resulting from the farm operation.

Not much is known about what shapes fouling communities in freshwater aquaculture. However, Dubost [14] showed that freshwater fouling is determined by characteristics of the water body such as temperature and currents.

19.2.2.3   *Farming supports fouling*

Aquaculture sites maintain a high level of biofouling. The reason for this follows from intensive production itself, resulting in leftover fish food and faeces. Non-selective filter-feeding fouling organisms including sponges, barnacles, ascidians and polychaetes thrive on such organic material; bivalves can also take profit from this additional food source [15, 16]. Growth of green algal species is stimulated by enhanced nutrient levels around farms [17].

19.2.3   *What does the fouling problem mean for aquaculture?*

For aquaculture organisms, care and environmental conditions need to be optimal to achieve a sellable product. For maximum growth, feeding needs to be at an optimum level too. This can only be accomplished if water flow through netting, trays or bags is as high as possible and therefore fouling minimal. If fouling is high, costs to production are increased due to reduced rate of stock growth leading to an overall increase in the production process as well as greater cleaning and maintenance costs.

19.2.3.1 Shellfish aquaculture

In shellfish aquaculture if flow through trays, bags or lantern nets is not maintained, plankton cannot reach the stock organisms [18] which are filter-feeders that depend on plankton as a food source. Consequently, growth rates fall [9, 19] leading to longer growing periods before the stock can go to market as well as lower survivorship. Current velocity through holding trays is an important growth factor for scallops and the right choice of mesh size is crucial [20]. In addition, fouling by filter-feeding organisms such as ascidians, bivalves or sponges means competition for food for filter-feeding stock species such as oysters, scallops or mussels. Some fouling organisms, such as sponges, can have negative impacts on bivalves as they may overgrow bivalve hinges, affecting their shell valve movements, thus reducing efficiency of escape from predators (scallops [21]) and their ability to feed properly. Sponge endobionts are also known to have negative effects on growth and meat condition of oysters [22]. López *et al.* [23] found that epibionts on the shells of the scallop *Argopecten purpuratus* resulted in higher mortality in the early culture of spat in pearl nets. Additionally, the visual aesthetics of the product perceived by the consumer are negatively influenced by shell fouling, which has to be removed before going to market if a low price is to be avoided. Biofouling has also been shown to act as a refuge for bivalve predators [24]. The authors showed that biofouling provided moisture for a polyclad turbellarian flatworm (*Stylochus frontalis*), thereby preventing desiccation of this predator whilst the bivalve was exposed to air. Furthermore, Enderlein *et al.* [25] showed that biofouling can assist some predatory crabs when it comes to handling their prey, thereby providing them with an easier meal and decimating stock (see also Chapter 7).

Overall, biofouling in shellfish aquaculture is therefore seen mainly as a negative issue; however, some positive effects of fouling have been documented. Some studies (e.g. [23, 26, 27]) have shown bivalve growth to be unaffected. Other studies, such as by Ross *et al.* [10], have shown that certain levels of biofouling on the equipment or infrastructure can improve growth or, in the case of sponge fouling, avoid subsequent fouling by hard foulers which are more difficult and costly to remove [11].

19.2.3.2 Finfish aquaculture

In finfish aquaculture, if flow is low or the netting even completely blocked, oxygen levels in the pen decrease dramatically. Fish cease feeding and therefore growing; mortalities will be high. In addition, removal of waste materials is prevented and the cage then acts as a reservoir or vector for disease and parasites. Fouling on netting can increase drag by up to three times [28] or the drag coefficient by up to 900% [29]. Therefore, wave and current loads lead to material fatigue of anchorage and cage systems.

19.2.3.3 Weight loads

One of the most important issues with fouling is the drastic increase in weight of structures, not only in finfish but also in shellfish culture. Fouling adds weight to submersed structures which can have implications for mechanical handling of equipment, particularly where nets are to be removed from the system for cleaning. Nets can gain up to 11 times their own weight by fouling within a few weeks depending on season and site, putting an extreme load on the structure and necessitating specialised lifting equipment (D. Fowler, personal communications).

19.2.3.4   Disease and parasites

Fouling may also serve as a stepping stone or safe harbour for disease or parasitic organisms. If the metabolic condition of any stock species is reduced, e.g. through stress or food availability (see above), then this may lead to greater susceptibility to attack from disease [22]. Fouling is suspected to transfer ectoparasites such as sea lice and enteroparasites [30] to the stock. Amoebic gill disease (AGD) is a major health issue in the salmon-farming sector caused by *Neoparamoeba pemaquidensis*. Despite occurring naturally in the water column, this paramoeba is known to be associated with certain fouling species (the bryozoan *Scrupocellaria bertholetti*, the solitary ascidian *C. intestinalis*) which act both as a vector and a reservoir [31]. Biofouling has been shown to concentrate plankton locally [10]. If this localisation involves plankton associated with disease, then implications arise relating to the health of the cultured stock and its suitability for consumption. For example, the dinoflagellate *Prorocentrum lima* (which causes diarrhetic shellfish poisoning (DSP)) has been shown to associate with the common fouling brown alga *Ectocarpus* sp. [32].

19.3   Major fouling groups

Fouling groups in aquaculture can include all locally present sessile organisms, but also include introduced algae or invertebrates. In addition, the aquaculture industry sometimes classifies fully mobile species such as sea stars, sea urchins or caprellid amphipods as 'fouling'. Obviously, these groups are not foulers, though they can create similar problems. Treatment measures against such mobile yet attached organisms will need to be quite different from commercial antifouling strategies; they will not be further discussed here.

19.3.1   Biofilm

Biofilm formation on netting in aquaculture follows normal succession (see Chapter 9). In a Scottish sea loch in winter, accrual of extracellular polymeric substance (EPS), bacteria and pennate diatoms, is followed by increased abundance of pennate and centric diatoms. After 6 weeks, biodiversity and abundance of diatoms is further increased and the first hydroids appeared after 8 weeks [33].

19.3.2   Macrofouling on infrastructure and equipment

Marine aquaculture fouling on infrastructure and equipment worldwide is dominated by mussels (*M. edulis*, *Mytilus galloprovincialis*), ascidians (*C. intestinalis*, *Botryllus* spp.), barnacles of the *Balanus* group and hydroids of the *Tubularia* genus (Table 19.1). Most of these species are non-encrusting and therefore pose a problem through their volume by reducing mesh sizes and therefore water flow. Not much is known about freshwater biofouling at aquaculture facilities. It may consist mainly of cyanobacteria, diatoms and other algae and bryozoans [14, 34]. It is clear that information on fouling groups in different regions is very limited and more research into defining the problem needs to done.

In Western Europe, most fouling consists of six groups: algae, hydroids, serpulids, mussels, barnacles and ascidians (S. Dürr *et al.*, unpublished data). Fouling in Western Europe can be

Table 19.1 Dominant fouling species found on different types of infrastructure surfaces at aquaculture farms in various oceanic regions of the world.

Oceanic region	Dominant species	Species group	Type of surface	Reference
NE Atlantic	Ulva spp.	GA	Net	[76]
	Chaetomorpha spp.	GA		
	Polysiphonia spp.	RA		
	Ceramium spp.	RA		
	Hincksia spp.	BA		
	Sphacelaria spp.	BA		
	Laminaria saccharina	K		
	Obelia spp.	HY		
	Mytilus edulis	BV		
	Ascidiella aspersa	AS		
	M. edulis	BV	Net cylinder	[40]
	A. aspersa	AS		
	Tubularia sp.	HY	Scallop nets	[41]
	Bougainvillia sp.	HY		
	Jassa falcata	STPA		
	Diplosoma listerianum	AS		
	Ascidiella scabra	AS		
	Ulva spp.	GA	PVC panels at fish and shellfish farms	S. Dürr et al., unpublished data
	Ectocarpus sp.	BA		
	Laminaria spp.	K		
	Alaria esculenta	K		
	Tubularia spp.	HY		
	M. edulis	BV		
	Mytilus galloprovincialis	BV		
	Balanus improvisus	BAR		
	Balanus crenatus	BAR		
	Balanus perforatus	BAR		
	Tube-building polychaetes and amphipods	STPA		
	Ciona intestinalis	AS		
	Diplosoma spp.	AS		
		HY	Timber panels at shellfish farms	[87]
	M. galloprovincialis	BV		
	Pomatoceros lamarcki	CP		
	Pomatoceros triqueter	CP		
	Elminius modestus	BAR		
	B. crenatus	BAR		
	B. perforatus	BAR		
	Balanus amphitrite	BAR		
	Psidia longicornis	BY		
	Electra pilosa	BY		
Adriatic	Polysiponia sp.	RA	Net cylinder	[40]
	Campanopsis sp.	HY		
	Schizobrachiella sanguinea	BY		
E Mediterranean	Tube-building polychaetes	STPA	Net cylinder	[40]

Table 19.1   (*Continued*)

Oceanic region	Dominant species	Species group	Type of surface	Reference
W Mediterranean		A	Oyster rope culture	[88]
		SP		
		BY		
	Phallusia mamillata	AS		
	Botryllus sp.	AS		
	C. intestinalis	AS		
	Mytilaster minimus	BV	Buoy at fish farm	[89]
	B. perforatus	BAR		
	Chthalamus stellatus	BAR		
	Soft-tube builders	STPA	PVC panels at oyster farm	S. Dürr *et al.*, unpublished data
Red Sea	*Jania* sp.	RA	Net cylinder	[40]
	Mycale fistulifera	SP		
White Sea	*Obelia longissima*	HY	Net	[90]
	Halichondria panicea	SP		
	Halisarca dujardini	SP		
	B. crenatus	BAR		
	Molgula sp.	AS		
	Styela rustica	AS		
NW Atlantic	*M. edulis*	BV	Mussel rope culture	[26]
	Hiatella arctica	BV		
	Semibalanus balanoides	BAR		
	E. pilosa	BY		
	Membranipora unicornis	BY		
	Bugula turrita	BY		
	Botrylloides aureum	AS		
	Botryllus schlosseri	AS		
	C. intestinalis	AS		
	Styela clava	AS		
		RA	Mussel sock	[91]
	B. turrita	BY		
	Molgula sp.	AS		
		A	Pearl net	[9]
	Tubularia larynx	HY		
	Oysters	BV		
	Mussels	BV		
		CP	Pearl net	[92]
		HY	Oyster tray	[61]
Tropical NW Atlantic		BY	Net	[12]
	Ostrea equetris	BV		
	Crassostrea rhizophorae	BV		
	Balanus trigonus	BAR		
		AS		
Tropical SW Atlantic	*M. edulis*	BV	Pearl net	[19]
	H. arctica	BV		
	Anomia sp.	BV		

(*Continued*)

Table 19.1   (*Continued*)

Oceanic region	Dominant species	Species group	Type of surface	Reference
NE Pacific	*Metridium senile*	AN	Net	[13, 28]
	Tubularia sp.	HY		
	Tubularia marina	HY		
	Hydroides ezoensis	HY		
	S. clava	AS		
NW Pacific	*Acropora* spp.	CO	Snail cage	[93]
	Clytia languida	HY	?	[94]
	O. longissima	HY		
	Bougainvillia ramosa	HY		
SW Pacific	*Cladophora* sp.	GA	Mussel ropes	[95]
	Hydroides elegans	CP		
	Watersipora subtorquata	BY		
	Bugula neritina	BY		
	C. intestinalis	AS		
	Bottryllus schlosseri	AS		
	Botrylloides leachii	AS		
	Corella eumyota	AS		
	Cnemidocarpa bicornuata	AS		
	Obelia australis	HY	Net	[30]
	M. edulis	BV		
	Scrupocellaria bertholetti	BY		
	Mogula ficus	AS		
	C. intestinalis	AS		
	Enteromorpha sp.	GA	Net	[84]
		SP		
	Ulva rigida	GA	Net	[38]
	Molgula ficus	AS		
	Asterocarpa humilis			
	Chondria fusifolia	BA	Net	[96]
	Brongniartella australis	RA		
	S. bertholettii	BY		
Tropical SW Pacific	*Dictyota* sp.	BA	Oyster trays	[97]
	Rock oysters	BV		
	Pteria spp.	BV		
	Pinctada fucata	BV		
		BY		

A, alga; AN, anemone; AS, ascidian; BA, brown alga; BAR, barnacle; BV, bivalve; BY, bryozoan; CB, cyanobacteria; CO, coral; CP, calcareous polychaete; GA, green alga; HY, hydroid; K, kelp; RA, red alga; SP, sponge; STPA, soft-tube-building polychaetes and amphipods.

split into two regions: northern and southern. The former is characterised by blue mussels and the ascidian *C. intestinalis*, the latter by soft-tube-building amphipods and polychaetes (Plate VI A). Problem algae include *Enteromorpha* sp. and *Ulva* sp. as well as the ever-present *Ectocarpus* sp., but also kelps such as *Laminaria saccharina* and *Alaria esculenta*. Within the hydroid group *Tubularia* sp. are common foulers. Calcareous tubeworms of the family Serpulidae are widely distributed and *Pomatoceros triqueter* (a serious problem for mussel farmers in the northern region) is only one example. Common mussel species include *M. edulis* and *M. galloprovincialis*. Amongst barnacles there are various subtidal *Balanus* species such as *Balanus improvisus*, *Balanus crenatus* and *Balanus perforatus*. Ascidians are a problem both in compound and in solitary form. The compound *Diplosoma* genus is generally present, while the solitary *C. intestinalis* can only be found in the north-western part of Europe.

19.3.3   Macrofouling on shellfish organisms

Fouling on shellfish organisms themselves is worldwide dominated by sponges, barnacles, spirorbids/serpulids and ascidians (Table 19.2). Bryozoans, hydroids, algae, other bivalves and shell-boring species can also be important. Biofouling of stock species can differ in type and intensity dependent on the stock in culture and even between the two valves of a single bivalve [35]. Rosell *et al.* [35] showed that oysters were more impacted by borers than were mussels when held at the same water depth. Mutualistic fouling relationships also differ between species with [21] showing sponge/bivalve mutualistic relationships occurring among scallops, but not oysters. Chernoff [36] found that such relationships improved scallop growth rates, while Pitcher and Butler [37] demonstrated an increase in the soft tissue levels present.

19.3.4   Weight loads

Published information on weight loads is very limited. In Tasmania, fouling wet weights can reach between 7.8 and 8.5 kg m^{-2} caused mainly by ascidians after only 163 days of immersion [38]. Pearl nets in Venezuela can have loads of approximately 350 g after 3 months [39]. Net samples in the north-western Atlantic accumulate approximately 400 g 0.01 m^{-2} in 1 year mainly in the form of blue mussels [13].

In Europe, Cook *et al.* [40] found average dry weights after 6 months to range between 100 and more than 250 g dry weight per net sample, with the highest values recorded in the Adriatic and the lowest in the eastern Mediterranean. Scallop nets at the Isle of Man, UK, can have loads between 400 and 600 g per net [41]. In western Europe, there is a separation between northern and southern regions with the border situated between Ireland and Spain (S. Dürr *et al.*, unpublished data). Fouling on 20 × 20 cm PVC panels exposed to the environment for 26 months added less than 400 g weight at aquaculture sites in southern Europe, while in northern Europe weight increase could exceed 5 kg. Weight of structures in southern countries such as Spain and Portugal was mainly due to soft-tube-building amphipods and polychaetes and associated algae and barnacles. The high loads in the north resulted from blue mussels, the ascidian *C. intestinalis* and large kelp species.

Weight of fouling on netting in European freshwater aquaculture can also be high. In one study [14], fouling reached a weight of up to 1400 g m^{-2} after only 3 weeks.

Relative fouling loads can impact on bivalve culture. Fouling on the shell of the pearl oyster *Pinctada imbricata* can reach loads of 1.53 g per shell in Venezuela [39]. However, the significance of the fouling weight for this bivalve is negligible [42].

19.3.5   Invasive fouling species

Better observation techniques in recent years may have led to an increase in numbers of invasive fouling species being identified that may have otherwise gone unnoticed. Equally, this increase

Table 19.2 Dominant fouling species found on the shells of different bivalve species cultivated at aquaculture farms in various oceanic regions of the world.

Oceanic region	Dominant fouling species	Species group	Fouled shellfish species	Reference
NE Pacific	*Lithophaga aristata*	BV	*Haliotis fulgens*	[98]
	Lithophaga plumula	BV	*Haliotis corrugata*	
	Penitella conradi	BV		
	Spionidae	CP	*Haliotis rufescens*	[99]
	Serpulidae	CP		
	Myxilla incrustans	SP	*Chlamys hastata hericia*	[100]
	Mycale adhaerans	SP	*Chlamys rubida*	
	Sabellid polychaetes	CP	*H. fulgens*	[101]
			H. corrugata	
	Various		*Chama pellucida*	[102]
Tropical NW Pacific	*Pinctada* spp.	BV	*Pinctada maxima*	[27]
	Pteria spp.	BV		
	Crassostrea spp.	BV		
	Polychaetes	CP		
	Barnacles	BAR		
SE Pacific	*Enteromorpha* sp.	GA	*Argopecten purpuratus*	[23]
	Hydroids	HY		
	Nevianipora milneana	BY		
	Schizoporella maulina	BY		
	Ciona intestinalis	AS	*A. purpuratus*	[103]
Tropical SW Pacific	*Pione* sp.	SP	*Pinctada margaritifera*	[86]
	Oysters	BV		
	Barnacles	BAR		
	Obelia sp.	HY	*Pinctada fucata*	[104]
	Saccostrea sp.	BV		
	Parasmittina sp.	BY		
	Didemnum sp.	AS		
	Cliona vastifica	SP	*Saccostrea commercialis*	[105]
	Cliona celata	SP		
SW Pacific	Algae	A	Various oyster species	[6]
	Sponges	SP		
	Oyster species	BV		
	Polydora websteri	STPA		
	Serpulids	CP		
	Barnacles	BAR		
	Bryozoans	BY		
	C. intestinalis	AS		
	Sponges	SP	*Chlamys asperrima*	[36]
	Algae	A	Oysters	[106]
	Spirorbid polychaetes	CP	*Pecten fumatus*	[107]
	Sponges	SP	*C. asperrima*	[37]
Baltic Sea	*Balanus improvisus*	CP	*Mytilus edulis*	[25]
White Sea	*Balanus crenatus*	CP	*M. edulis*	[90]
Mediterranean Sea	*Cliona viridis*	SP	*Ostrea edulis*	[35]
	Cliona celata	SP		

Table 19.2   (*Continued*)

Oceanic region	Dominant fouling species	Species group	Fouled shellfish species	Reference
NE Atlantic	*Suberites ficus* ssp. *rubrus*	SP	*Chlamys opercularis*	[11]
	Pomatoceros spp.	PW	*M. edulis*	[5]
	Hydroids	HY	*Pecten maximus*	[41]
	Anomia spp.	BV		
	Ascidians	AS		
NW Atlantic	*M. edulis*	BV	*M. edulis*	[54]
	C. intestinalis	AS	Mussels	[57]
			Oysters	
	C. vastifica	SP	*Placopecten magellanictis*	[108]
	P. websteri	STPA		
	Polydora concharum	STPA		
	Halichondria panicea	SP	*Chlamys varia*	[21]
	Polydora sp.	STPA	*Crassostrea virginica*	[22]
Tropical NW Atlantic	*Balanus eburneus*	BAR	*C. virginica*	[109]
SW Atlantic	*Balanus* sp.	BAR	*Aequipecten tehuelchus*	[110]
	Ascidians	AS		
SE Atlantic	Cyanobacteria	CB	*Perna perna*	[55]

A, alga; AN, anemone; AS, ascidian; BA, brown alga; BAR, barnacle; BV, bivalve; BY, bryozoan; CB, cyanobacteria; CO, coral; CP, calcareous polychaete; GA, green alga; HY, hydroid; K, kelp; RA, red alga; SP, sponge; STPA, soft-tube-building polychaetes and amphipods.

may simply represent the accumulative invasive species transmitted either by aquaculture itself or by shipping. These introduced species often dominate the fouling assemblage. Movements of invasive species from one farm to another can in shellfish aquaculture often happen through transportation of seed stock. In southern Portugal, the introduced barnacles *Balanus amphitrite* and *Elminius modestus* can be found on oyster cages. In Norway and Scotland, the hydroid *Tubularia larynx* and the caprellid *Caprella mutica* are present at salmon farms ([40]; S. Dürr *et al.*, unpublished data). For the New Zealand mussel farming industry, the introduced alga *Undaria pinnatifida* is a serious problem ([43]; Chapter 24).

19.4   Antifouling strategies

Currently, fouling control in aquaculture is limited to cleaning, husbandry and use of some antifouling coatings. The various approaches are not mutually exclusive, but are usually combined to yield an antifouling strategy that has proven successful for a particular site. For example, the application of an antifouling coating on a net, followed by air-drying to kill foulers and cleaning in a net washer can be combined. Olafsen [44] found, in a study based on industrial interviews, that the Norwegian finfish industry basically follows three different combinatory strategies: Strategy 1: copper-based coating on nets in combination with drying of nets on site; Strategy 2: a copper-based coating in combination with washing and even sometimes air-drying (Plate VI B); Strategy 3: using uncoated nets that are cleaned frequently.

19.4.1   Husbandry/cleaning

19.4.1.1   Natural self-cleaning in bivalves

One particular difference from biofouling in other industries is that in aquaculture the affected surfaces sometimes have a degree of natural self-cleaning ability. Often the stock species, particularly bivalves, have their own techniques to maintain a clean surface without needing any human intervention. Wahl *et al.* [45] proposed that the periostracum of bivalve shells acts as a defensive barrier against fouling organisms (see also [46]), and Harper and Skelton [47] showed that this barrier property prevented attack by boring organisms. These techniques involve shell surface texture and chemistry ([48, 49]; Chapter 8). Certain bivalves also have the ability to clean themselves. For example, mussels can sweep their foot across their shell, removing newly settled organisms [50].

19.4.1.2   Human-induced techniques

There are a number of methods which can be employed to prevent or reduce fouling [51]. Methods can be as simple as reducing the level of fouling through exposure of nets etc. to air [52] (Plate VI B), the use of disk cleaners in situ or nets can be removed and cleaned in net washers on shore. Fouling has also been shown to decrease with depth [9]; therefore, by lowering gear during times of heavy spatfalls, reductions in the levels of fouling can be achieved [53]. The dropping of gear has produced positive results in the Canadian longline mussel industry as it assists in avoiding secondary spatfalls and allows predators to control the numbers of smaller recruits [54].

Shell strength in farmed shellfish is often affected by boring organisms [55] leading to difficulties when it comes to the processing of the bivalves [56]. Regular cleaning cannot control such endobionts, but is effective against epibiotic fouling organisms and can help limit any reduction in shell quality – particularly important in maintaining the market value of certain species. However, if the fouling organisms are hard in nature (e.g. barnacles, serpulid worms) and allowed to grow unchecked then they may need to be chipped off of the bivalve shell – a process that can result in shell damage. Chemical cleaning techniques such as dipping in caustic solutions and/or more benign solutions (such as hot water, brine, acid solutions [57]) has the ability to kill both endobiotic [58] and epibiotic organisms. Such dipping techniques can only be utilised for bivalves that have the ability to completely close their valves (e.g. oysters, mussels) and have good thermal tolerances. They are inappropriate for scallops, which cannot close fully.

Cleaning techniques should be targeted to the cultured organisms. Certain species need only be cleaned once prior to sale, whilst others are regular cleaned throughout their life cycle. Countries where labour costs are high generally use expensive machines (e.g. tumblers and/or brush graders) to carry out the cleaning. Cleaning by hand is only possible in countries where labour costs are low [53]. Despite cleaning being necessary for the prevention of biofouling accumulation, the repetition of handling the stock organisms can lead to reduced bivalve growth and increased mortality levels [59].

The recent CRAB study [8] revealed, through the use of industrial questionnaires, the different cleaning techniques employed by a wide variety of European aquaculturists, and determined the costs involved with each (D.I. Watson and S. Dürr, unpublished data). Despite major differences in approach, such as reliance on machines for mechanical cleaning in certain regions, but use of manual labour in others, the costs, in relation to total expenditure, were largely similar at up to 30% (e.g. for oyster producers; E. Bergtun, personal communications). Other costs in the industry are often incurred by cleaning-related losses (broken shells etc.), which can be significant in number. The CRAB study [8] showed typical levels of losses during the mussel cleaning process of 20% (in relation to total operating costs; J. Maguire, personal communications).

19.4.2 Biological control

Biological control of fouling can be natural or initiated by the aquaculturist. Mutualisms can assist fouled organisms by reducing their susceptibility to predation and/or additional fouling pressure [11,21]. Additionally, the mere presence of the stock species filtering the surrounding water can reduce the level of biofouling pressure [45,60].

Grazers and predators can also be introduced into both finfish and shellfish aquaculture in order that fouling on equipment and stock may be controlled. Hidu *et al.* [61] noted the benefits of crabs at controlling mussel fouling in oyster culture. A further study by Enright *et al.* [7] showed that the presence of crabs in oyster culture resulted in the improvement in shell shape and quality of the final product, whilst reducing fouling on the shells by 76–79% and increasing growth by 10–60%. Other studies have also shown the benefits of using grazers in the shellfish sector with Enright *et al.* [62] demonstrating a 30% increase in growth rates of oysters, while Minchin and Duggan [63] showed that dogwhelks were successful in controlling mussel fouling in scallop and oyster culture. Sea urchins have also been used successfully at controlling fouling in both scallop and oyster culture [39,41] (Plate VI C), reducing fouling levels by 74% on equipment and 71% on stock. Biological control organisms may be able to reduce the levels of hard foulers (e.g. barnacles and serpulids [63]). Even fish may decrease levels of ascidians in bivalve culture [64]. Such positive effects on growth rates can reduce costs by reducing time to market [65].

Grazers such as sea urchins have also been utilised in the finfish sector for the control of fouling on nets. However, the most common organisms utilised in the finfish sector are various wrasse species as they have the additional ability to control sea lice on stock species such as salmon [66] as well as affecting the level of fouling on the netting. Use of wrasse has allowed up to a 50% reduction in the requirement for net changes [67]. Although wrasses have been shown to be effective biological control agents, there are issues involved in their deployment. These include availability (local wrasse populations need to be large and the fishery for them sustainable), husbandry (particularly wrasse:stock ratio and the provision of overwintering shelters). It is also the case that they are only effective at controlling sea lice on salmon less than 2 kg in body weight [67].

19.4.3 Coatings

Tributyltin (TBT) coatings are dangerously ecotoxic in the marine ecosystem (Chapters 16 and 17) and therefore also to aquaculture production and human consumption [68–70]. Despite its toxicity, TBT has even been used directly in aquaculture to coat nets and other structures (e.g. [71,72]). Until 1987, TBT was used in UK aquaculture operations and low-level concentrations were still present years afterwards leading to imposex in dogwhelks in the impacted areas [73].

In Japan and Taiwan, TBT concentrations at aquaculture sites are still an ecological hazard [74,75]. Tests in the 1990s clearly indicated continued usage of TBT in the local aquaculture industry in Thailand, India and the Philippines [76]. Nowadays, in the western world, common coatings are based on copper, copper oxide, zinc oxide and halogenids. Antifouling products are produced worldwide by various coating companies and mainly used in finfish aquaculture [77]. Even so, coatings are only applied by a small percentage of non-organic producers due to their cost of approximately €2000 per cage and short life span (D. Fowler, personal communications).

Copper-based antifouling coatings specially formulated for aquaculture are now the most commonly used coatings. They are effective against fouling. Braithwaite *et al.* [78] applied a specific copper product on net panels in the Shetland Islands. The study ran for 10 months but was only started in August after the main fouling season had occurred. Fouling on copper-treated nets was inhibited for 150 days longer than on the control nets. The mesh of the control netting was blocked by as much as 100%. After 10-month immersion, netting mesh was blocked by about 55% in all net treatments while the wet weight of each sample and the number of species present on the netting depended on how the net sample was treated. The control net weighed approximately 5 kg m^{-2} and had 15 species present per sample, whereas the copper-treated net weighed approximately 1.6 kg m^{-2} and 10 species were present per sample. Therefore, at this particular site the copper-treated net was of advantage in terms of weight load, but not in terms of blocking of the mesh.

Despite their advantages, copper coatings can be disadvantageous to production. Douglas-Helders *et al.* [79] found a higher risk of amoebic gill disease in fish held in copper-treated nets. Copper leakage by antifouling coatings should not be neglected either. Ellingsen *et al.* [80] estimated that in Norway for each 2 kg of salmon produced 1 g of copper leaks into the environment. However, Solberg *et al.* [81] found no difference in copper content of salmon muscle or blue mussel tissues between sites where nets were treated with copper or left untreated. Davies and Paul [82] investigated copper and nickel accumulation in scallops and oysters grown in nets coated with a copper oxide-based antifouling paint and in trays of copper-nickel alloy mesh. Nickel was not significantly accumulated by either species. However, copper was highly accumulated in both culturing techniques and in both species. While scallops gradually lost their copper burden after being removed from the source, oysters retained copper indefinitely. Consequently, Norway and many other countries aim to reduce the use of copper coatings in aquaculture [83]. Due to these developments, non-toxic antifouling strategies are being more openly and urgently discussed in the finfish as well as shellfish aquaculture industries (e.g. [8]).

Other solutions to the use of copper as biocide are being investigated with some products already in use in aquaculture. Svane *et al.* [84] tested one of these products, a water-based synthetic latex coating (active ingredients are 2-isothiazolinones) at a southern bluefin tuna farm in South Australia under real working conditions. After 5.5 months, the total area covered by fouling on the untreated netting was 80.6% and on the treated netting 65.9% showing a relatively poor antifouling performance for the coating. Unfortunately, cleanability, other material characteristics and stock quality and welfare were not assessed.

It is clear that available general purpose non-toxic antifouling coatings need to be further adapted for use in aquaculture. A promising technology already used in shipping is silicone coatings. However, other nanotechnological material solutions (e.g. [85]) may equally be the way forward if food safety is further researched and can be guaranteed.

Silicone coatings have shown effective antifouling performance. In one study [38], fouling wet weight on a silicone-treated net was reduced to approximately 25% of the uncoated netting. Interestingly, community composition was affected by the coating leading to dominance by the alga *Ulva rigida*, while fouling on the uncoated net was dominated by solitary ascidians. Cleaning proved to be very easy, but stock quality and welfare were not assessed.

Tests of other commercial silicone coatings from four international paint companies on nettings and trays were conducted by S. Dürr *et al.* (unpublished data). Antifouling efficacy as well as cleanability proved to be very good in general, especially for the net samples in the first season. However, application of coatings was not completely successful as applied coatings and base coats were not specifically developed for netting and tray material requirements. Net samples showed cracking after cleaning with low water pressure systems and coatings started to peel off tray samples soon after immersion. This impaired antifouling performance drastically and further application development by the paint companies needs to be done. One of the silicone coatings was tested on a net cage in a tank stocked with Atlantic salmon (*Salmo salar*). Antifouling performance was again very good. However, the weight of the net was estimated to be twice that of the uncoated net, hard to handle, very inflexible and showed failures rapidly in the form of cracking at net joints. The gravest issue was that fish showed abrasions to noses, higher mortality and weight loss. This was not observed in a tank without net or in a tank with uncoated netting. For a full appraisal of effects on stock, full-scale silicone-coated netting needs to be tested in sea cages.

As well as non-toxic coatings being developed for infrastructure (see above), the natural barrier of the bivalve periostracum has been seen as a model for the creation of an antifoulant coating for the bivalves themselves. De Nys and Ison [86] have developed such a synthetic but biocidal coating for the Australian pearl oyster industry. Similar coatings, plus waxy coatings, have been applied directly to bivalve species and have the ability to smother endobiotic organisms (see [58] and references therein) as well as preventing fouling build-up [86].

19.5 Future solutions

Any new antifouling technique must have the ability to work at an industrial scale, be cost-effective and simple to apply. Prevention techniques through the identification of spatfall timing and measures to avoid such events are simple but effective, negating the requirement for extensive cleaning technologies and/or labour costs later on in the production cycle. Add to this potentially cheap and effective non-toxic coatings and the reduction in manpower costs will likely be significant. However, on top of this there are additional new challenges. With global climate change and apparent increases in the prevalence of invasive species, the biofouling problems faced by the aquaculture industry are highly likely to change in the levels and number of species groups present. How the problem will change is not clear, but it may necessitate changes in antifouling strategies. Investigations into sustainability of aquaculture, especially shellfish aquaculture, including conservation of the impacted benthic communities are becoming more important in light of the aforementioned challenges.

19.6   Conclusions

- Costs of biofouling to the aquaculture industry are great, typically 20–30% of total operating costs.
- Fouling affects infrastructure, equipment and even stock in aquaculture. It impacts on a wide range of materials. Aquaculture-specific problems include reduction of water flow (leading to reductions in oxygen levels and food limitation), increase of weight and drag of structures, overgrowth of stock and disease. The type of fouling, and therefore problem, varies not only with material and type of equipment or stock but also with site, season and immersion time.
- Major fouling groups worldwide on infrastructure and equipment are hydroids, mussels, barnacles and ascidians. On shells of cultured bivalves, they are sponges, barnacles, spirorbids/serpulids and ascidians.
- Antifouling strategies are usually combinations of various techniques and technologies. Commonly used are husbandry, cleaning, biological control and antifouling coatings.
- Future solutions in antifouling strategies need to be cost-effective and simple to apply, and must be compatible with aquaculture's status as a globally vital food industry.

Acknowledgements

We thank the partners in the project CRAB (Collective Research on Aquaculture Biofouling – EU Project COLL-CT-2003-500536-CRAB), especially all the aquaculture workers of the SME partners who showed unrelenting commitment to data collection. Special thanks to John Davenport whose comments helped to improve the draft manuscript.

References

1. Subasinghe, R. (2005) Fisheries topics: resources – the state of world aquaculture. In: *FAO Fisheries and Aquaculture Department* (ed. FAO). Food and Agriculture Organisation of the United Nations, Rome. Available from http://www.fao.org/fi/website/FIRetrieveAction.do?dom =topic&fid=13540. Accessed 6 July 2009.
2. Lane, A. & Willemsen, P. (2004) Collaborative effort looks into biofouling. *Fish Farming International*, September, 34–35.
3. Olafsen, T. (2006) *Kostnadsanalyse av ulike begroingshindrende strategier*. SFH80 A066041 – Åpen, SINTEF, Trondheim, Norway.
4. Rana, K.J. (2005) *Regional Review on Aquaculture Development. 6. Western European Region – 2005*. FAO Fisheries Circular No. 1017/6 FIMA/C1017/6, Food and Agriculture Organisation of the United Nations, Rome.
5. Campbell, D.A. & M.S. Kelly (2002) Settlement of *Pomatoceros triqueter* (L.) in two Scottish lochs, and factors determining its abundance on mussels grown in suspended culture. *Journal of Shellfish Research*, **21** (2), 519–527.
6. De Nys, P.C., Steinberg, P., Hodson, S. & Heasman, M. (2002) *Evaluation of Antifoulants on Overcatch, Other Forms of Biofouling and Mudworms in Sydney Rock Oysters*. Final Report Project No. 98/314, Fisheries Research and Development Corporation, Deakin, Australia.
7. Enright, C.T., Elner, R.W., Griswold, A. & Borgese, E.M. (1993) Evaluation of crabs as control agents for biofouling in suspended culture of European oysters. *World Aquaculture*, **24** (4), 49–51.

8. CRAB (2004–2007) *Collective Research in Aquaculture Biofouling.* EU FP6 COLL-CT-2003-500536-CRAB. Available from http://www.crabproject.com. Accessed 6 July 2009.

9. Claereboudt, M.R., Bureau, B., Côté, J. & Himmelman, J.H. (1994) Fouling development and its effect on the growth of juvenile giant scallops (*Placopecten magellanicus*) in suspended culture. *Aquaculture*, **121**, 327–342.

10. Ross, K.A., Thorpe, J.P., Norton, T.A. & Brand, A.R. (2002) Fouling in scallop cultivation: help or hindrance? *Journal of Shellfish Research*, **21** (2), 539–547.

11. Armstrong, E., McKenzie, J.D. & Goldsworthy, G.T. (1999) Aquaculture of sponges on scallops for natural products research and antifouling. *Journal of Biotechnology*, **70**, 163–174.

12. Hincapié-Cárdenas, C. (2007) *Macrobiofouling on Open-Ocean Submerged Aquaculture Cages in Puerto Rico.* MSc thesis, University of Puerto Rico.

13. Greene, J.K. & Grizzle, R.E. (2007) Successional development of fouling communities on open aquaculture fish cages in the western Gulf of Maine, USA. *Aquaculture*, **262**, 289–301.

14. Dubost, N., Masson, G. & Moreteau, J.C. (1996) Temperate freshwater fouling on floating net cages: method of evaluation, model and composition. *Aquaculture*, **143** (3–4), 303–318.

15. Lojen, S., Spanier, E., Tsemel, A., Katz, T., Eden, N. & Angel, D.L. (2005) δ^{15}N as a natural tracer of particulate nitrogen effluents released from marine aquaculture. *Marine Biology*, **148**, 87–96.

16. Sarà, G., Martire, M.L., Buffa, G., Mannino, A.M. & Badalamenti, F. (2007) The fouling community as an indicator of fish farming impact in Mediterranean. *Aquaculture Research*, **38**, 66–75.

17. Ruokolahti, C. (1988) Effects of fish farming on growth and chlorophyll α content of *Cladophora*. *Marine Pollution Bulletin*, **19** (4), 166–169.

18. Paul, J.D. & Davies, I.M. (1986) Effects of copper- and tin-based anti-fouling compounds on the growth of scallops (*Pecten maximus*) and oysters (*Crassostrea gigas*). *Aquaculture*, **54**, 191–203.

19. Lodeiros, C.J.M & Himmelman, J.H. (1996) Influence of fouling on the growth and survival of the tropical scallop, *Euvola* (*Pecten*) *ziczac* (L. 1758) in suspended culture. *Aquaculture Research*, **27** (10), 749–756.

20. Brake, J. & Parsons, G.J. (1999) Flow rate reduction in scallop grow-out trays. *Bulletin of the Aquaculture Association of Canada*, **2**, 62–64.

21. Forester, A.J. (1979) The association between the sponge *Halichondria panicea* (Pallas) and scallop *Chlamys varia* (L.): a commensal-protective mutualism. *Journal of Experimental Marine Biology and Ecology*, **36**, 1–10.

22. Wargo, R.N. & Ford, S.E. (1993) The effect of shell infestation by *Polydora* sp. and infection by *Haplosporidium nelsoni* (MSX) on the tissue condition of oysters, *Crassostrea virginica*. *Estuaries*, **16** (2), 229–234.

23. López, D.A., Riquelme, V.A. & González, M.L. (2000) The effects of epibionts and predators on the growth and mortality rates of *Argopecten purpuratus* cultures in southern Chile. *Aquaculture International*, **8**, 431–442.

24. Littlewood, D.T.J. & Marsbe, L.A. (1990) Predation on cultivated oysters, *Crassostrea rhizophorae* (Guilding), by the polyclad turbellarian flatworm, *Stylochus Stylochus) frontalis* Verrill. *Aquaculture*, **88**, 145–150.

25. Enderlein, P., Moorthi, S., Röhrscheidt, H. & Wahl, M. (2003) Optimal foraging *versus* shared doom effects: interactive influence of mussel size and epibiosis on predator preference. *Journal of Experimental Marine Biology and Ecology*, **292**, 231–242.

26. Lesser, M.P., Shumway, S.E., Cucci, T. & Smith, J. (1992) Impact of fouling organisms on mussel rope culture: interspecific competition for food among suspension-feeding invertebrates. *Journal of Experimental Marine Biology and Ecology*, **165** (1), 91–102.

27. Taylor, J.J., Southgate, P.C. & Rose, R.A. (1997) Fouling animals and their effect on the growth of silver-lip pearl oysters, *Pinctada maxima* (Jameson) in suspended culture. *Aquaculture*, **153**, 31–40.

28. Swift, M.R., Fredriksson, D.W., Unrein, A., Fullerton, B, Patursson, O. & Baldwin, K. (2006) Drag force acting on biofouled net panels. *Aquacultural Engineering*, **35**, 292–299.

29. Baldwin, K.C., Celikkol, B., Swift, M.R., Fredriksson, D. & Tsukrov, I. (2003) Open ocean aquaculture engineering II. *OCEANS 2003*, **3**, 1454–1464.

30. Gonzalez, L. (1998) The life cycle of *Hysterothylacium aduncum* (Nematoda: Anisakidae) in Chilean marine farms. *Aquaculture*, **162**, 173–186.

31. Tan, C.K., Nowak, B.F. & Hodson, S.L. (2002) Biofouling as reservoir of *Neoparamoeba permaquidensis* (Page, 1970), the causative agent of amoebic gill disease in Atlantic salmon. *Aquaculture*, **210** (1–4), 49–58.

32. Morton, S.L., Leighfield, T.A., Haynes, B.L., *et al.* (1999) Evidence of Diarrhetic Shellfish Poisoning along the coast of Maine. *Journal of Shellfish Research*, **18** (2), 681–686.

33. Corner, R.A., Ham, D., Bron, J.E. & Telfer, T.C. (2007) Qualitative assessment of initial biofouling on fish nets used in marine cage aquaculture. *Aquaculture Research*, **38**, 660–663.

34. Norberg, J. (1999) Periphyton fouling as a marginal energy source in tropical tilapia cage farming. *Aquaculture Research*, **30**, 427–430.

35. Rosell, D., Uriz, M-J. & Martin, D. (1999) Infestation by excavating sponges on the oyster (*Ostrea edulis*) populations of the Blanes littoral zone (north-western Mediterranean Sea). *Journal of the Marine Biological Association of the United Kingdom*, **79**, 409–413.

36. Chernoff, H. (1987) Factors affecting mortality of the scallop *Chlamys asperrima* (Lamarck) and its epizooic sponges in South Australian waters. *Journal of Experimental Marine Biology and Ecology*, **109**, 155–171.

37. Pitcher, C.R. & Butler, A.J. (1987) Predation by asteroids, escape response, and morphometrics of scallops with epizoic sponges. *Journal of Experimental Marine Biology and Ecology*, **112**, 233–249.

38. Hodson, S.L., Burke, C.M. & Bissett, A.P. (2000) Biofouling of fish-cage netting: the efficacy of a silicone coating and the effect of netting colour. *Aquaculture*, **184**, 277–290.

39. Lodeiros, C. & García, N. (2004) The use of sea urchins to control fouling during suspended culture of bivalves. *Aquaculture*, **231**, 293–298.

40. Cook, E.J., Black, K.D., Sayer, M.D.J., *et al.* (2006) The influence of caged mariculture on the early development of sublittoral fouling communities: a pan-European study. *ICES Journal of Marine Science*, **63**, 637–649.

41. Ross, K.A., Thorpe, J.P. & Brand, A.R. (2004) Biological control of fouling in suspended scallop cultivation. *Aquaculture*, **229**, 99–116.

42. Lodeiros, C., Galindo, L., Buitrago, E. & Himmelman, J.H. (2007) Effects of mass and position of artificial fouling added to the upper valve of the mangrove oyster *Crassostrea rhizophorae* on its growth and survival. *Aquaculture*, **262**, 168–171.

43. Forrest, B.M. & Blakemore, K.A. (2006) Evaluation of treatments to reduce the spread of a marine plant pest with aquaculture transfers. *Aquaculture*, **257**, 333–345.

44. Olafsen, T. (2005) *Antifouling Strategies in the Norwegian Aquaculture Industry – A Survey*. Report, SINTEF, Trondheim, Norway.

45. Wahl, M., Kröger, K. & Lenz, M. (1998) Non-toxic protection against epibiosis. *Biofouling*, **12**, 205–226.

46. Bers, A.V., D'Souza, F., Klijnstra, J.W., Willemsen, P.R. & Wahl, M. (2006) Chemical defence in mussels: antifouling effect of crude extracts of the periostracum of the blue mussel *Mytilus edulis*. *Biofouling*, **22** (4), 251–259.

47. Harper, E. & Skelton, P. (1993) A defensive value of the thickened periostracum in the Mytiloidea. *Veliger*, **36**, 36–42.

48. Scardino, A., de Nys, R., Ison, O., O'Connor, W. & Steinberg, P. (2003) Microtopography and antifouling properties of the shell surface of the bivalve molluscs *Mytilus galloprovincialis* and *Pinctata imbricata*. *Biofouling*, **19** (Suppl.), 221–230.

49. Bers, A.V. & Wahl, M. (2004) The influence of natural surface microtopographies on fouling. *Biofouling*, **20** (1), 43–51.

50. Thiesen, B.F. (1972) Shell cleaning and deposit feeding in *Mytilus edulis* L. (Bivalvia). *Ophelia*, **10**, 49–55.

51. Southgate, P.C. & Beer, A.C. (2000) Growth of blacklip pearl oyster (*Pinctada margaritifera*) juveniles using different nursery culture techniques. *Aquaculture*, **187**, 97–104.

52. Crawford, C.M., Lucas, J.S. & Nash, W.J. (1988) Growth and survival during the ocean-nursery rearing of Giant Clams, *Tridacna gigas*: 1. Assessment of four culture methods. *Aquaculture*, **68**, 103–113.

53. Enright, C. (1993) Control of fouling in bivalve culture. *World Aquaculture*, **24** (4), 44–46.

54. Bourque, F. & Myrand, B. (2006) Sinking of mussel (*Mytilus edulis*) longlines as a strategy to control secondary set in Îles-de-la-Madeleine. *Aquaculture Canada, AAC Special Publication*, **10**, 64–66.

55. Kaehler, S. & McQuaid, C.D. (1999) Lethal and sub-lethal effects of phototrophic endoliths attacking the shell of the intertidal mussel *Perna perna*. *Marine Biology*, **135**, 497–503.

56. Wesche, S.J., Adlard, R.D. & Hooper, J.N.A. (1997) The first incidence of clionid sponges (Porifera) from the Sydney rock oyster *Saccostrea commercialis* (Iredale and Roughly, 1933). *Aquaculture*, **157**, 173–180.

57. Carver, C.E., Chisholm, A. & Mallet, A.L. (2003) Strategies to mitigate the impact of *Ciona intestinalis* (L.) biofouling on shellfish production. *Journal of Shellfish Research*, **22** (3), 621–631.

58. Leighton, D.L. (1998) Control of sabellid infestation in green and pink abalones, *Haliotis fulgens* and *H. corrugate*, by exposure to elevated water temperatures. *Journal of Shellfish Research*, **17** (3), 701–705.

59. Parsons, G.J. & Dadswell, M.J. (1992) Effect of stocking density on growth, production, and survival of the giant scallop, *Placopecten magellanicus*, held in intermediate suspension culture in Passamaquoddy Bay, New Brunswick. *Aquaculture*, **103**, 291–309.

60. Wildish, D.J. & Kristmanson, D. (1985) Control of suspension-feeding bivalve production by current speed. *Helgoländer Wissenschaftliche Meeresuntersuchungen*, **39**, 237–243.

61. Hidu, H., Conary, C. & Chapman, S.R. (1981) Suspended culture of oysters: biological fouling control. *Aquaculture*, **21**, 189–192.

62. Enright, C., Krailo, D., Staples, L., *et al.* (1983) Biological control of fouling algae in oyster aquaculture. *Journal of Shellfish Research*, **3** (1), 41–44.

63. Minchin, D. & Duggan, C.B. (1989) Biological control of the mussel in shellfish culture. *Aquaculture*, **81**, 97–100.

64. Flimlin, G.E., Jr. & Mathis, G.W., Jr. (1993) Biological biofouling control in a field based nursery for the hard clam, *Mercenaria mercenaria*. *World Aquaculture*, **24** (4), 47–48.

65. Cigarría, J., Fernández, J. & Magadán, L.P. (1998) Feasibility of biological control of algal fouling in intertidal oyster culture using periwinkles. *Journal of Shellfish Research*, **17** (4), 1167–1169.

66. Treasurer, J.W. (1996) Wrasse (Labridae) as cleaner-fish of sea lice on farmed Atlantic salmon in west Scotland. In: *Wrasse: Biology and Use in Aquaculture, Fishing News Books* (eds M.D.J. Sayer, J.W. Treasurer & M.J. Costello), pp. 185–195. Blackwell Science, Oxford.

67. Kvenset, P.G. (1996) Large-scale use of wrasse to control sea lice and net fouling in salmon farms in Norway. In: *Wrasse: Biology and Use in Aquaculture, Fishing News Books* (eds M.D.J. Sayer, J.W. Treasurer & M.J. Costello), pp. 196–203. Blackwell Science, Oxford.

68. Alzieu, C. (1998) Tributyltin: case study of a chronic contaminant in the coastal environment. *Ocean & Coastal Management*, **40**, 23–36.

69. Bruno, D.W. & Ellis, A.E. (1988) Histopathological effects in Atlantic salmon, *Salmo salar* L., attributed to the use of tributyltin antifoulant. *Aquaculture*, **72** (1–2), 15–20.

70. Davies, I.M., McKie, J.C. & Paul, J.D. (1986) Accumulation of tin and tributyltin from antifouling paint by cultivated scallops (*Pecten maximus*) and Pacific oysters (*Crassostrea gigas*). *Aquaculture*, **55** (2), 103–114.

71. Short, J.W. & F.P. Thrower (1986) Accumulation of butyltins in muscle tissue of Chinook salmon reared in sea pens treated with tri-n-butyltin. *Marine Pollution Bulletin*, **17** (12), 542–545.

72. Hao, Y., Sun, B. & Zhu, C. (1990) The characteristics and control of fouling organisms on the lantern nets for cultivating scallops (*Pecten maximus*) in Penglai and Changdao cultivating farms. *Journal of Oceanography of Huanghai and Bohai Seas*, **8** (1), 57–62.

73. Bailey, S.K. & Davies, I.M. (1991) Continuing impact of TBT, previously used in mariculture, on dogwhelk (*Nucella lapillus* L.) populations in a Scottish sea loch. *Marine Environmental Research*, **32** (1–4), 187–199.

74. Hung, T.C., Hsu, W.K., Mang, P.J. & Chuang, A. (2001) Organotins and imposex in the rock shell, *Thais clavigera*, from oyster mariculture areas in Taiwan. *Environmental Pollution*, **112** (2), 145–152.

75. Murai, R., Takahashi, S., Tanabe, S. & Takeuchi, I. (2005) Status of butyltin pollution along the coasts of western Japan in 2001, 11 years after partial restrictions on the usage of tributyltin. *Marine Pollution Bulletin*, **51**, 940–949.

76. Tanabe, S., Prudente, M.S., Kan-atireklap, S. & Subramanian, A. (2000) Mussel watch: marine pollution monitoring of butyltins and organochlorines in coastal waters of Thailand, Philippines and India. *Ocean & Coastal Management*, **43**, 819–839.

77. SEPA (2003) *Regulation and Monitoring of Marine Cage Fish Farming in Scotland – A Procedures Manual*. Scottish Environmental Protection Agency, Stirling, UK.

78. Braithwaite, R.A., Cadavid Carrascosa, M.C. & McEvoy, L.A. (2007) Biofouling of salmon cage netting and the efficacy of a typical copper-base antifoulant. *Aquaculture*, **262**, 219–226.

79. Douglas-Helders, G.M., Tan, C., Carson, J. & Nowak, B.F. (2003) Effects of copper-based antifouling treatment on the presence of *Neoparamoeba pemaquidensis* Page, 1987 on nets and gills of reared Atlantic salmon (*Salmo salar*). *Aquaculture*, **221** (1–4), 13–22.

80. Ellingsen, H. & Aanondsen, S.A. (2006) Environmental impacts of wild caught cod and farmed salmon – a comparison with chicken. *The International Journal of Life Cycle Assessment*, **1**, 60–65.

81. Solberg, C.B., Saethre, L. & Julshamn, K. (2002) The effect of copper-treated net pens on farmed salmon (*Salmo salar*) and other marine organisms and sediments. *Marine Pollution Bulletin*, **45** (1–12), 126–132.

82. Davies, I.M. & Paul, J.D. (1986) Accumulation of copper and nickel from antifouling compounds during cultivation of scallops (*Pecten maximus* L.) and Pacific oysters (*Crassostrea gigas* Thun.). *Aquaculture*, **55** (2), 93–102.

83. Sandberg, M.G. & Olafsen, T. (2006) *Overview of Laws and Regulations Regarding Antifouling Methods in Fish Farming*. SFH80 AO66001 – Open Report, SINTEF, Trondheim, Norway.

84. Svane, I., Cheshire, A. & Barnett, J. (2006) Test of an antifouling treatment on tuna fish-cages in Boston Bay, Port Lincoln, South Australia. *Biofouling*, **22** (4), 209–219.

85. AMBIO (2005–2009) *Advanced Nanostructured Surfaces for the Control of Biofouling*. EU FP6 NMP4-CT-2005–011827. Available from http://www.ambio.bham.ac.uk/. Accessed 6 July 2009.

86. de Nys, R. & Ison, O. (2004) *Evaluation of Antifouling Products Developed for the Australian Pearl Industry*. Fisheries Research and Development Corporation Project No. 2000/254. James Cook University, Townsville, Australia.

87. Brown, C.J., Eaton, R.A., Cragg, S.M., *et al.* (2003) Assessment of effects of chromated copper arsenate (CCA)-treated timber on nontarget epibiota by investigation of fouling community development at seven European sites. *Archives of Environmental Contamination and Toxicology*, **45**, 37–47.

88. Mazouni, N., Gaertner, J.-C. & Deslous-Paoli, J.-M. (2001) Composition of biofouling communities on suspended oyster cultures: an *in situ* study of their interactions with the water column. *Marine Ecology Progress Series*, **214**, 93–102.

89. Sarà, G., Lo Martire, M., Buffa, G., Mannino, A.M. & Badalamenti, F. (2007) The fouling community as an indicator of fish farming impact in Mediterranean. *Aquaculture Research*, **38**, 66–75.

90. Khalaman, V.V. (2001) Succession of fouling communities on an artificial substrate of a mussel culture in the White Sea. *Russian Journal of Marine Biology*, **27** (6), 345–352.

91. LeBlanc, N., Landry, T. & Miron, G. (2003) *Identification of Fouling Organisms Covering Mussel Lines and Impact of a Common Defouling Method on the Abundance of Foulers in Tracadie Bay, Prince Edward Island*. Report no. 2477 0706-6457, Canadian Technical Reports in Fisheries and Aquatic Sciences, Moncton, Canada.

92. Heffernan, P.B., Walker, R.L., Crenshaw, J.W., Hoats, J. & Vaughan, D.E. (1988) Observations on grow out systems for the southern bay scallop, *Argopecten irradians concentricus*. *Journal of Shellfish Research*, **7** (3), 547.

93. Omori, M., Kubo, H., Kajiwara, K., Matsumoto, H. & Watanuki, A. (2006) Rapid recruitment of corals on top shell snail aquaculture structures. *Coral Reefs*, **25**, 280.

94. Chaplygina, S.F. (1994) Hydroids in the fouling of mariculture installations in Peter the Great Bay, Sea of Japan. *Russian Journal of Marine Biology*, **19** (2), 85–89.

95. Forrest, B.M., Hopkins, G.A., Dodgshun, T.J. & Gardner, J.P.A. (2007) Efficacy of acetic acid treatments in the management of marine biofouling. *Aquaculture*, **262**, 319–332.

96. Hodson, S.L., Lewis, T.E. & Burke, C.M. (1997) Biofouling of fish-cage netting: efficacy and problems of *in situ* cleaning. *Aquaculture*, **152**, 77–90.

97. Pit, J.H. & Southgate, P.C. (2003) Fouling and predation; how do they affect growth and survival of the blacklip pearl oyster, Pinctada margaritifera, during nursery culture. *Aquaculture International*, **11** (6), 545–555.

98. Alvarez-Tinajero, del C., Càceres-Martínez, J. & Gonzàlez-Avilís, J.G. (2001) Shellboring clams in the blue abalone *Haliotis fulgens* and the yellow abalone *Haliotis corrugata* from Bahia California, Mexico. *Journal of Shellfish Research*, **20** (2), 889–893.

99. Caceres-Martinez, J. & Tinoco-Orta, G.D. (2001) Symbionts of cultured red abalone *Haliotis rufescens* from Bahia California, Mexico. *Journal of Shellfish Research*, **20** (2), 875–881.

100. Bloom, S.A. (1975) Motile escape response of a sessile prey, a sponge – scallop mutualism. *Journal of Experimental Marine Biology and Ecology*, **17**, 311–321.

101. Leighton, D.L. (1998) Control of sabellid infestation in green and pink abalones, *Haliotis fulgens* and *H. corrugate*, by exposure to elevated water temperatures. *Journal of Shellfish Research*, **17** (3), 701–705.

102. Vance, R.R. (1978) A mutualistic interaction between a sessile marine clam and its epibionts. *Ecology*, **59** (4), 679–685.

103. Uribe, E. & Etchepare, I. (2002) Effects of biofouling by *Ciona intestinalis* on suspended culture of *Argopecten purpuratus* in Bahia Inglesa, Chile. *Bulletin of the Aquaculture Association Canada*, **102** (3), 93–95.

104. Guenther, J., Southgate, P.C. & De Nys, R. (2005) The effect of age and shell size on accumulation of fouling organisms on the Akoya pearl oyster *Pinctada fucata* (Gould). *Aquaculture*, **253** (1–4), 366–373.

105. Wesche, S.J., Adlard, R.D. & Hooper, G.N.A. (1997) The first incidence of clionid sponges (Porifera) from the Sydney rock oyster *Saccostrea commercialis* (Aredale & Roughley, 1933). *Aquaculture*, **157**, 173–180.

106. Cigarria, J., Fernandez, J. & Magadan, L.P. (1998) Feasibility of biological control of algal fouling in intertidal oyster culture using periwrinkles. *Journal of Shellfish Research*, **17** (4), 1167–1169.

107. O'Connor, S.J., Heasman, M.P. & O'Connor, W.A. (1999) Evaluation of alternative suspended culture methods for the commercial scallop, *Pecten fumatus* Reeve. *Aquaculture*, **171** (3–4), 237–250.

108. Evans, J.W. (1969) Borers in the shell of the sea scallop, *Placopecten magellnnicus*. *American Zoologist*, **9** (3), 775–782.

109. Athanas, S.B. & Rouse, D.B. (1996) Incidence of fouling at two mariculture sites in Bon Secure Bay, Alabama. *Journal of Shellfish Research*, **15** (2), 524.

110. Narvarte, M.A. (2003) Growth and survival of the tehuelche scallop *Aequipecten tehuelchus* in culture. *Aquaculture*, **216**, 127–142.

Chapter 20

Fouling and Antifouling in Other Industries – Power Stations, Desalination Plants – Drinking Water Supplies and Sensors

Peter Henderson

This chapter reviews biofouling of water intakes, desalination plants and power stations by microorganisms, plants and animals. The main groups of fouling organisms are introduced and the serious effects these organisms can have are summarised. There is a small number of particularly problematical, widely distributed, fouling organisms, which are identified and discussed. While there are many methods to control fouling, the most important of which are reviewed below, these are often expensive and harmful to the environment. It is argued that the key to minimising biofouling problems is good plant design, which limits scope for fouling organisms to colonise and allows efficient application of antifouling methods. Good design requires an understanding of local ecology, the potential fouling organisms present and their biology. Consideration of water extraction systems with biofouling problems often reveals that they were designed without any analysis of the local biofouling threats and the appropriateness of the antifouling methods installed. Finally, the requirement on all plant operators for continual vigilance is discussed. It is common for plant operators to be surprised by the development of a new fouling problem. Over the last 50 years, there has probably been a rise in the frequency of new biofouling problems as movements of ships aids the dispersal of organisms and environmental change allows some fouling species to rapidly increase their range and abundance. This trend is likely to continue.

20.1 Microfouling organisms of water intakes and cooling systems

Bacteria and other microorganisms rapidly colonise any surface placed in water (see Chapters 4 and 7). The microbial community that develops depends on the physicochemical proprieties of both the water and the surface. The smoothness of the surface and its chemical properties affects the rate of attachment and development of microbial colonies. Many materials including plastics and metals will initially leach toxic compounds that inhibit colonisation. Colonisation progresses via the acquisition of colonisers resistant to any toxins present; these resistant colonisers then reduce the release of toxins allowing non-resistant forms to colonise.

By far the most important microbial fouling communities are termed microbial slimes formed by communities of *Pseudomonas*, *Flavobacterium* and *Bacillus* bacteria [1]. Bacterial slimes are problematical in condensers where they can reduce heat exchange efficiency. They also form on the packing in cooling towers and can cause operational problems when they result in the accumulation of silt or scale increasing the weight to the point where the pack collapses. The changeover from wood and asbestos to plastic packing at British power stations cooling towers in the 1980s did not bring the benefits in heat exchange expected because of the rapid development of a slimy biofilm [2]. At Cottam power station, the retrofitting of a Munters C10/19 pack resulted in a biofilm growth of 2.5 kg m^{-2} day^{-1} giving an increase in weight over a 3-month period of 55.2 kg m^{-3}. This example demonstrates the need for biofouling tests before new materials are brought into general use [2].

Biofilms are particularly problematical in desalination plants where they form on the high-pressure filtration membranes [3]. A study of the biofilms on nanofiltration membranes found the community to be dominated by *Flavobacterium* sp., *Pseudomonas*, *Ralstonia* and *Cytophaga* bacteria over a range of conditions; protozoa were also present [4]. Additional operational problems caused by microbial fouling include damage to sensors, fouling of fine filters and the blocking of aeration and bubble curtain systems. The rate of development of slimes depends on temperature, nutrient availability and suitable surface area [5]. Heavy slime growth requires flowing water to supply nutrients and oxygen and remove waste products. However, at high flows a thick biofilm cannot develop as it is continuously sheared from the surface. Where light is present, sheet-like biofilms can develop. These have been found to grow within cooling towers resulting in the shedding of sheets when the tower is taken out of commission for maintenance [6, 7].

20.2   Macrofouling organisms

20.2.1   *Passive clogging of filter screens and pipes*

Whenever water is pumped into pipework, there is a risk that material in the water will physically block the pipework. To protect the plant it is common practice to install filter screens, which can themselves become fouled. Filter screens are frequently fouled by rubbish such as plastic bags, bottles, rope and driftwood, which will not be considered further as this chapter focuses on problems generated by the living world other than humans. The natural world does, however, produce inert debris which in many ways acts like anthropogenic rubbish.

Inert plant material including leaves, sticks, tree trunks and seaweed is normally pulled onto filter screens. It is not unusual for filter screens to be blocked by plant material, particularly after unusual weather events. For example, marine seaweed ingress into power station intakes can become problematical after gales when large quantities of weed can become detached from rocks in the intertidal in the vicinity of the intakes. The following quotation is typical

Millstone Unit 1, October 4, 1990. The licensee, Northeast Nuclear Energy Company, manually tripped the reactor from 45 percent of full power because of circulating water system and service water system fouling that resulted in degraded SWS cooling, which resulted in increased

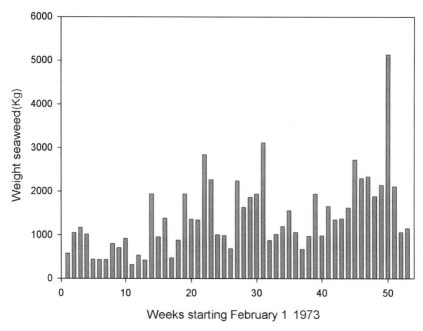

Figure 20.1   The seasonal pattern of weed ingress at Fawley Power Station, England, 1973–1974. The wet weight of seaweed caught on the drum screens for each 7-day period is plotted. The majority of the weed at this site comprised *Griffithsia flosculosa* and *Ulva* species. (Data from Reference [9].)

containment temperature and pressure. Storm-induced high winds and seas caused an excessive amount of seaweed to accumulate on the travelling screens of the circulating water system. [8]

Clogging by plant debris is seasonal and to a great extent can be anticipated. A typical pattern is the seaweed ingress at Fawley Power Station shown in Figure 20.1 [9]. The minimum of 313 kg week^{-1} was recorded in April and a maximum of 5133 kg week^{-1} in January. The high between-week variability is typical and can be related to wind strength and direction, sea state and the spring–neap tidal cycle.

While macroalgae such as kelp or *Fucus* frequently contribute the greatest impinged mass on filter screens at marine and estuarine intakes, they may not be the most problematical plants. For small mesh screens and wedgewire filters, plants in the form of long strands can wrap round the screen and become difficult to wash off. For example, in marine habitats, seagrass debris has blocked wedgewire screens. Many freshwater algae and higher plants also produce long strands that have been found to block intakes. At power stations, filamentous strands able to pass through the filter screens can cause a blockage at the condenser end plates where they bridge two condenser tubes and cause a build-up of material. *Griffithsia* spp. have been known to form mats across condenser end plates.

Animals can also act like plants on filter screens. For example, the hydrozoan *Sertularia argentea* has a branching form and grows on shelly sand in the outer reaches of estuaries. It is so plant-like that it is commonly called white weed. In the autumn, it sheds side branches and this free material can block intakes, and on one occasion caused a sudden shutdown of Tilbury power station, London, by suddenly covering five band screens in a dense mat [10].

Other animals that have blocked filter screens include jellyfish, starfish and fish. Clogging incidences with jellyfish seem to be increasing and are particularly commonly reported from the Pacific region. For example, in July 2006, power from two reactors at Chubu Electric Power Co.'s nuclear plant in Hamaoka, Japan, had to be reduced after the sea water intake system shut down automatically because of jellyfish clogging. Screen clogging by fish occurs at localities where intakes have been placed in waters through which large shoals of fish pass. A particularly serious example of fish inundation occurred at Dungeness Nuclear Power Station, England, where a number of emergency shutdowns occurred following the loss of condenser cooling water when the band screens were blocked and damaged by the weight of sprats, *Sprattus sprattus*, sucked into the intakes [11]. Massive ingresses resulting in clogging with fish are only likely to occur in regions seasonally occupied by shoaling fish and are most likely to occur with highly abundant members of the herring family such as sprat, sardine or shad.

20.2.2   Fouling by attachment

Animals and plants that readily colonise man-made hard surfaces such as those in screens, culverts, pipework and cooling towers cause macrofouling. The potential range of colonising organisms at any site can usually be anticipated by examining local hard surfaces such as rocks or jetty pilings. Most man-made surfaces vulnerable to fouling are in dark or low light areas so that algal fouling is rarely important. Typical fouling organisms enter the system as larval or juvenile stages, which attach to a surface. They obtain their food from the water flowing through the system.

20.2.3   Principle freshwater fouling species

In water extraction and cooling systems, where sufficient light is available, algae can cause appreciable fouling. Blanket weed, *Cladophora glomerata*, has fouled the packing in evaporative cooling towers resulting in reduced flow and heat exchange [2]. The growth of *C. glomerata* became particularly problematical in the cooling towers at High Marnham Power Station, England, where the weight of algae on this material became so great that it caused the pack to collapse (Plate VI D). Algae can also grow on fine screens.

Colonial bryozoans such as *Plumatella* spp. may form colonies on screens, culverts and cooling towers. Generally, bryozoans are most abundant in waters with low sediment loads.

Bivalve molluscs such as the zebra mussel, *Dreissena polymorpha*, the quagga mussel, *Dreissena bugensis*, and the Asiatic clam, *Corbicula fluminea*, are amongst the most damaging fouling organisms in freshwater (for information on zebra mussel see [12] and for quagga mussel see [13]). The zebra mussel has been the subject of extensive study. This species is a particularly effective fouling organism because of the planktonic larval stage and the ability of adults to continue to make byssus threads. These threads anchor the animal to a surface and zebra mussels are able to attach as adults. Following their recent introduction, zebra mussels have become so abundant and dominant in habitats such as the Great Lakes and the Hudson River that they have had a major impact on the functioning of ecosystem (see [14] for information on the zebra mussel invasion of the River Hudson). Zebra mussels are also able to withstand the low salinities in upper estuarine tidal freshwaters. Unlike zebra mussels, adult Asiatic clams are not attached to the substratum and the shells can become entrained into cooling water intakes. The maximum size of the shells is about 60 mm in the USA and 40 mm in France [15].

In areas of low flow or within fine filters, colonies of the tube-building amphipod *Corophium* spp. can develop. *Corophium* colonies can also cause fouling in estuarine waters.

20.2.4   Principal marine and estuarine fouling species

Hydroids can form dense colonies on suitable surfaces and are most likely to be problematical in estuarine waters. In Europe, *Cordylophora caspia* is the main fouling species [1]. Colonies develop in brackish waters of between 0 and 10 salinity and are propagated asexually by budding. Hydroids are often early colonisers of new surfaces.

The tubes of serpulid polychaete worms are frequently observed encrustations. *Ficopomatus enigmaticus* is the species most frequently recorded as a fouling species. It is a euryhaline species able to tolerate salinities between 1 and 35 and lives in a calcareous tube about 20 mm long and 2 mm in diameter. These tubes can form into a dense mass 10 cm deep. This species is believed to have originated in the southern hemisphere but is now widespread and is probably dispersed by ships.

Barnacles are sessile crustaceans that may settle on almost any hard surface in sea water. Their cypris larvae are seasonally one of the most abundant members of the coastal plankton. These larvae attach by their head to a hard surface and develop the familiar barnacle form. Barnacles can rapidly colonise clean surfaces placed in the sea when cypris larvae are abundant during the spring or summer. The prevention of barnacle fouling is discussed by Christie and Dalley [16].

A number of species of mussel will colonise water extraction systems and if not checked will form massive colonies resulting in serious operational effects. They particularly favour concrete culverts. The young stages are planktonic and newly settled spat can reach natural densities in the range 1 to 6×10^5 m^{-2} [17]. Spat settlement is seasonal and varies between localities. Mussel growth is determined by water temperature and food; they can reach 20 mm or more in shell length within 6 months of settlement. While living mussel colonies can reduce flow, the dead mussel shells from fouled culverts can also be damaging. In power stations, these can pass across the filter screens and subsequently lodge in the condenser tubes where they cause erosion failure. Oysters can also attach to culverts and pipes. Generally, oysters pose less of an operational problem than mussels.

Tunicates are soft-bodied animals that are an important component of many fouling communities. They can grow on wedgewire screens and other fine mesh filters reducing flow. Within culverts and pipework, they are rarely an operational problem because of their soft bodies. A number of introduced species have become damaging fouling organisms of buoys and fish farming structures.

20.2.5   The development of fouling communities

Like all biological systems, fouling communities develop by the arrival of a succession of species. During the initial phase, the clean surface is occupied by early colonisers, which are able to attach to a clean surface. As the community develops and forms a three-dimensional structure, climax community species arrive. Some of these climax forms are fouling organisms that are particularly able to compete in a crowded environment, while others are mobile forms utilising the space and resources offered by the attached organisms. The existence of an ordered species succession does not mean that the final form of a fouling community is fixed. The actual fouling sequence will be determined by the surface and also by the time of year at which the surface was first made available for colonisation. For example, if a hard surface is placed in sea water during spring, when barnacle cypris larvae are highly abundant, almost the entire surface may be rapidly colonised by barnacles which subsequently restrict and to some extent determine the entry of other species. At another time of year, mussels may be the initial colonisers that influence subsequent development.

It is frequently assumed that fouling commences with the arrival of microorganisms conditioning the surface prior to the arrival of high organisms such as barnacles or molluscs. This does not seem to be essential, as barnacles and mussels have been observed to attach to clean surfaces.

As a first example of a typical fouling sequence, the initial colonisation of a geotextile filter drilled with 1 mm holes and placed in an estuarine environment is described [18]. Panels of the filter were placed in Haverstraw Bay in the Hudson estuary, New York. Water temperature was about 80°F, salinities ranged between 0.1 and 10 and dissolved oxygen concentration was 6–8 mg L^{-1}. The fouling community was examined after 11, 20 and 29 days. A summary of the fouling sequence is given below.

Day 11: the fabric had been colonised by *Corophium* and *Gammarus* spp. with about 5% of the surface showing evidence of colonisation. Tube-building *Corophium* spp. had colonised the 1 mm holes in the fabric. Many holes were completely filled. Smaller *Corophium* used part of a hole as a base for building a tube. Mean bacterial count, 7000 cells mL^{-1} of water washed from the fabric. Rod-shaped bacterial count 4806 mL^{-1}. Zebra mussel count per panel $= 0$.

Day 20: the fabric had started to be colonised by several additional organisms including zebra mussels, chironomids and small amounts of filamentous algae. Approximately 30% of the fabric surface showed evidence of colonisation. *Corophium* continued to colonise the 1 mm holes and their tubes were also widely dispersed over the surface of the panels. These surface tubes were bound under the outer filaments of the fabric and the surface of the fabric was becoming looser. Some chironomid tubes were found. Copepods were observed moving across the surface of the fabric. Mean bacterial count, 220 000 cells mL^{-1} of water washed from the fabric. Rod-shaped bacterial count 3922 mL^{-1} of water. Zebra mussel count per panel $= 3.5$.

Day 29: obvious colonisation had increased to 70% of the available surface area. Many *Corophium* surface tubes were observed and most of the 1 mm holes in the fabric were occupied. Holes were now only occupied by large *Corophium*, which completely filled the holes with their tubes. A large number of chironomid tubes were present on the upper panels exposed at a depth of approximately 1 m. The community had increased in diversity to include several predatory organisms including ostracods and ciliates. Other groups included the vorticellids, hydroids and bryozoans. Mean bacterial count, 2 020 000 cells mL^{-1} of water washed from the fabric. Rod-shaped bacterial count 10 006 mL^{-1} of water. Zebra mussel count per panel $= 16.33$.

An important point to note about the above fouling sequence is that the microbial and macrofouling communities are both developing over time. Bacterial counts increase through time and some protozoans are not recorded before day 29. Early colonisers include crustaceans and insects and zebra mussels were detected at day 20.

The second example is from a study of the long-term colonisation of a variety of panels placed in sea water at the entrance to Southampton Water, England [19]. The tests were designed to study the antifouling properties of the Johnson wedgewire screen 715 alloy which leaches copper. The study was undertaken between September 1982 and January 1984. Table 20.1 shows a summary of the observed fouling development over a 16-month period. Initial colonisation was by benthic diatoms, which formed a slimy layer, rarely more than 1 mm thick. These were succeeded during the winter by seaweeds, which did not grow on the 715 alloy. During the following warm summer months, animal colonisers replaced the plants. The two main features of this successional study are that copper leaching from the surface had a major impact on the fouling community and, secondly, the timing of the succession from plants to animals was determined by the seasonality. The colonisation by the main animal groups was delayed from September until the following May when the main period of animal settlement commenced.

20.3   Consequences of fouling for plant operation and safety

The economic and safety costs linked to fouling are considerable and fouling issues need to be considered during the design of all water extraction systems. The main operational problems can be categorised as (1) sudden loss of water supply, (2) progressive reduction in flow, (3) erosion damage, (4) corrosion damage, (5) loss of heat transfer efficiency, (6) loss of monitoring data and (7) risk of infection with pathogens and reduced safety in the work place. Additional problems, which can be categorised under the heading of antisocial effects, include the disposal of biofouling organisms, smell and discharge pollution effects linked to fouling. All of the above will be considered in turn.

20.3.1   Sudden loss of water supply

When intakes or filter screens become suddenly blocked with large quantities of plants or animals, a sudden loss of water supply can occur resulting in an emergency shutdown of the plant. Globally, this is a frequent occurrence at large direct-cooled power stations using marine water. In addition to the immediate loss of power generation, these blockages often also cause damage to screens, pumps and electric motors driving drum and band screens. Safety systems can also be degraded, for example it has been known for the firefighting water supply to be lost through fouling.

20.3.2   Progressive reduction in flow

The presence of large quantities of organisms such as mussels on the walls of culverts and tubes can increase the head loss across pumps and reduce water flow. Fouling can also block or jam valves and gates so that it becomes impossible to isolate pipes or culverts. Mussel fouling is particularly prevalent in marine cooling water culverts where the intake is situated offshore but the injection of chlorine for biofouling control occurs downstream at the filter screens

Table 20.1   Summary of fouling development of panels placed in Southampton Water [19].

Date	Johnson 715 alloy	Other panels made of aluminium, stainless steel, shot-blasted Tufnol and not shot-blasted Tufnol
September 82		
October 82		
November 82	Diatomaceous slime developing	Diatomaceous slime developing
December 82		
January 83		Red weeds (*Griffithsia*)
Februray 83		Weed growth becoming denser; *Laminaria* settling
March 83		
April 83		90% weed cover: *Laminaria* growing fast, *Ulva* appearing
May 83		
June 83	Few hydroids	Animal fouling developing, dense colonies of hydroids, bryozoans, tunicates; weeds dying back. Crevices blocking
July 83	More hydroids	
August 83		
September 83		
October 83	Single serpulid worm and growing tufts of *Griffithsia*	Dense fouling cover on all panels. Only *Griffithsia* remains of the weeds
November 83		
December 83	Additional serpulids (*Hydroides*)	Barnacles, serpulids and encrusting bryozoans apparent
January 84		

Note: The study was undertaken to compare the fouling resistance of a Johnston wedgewire alloy 715 against a range of other common materials.

onshore. This extraordinarily incompetent design is commonly encountered at recently built power stations.

20.3.3   Erosion damage

Damage to condenser tubes is frequently caused by the jamming of mussels and other hard-bodied fouling organisms in the tubes. This results in increased water velocity and erosion of the tube wall. Often it is found that the erosion is caused by dead shells probably derived from the fouling community growing on the intake structure or within the intake culverts. Condenser tube leaks in power stations result in the salt contamination of the boiler-feed water resulting in corrosion of the boiler tubes and damage to turbine blades.

20.3.4   Corrosion damage

Microbial slimes increase metal corrosion rates by creating anaerobic conditions under the slime. There is also a tendency for macrofouling communities to generate increased localised corrosion.

20.3.5   Loss of heat transfer efficiency

It is well known that the development of microbial slimes in condenser tubes and heat exchangers results in reduced heat exchange.

20.3.6   Loss of monitoring data

Fouling can result in the malfunctioning of pressure, corrosion or temperature probes placed within cooling water systems.

20.3.7   Risk of infection with pathogens and reduced safety in the work place

Microbial fouling can also include dangerous pathogens, including the bacteria *Legionella pneumophila* and the protozoan *Naegleria fowleri*. *L. pneumophila* is the causative agent of deadly Legionnaires disease. Both of these organisms are associated with water-cooling systems as they are thermophilic. *N. fowleri* has been detected in lakes receiving the heated discharge of power stations [20] and steel works [21]. Special care is needed when entering heavily fouled culverts, cooling towers and other structures that have been drained for maintenance. In addition to the risks from inhalation of microorganisms, there are also infection risks from rotting organisms. Fouling often produces slippery surfaces and steps and increases the risk of general workplace accidents.

20.3.8   Antisocial effects

Fouling of screens and the cleaning of fouling organisms of culverts and cooling towers can generate a waste disposal problem. Under some circumstances such as when a system is drained down, the rotting organism can produce unacceptable smells and potentially dangerous levels of hydrogen sulphide gas. Finally, the destruction of large numbers of fouling organisms or animals sucked into intakes can result in the discharge of water enriched with dead organic matter. In exceptional cases, this has been found to produce excessive foaming of discharges.

20.4   Control and mitigation of fouling

Control methods can be categorised under (1) physical, (2) mechanical and (3) chemical methods. In water extraction systems, antifouling paints and surface treatments are generally not practical and are not considered here.

20.4.1   Physical methods

Potentially, the simplest control method is to use the flow of the water to suppress fouling growth. Many organisms cannot settle and attach above a certain flow rate and it is recommended that flow in the main cooling water circuits of power stations should always be in excess of 1.5 m s^{-1} [22]. It is frequently found that mussels and barnacles grow in corners and other dead zones within culverts. Water extraction systems should be designed to minimise regions of low flow.

The control of macrofouling including mussels by heat treatment has been successfully practiced for more than 50 years at power stations. Fouling organisms in the cooling water circuits are killed by re-circulating the cooling water until a lethal temperature exposure time is achieved. The method has been used routinely about every 6 weeks by Californian direct sea water-cooled power stations. This method is less frequently used in Europe, but one exception is at Ems power station in the Netherlands. At Ems, three to four heat treatments a year are usually undertaken. The first, at the end of May, kills the spring settlement of barnacles, the second and third in July and August kill mussel spat and a final treatment in the autumn may be used to clean the system for the winter [22].

UV light can be used to kill microorganisms in drinking water, but is not used for biofouling control because in most waters the suspended solid levels are too high to allow sufficient light penetration.

20.4.2   Mechanical methods

Microbial slime in tubes can be removed by passing sponge rubber balls with an abrasive strip along the pipework. Once heavy macrofouling by mussels has occurred, the only way to clean culverts is by manual scraping.

Wedgewire screens and fine mesh filters are typically cleared of clogging debris and lightly attached animals and plants using airbursts. These methods are usually only effective when the screens are placed in flowing waters so that the material displaced from the screen surface is swept away with the current. Some protection of intakes from clogging with jellyfish and other aquatic organisms can be provided at suitable intakes by the use of bubble curtains and booms in front of the intakes.

20.4.3   Chemical methods

20.4.3.1   Chlorination

Chlorination is the most commonly used antifouling technique for reasons of effectiveness and cost. It is used to control both microbial slimes and macrofouling. Chlorination of water is achieved by (1) dissolution of chlorine gas, (2) addition of sodium hypochlorite solution or (3) at marine sites by the generation of sodium hypochlorite by electrolysis. The danger of accidents during transportation has acted against the use of chlorine gas.

When added to water, chlorine reacts to produce a wide range of oxidising compounds, many of which are short-lived and difficult to measure. To measure the amount of active chlorine in water, it is common practice to measure the oxidant capacity expressed as total

Table 20.2 Concentrations of bromoform from the chlorinated discharges of European power stations [23].

Power station	Bromoform (μg L^{-1})
Heysham	23–29.2
Dunguness	5.75
Wylfa	27–27.5
Bradwell	25
Hartlepool	3.5
Sizewell	14.5
Penly	13.4–15.0
Paluel	3.1–9.65
Gavelines	18.6
Maasvlakte	8.4–11.5

residual oxidant (TRO). This measure is more useful than total residual chlorine as in saline waters chlorination produces brominated oxidants.

When chlorine is added to sea water, the bromide is oxidised to hypobromous acid:

$$Br^- + OCl^- \rightarrow Cl^- + OBr^-$$

so that in sea water free oxidants are predominantly composed of hypobromous acid. When ammonia levels are high relative to the chlorine dose added, chlorine reacts with ammonia to produce monochlorororamine (NH_2Cl) [23]. The hypobromous acid reacts with NH_2Cl producing dibromamine and a subsequent range of bromine-based compounds. Experimental data have shown that bromoform ($CHBr_3$) is the major organohalogenated compound finally produced by chlorination. Concentrations of bromoform at a number of European power stations are presented in Table 20.2.

In freshwaters, the addition of chlorine or hypochlorite results in the almost instantaneous production of an equilibrium mixture of hypochlorous acid (HOCl) and hypochlorite ions.

$$Cl_2 + H_2O \leftrightarrow HCl + HOCl$$
$$HOCl \leftrightarrow H^+ + OCl^-$$

Of the compounds produced, HOCl is the most effective biocide. The hypochlorous acid reacts with nitrogenous compounds present in the water to potentially produce a wide range of chlorinated compounds. Of particular importance is the reaction with ammonia which is present in low concentrations in most freshwaters as it is excreted by fish and many other active organisms. The chlorine sequentially substitutes the hydrogen atoms producing monochloramine, dichloramine and finally trichloramine.

Because of the rapid reaction of chlorine with dissolved compounds, it is common practice to define the chlorine demand as the difference between the amount of chlorine added and the residual chlorine (actually oxidant) remaining after a specified period. It is the TRO that measures the biocidal, antifouling, ability of the water.

For the control of slimes, regular 'shock' injections of chlorine are used. Effective treatment may require the addition of chlorine to produce a TRO of between 2 and 10 mg L^{-1} at the point of injection every 4–8 hours [22]. However, in a study in Lake Erie between July and November, continuous low-level chlorination with a TRO as low as 0.1 mg L^{-1} was found to be effective. In this study, biofilm growth was not inhibited by intermittent treatment for 30 minutes every 12 hours at dosages of between 0.5 and 1.5 mg L^{-1}.

Considerable attention has been paid to the use of chlorination for the control of freshwater and marine molluscs and mussels in particular [1]. Both low level continuous and higher concentration pulsed chlorination are used. The effectiveness of chlorination is highly temperature dependent. To obtain a 95% mortality of zebra mussels with a TRO of 0.5 mg L^{-1} required chlorination for 42 days at 10°C but only 7 days at 25°C [24]. In saline waters, mussels are best controlled by continuous low-level chlorination at TRO concentrations of less than 1.0 mg L^{-1}. It is believed that mussel fouling is suppressed by two processes: (1) reduction in feeding and (2) chronic toxicity through chemicals absorbed from the water [25]. It has generally been found that a chlorination treatment regime that suppresses mussel fouling will also suppress other macrofouling organisms such as barnacles.

20.4.4   *Copper*

Copper in a number of forms has been used to control macrofouling. Wedgewire screens can be constructed from cupro-nickel alloys which leach copper and suppress biofouling. The dosing of cooling water with copper ions using electrolysis systems has also been used and has not generally found favour because of environmental concerns. In a review of the available methods for controlling biofouling of optical sensors placed in the sea, it was concluded that copper-based methods were superior to previously used methods based on highly toxic compounds [26].

A combination of low-level copper (5 µg L^{-1}) and chlorine (50 µg L^{-1}) dosing has been found to be effective for the control of both microbial and macrofouling at desalination plants in the Persian Gulf [27].

20.4.5   *Other methods*

In some situations, ozone disinfections can be used to maintain control of microbial populations. When added to sea water, ozone reacts with bromide ions so the chemical effect is similar to the addition of chlorine. Cloete *et al.* [28] review the oxidising and non-oxidising chemical treatments that can be used in industrial plant to remove biofilms. In closed systems, where the cleaning water will not be released to the environment detergents can be used. In some smaller scale applications enzymes may also be practical. Ultrasonic cleaning methods have been tested with limited success [29]. These methods are probably most applicable for sensor cleaning. Abarzua and Jakubowshi [30] review potential biogenic and biochemical control agents.

20.5   Consequences of treatments to the environment

Both heat and chemical biofouling treatments impact the environment. Heat treatment can result in the release in the cooling water discharge of warm water that will kill organisms in the vicinity of the discharge. This impact will be local and is not considered further. When chlorination is used in once-through power station cooling water systems, the discharge is warmed by typically 10°C and chlorinated. In practice, it can be difficult to determine if the damage observed in local organisms is caused by the chlorination or the heat or, most likely, a combination of the two.

Following chlorination, the amount of TRO in water declines through time, typically in an approximately exponential fashion. In freshwater chloroform, the main halogenated compound formed, disappears mainly by passing into the atmosphere with a half-life of between 36 hours for a large river and 9–10 days for a lake [1]. The speed at which it declines depends on the chemicals present in the water and the amount of organic matter in particular. Empirical studies have shown residual chlorine concentration decay in sea water to be a two-stage process. There is a very rapid initial loss of residual chlorine, termed the 'instantaneous chlorine demand' followed by a slower, approximately exponential, decay. To encompass both the chlorine demand and the decay, Davis and Coughlan [31] developed an empirical mathematical model for the loss of chlorine in a power station discharge plume, of the form:

$$C_t = \frac{(C_{in} - C_{id})e^{-kt}}{(D + k'D)^t}$$

where C_t is the residual chlorine concentration at time t after discharge (mg L^{-1}), D is the dilution rate, C_{in} is the chlorine concentration at the point of injection (mg L^{-1}), C_{id} is the instantaneous chlorine demand (mg L^{-1}), k is the chlorine decay constant (per minute), k' is the chlorine demand constant linked to dilution and t is time.

The decay constant of residual chlorine in sea water was determined by Davis and Coughlan [31] to range between 0.012 and 0.042 min^{-1} for temperatures between 0 and 33°C (see Figure 20.2). The best function to describe the relationship of the decay constant, k, to temperature, T, is unknown, but a simple linear relationship of the form

$$k = 0.0011T + 0.0097$$

gives a good fit to the data ($R^2 = 0.956$).

The initial fast loss of residual chlorine from water is termed the instantaneous demand. Davis and Coughlan [31] measured the instantaneous demand following the introduction of 2.5 mg L^{-1} NaOCl. As shown in Figure 20.3, they found that the instantaneous demand as determined 1 min after the addition of NaOCl ranged between 0.1 and 1 mg L^{-1} for temperatures between 0 and 33°C.

Using non-linear regression, an exponential model of the form

$$C_{id} = a(1 - e^{-bT})$$

gave a good fit to the data ($R^2 = 0.891$) with $a = 0.972$, $b = 0.0888$ and T, the temperature.

20.5.1   Chlorination impacts

It is quite difficult to measure accurately the biologically active chlorination products in water, but it is clear that they are toxic to aquatic life at very low levels. Summarised below are the TRO concentrations at which different groups of organisms are impacted arranged in taxonomic order of increasing complexity. It is notable that it is the simpler organisms at the base of the food chain that are most sensitive to chlorination.

Bacterial activity is suppressed at chlorine levels below those that were chemically detectable [31]. A TRO of less than 0.01 mg L^{-1} causes a complete failure of heterotrophic bacterial activity [32]. Marine phytoplankton photosynthesis, as measured by the ration of carbon

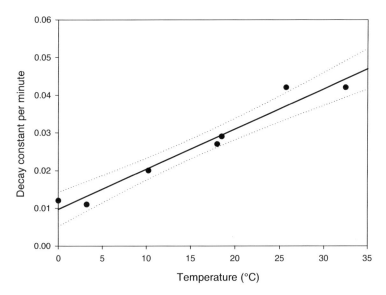

Figure 20.2 The change in the decay constant of chlorine with temperature in sea water. (Data from Reference [31].)

fixation pre- and post-chlorination, is also known to be reduced at a TRO of less than 0.1 mg L^{-1} (see Figure 20.4). Zooplankton also shows severe metabolic and reproductive suppression after exposure to levels as low as 0.01 mg L^{-1} in sea water [33].

Molluscs, particularly the young stages, are also highly sensitive to chlorine. Chlorine concentrations of only 0.05 mg L^{-1} caused about 50% of Pacific oyster, *Crassostrea gigas*,

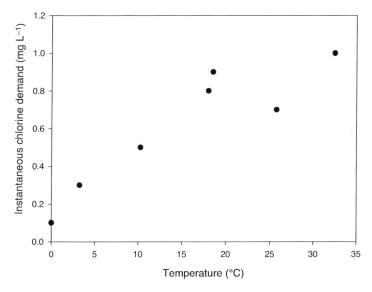

Figure 20.3 The instantaneous chlorine demand of sea water between 0 and 35°C. The initial chlorine concentration was 2.5 mg L^{-1} and the demand was measured 1 minute after addition of the NaOCl.

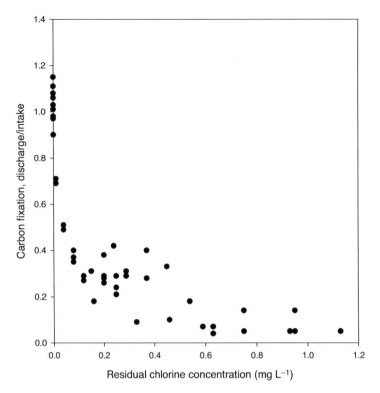

Figure 20.4 The reduction in carbon fixation following chlorination at British marine power stations [22]. The ratio of carbon fixation is calculated between the intake water prior to chlorination and the chlorinated discharge water.

larvae to develop abnormally [34]. The larvae of American oysters have a 48-hour LC50 of less than 0.005 mg L^{-1} [35].

Lethal concentrations for fish occur in the concentration range 0.008–0.01 mg L^{-1} [36]. Fish show many sub-lethal responses. Fathead minnow egg production is reduced by concentrations of 0.043 mg L^{-1}. Fish will avoid chlorinated water; the avoidance threshold for trout may be as low as 0.001 mg L^{-1} [37, 38]. At R. L. Hearn Power Station, Toronto, fish avoided areas with more than 0.035 mg L^{-1}. From fish aquaculture experiments, it has been found that eels, mullet, black bass and sea bream can survive at 0.05 mg L^{-1} for extended periods [39].

The available data indicate that chlorination products are damaging to the environment in low concentrations that may be difficult to measure. The lower organisms at the base of the food chain are particularly sensitive. The main area of impact will be close to discharges prior to dilution in receiving waters. Further, in sea water, toxicity declines as bromoform is generated, which is known to have a low toxicity for bacteria, algae and fish. Generally, TRO levels decline most rapidly in warmer waters with high organic matter content.

20.6   Changing biofouling threats

The risk of serious biofouling does not stay constant and can change dramatically over the typical life of an industrial facility of 40–50 years. The main reasons for change are (1) the introduction of alien species and (2) changes in water quality. The zebra mussel, *D. polymorpha*, gives an example of fouling following accidental introduction. Zebra mussels were introduced into the Great Lakes in the 1980s. They rapidly colonised the lakes radically changing planktonic and benthic communities [40]. First seen in the Hudson at Catskill in May 1991, zebra mussels now inhabit the Mohawk River and the Hudson River from Albany to Haverstraw Bay where they are a constant threat to intakes [41] and fine filter screens. Changes in water quality can considerably affect biofouling incidence. On a global scale, recent warming may allow some problem organisms to extend their geographical range. Recent large-scale increases in the abundance of jellyfish have been related to environmental change [42]. Within rivers and estuaries, water quality improvements can actually lead to an increase in fouling as oxygen levels improve and concentrations of toxic chemical decrease. While ecosystems can change quite rapidly, it is normally the case that a new biofouling problem develops over a number of months or years. Operators who are suddenly surprised by a failure caused by biofouling are probably not carrying out regular inspections of their plant and the local environment.

20.7   Conclusions

- Biofouling is a constant threat to the operation and integrity of power station and water supply water extraction and discharge systems.
- Passive blocking of filter screens by weed, jellyfish and occasionally by huge shoals of fish is regularly reported and increasingly problematical.
- Both freshwater and marine underwater surfaces become coated in microbial communities that reduce water flow, impede heat transference and damage surfaces and monitoring equipment. By far the most important microbial fouling communities are termed microbial slimes formed by communities of *Pseudomonas*, *Flavobacterium* and *Bacillus* bacteria.
- Macrofouling occurs in both fresh and marine waters, but is particularly problematical and almost universal in marine and estuarine environments. Bivalve molluscs, particularly mussels, are a particularly damaging group of fouling organisms.
- Numerous methods for biofouling control have been tested. At present, heat treatment and chlorination are the most favoured, although the slow release of low concentrations of copper ions is also frequently used. Many of these treatments result in damage to the environment via the release of low levels of toxins.
- Biofouling threats are constantly changing and there is a need for constant vigilance.
- Biofouling threats can be minimised by good design and construction. It can be difficult and costly to reduce or eliminate biofouling in a poorly designed system.

References

1. Jenner, H.A, Whitehouse, J.W., Taylor, C.J.L. & Khalanski, M. (1998) Cooling water management in European power stations. Biology and control of fouling. *Hydroecologie Appliquee*, **10** (1–2), 1–225.
2. Whitehouse, J.W. & Stubberfield, L.C.F. (1989) *Studies of Biofouling Growth on Plastics Packing Using Miniature Cooling Towers*. Report RD/L/3423/R88, Central Electricity Generating Board, UK.

3. Flemming, H.C., Schaule, G., Griebe, T., Schmitt, T. & Tamachkiarowa, A. (1997) Biofouling – the Achilles heel of membrane processes. *Desalination*, **113**, 215–225.

4. Ivnitskya, H., Katza, I., Minzc, D., *et al.* (2005) Characterization of membrane biofouling in nanofiltration processes of wastewater treatment. *Desalination*, **185**, 255–268.

5. Characklis, W.G. (1981) Fouling biofilm development. A process analysis. *Biotechnology & Bioengineering*, **23**, 1923–1960.

6. Aprosi, G. & Nepveu de Villemarceau, C. (1988) French operational experience on fouling problems (Algae, Bryozoa) in cooling towers. Paper presented at *the 6th IAHR Cooling Tower Workshop*, Pisa, Italy, 4–7 October 1988.

7. Aprosi, G. & Costaz, J.L. (1989) *La formation des Algues sur les Coques des Refrigerants*. Rapport EFD/DER HE/31–89.14. The International Association of Hydraulic Engineering and Research, Madrid, Spain.

8. NRC (1992) *Recent Loss or Severe Degradation of Service Water Systems*. NRC information notice 92–49, 2 July 1992. Nuclear Regulatory Commission, USA.

9. Holmes, R.H.A. (1975) *Fish and Weed on Fawley Generating Station Screens February 1973–January 1974*. Report RD/L/N 129/75, Central Electricity Generating Board, UK.

10. Board, P. (1975) *White-Weed Fouling of Tilbury 'A' Power Station*. Report LM/BIOL/001, Central Electricity Generating Board, UK.

11. Robinson, N.E., Wickett, T.A. & Mansfield, S.E. (1981) *Fish inundation at Dungeness 'A' Power Station – Cause and Counteraction*. Report SE/SSD/R/81/046, Central Electricity Generating Board, UK.

12. USGS (2000) *Zebra Mussels Cause Economic and Ecological Problems in the Great Lakes*. GLSC Fact sheet 2000–6, US Geological Survey, Great Lakes Science Center, USA. Available from http://www.glsc.usgs.gov/_files/factsheets/2000-6%20Zebra%20Mussels.pdf. Accessed 8 July 2009.

13. Benson, A.J., Richerson, M.M. & Maynard, E. (2008) *Dreissena rostriformis bugensis*. USGS Nonindigenous Aquatic Species Database, Gainesville, USA. Available from http://nas.er.usgs.gov/queries/FactSheet.asp?speciesID=95. Revised on 10 October 2008.

14. Strayer, D.L. (2008) *Zebra Mussels and the Hudson River*. Cary Institute of Ecosystem Studies, New York. Available from http://www.ecostudies.org/people_sci_strayer_zebra_mussels.html. Accessed 30 September 2008.

15. Dubois, C. (1995) *Biologie et Demo-Ecologie d'une Espece Invasive,* Corbicula fluminea *(Mollusca, Bivalva) Originaire d'Asia. Edude en Milieu Naturel (Canel Lateral a la Geronne) et en Canal Experimental*. PhD thesis, University of Toulouse.

16. Christie, A.O. & Dalley, R. (1987) Barnacle fouling and its prevention. In: *Barnacle Biology: Crustacean Issues V* (eds A.J. Southward & D.J. Crisp), pp. 419–433. A.A. Balkema, Rotterdam.

17. Dare, P.J. (1973) *The Stocks of Young Mussels in Morecambe Bay, Lancashire*. Shellfish Information Leaflet 28, Ministry of Agriculture and Food (MAFF), UK.

18. Henderson, P.A., Seaby, R.M.H., Cailes, C. & Somes, J.R. (2001) *Gunderboom Fouling Studies in Bowline Pond*. Report by Pisces Conservation Ltd. for Riverkeeper Inc., USA. Available from http://www.powerstationeffects.co.uk/reports/gunderboom.pdf. Accessed 8 July 2009.

19. Bamber, R.N & Turnpenny, A. (1986) *Trial of a Johnson 715 Alloy Wedge-Wire Screen at a UK Coastal Site*. Report TPRD/L/3012/R86, Central Electricity Generating Board, UK.

20. Delattre, J.-M. & Oger, C. (1981) *Naegleria* pathogens et pollution thermique. *Journal of French Hydrology*, **12**, 239–244.

21. Tyndall, R.L., Ironside, K.S. Metler, P.L., Tan, E.L., Haezn, T.C. & Fliermans, C.B. (1989) Effect of thermal addition on the density and distribution of *Naegleria fowleri* thermophilic amoebae and pathogenis in a newly created cooling lake. *Applied Environmental Microbiology*, **55**, 722–732.

22. Whitehouse, J.W., Khalanski, M., Saroglia, M.G. & Jenner, H.A. (1985) *The Control of Biofouling in Marine and Estuarine Power Stations*. Joint Report of CEGB, EDF, ENEL & KEMA. Central Electricity Generating Board (UK), Electricité de France (France), Ente Nazionale Energia Elettrica Joint Laboratories (Italy), Consulting Services of the Dutch Electricity Supply Companies (The Netherlands).

23. Jenner, H.A, Taylor, C.J.L., Van Dork, M. & Khalanski, M. (1997) Chlorination by-products in chlorinated cooling water of some coastal power stations. *Marine Environment Research*, **43**, 279–293.

24. Van Benschoten, J.E., Jensen, J.N., Lewis, D. & Brady, T.J. (1993) Chemical oxidants for controlling Zebra Mussels (*Dreissena polymorpha*): a synthesis of recent laboratory and field studies. In: *Zebra Mussels Biology, Impacts and Control* (eds T.F. Nalepa & D.W. Schloesser), pp. 599–619. Lewis Publishers, London.

25. Khamanski, M. & Bordet, F. (1981) Modalite d'action du chlore sur la moule marine. Paper presented at *2nd Journees de la Thermo-ecologie Institut Scientifique et Technique des Peches Maritimes, 14–15 November 1979, Paris*.

26. Manov, D.V., Chang, G.C. & Dickey, T.D. (2004) Methods for reducing biofouling of moored optical sensors. *Journal of Atmospheric and Oceanic Technology*, **21**, 958–968.

27. Knox-Holmes, B. (1993) Biofouling control with low levels of copper and chlorine. *Biofouling*, **7**, 157–166.

28. Cloete, T.E., Jacobs, L. & Brözel, V.S. (1998) The chemical control of biofouling in industrial water systems. *Biodegradation*, **9**, 23–37.

29. Bott, T.R. (2000) Biofouling control with ultrasound. *Heat Transfer Engineering*, **21**, 43–49.

30. Abarzua, S. & Jakubowshi, S. (1995) Biotechnological investigation for the prevention of biofouling. I. Biological and biochemical principles for the prevention of biofouling. *Marine Ecology Progress Series*, **123**, 301–312.

31. Davis, M.H. & Coughlan, J. (1983) Model for predicting chlorine concentration within marine cooling circuits and its dissipation at outfalls. In: *Water Chlorination, Environmental Impact and Health Effects*, Vol. 4 (eds R.L. Jolley, W.A. Brungs, J.A. Cotruvo, et al.), pp. 347–357. Ann Arbor Science, Ann Arbor.

32. Davis, M.H. & Coughlan, J. (1978) Response of entrained plankton to low-level chlorination at a coastal power station. In: *Water Chlorination: Environmental Impact and Health Effects*, Vol. 2 (eds R.L. Jolley, W.A. Brungs, J.A. Cotruvo, et al.), pp. 369–376. Ann Arbor Science, Ann Arbor.

33. Goldman, J.C, Capuzzo, J.H. & Wong, G.T.F. (1978) Biological and chemical effects of chlorination at coastal power plants. In: *Water Chlorination: Environmental Impact and Health Effects*, Vol. 2 (eds R.L. Jolley, W.A. Brungs, J.A. Cotruvo, et al.), pp. 291–306. Ann Arbor Science, Ann Arbor.

34. Bamber, R.N., Seaby, R.M., Fleming, J.M. & Taylor, C.J.L. (1994) The effects of entrainment passage on embryos of the Pacific oyster, *Crassostrea gigas*. *Nuclear Energy*, **33**, 353–357.

35. Mattice, J.S. & Zittel, H.E. (1976) Site-specific evaluation of power plant chlorination. *Journal of the Water Pollution Control Federation*, **48**, 2284–2308.

36. Langford, T.E. (1983) *Electricity Generation and the Ecology of Natural Waters*. Liverpool University Press, Liverpool.

37. Sprague, J.B. & Drury, D.E. (1969) Avoidance reactions of salmonid fish to representative pollutants. In: *Advances in Water Pollution Research* (ed. S.H. Jenkins), pp. 169–179. Pergamon Press, Oxford.

38. White, G.C. (1972) *Handbook of Chlorination*. Van Nostrand, New York.

39. Saroglia, M.G., Quierazza, G. & Scarano, G. (1981) Water quality criteria for aquaculture in thermal effluents, heavy metals and residual anti-fouling products. In: *Aquaculture in Heated Effluents and Circulation Systems*, Vol. 1 (ed. K. Tiews), *Proceedings of a World Symposium*, Stavanger, 1980. Heinemann, Berlin.

40. Caraco, N.F., Cole, J.J., Raymond, P.A., *et al.* (1997) Zebra mussel invasion in a large, Turbid River: phytoplankton response to increased grazing. *Ecology*, **78**, 588–602

41. Strayer, D.L., Powell, J., Ambrose, P., Smith, L.C., Pace, M.L. & Fischer, D.T. (1996) Arrival, spread, and early dynamics of a zebra mussel (*Dreissena polymorpha*) population in the Hudson River estuary. *Canadian Journal of Fisheries and Aquatic Sciences*, **53**, 1143–1149.

42. Purcell, J.E. (2005) Climate effects on formations of jellyfish and ctenophore blooms: a review. *Journal of the Marine Biological Association of the United Kingdom*, **85**, 461–476.

Chapter 21
Regulation of Marine Antifouling in International and EC Law

Ilona Cheyne

The regulation of marine antifouling is a relatively recent legal development and must be understood within the wider context of the international law of the sea and national environmental and safety legislation. This chapter examines the international law background to antifouling restrictions and the way in which they have been introduced in EC law.

21.1   The international law of the sea

In the eighteenth and nineteenth centuries, use of the sea was largely unrestricted outside narrow coastal areas. The emphasis was on freedom of navigation and free use of resources. This began to change in the twentieth century as technological developments offered greater opportunities to exploit both living and non-living resources. Coastal States began to make more expansive claims over areas such as the continental shelf and exclusive fishing zones. Developing countries became anxious that they would be excluded from exploiting the mineral resources of the deep seabed. Concerns grew about the impact of human activities on the marine environment, including over-fishing, pollution from accidental oil spills and pollution from deliberate acts such as dumping waste and washing out oil tanks. A series of attempts were made to negotiate written rules which would help to regulate the growing uses of maritime areas and help prevent or reduce the conflicts that occurred as a result. The most recent and ambitious attempt began in 1973. After long and difficult negotiations, the United Nations Law of the Sea Convention (UNCLOS) was adopted in 1982[1]. Although UNCLOS did not come into effect until 16 November 1994, it now has 159 parties[2]. It has, through its links with the International Maritime Organisation (IMO), become a powerful agent for promoting and enforcing environmental protection with regard to shipping.

The extent to which activities in marine areas are controlled depends on which States have the power to regulate them. The sea is divided into a number of jurisdictional zones which determine the nature of those powers and the identity of the States that are entitled to exercise them. The relevant marine areas are the territorial sea, the exclusive economic zone (EEZ) and the high seas[3]. The territorial sea allows States to exercise sovereign powers over a belt of water up to 12 miles from their coast. However, their sovereign control is limited by the right of innocent passage for foreign vessels[4]. This right allows foreign shipping to pass through the territorial sea, subject to certain navigation safety restrictions and the limited exercise of criminal and civil jurisdiction. The coastal State has an obligation not to hamper innocent passage, which includes an obligation not to impose requirements that have the practical effect of denying or impairing the right[5]. Passage means navigation for the purpose of crossing the territorial sea without entering internal waters or port, or proceeding to or from internal waters or port[6]. Passage must be continuous and expeditious, although it includes stopping and anchoring where incidental to ordinary navigation or necessary because of force majeure, distress or the need to assist persons, ships or aircraft in danger or distress[7]. In order to be innocent, passage must not be 'prejudicial to the peace, good order or security of the coastal State'[8]. This definition covers obviously threatening activities such as the use of force or testing weapons, but is limited as it affects environmental protection. The only explicit environmental reference excludes 'any act of wilful and serious pollution contrary to this Convention' from the exercise of innocent passage[9]. It is therefore only in rather extreme circumstances that passage can be prevented, given the coastal State's obligation not to hamper it when innocent. This also means that national laws that might affect the ability of foreign shipping to exercise their right of innocent passage must be of a type that is permitted by the Convention. For example, coastal States are entitled to introduce laws and regulations in respect of the conservation of marine living resources and the preservation of the environment of the coastal State, including the prevention, reduction and control of pollution[10]. However, they may not impose laws and regulations applying to the design, construction, manning or equipment of foreign ships unless they follow generally accepted international rules or standards[11].

The EEZ is an area extending from the territorial sea up to 200 miles from the coast[12]. The coastal State has, inter alia, sovereign rights to explore and exploit, conserve and manage the natural resources, living and non-living, in the water, seabed and subsoil. It also has jurisdiction with regard to the protection and preservation of the marine environment[13]. Foreign shipping continues to enjoy freedom of navigation and normal lawful operation of ships[14]. However, other States must have due regard to the rights and duties of the coastal State and must comply with its laws and regulations[15]. The coastal State has the right to take measures to ensure compliance with its laws and regulations, including boarding, inspection, arrest and judicial proceedings[16].

The high seas cover the area beyond the limits of the EEZ, and all States have the freedom to use them including for the purposes of navigation[17]. With limited exceptions, the only States entitled to exercise jurisdiction over ships while they are on the high seas are flag States[18]. This is based on the fact that ships have the nationality of the State in which they are registered and whose flag they fly; there should be a 'genuine link' between a flag State and its ships[19]. However, the exclusive jurisdiction of flag States has caused significant problems because of the variability with which different States have imposed safety and other standards[20]. This is despite the rule, reaffirmed in the 1982 UNCLOS, that flag States have a duty to exercise effective jurisdiction and control over all matters concerning their registered ships[21].

The law of the sea demonstrates an uneasy balance between the rights of coastal States to control activities which might threaten their interests and the right of other States to have their registered ships navigate as freely as possible through all marine areas. Protection of the marine environment has historically been somewhat neglected, although UNCLOS does contain a general obligation to protect and preserve the marine environment[22]. However, the provision which deals directly with the prevention, reduction and control of pollution is disappointingly vague. Thus, Article 194(1) states:

> States shall take, individually or jointly as appropriate, all measures consistent with this Convention that are necessary to prevent, reduce and control pollution of the marine environment from any source, using for this purpose the best practicable means at their disposal and in accordance with their capabilities, and they shall endeavour to harmonise their policies in this connection.

The use of words such as 'best practical means' and 'in accordance with their capabilities' indicates that this is a soft law provision, that is to say, one that would be difficult to apply as a firm and precise legal obligation. More specifically, however, Article 211 of UNCLOS provides that Member States shall establish international rules and standards to prevent, reduce and control pollution of the marine environment from vessels, and obliges flag States to legislate in order to implement those rules and standards with regard to their registered ships[23]. Compliance with these international rules and standards may be a condition for the entry of foreign shipping into ports[24]. It is also possible for coastal States to submit evidence to IMO in support of special measures in particular areas of their EEZ which are needed because of oceanographical and ecological conditions, and the need to protect resources[25].

The effectiveness of legal provisions depends to a large extent on whether enforcement is possible. It can be seen that the coastal State's powers are quite limited even within its own territorial sea and EEZ because of the need to safeguard the right of innocent passage and freedom of navigation, respectively. The position in the high seas is even less promising, given the almost exclusive reliance on flag State jurisdiction. To this end, UNCLOS imposes a specific obligation on flag States to ensure that ships flying their flag or of their registry comply with international rules and standards, including provision for effective enforcement, prevention of non-compliant vessels from sailing and requirements to carry appropriate documentation[26]. Where a violation of an international rule or standard does occur, the flag State shall immediately investigate and, where appropriate, institute proceedings[27]. However, it was not considered sufficient to leave all enforcement proceedings to flag States, and UNCLOS also provides for coastal State and port State enforcement. Coastal States have enforcement powers where a violation of an international rule or standard for the prevention, reduction or control of pollution has occurred in their territorial sea or the EEZ. The ship must have voluntarily entered its port but, once this occurs, the coastal State has the power to investigate and institute proceedings[28]. More interesting, perhaps, is port State control. This allows a port State to exercise enforcement powers even where the violation of the international rule or standard has occurred outside the territorial sea or EEZ of that State. The only condition of jurisdiction is that the ship has entered the port voluntarily[29]. The exercise of these enforcement powers is subject to certain safeguards, including the obligation not to delay a vessel longer than is necessary, not to inspect documents other than those required by international rules and standards and only to carry out physical inspections where there are clear grounds for believing that the condition of the vessel or its equipment is substantially different from the documents, or the documents are insufficient or missing[30]. Proceedings by the coastal or port State should be suspended if the flag State chooses to bring corresponding charges, but this will not apply if the coastal State has suffered major damage or if the flag State has repeatedly failed to enforce international rules and standards against its own vessels[31].

Thus, flag State jurisdiction and enforcement remain fundamentally important but port States in particular have been given significant powers to correct failures of compliance by ships that voluntarily enter their jurisdiction. The strength of port State control lies in the fact that it allows international rules and standards to be applied to all shipping, even vessels registered with States that are not party to the international agreement provided the rules and standards have been 'generally accepted'. The effectiveness of port State control has been enhanced by the establishment of regional cooperation schemes. There are currently nine of these regional schemes set up under memoranda of understanding (MOU): Europe and the North Atlantic (Paris MOU), Asia and the Pacific (Tokyo MOU), Latin America (Acuerdo de Viña del Mar), Caribbean (Caribbean MOU), West and Central Africa (Abuja MOU), the Black Sea region (Black Sea MOU), the Mediterranean (Mediterranean MOU), the Indian Ocean (Indian Ocean MOU) and the Arab States of the Gulf (GCC MOU/Riyadh MOU). The purpose of these schemes is to ensure that ships are inspected on a regular basis, and information about inspections, detentions and bans is shared. The advantage is that ships are not unnecessarily inspected but neither are they easily able to avoid inspection by varying the ports that they visit within a region[32].

The link between UNCLOS and the international rules and standards developed by specialised agencies makes UNCLOS a framework or umbrella treaty. This means that it lays down general principles but can be implemented in detail through other legal instruments. UNCLOS provisions clearly delineate the jurisdiction of States over geographically defined areas and the types of legislation that they can enact and enforce over activities in those areas. The actual content of that legislation is not specified in detail, although it may be limited, for example, by the rule against discrimination or against regulating design, construction, manning or equipment of foreign ships[33]. However, the specific scope and content of coastal State and port State regulation are tied to international rules and standards. This has the beneficial effect of ensuring that Member States apply the same rules, thus easing the burden of compliance for transboundary maritime transport. It also ensures that the development of detailed technical standards is carried out by competent and specialised organisations. IMO is one of those organisations, even though it is not always explicitly mentioned in the text of UNCLOS[34]. It currently comprises 168 Member States, slightly more than the 156 parties of UNCLOS. It is the competent international organisation with regard to the protection and preservation of the marine environment[35].

IMO has adopted a variety of instruments, including resolutions, recommendations, codes and guidelines, which are not mandatory in themselves. However, they may become mandatory by being incorporated into IMO treaties and by being implemented in national legislation. National legislation is binding on domestic ships and may be enforced against foreign shipping entering the jurisdiction of that State. The UNCLOS provisions which refer to these international rules and standards do not always require compliance, but it is expected and, in any case, adoption into national legislation has much the same effect[36].

21.2 The International Convention on the Control of Harmful Antifouling Systems on Ships, 2001[37]

The International Convention on the Control of Harmful Antifouling Systems on Ships, 2001 (ICAFS) came into force on 17 September 2008[38]. The purpose behind the ICAFS is to prohibit the use of organotins in antifouling paints and to enable other antifouling substances to be prohibited or restricted in the future if circumstances warrant[39]. It defines an antifouling system as 'a coating, paint, surface treatment, surface, or device that is used on a ship to control or prevent attachment of unwanted organisms'[40]. A ship for these purposes is widely defined as 'a vessel of any type whatsoever operating in the marine environment and includes hydrofoil boats, air-cushion vehicles, submersibles, floating craft, fixed or floating platforms, floating storage units (FSUs) and floating production storage and off-loading units (FPSOs)'[41]. The obligations of the ICAFS apply to ships flying the flag or under the authority of a State party, and to other ships that enter a port, shipyard or offshore terminal of a party[42].

The ICAFS requires each party to legislate in order to prohibit or restrict any application, re-application, installation or use of harmful antifouling systems on their own ships and on foreign ships while in their ports, shipyards or offshore terminals, according to the requirements listed in Annex 1[43]. If the flag State is informed that a violation has occurred and there is sufficient evidence to bring proceedings, it is under an obligation to do so[44]. The State in whose jurisdiction the violation occurs may either inform the flag State or institute its own proceedings[45]. At present, Annex 1 provides that organotin compounds which act as biocides in antifouling systems must not be applied or re-applied to any ships after 1 January 2003, and they must either not be borne by their hulls or external parts or surfaces or must be under a coating which acts as a barrier to prevent them from leaching by 1 January 2008[46]. Small quantities of organotin compounds acting as a chemical catalyst are allowed, provided they are at a level that does not provide a biocidal effect to the coating[47]. If Annex I is amended in the future, ships with a non-compliant antifouling system may retain that system until the next scheduled renewal of the system provided the period is no longer than 60 months, unless the Marine Environment Protection Committee of IMO decides that exceptional circumstances exist to warrant earlier implementation[48]. In addition, parties must ensure that wastes resulting from the application or the removal of an antifouling system are collected, handled, treated and disposed of safely and in an environmentally sound manner in order to protect human health and the environment[49].

The ICAFS lays down several provisions for implementation. First, flag States must ensure that their ships are surveyed and certified if they are of or above 400 gross tonnage and engaged in international voyages[50]. The need for survey occurs when a ship is put into service or when the International Antifouling System (IAFS) Certificate is to be issued for the first time or after the antifouling system is changed or replaced[51]. The Certificate issued after survey must be accepted as valid by the other parties[52]. Flag States must ensure that their ships which are less than 400 gross tonnage but more than 24 m in length carry a declaration signed by the owner or the owner's authorised agent which is appropriately endorsed or accompanied by documentation such as a paint receipt or a contractor's invoice[53]. There is no formal requirement of certification or self-certification for vessels of less than 24 m in length, an omission which is understandable given that they are unlikely to undertake many international voyages.

Secondly, inspections may be undertaken by port States when ships are in their ports, shipyards or offshore terminals to determine whether they are compliant with the ICAFS. However, the inspection is limited to verifying, where required, that the ship carries an IAFS Certificate or declaration and a brief sampling of the antifouling system that does not affect its structure, integrity or operation[54]. If there is no IAFS Certificate, for example where the ship is not registered with a party to the ICAFS, other documentation such as a declaration of compliance from the antifouling system manufacturer may be sufficient[55]. If there are clear grounds for believing that the ship is not compliant with the ICAFS, a thorough inspection may be carried out[56]. This may include inspection of additional documentation, such as the ship's log, or requests for information such as the date of the last antifouling system application or the name of the facility in which the work was performed, and more comprehensive sampling and analysis of the antifouling system[57]. Inspections may also be carried out if a request for an investigation is received from another party, along with sufficient evidence that the ship is operating or has operated in violation of the ICAFS[58]. The port State has the power to warn, detain, dismiss or exclude from its ports any ship which it detects to be in violation of the ICAFS[59]. However, it must make all possible efforts to avoid unduly detaining or delaying a ship by inspection or instituting proceedings against it[60].

21.3   Implementation of the International Antifouling Convention by the EC

The EC cannot become a party to the ICAFS in its own right, but it was decided that a harmonised approach within the Community was needed to ensure that all the Member States introduced the same rules at the same time. The EC therefore implemented the ICAFS prior to its coming into force by Regulation 782/2003/EC on the prohibition of organotin compounds on ships[61]. It characterises the Convention as a framework agreement which allows specific implementation by States parties, and notes that it pays 'due regard to the precautionary principle'[62]. The Regulation uses the same definitions of antifouling systems and ship as the ICAFS, and adopts the same mechanisms of certificate, declaration and statement of compliance[63]. It also makes the same distinction between ships flying the flag or operating under the authority of Member States, and other ships which enter a port or offshore terminal of a Member State[64].

The Regulation prohibits organotin compounds that act as biocides in antifouling systems from being applied or re-applied on ships after 1 July 2003, but only if they fly the flag or operate under the authority of a Member State[65]. According to the wording of the Regulation, these ships may not bear organotin compounds on their hulls or external parts or surfaces unless covered by a barrier coating if their antifouling system has been applied, changed or replaced since 1 July 2003[66]. This was rather ambiguous in meaning, but a subsequent document issued by the Commission interpreted the provision to mean that the EC and IMO schemes were identical, so that during the interim period between 1 July 2003 and the entry into force of the Convention, ships registered with EC Member States were allowed to keep their existing antifouling system but any new paint used needed to be tributyltin-free. The obligation to remove or seal off did not come into effect until 1 January 2008[67]. The prohibition on bearing organotin compounds without a barrier coating was applied to non-Member State ships on 1 January 2008, even though the Convention had not come into force by then[68]. The Commission seems to have acted on the assumption that the shipping industry would introduce its own voluntary restraints, quoting the International Chamber of Shipping statement made shortly after the ICAFS Conference: 'whether or not the Convention enters into force by 1 January 2003 is perhaps somewhat academic as the fixed dates of 1 January 2003 and 1 January 2008 should be regarded as firm for any ship operating in international trade'[69].

Ships belonging to EC Member States are subjected to the same survey and certification requirements as provided for under the Convention[70]. Thus ships over 400 gross tonnage are subject to a survey and certification scheme which is equivalent to that of the ICAFS, although the Regulation goes further than the Convention by applying to ships even if they do not undertake international voyages. Ships under 400 gross tonnage weight but more than 24 m in length are also subject to equivalent requirements in the shape of a declaration of compliance. The Commission noted in its proposal that no specific survey or certificate was proposed but a harmonised regime could be introduced in the future[71]. Ships under 24 m in length are not subject to specific survey or certification requirements, which is again equivalent to the ICAFS.

The normal port State control directive is only effective with regard to international conventions that are in force, and therefore an interim procedure originally had to be put in place[72]. The Regulation therefore provides that control provisions equivalent to the normal port State control legislation shall be applied to ships above 400 gross tonnage and flying the flag of a Member State[73]. Appropriate compliance procedures for third State ships were to be developed once the ICAFS came into force[74].

21.4   Authorisation of biocidal substances

Much of the EC's control over biocidal substances is exercised through obligatory authorisation procedures which must be successfully completed before such products can be marketed or used. In some cases, products are simply banned. Thus, Directive 76/769/EEC banned organotin compounds from being placed on the market for use as biocides in free association antifouling paint in 1976. Originally, the ban covered only antifouling for boats of less than 25 m or on any kind of vessel used predominantly on inland waterways and lakes[75]. The ban has now been extended to all tall craft irrespective of their length intended for use in marine, coastal, estuarine and inland waterways and lakes[76]. Neither can they be used on cages, floats, nets and any other appliances or equipment used for fish or shellfish farming or on any totally or partly submerged appliance or equipment. In addition, they may not be used as substances and constituents of preparations intended for use in the treatment of industrial waters[77]. This restriction obviously has an effect on aquaculture practices, and potentially on other activities such as lobster fishing or industries using outlets into water, such as power stations (see Chapters 19 and 20).

Apart from these specific bans, however, biocidal products are subject to an authorisation process under Directive 98/8/EC (the Biocides Directive)[78]. This Directive imposes certain requirements before products can be placed on the market or used for biocidal purposes[79]. Its purpose is to harmonise the release of biocidal products onto the market by imposing common review procedures and mutual recognition of national authorisation, and to provide a high level of protection for people, animals and the environment. Member States have the power to authorise products consisting of, or containing, one or more active substances intended to 'destroy, deter, render harmless, prevent the action of or otherwise exert a controlling effect on any harmful organism by chemical or biological means'[80]. For the purposes of the Directive, an active substance is defined as a 'substance or micro-organism including a virus or a fungus having general or specific action on or against harmful organisms'[81]. A harmful organism is any organism which has 'an unwanted presence or a detrimental effect for humans, their activities or the products they use or produce, or for animals or for the environment'[82].

The power to authorise biocidal products is subject to conditions contained in the Directive, including the need to appraise a dossier of information submitted by the applicant[83]. The active substance or substances contained in the product must be listed in Annex I of the Directive, and any accompanying conditions must be included in the authorisation[84]. The type of conditions that might be imposed when a substance is included in Annex I include the minimum degree of purity, the product type in which the substance may be used, the manner and area of use, and the category of user, such as professional or industrial users[85]. Once authorisation is given by a Member State, it must be recognised as valid by other Member States[86]. Authorisations can be valid for up to 10 years, and may be renewed, reviewed, cancelled or modified[87]. The criteria for inclusion of substances in Annex I are broadly similar to the criteria to be used by Member State when authorising products. Most importantly, the product must be sufficiently effective and have no unacceptable effects on the target organisms, such as resistance or unnecessary suffering, or on human or animal health, or on the environment[88]. However, a substance cannot be included in Annex I if it is carcinogenic, mutagenic, toxic for reproduction, sensitising or is bioaccumulative and does not readily degrade[89].

The aim is therefore to develop a list of acceptable substances which will be listed in Annex I. In order for this to happen, biocidal substances are divided into two groups. The first comprises existing biocidal substances, namely, those already on the market when the Biocides Directive came into force on 14 May 2000[90]. Existing substances are subject to a programme of review which is expected to last 10 years, and antifouling biocides are covered by the second phase[91]. During the review period, the marketing of biocidal products will be governed by national legislation until a decision under the Biocides Directive is made[92]. The second group comprises new biocidal substances, that is to say, those substances put forward for sale and use after 14 May 2000. Inclusion of new active substances requires the submission of a dossier of information on the active substance and on at least one product containing that substance to the competent authority of a Member State[93]. The competent authority evaluates the information contained in the dossiers and sends its findings to the Commission and other Member States, and the Commission is then responsible for proposing a decision to include or not to include the substance in Annex I. For this purpose, the Commission is assisted by the Standing Committee on Biocidal Products[94].

21.5   Conclusions

- Legal regulation of marine antifouling is relatively new.
- Most international law designed to protect the marine environment from shipping activity has focused on safety or operating procedures. Regulating the use of antifouling paints and biocides does not fit neatly within the existing legal framework.
- It is not clear whether the use of antifouling paint falls under the phrase 'design, construction, manning or equipment' or whether the use of biocidal antifouling within the territorial sea of coastal States might constitute 'wilful and serious pollution' for the purposes of defining innocent passage in the territorial sea.
- It is not clear whether the use of antifouling paint would contravene the general obligation to protect the marine environment contained in UNCLOS.
- The introduction of the ICAFS will clarify the legal picture by providing a clear process by which biocidal substances may be evaluated and, if necessary, restricted.

- The effectiveness of the ICAFS will depend on the participation and cooperation of the shipping industry, and on the power of port States to enforce its provisions against ships whose flag States have failed to take the necessary action.
- Restrictions on biocidal substances will be developed on a case-by-case basis under the aegis of the ICAFS.
- The ICAFS came into force in September 2008. However, it was already being implemented in national and regional legislation, for example in the EC, and through voluntary action by the industry.

Notes

1. UNCLOS, 1982 (1994) 1833 UNTS 3.
2. As at 16 July 2008. The UK acceded on 25 July 1997. The USA has not yet acceded.
3. Others include the contiguous zone (customs, fiscal, immigration or sanitary laws and regulations), and the Area (seabed and ocean floor and subsoil beyond the limits of national jurisdiction).
4. UNCLOS, Article 17.
5. UNCLOS, Articles 24(1) and 25.
6. UNCLOS, Article 18(1).
7. UNCLOS, Article 18(2).
8. UNCLOS, Article 19.
9. UNCLOS, Article 19(2)(h).
10. UNCLOS, Article 21(1)(d) and (f). Foreign ships exercising the right of innocent passage have the corresponding duty to comply with the coastal State's regulations, see Article 21(4).
11. UNCLOS, Article 21(2).
12. UNCLOS, Article 57.
13. UNCLOS, Article 56. However, much of the environmental powers in this section relate to conservation and exploitation of the natural resources present in the EEZ, see Articles 61–70.
14. UNCLOS, Article 58.
15. UNCLOS, Article 58(3).
16. UNCLOS, Article 73.
17. UNCLOS, Articles 87 and 90.
18. UNCLOS, Article 92. Exceptions include piracy, slavery or unauthorised broadcasting, see Articles 105, 109, 110.
19. UNCLOS, Article 91.
20. See, for example, the International Transport Workers' Federation campaign against flags of convenience, available from http://www.itfglobal.org/flags-convenience/index.cfm. Accessed 8 July 2009.
21. UNCLOS, Article 94.
22. UNCLOS, Article 192.
23. UNCLOS, Article 211(1)-(2).
24. UNCLOS, Article 211(4) and see also Article 21. Note, however, that this right is subject to the obligation not to hamper the exercise of the right of innocent passage discussed above.
25. UNCLOS, Article 211(6). Two types of areas may be designated under IMO procedures. The first is a particularly sensitive sea area which may qualify because of characteristics such as a unique or rare ecosystem, diversity or vulnerability, or economic, scientific or social value. See Assembly Resolution A.982(24) *Revised guidelines for the identification and designation of Particularly Sensitive Sea Areas (PSSAs)*. The second type of area may fall under Annex I on prevention of pollution by oil, Annex II on control of pollution by noxious liquid substances or Annex V on prevention of pollution by garbage from ships of the International Convention for the Prevention of

Pollution from Ships, 1973, as modified by the Protocol of 1978 relating thereto (MARPOL 73/78). This means that areas which, for technical reasons relating to their oceanographical and ecological condition and to their sea traffic, require special mandatory methods for their protection, such as restricted oil tanker navigation routes, can be classified as 'special areas'. The two types are not necessarily mutually exclusive.

26. UNCLOS, Article 217(1)-(3).
27. UNCLOS, Article 217(4).
28. UNCLOS, Article 220.
29. UNCLOS, Article 218.
30. UNCLOS, Article 226. There is also an obligation not to discriminate 'in form or in fact' against vessels from a particular State, see Article 227.
31. UNCLOS, Article 228.
32. See, for example, the Annual Report 2006 of the Paris MOU, available from http://www. parismou.org/upload/anrep/anrep2006low.pdf. Accessed 8 July 2009.
33. See, for example, Articles 24(1)(b) and 21(2), respectively.
34. The United Nations Division for Ocean Affairs and the Law of the Sea lists 18 relevant international organisations, including the Food and Agricultural Organisation (FAO), the International Labour Organisation (ILO), the United Nations Environment Programme (UNEP), and the World Health Organisation (WHO). For a breakdown of which organisations are relevant for specific provisions, see the *Law of the Sea Bulletin No. 31*, pp. 79–95. And see generally, *Implications of the United Nations Convention on the Law of the Sea for the International Maritime Organization*, 31 January 2007, LEG/MISC.5.
35. Ibid. Key IMO environmental treaties include the International Convention for the Prevention of Pollution from Ships, 1973, as modified by the 1978 Protocol (MARPOL), the International Convention on Oil Pollution Preparedness, Response and Co-operation, 1990 (OPRC) and the International Convention Relating to Intervention on the High Seas in Cases of Oil Pollution Casualties, 1969 (the Intervention Convention).
36. For example, a provision may state that the Member State should 'give effect to' or 'implement' international rules and standards, but in other cases merely 'take account of' them. In some cases, the provision may say that the international rule or standard is 'applicable' or just 'generally accepted'. On the whole, the minimum percentage of the world merchant fleet requirements for an agreement to come into force suggests that IMO instruments fulfil the applicability or acceptance criteria. More formally, Article 311(2) of UNCLOS ensures that rights and obligations arising from other agreements shall not be affected provided that they are compatible with UNCLOS and do not affect the application of its basic principles. Article 237(1) provides with regard to the protection of the marine environment that previously concluded agreements and future agreements concluded in furtherance of the general principles of UNCLOS will be unaffected. Article 237(2) provides that specific obligations under conventions relating to the protection and preservation of the marine environment should be carried out in a manner consistent with the general principles and objectives of UNCLOS. See also *Implications*, footnote 34, pp. 7–8.
37. Adopted on 18 October 2001, see AFS/CONF/26. The issue of antifouling paint was first considered by the Marine Environment Protection Committee (MEPC) in 1988. See also MEPC Resolution 46(30) on Measures to Control Potential Adverse Impacts Associated with Use of Tributyl Tin Compounds in Antifouling Paints, adopted 16 November 1990; Assembly Resolution A.895(21), adopted 25 November 1999.
38. This is 12 months after ratification by 25 States representing 25% of the world's merchant shipping tonnage, see Article 18. In other conventions, such as MARPOL, the required tonnage would be 50% but it was thought that it would take too long for the ICAFS to come into force.
39. It will also establish a mechanism to restrict the use of other harmful substances in antifouling systems in the future. This is discussed further in Chapter 26.
40. Article 2(2).

41. Article 2(9).
42. Article 3(1). The Convention does not apply to warships or other state non-commercial ships, although parties should ensure that such ships comply with it in so far as reasonable and practicable, see Article 3(2). For ease of reference, term 'ship flying the flag' will be used to include ships operating under the authority of a State, a phrase which is used to cover offshore platforms.
43. Article 4(1). The EC legislation is discussed below. See, as another example of domestic implementation, the Australian Protection of the Sea (Harmful Antifouling Systems) Act 2006.
44. Article 12(1).
45. Article 12(2).
46. The requirement to remove altogether or cover with a barrier coating does not, however, apply to fixed and floating platforms, FSUs and FPSOs constructed prior to 1 January 2003 and that have not been in dry-dock on or after 1 January 2003. Since the Convention did not come into force until after 2003, the obligation did not apply until the Convention had become legally applicable. Despite this, it appears that the majority of shipowners have already implemented the deadline contained in the Convention; see International Chamber of Shipping and International Shipping Federation, Annual Review 2007, p. 5.
47. This should mean in practice no more than 2500 mg total tin per kg of dry paint. See Guidelines for Survey and Certification of Antifouling Systems on Ships, Resolution MEPC.102(48), adopted 11 October 2002, Appendix. When sampled, no sample must exceed the concentration 3000 mg SN per kg dry paint, see Guidelines for Brief Sampling of Antifouling Systems on Ships, Resolution MEPC.104(49), adopted 18 July 2003, p. 19.
48. Article 4(2).
49. Article 5.
50. Article 10 and Annex 4. Fixed or floating platforms, FSUs and FPSOs are excluded from this requirement, see Annex 4, Regulation 1(1). And see Guidelines for Survey and Certification of Antifouling Systems on Ships, Resolution MEPC.102(48), adopted 11 October 2002. Even before the Convention entered into force, the flag State could conduct surveys and issue a Statement of Compliance which would be sufficient.
51. Ibid.
52. Annex 4, Regulation 2(1). The flag State may request that another party carries out the survey and issues the Certificate, see Annex 4, Regulation 3. The validity of the Certificate ceases if the antifouling system is changed or replaced, unless endorsed after a fresh survey, or if the ship is transferred to another flag State, see Annex 4, Regulation 4. A record of antifouling systems used on the ship must be attached to the Certificate.
53. Annex 4, Regulation 5. The Declaration must at least be in English, French or Spanish.
54. Article 11(1), and see Guidelines for Brief Sampling of Antifouling Systems on Ships, Resolution MEPC.104(49), adopted 18 July 2003.
55. Guidelines for Inspection of Antifouling Systems on Ships, Resolution MEPC.105(49), adopted 18 July 2003, p. 3.
56. Article 11(2).
57. Guidelines for Inspection of Antifouling Systems on Ships, Resolution MEPC.105(49), adopted 18 July 2003, pp. 4–5.
58. Article 11(4).
59. Article 11(3). The port State must immediately notify the flag State of the action it is taking.
60. Article 13. Compensation shall be paid for any loss or damage suffered through undue detention or delay.
61. OJ L 115/1.
62. Preamble, recital 3.
63. Articles 2 and 6.
64. Article 3.
65. Article 4.
66. Article 5(1).

67. *TBT Regulation Consequences*, DG for Energy and Transport, 3 September 2003.
68. Article 5(2). None of these obligations apply to fixed and floating platforms, FSUs or FPSOs if they were constructed before 1 July 2003 and have not been in dry-dock on or after that date, see Article 5(3).
69. Commission Proposal for a Regulation of the European Parliament and of the Council on the Prohibition of Organotin Compounds on Ships, COM(2002) 396 final, p. 7.
70. Article 6(1)-(2); Annexes I, II and III.
71. Commission Proposal, see footnote 69, p. 10.
72. Council Directive 95/21/EC of 19 June 1995 concerning the enforcement, in respect of shipping using Community ports and sailing in the waters under the jurisdiction of the Member States, of international standards for ship safety, pollution prevention and shipboard living and working conditions, OJ L 157/1. And see also Council Directive 97/70/EC of 11 December 1997 setting up a harmonised safety regime for fishing vessels of 24 m in length and over, OJ L 34/1. And see Commission Proposal, footnote 69, p. 10.
73. Article 7, referring to Directive 95/21/EC concerning the enforcement, in respect of shipping using Community ports and sailing in the waters under the jurisdiction of the Member States, of international standards for ship safety, pollution prevention and shipboard living and working conditions, OJ L 157/1.
74. See European Maritime Safety Agency, Workshop Report on Implementation of Regulation (EC) 782/2003 on the Prohibition of Organotin Compounds on Ships, Lisbon, 14 May 2007, and Work Programme 2008, p. 33. Commission Regulation 536/2008 of 13 June 2008 giving effect to Article 6(3) and Article 7 of Regulation (EC) No 782/2003 on the prohibition of organotin compounds on ships and amending that Regulation, OJ L 156/10.
75. Commission Directive 1999/51/EC of 26 May 1999 adapting to technical progress for the fifth time Annex I to Council Directive 76/769/EEC on the approximations of the laws, regulations and administrative provisions of the Member States relating to restrictions on the marketing and use of certain dangerous substances and preparations (tin, PCP and cadmium), OJ L 142/22.
76. Annex I, Section 21 of Council Directive 76/769/EEC of 27 July 1976 on the approximation of the laws, regulations and administrative provisions of the Member States relating to restrictions on the marketing and use of certain dangerous substances and preparations, OJ L 262/201, amended by Commission Directive 2002/62/EC of 9 July 2002 adapting to technical progress for the ninth time Annex I to Council Directive 76/769/EEC on the approximation of the laws, regulations and administrative provisions of the Member States relating to restrictions on the marketing and use of certain dangerous substances and preparations, OJ L 183/58.
77. Ibid.
78. Directive 98/8/EC of the European Parliament and of the Council of 16 February 1998 concerning the placing of biocidal products on the market, OJ L 123/1.
79. Placing on the market includes supplying the product for payment or free of charge, and importing into the EC. See Article 2(1)(h).
80. Article 2(1). 23 product groups are covered by the Directive of which number 21 covers antifouling substances, see Directive Annex V. The Biocides Directive does not cover substances which are already governed by other legislation, such as plant protection products or veterinary medicines. See Article 1(2). For decisions on the scope of the Directive, the Commission has issued guidelines and the Manual of Decisions, available from http://ec. europa.eu/environment/biocides/index.htm. Accessed 8 July 2009. However, any substantive disagreement can only be resolved by the European Court of Justice.
81. Article 2(1)(d).
82. Article 2(1)(f).
83. Article 8, laying down specific requirements for the information to be contained in the dossier. Where the information is already held by the competent authority, it cannot be used for another applicant without the permission of the original applicant. In that case, the applicant may submit a letter of access. See Article 12.

84. Article 5. Annex I lists active substances. Annex IA will list active substances that can be used in 'low-risk biocidal products'. These are products which do not contain substances of concern, that is to say, substances which have an inherent capacity to cause an adverse effect on humans, animals or the environment (Article 2(1)(b) and (e)). Annex IB will list basic substances, that is to say, substances which are used primarily for non-pesticidal purpose but have minor use as a biocide, for example carbon dioxide, nitrogen, ethanol and acetic acid (Article 2(1)(c)). Low-risk biocidal products can be authorised by Member States under a less onerous procedure, provided that they are registered (Articles 3(2) and 8(3)). Member States may also employ frame-formulations, i.e., specifications for a group of biocidal products having the same use and user type (Article 3(4)).

85. Article 10(2).

86. Article 4. This is implemented in the UK through the Biocidal Products Regulations 2001, SI 2001 No. 880, amended by SI 2003 No. 429, SI 2005 No. 2451 and SI 2007 No. 293.

87. Articles 3(6), 6 and 7.

88. Article 5(1)(b), applied to substances to be included in Annex I through Article 10.

89. Article 10(1).

90. Producers, formulators or associations must also have identified them as active substances or notified their intention to support the active substance in specific product types in the review programme. See Commission Regulation (EC) No 1896/2000 on the first phase of the programme referred to in Article 16(2) of Directive 98/8/EC of the European Parliament and of the Council on biocidal products, OJ L 228/6, Articles 1–4. For up-to-date information on identified or notified substances, see the European Chemicals Bureau website at http://ecb.jrc.it/biocides. Accessed 8 July 2009.

91. Commission Regulation (EC) No 2032/2003 on the second phase of the 10-year work programme referred to in Article 16(2) of Directive 98/8/EC of the European Parliament and of the Council concerning the placing of biocidal products on the market, and amending Regulation (EC) No 1896/2000, OJ L 307/1. A process of identification and notification of substances is laid out in the first review regulation, Commission Regulation (EC) No 1896/2000 on the first phase of the programme referred to in Article 16(2) of Directive 98/8/EC of the European Parliament and of the Council on biocidal products, OJ L 228/6. This has been amended by the third phase review regulation, Commission Regulation (EC) No 1896/2000 on the first phase of the programme referred to in Article 16(2) of Directive 98/8/EC of the European Parliament and of the Council on biocidal products, OJ L 228/6.

92. Article 16.

93. Article 11. The information requirements for the active substance are contained in Annex II, III and IVA. The competent authority in the UK is the Biocides & Pesticides Unit of the Health and Safety Executive.

94. Articles 11 and 28. Member States may temporarily authorise the marketing of a biocidal product for up to 120 days for limited and controlled use if it appears necessary because of an unforeseen danger which cannot be contained by other means. See Article 15(1). Member States may also give provisional authorisation for a new active substance for up to 3 years if it is not listed in Annex I or IA. See Article 15(2).

Chapter 22
Techniques for the Quantification of Biofouling

*Alan J. Butler, João Canning-Clode, Ashley D.M. Coutts,
Phillip R. Cowie, Sergey Dobretsov, Simone Dürr, Marco
Faimali, John A. Lewis, Henry M. Page, Jonathan Pratten,
Derren Ready, Dan Rittschof, David A. Spratt,
Antonio Terlizzi and Jeremy C. Thomason*

(Authors are in alphabetical order.)

For this chapter, we have collated from the authors in this book approaches and techniques for the quantification and qualification of biofouling and have presented most of it in referenced tabular form to permit easy access to the information (Table 22.1). There are also two more thorough sections neither of which has been published before: one on the sampling of hulls of large vessels and another which details a broadly applicable advanced stereological technique for the quantification of fouling. The collection is unlikely to be fully comprehensive, but we have aimed to give the reader techniques for quantification of different levels of the biofouling community, from biofilm to macroorganisms, and from single species to the whole community. The techniques are for use in the office, laboratory or the field and range from cheap to costly. All of these techniques have proven their usefulness and with this chapter we would like to improve the exchange of approaches to biofouling quantification between the different disciplines. For this reason, the techniques come from across the basic and applied sciences in industry, biology and medicine.

MEASURING BIOFOULING

Ashley D. M. Coutts

This section describes a stratified sampling design to ensure that biofouling on vessels is representatively and accurately sampled.

22.1   Vessel hull

A single sampling design was used to determine the distribution and abundance of fouling organisms upon a vessel's hull. The hull was split into three lengths (bow, midship and stern) and three depths (top, middle and bottom). Bottom sampling sites were split into two due

Table 22.1 Summary of approaches and techniques for the quantification and qualification of biofouling.

	Parameter	Normal application	Technique	Equipment	Limitations	Suitability for deep sea studies	Industrial usefulness	Cost (cheap: <€1000, expensive: >€10 000)	Time (quick: 1 day, intermediate: 1–14 days, long: >14 days)	Comments	References
Biofilm	Community composition and richness	All	Colony-forming units	Anaerobic cabinet	Identification of cultivable organisms only	No	Yes	Cheap	Quick		[8]
Biofilm	Community composition and richness	All	Quantitative PCR	Quantitative PCR machine	Suitable primers	Yes	Yes	Expensive	Quick		[9]
Biofilm	Community composition and richness	All	Metagenomics	Cloning kits and sequencing apparatus	Extensive methodology	No	Yes	Expensive	Long		[10]
Biofilm	Function	All	Metabolic diversity	Biological plates, specification	Interpretation of results	No	Yes	Expensive	Quick		[11]
Biofilm	Function	Small samples	RNA/DNA stable isotope probing	Ultracentrifuge and sequencing equipment	Extensive methodology	No	Yes	Expensive	Long		[12]
Biofilm	Richness	All	16S rRNA comparative sequencing	PCR and sequencing apparatus	Gives no indication of the number of organisms present	Yes	Yes	Cheap	Intermediate		[8]
Biofilm	Structure	All	Confocal microscopy	Viability or lectin fluorescent staining, confocal microscope	Resolution limitations due to wavelength of light and small field of view	No	Yes	Expensive	Quick		[13]
Biofilm	Structure	All	Environmental electron microscopy	Environmental electron microscope	Small samples in environmental mode	No	Yes	Expensive	Quick		[14]
Biofilm	Comparison of community composition	Small samples, laboratory, field	Denaturing gradient gel electrophoresis (DGGE)	PCR, DGGE, electrophoresis, gel scanner, image analysis software	DNA extraction bias, PCR bias. Cannot estimate accurately the density of microorganisms. Separation of some bands in complex samples is difficult	Yes	Yes	Expensive, free software available	Intermediate	Some bands can be cut and sequenced to get identity of an organism.	[15–17]

Environment	Purpose	Scale	Method	Equipment	Limitations			Cost	Time	Notes	Reference
Biofilm	Comparison of community composition	Small samples, laboratory, field	Restriction length polymorphism (RFLP)	PCR, electrophoresis, gel scanner, image analysis software	DNA extraction bias, PCR bias. Cannot provide identity of the strains. Limited by specificity of enzymes used	Yes	Yes	Expensive	Intermediate	In combination with sequencer can be used for an automatic comparison of many samples.	[18]
Biofilm	Community composition and species richness	Small samples, laboratory, field	Isolation	Incubator, autoclave	Only small proportion of microorganisms can be cultivated	Yes	Yes	Cheap	Long	Widely used for isolation of microorganisms that produce bioactive compounds. Different media may result in isolation of different microorganisms	[19]
Biofilm	Community composition	Small samples, laboratory, field	Cloning	PCR, electrophoresis, water bath, hood	DNA extraction bias, PCR bias. Cannot estimate accurate the density of microorganisms. Very slow way to characterise a biofilm community. It is not suitable for large amount of samples	Yes	Yes	Expensive	Long	One can identify even non-cultivable microbes according to their DNA. Accuracy of this method depends on the size of the clone library.	[20, 21]
Biofilm	Composition and density	Small samples, laboratory, field	Laboratory-based, fluorescent in situ hybridization (FISH)	Epifluorescent microscope, water bath, incubator	Limited by specificity of fluorescent probes. Identify only known microorganisms. Cannot be used for thick and dense biofilms	Yes	Yes	Expensive	Intermediate	DAPI staining widely used for counting bacteria. Fluorescent probes can be group-, gene- or species-specific	[22]
Biofilm	Density of microorganisms	Small samples, laboratory, field	Scanning electron microscopy (SEM)	Scanning electron microscope	Destructive. Include fixation and dehydration	Yes	Yes	Expensive	Intermediate	Maximum magnification of the order of 100 000 times	[23]

(*Continued*)

Table 22.1 (*Continued*)

	Parameter	Normal application	Technique	Equipment	Limitations	Suitability for deep sea studies	Industrial usefulness	Cost (cheap: <€1000, expensive: >€10 000)	Time (quick: 1 day, intermediate: 1–14 days, long: >14 days)	Comments	References
Biofilm	Microbial cells and cell structures	Small samples, laboratory, field	Transmission electron microscopy (TEM)	Electron microscope, microtome	Destructive method include fixation and dehydration	Yes	Yes	Expensive	Intermediate	Maximum magnification of the order of 90 000 000 times	[23]
Biofilm	Cell structures and proteins	Single cells	Magnetic resonance force microscopy (MRFM)	Magnetic resonance force microscope	Only high magnification	No	Yes	Expensive	Quick	Non-destructive three-dimensional imaging with angstrom-scale resolution	[23]
Biofilm	Surface topography	Small samples, laboratory, field	Atomic force microscopy (AFM)	Atomic force microscope	Only high-resolution surface topography	Yes	Yes	Expensive	Quick	Maximum magnification of the order of 10^7 times	[23]
Biofilm	Three-dimensional structure	Small samples, laboratory, field	Laboratory-based – confocal scanning laser microscope (CSLM)	Confocal scanning laser microscope and image analysis software	Cannot use for thick biofilms	Yes	Yes	Expensive	Quick	Can be used with combination of other techniques, like FISH, for identification of different biofilm components. Not destructive and do not require biofilm fixation.	[24]
Biofilm	Isolation of whole DNA and genes	Small samples, laboratory, field	Metagenomics	RCR, sequencers, DNA alignment software	Require recovery of DNA sequences longer than 3000 bp from environment and their alignment	Yes	Yes	Extremely expensive	Long	Help to identify new types of bacteria and Archea, discover new genes and their products	[25]

	Parameter	Scale	Method	Equipment	Limitations			Cost	Time	Notes	Reference
Macrofouling	Composition and cover	Medium samples, field	Photoquadrats or surveys	SCUBA or ROV with camera, quadrat frame, manual or computer-assisted image analysis	Layering of taxa, standardisation, species identification	Yes	Yes	Moderate depending on depth, high below SCUBA depths	Intermediate, quick field, longer laboratory analyses	Useful for detection of exotic species; oil platforms	[26-27]
Macrofouling	Recruitment	Medium samples, field	Settlement surfaces, field-based	SCUBA, ceramic tiles, scrub pads, scrub brushes, direct/computer image analysis	Species identification, attachment to structure, field conditions, depth limitations	Yes	Yes	Moderate, depending on depth, high below SCUBA depths	Intermediate, long	Useful for detection of exotic species; oil platforms	[27–29]
Macrofouling	Biomass	Medium samples, destructive sampling	Scrape/vacuum small quadrats, field/laboratory-based	SCUBA, airlift vacuum, dissecting scope, analytical balance, drying oven	Field conditions, depth limitations	No	Yes	Moderate, depending on depth, high below SCUBA depths	Quick	Oil platforms	[29, 30]
Macrofouling	Rate of growth/biomass accumulation of individual fouling taxa	Medium samples, field	Growth of caged individuals, field-based	Field-based, SCUBA, cages	Field conditions	No	Yes	Moderate	Intermediate, long	Can include harvestable taxa (e.g. mussels); oil platforms	[31]
Macrofouling	Rate of biomass accumulation	Medium samples, field	Change in biomass over time on cleaned quadrats	Airlift vacuum, SCUBA	Field conditions, depth limitations	No	Yes	Moderate, depending on depth, high below SCUBA depths	Intermediate, long	Oil platforms	[32]
Macrofouling	Faunal litter fall	Medium samples, field	Collect sloughed macrofouling organisms in traps	Suspended hoops with mesh bags, SCUBA	Field conditions, suitable structure needed, depth limitations	No	Yes	Moderate, depending on depth	Intermediate, long	Oil platforms	[33]
Macrofouling	Percentage cover	Medium samples, field	Digital imaging, thresholding	Camera, computer, image analysis programme	May miss cryptic species or multilayered communities. Dependent on camera resolution	Yes	Yes	Cheap, free software available	Quick		[6]
Macrofouling	Percentage cover	Medium samples, field	Digital imaging, stereology	Camera, computer, image analysis programme	May miss cryptic species or multilayered communities. Dependent on camera resolution	Yes	Yes	Cheap, free software available	Quick		See text: S. Dürr, J.C. Thomason

(Continued)

Table 22.1 (Continued)

Parameter		Normal application	Technique	Equipment	Limitations	Suitability for deep sea studies	Industrial usefulness	Cost (cheap: <€1000, expensive: >€10 000)	Time (quick: 1 day, intermediate: 1–14 days, long: >14 days)	Comments	References
Macrofouling	Percentage cover	Ship-scale, field	Visual estimation	None	Cross standardisation required if more than one person	No	Yes	Cheap	Quick		[7, 34]
Macrofouling	Percentage cover	Medium (15 × 15 cm) samples	Visual estimation (in 5% steps)	15 × 15 cm grid from plastic wire	Cross standardisation required if more than one person	No	Yes	Cheap	Quick		[35, 36]
Macrofouling	Recruitment abundance	Test raft control panels	Laboratory, frequency assessment	Stereo microscope	Only suitable for monitoring early settlement; before macrofouling cover approaches 100%	No	Yes	Cheap	Intermediate		[37, 38]
Macrofouling	Biomass	Medium samples, field	Wet weight, dry weight, scraped samples, samples scraped from surfaces, quadrats, samples intact on fouling panels	Weighing scales	Wet weight is crude but good estimator, dry weight more precise but needs more time and may smell considerably	Yes	Yes	Cheap	Quick		[39]
Macrofouling	Settlement behaviour	Small scale, field	Direct observation by divers	Direct observation, timer, camera	Suitable only for large larvae; detects settlers close to release but longer dispersers are lost	No	Yes	Cheap to expensive	Intermediate	Useful for understanding natural dynamics of fouling communities	[40, 41]
Macrofouling	Larval behaviour	Small samples laboratory	Larval chromatography	Glass tubes, test tubes, flow meter, dissecting microscope	Supply of larvae and their handling	No	Yes	Cheap	Quick	Useful for additives, basic research, and testing surface properties.	[42–45]

Macrofouling	Toxicity	Small samples, laboratory	Second-stage barnacle nauplii	Test tubes, temperature control supply of larvae	Supply of larvae	No	Yes	Cheap	Quick	Useful for additives, natural products, testing toxicity of leachates	[46]
Macrofouling	Settlement assay	Small samples, laboratory	Barnacle cyprids	Containers and incubator	Supply of larvae	No	Yes	Cheap	Quick	Useful for additives, natural products, testing toxicity of leachates	[42, 46]
Macrofouling	Anti-settlement and foul release	Small samples, field	Field settlement arrays	Field site and means of deployment. Dissecting microscope	Secure field site	Yes	Yes	Cheap	Quick to long	An inexpensive replicated way to compare series for coatings with relatively small amounts of material	[47, 48]
Macrofouling	Barnacle strength of adhesion after re-attachment	Small samples, laboratory	Re-attachment of 0.5 cm basal diameter barnacles	Facilities to settle and grow on barnacles	Requires continued care of re-attaching barnacles	No	Yes	Cheap	Long	Useful as a screening device to down-select from large numbers of coatings for field testing.	[49]
Macrofouling	Cyprid strength of adhesion	Laboratory samples	Removal of individual cyprids form a surface	Microbalance, high-quality microscope	Supply of cyprids (wild caught or laboratory raised). Handling very small larvae	No	Yes	Expensive	Intermediate	Not often used, but may be a good screening assay	[50, 51]
Macrofouling	Barnacle strength of adhesion	Field samples	Naturally settled barnacles on coatings	Hand-held force gauge and optical scanner	Only really useful for foul release coatings as barnacles need to be removed intact	Yes	Yes	Cheap	Intermediate	A good screening assay for field samples. Widely used	[52]

(Continued)

Table 22.1 (*Continued*)

	Parameter	Normal application	Technique	Equipment	Limitations	Suitability for deep sea studies	Industrial usefulness	Cost (cheap: <€1000, expensive: >€10 000)	Time (quick: 1 day, intermediate: 1–14 days, long: >14 days)	Comments	References
Macrofouling	Barnacle strength of adhesion	Small samples, laboratory	Barnacles grown in laboratory	Automated fouling-release tester; facilities to settle and grow on barnacles	Only really useful for foul release coatings as barnacles need to be removed intact	No	Yes	Expensive	Long	A good screening assay for laboratory samples but needs made-to-order push-off machine	[51]
Macrofouling	Nature and extent, and biosecurity risk	Laboratory	Utilisation of video footage from in-water hull inspection surveys	Cameras, image processing software	Cannot confirm species present, only use higher taxonomic groupings	Yes	Yes, could be used by industry to assess the patterns of fouling on different vessels and performance of antifouling coatings at various hull locations.	Cheap	Long	Cost-effective method for quantifying levels of fouling on vessels at various hull locations	[53]
Macrofouling	Nature and extent, biosecurity risk	Generally small samples	Field, in situ, samples, laboratory	Cameras, quadrats, scrapers, collectors	Occupational health and safety issues	No	As above	Expensive	Long	Most accurate method for quantifying nature and extent of fouling on artificial structures	See text: A. Coutts

to the presence of dry-docking support strips (DDSS). Sampling occurred both inside and outside of the DDSS. All merchant vessels possess an intertidal zone, similar to a seashore. This zone refers to the area between the vessel's highest and lowest watermark. Top (shallow) sampling sites were empirically determined as a point just below the highest marine organism sighted on the side of the vessel's hull. Middle sampling positions were defined as the half way point between the top sampling sites on the sides of vessels and the bilge keels. Bottom sampling sites were underneath the vessel around the keel region in the middle of the vessel, inside and out of DDSS. Sampling sites at the bow were within 5–10% of the total length of the vessel. For instance, sampling commenced between 10 and 20 m from the bow of a vessel 200 m in length. Stern sampling sites commenced just forward of the super structure of vessels to avoid ballast intake vents (sea chests) and propeller regions. Midship sampling sites were identified as the area just forward of the vessel's load limit markings. Sampling was conducted during daylight hours at slack tide. Only one side (port or starboard) was surveyed per vessel. For safety reasons, all vessels were sampled on the riverside (away from the wharf) which was determined by the berthing side of the vessel. Sampling commenced at either the bow or stern depending on tidal movements and commenced at the deepest sampling sites (inside/out of DDSS) progressing towards the surface to ensure a safer diving profile. Each of the 12 locations was sampled six times using a quadrat measuring 0.225 m^2. A total of three DDSS were sampled within each location, i.e. bow, midship and stern by placing two quadrats inside each DDSS. The orientation and placement of quadrats was haphazard. Six quadrats were also haphazardly placed immediately outside the DDSS within each location. Quadrats were temporarily attached to the hull using neodymium iron boron magnets. Upon placement, a Nikonos V underwater camera with a 28 mm lens plus Nikonos overlens and strobe attached was used with 100 SA Fujichrome film to photograph each quadrat. After photographing, a paint scraper was used to remove and transfer all fouling organisms present within the quadrat area into 330 × 330 mm plastic snaplock bags. Seventy-two photographs and samples were required for each vessel to satisfy the sampling design.

22.2   Steel Mariner barge

An initial inspection of the 'Steel Mariner' revealed that heavy biofouling was distributed over the entire submerged area of the hull. Four 25 m transects were therefore randomly placed underneath the barge (port to starboard). Eight random 0.25 m^2 quadrats, two on each of the vertical sides and four underneath the barge along each of the four transects, were selected to assess the type and extent of biofouling on the hull. A Nikonos III underwater camera with a 28 mm lens and 200 ASA Fujichrome film was used to photograph all 32 quadrats to assist with identifying various fouling organisms. Two divers, one using a paint scraper and the second to hold the collection bag, removed and collected all of the biofouling in each quadrat. Biofouling from each quadrat was transported to the surface inside the collection bags. Samples were transferred to 500 μm bags then allowed to drain for 5 minutes before being weighed using hand-held scales. Given the excessive amount of biofouling, the wet weight of all material collected from each quadrat was recorded in kilogrammes. All biofouling from each quadrat was preserved in separate containers using 5% formaldehyde and 95% seawater for taxonomic analysis. Estimates of the quantity of the biofouling on the barge were determined by calculating the mean and standard error of wet biomass weight of biofouling per square metre for each side (port and starboard) and the bottom of the barge. These were then scaled by the submerged area of the hull for each stratum (given the marked differences in the level of biofouling between strata). An overall estimate of wet biomass biofouling on the barge was also determined by calculating the mean and standard error of wet biomass weight of biofouling per square metre amongst all 32 quadrats and then scaling these by the total submerged area of the hull.

STEREOLOGICAL ANALYSIS OF BIOFOULING

Simone Dürr and Jeremy C. Thomason

This section describes a digital image analysis technique for the quantification of biofouling which incorporates best practice stereology from medical science to reduce the measurement error and therefore improve both the accuracy and precision of data and to obtain an unbiased estimation. Additionally, the analysis process we portray was implemented as a partially automated process using self-written plug-ins for the freeware ImageJ [1] in order to facilitate a high throughput of photographs ($>10\,000$ photographs). This software has great advantages of no set-up cost, an extensive user base ($>80\,000$ downloads), online support and platform independence. User-compiled plug-ins can be directly downloaded from a large online collection on the developer's website. These can easily be adapted to the user's own needs or they can be easily written with a little practise in Java. This makes ImageJ very adaptable to the individual user, connectable to other programmes and file formats, and at the same time independent of commercial developers. All of this is available without any costs. These are advantages that are not substantially covered by other free image analysis programmes such as Photogrid [2] and NIH Image [3].

The technique we describe was successfully used to analyse digital photographs taken by field workers of non-scientific background in a pan-European project studying biofouling in the aquaculture industry [4]. Photographs were taken by workers at 11 sites using standard protocols and equipment. They were edited in batches using Adobe Photoshop® and analysed in ImageJ using an advanced stereological technique known as random systematic sampling (RSS) combined with a fractionator [5,6]. In RSS, the photograph is apportioned into identically sized fields of view, with the number of fields and size of fields depending both on organism size and the area that is to be analysed. A single field should be big enough not to be filled by a single organism but still small enough to represent a sample of the whole area. The size of the field has to remain constant within one analysis. Furthermore, only some of these fields need to be analysed: this is determined a priori. This is the principle of the fractionator. Only a fraction of the whole area is analysed and thus not all of the defined fields need analysing. The number of fields to be analysed depends on the measurement error desired:

$$\text{Measurement error} = \frac{\sqrt[2]{n}}{n}$$

where n is the number of dots enumerated (Figure 22.1).

The number of fields also depends on the distribution of the organism(s) to be investigated such that the more uniform the distribution the fewer fields are necessary. If the selection of fields and placement of dots is automated then the time needed for analysis will not increase

(a)   (b)

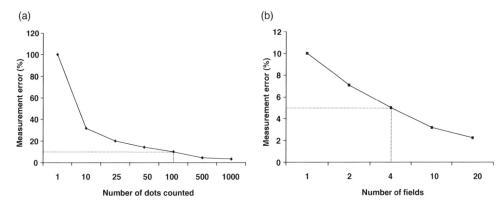

Figure 22.1   Relationship between measurement error and the number of dots counted (a) and the number of fields investigated (b), respectively. The measurement error for the number of dots counted (a) is for the analysis of one field only. The measurement error for the number of fields used is with the number of dots held constant at 100. Dotted lines show parameters used in the CRAB project [4].

concomitantly with an increase in the number of fields analysed. Identification of the species under the dot is done by the user and therefore incurs a time penalty. However, if the total number of dots is not increased then the time required for analysis will also not increase. In fact by increasing the number of fields, the number of dots can be decreased whilst not altering the measurement error (Figure 22.1).

We based our automated process on the use of nine fields. The first field of view to be analysed was selected randomly by ImageJ. Starting from this field, the other fields were determined uniformly and continuously. We decided to select every 5th field and selected 4 fields in total, thus giving a fraction of 4/9. As in the basic stereological approach used for sessile community analysis [7], we analysed organisms under 100 virtual dots, but in this case the dots were dispersed equally between the four selected fields of view. Placement of dots was determined by a uniform grid. For every field, the number of dots overlaying each species was determined and then summed for all the fields of view and every species. The result was percentage cover for every species in the photograph. Field selection and placement of grids were automated using plug-ins. Values for every species were input into Microsoft Excel and automatically summed. Therefore, there was no increase in time compared to the conventional random dot stereological technique. Furthermore, this advanced stereological technique combining random systematic sampling with a fractionator decreases the measurement error of the results compared to the standard random dot stereological technique by at least a factor of two [5, 6]. If the same measurement error is acceptable then less time is needed to achieve the same results. This is an improvement when compared to visual estimation techniques [7] or less sophisticated software as it provides an increase in accuracy which leads to a reduction in the variance of biofouling caused by the measurement process and hence more powerful statistics.

References

1. Rasband, W.S. (1997–2008) *ImageJ*. US National Institutes of Health, Bethesda. Available from http://rsb.info.nih.gov/ij/. Accessed 8 July 2009.

2. Bird, C.E. (2003) *Photogrid – Ecological Analysis of Digital Photographs*. Hawaii. Available from http://www.photogrid.netfirms.com/. Accessed 8 July 2009.
3. Rasband, W.S. (2002) *NIH Image*. US National Institutes of Health, Bethesda. Available from http://rsbweb.nih.gov/nih-image/index.html. Accessed 8 July 2009.
4. CRAB (2004–2007) *Collective Research in Aquaculture Biofouling*. EU FP6 COLL-CT-2003–500536-CRAB. Available from http://www.crabproject.com. Accessed 8 July 2009.
5. Howard, C.V. & Reed, M.G. (1998) *RMS Handbook 41: Unbiased Stereology – Three-Dimensional Measurement in Microscopy*. Bios, Oxford.
6. Russ J.C. & Dehoff R.T. (2000) *Practical Stereology*, 2nd edn. Kluwer Academic/Plenum, New York.
7. Meese, R.J. & Tomich, P.A. (1992) Dots on the rocks – a comparison of percent cover estimation methods. *Journal of Experimental Marine Biology and Ecology*, **165** (1), 59–73.
8. Pratten, J, Wilson, M. & Spratt, D.A. (2003) Characterization of *in vitro* oral bacterial biofilms by traditional and molecular methods. *Oral Microbiology and Immunology*, **18**, 45–49.
9. Dalwai, F., Spratt, D.A. & Pratten, J. (2007) The use of quantitative PCR and culture methods to characterise ecological flux's in bacterial biofilms. *Journal of Clinical Microbiology*, **45** (9), 3072–3076.
10. Diaz-Torres, M.L., Villedieu, A., Hunt, N., *et al.* (2006) Determining the antibiotic resistance potential of the indigenous oral microbiota of humans using a metagenomic approach. *FEMS Microbiology Letters*, **258** (2), 257–262.
11. Spratt, D.A. & Pratten, J. (2005) Carbon substrate utilisation as a method of studying biofilm development. *Biofilms*, **2**, 239–243.
12. Neufeld, J.D., Dumont, M.G., Vohra, J. & Murrell, J.C. (2007) Methodological considerations for the use of stable isotope probing in microbial ecology. *Microbial Ecology*, **53** (3), 435–442.
13. Hope, C.K. & Wilson, M. (2003) Measuring the thickness of an outer layer of viable bacteria in an oral biofilm by viability mapping. *Journal of Microbiological Methods*, **54** (3), 403–410.
14. Muscariello, L., Rosso, F., Marino, G., *et al.* (2005) A critical overview of ESEM applications in the biological field. *Journal of Cell Physiology*, **205** (3), 328–334.
15. Muyzer, G., de Waal, E.C. & Uitterlinden, A.G. (1993) Profiling of complex microbial population by DGGE analysis of polymerase chain reaction amplified genes encoding for 16S rRNA. *Applied Environmental Microbiology*, **62**, 2676–2680.
16. Zang, T. & Fang, H.H.P. (2000) Digitization of DGGE (denaturing gradient gel electrophoresis) profile and cluster analysis of microbial communities. *Biotechnology Letters*, **22**, 399–405.
17. Gafan, G.P., Lucas, V.S., Roberts, G.J., Petrie, A., Wilson, M. & Spratt, D.A. (2005) Statistical analyses of complex denaturing gradient gel electrophoresis profiles. *Journal of Clinical Microbiology*, **43** (8), 3971–3978.
18. Liu, W.T., Marsh, T.L., Cheng, H. & Forney, L.J. (1997) Characterization of microbial diversity by determining terminal restriction fragment length polymorphisms of genes encoding 16S rRNA. *Applied Environmental Microbiology*, **63**, 4516–4522.
19. Paul, H. (2001) *Marine Microbiology. Methods in Microbiology*, Vol. **30**. Academic Press, San Diego.
20. Fuhrman, J.A., Griffith, J.F. & Schwalbach, M.C. (2002) Prokaryotic and viral diversity pattern in marine plankton. *Ecological Research*, **17**, 183–194.
21. Sekiguchi, Y., Kamagata, Y., Syutsubo, K., Ohashi, A., Harada, H. & Nakamura, K. (1998) Phylogenetic diversity of mesophilic and thermophilic granular sludge determined by 16S rRNA gene analysis. *Microbiology*, **144**, 2655–2665.
22. Amann, R.I., Krumholz, L. & Stahl, D.A. (1990) Fluorescent-oligonucleotide probing of whole cells for determinative, phylogenetic, and environmental studies in microbiology. *Journal of Bacteriology*, **172**, 762–770.
23. Beech, J.B., Tapper, R.C. & Gubner, R.J. (2000) Microscopy methods for studying biofilms. In: *Biofilms: Recent Advances in their Study and Control* (ed. L.V. Evans), pp. 51–70. Harwood Academic Publishers, Amsterdam.

24. Barranguet, C., Beusekom, S.A.M., Veuger, B., *et al.* (2004) Studying undisturbed autotrophic biofilms: still a technical challenge. *Aquatic Microbiology and Ecology*, **34**, 1–9.
25. Handelsman, J. (2004) Metagenomics: application of genomics to uncultured microorganisms. *Microbiology and Molecular Biology Reviews*, **68**, 669–685.
26. Gallaway, B. & Lewbel, G. (1982) *The Ecology of Petroleum Platforms in the Northwestern Gulf of Mexico: A Community Profile*. Biological Services Program FWS/OBS-82/72, BLM Open File Report 82–03, US Department of the Interior, Fish and Wildlife Service, USA.
27. Carney, R.S. (2005) *Characterization of Algal-invertebrate Mats at Offshore Platforms and the Assessment of Methods for Artificial Substrate Studies*. OCS Study MMS 2005–038, US Department of the Interior, Minerals Management Service, New Orleans.
28. Page, H.M., Dugan, J.E., Culver, C.S. & Hoesterey, J. (2006) Exotic invertebrate species on offshore oil platforms. *Marine Ecology Progress Series*, **325**, 101–107.
29. Bram, J.B., Page, H.M. & Dugan, J.E. (2005) Spatial and temporal variability in early successional patterns of an invertebrate assemblage at an offshore platform. *Journal of Experimental Marine Biology and Ecology*, **317**, 223–237.
30. Page, H.M., Dugan, J.E., Schroeder, D.M., Nishimoto, M.M., Love, M.S. & Hoesterey, J. (2007) Ecological performance and trophic links: comparison among offshore oil platforms and natural reefs for a selected fish and their prey. *Marine Ecology Progress Series*, **344**, 245–256.
31. Page, H.M. & Hubbard, D.M (1987) Temporal and spatial patterns of growth in mussels, *Mytilus edulis*, on an offshore platform: relationships to water temperature and food availability. *Journal of Marine Biology and Ecology*, **111**, 159–179.
32. Bram, J.B., Page, H.M. & Dugan J.E. (2004) *Spatial and Temporal Variability in Early Successional Patterns of an Invertebrate Assemblage at an Offshore Oil Platform*. Technical Report MMS OCS Study 2005–003, US Department of the Interior, Minerals Management Service, New Orleans.
33. Page, H.M., Dugan, J.E., Dugan, D. & Richards, J. (1999) Effects of an offshore oil platform on the distribution and abundance of commercially important crab species. *Marine Ecology Progress Series*, **185**, 47–57.
34. Drummond, S.P. & Connell, S.D. (2005) Quantifying percentage cover of subtidal organisms on rocky coasts: a comparison of the costs and benefits of standard methods. *Marine and Freshwater Research*, **56** (6), 865–876.
35. Canning-Clode, J., Kaufmann, M., Molis, M. Wahl, M. & Lenz, M. (2008) Influence of disturbance and nutrient enrichment on early successional fouling communities in an oligotrophic marine system. *Marine Ecology*, **29** (1), 115–124.
36. Valdivia, N., Heidemann, A., Thiel, M., Molis, M. & Wahl, M. (2005) Effects of disturbance on the diversity of hard-bottom macrobenthic communities on the coast of Chile. *Marine Ecology Progress Series*, **299**, 45–54.
37. Lewis, J.A. (1981) *A Comparison of Possible Methods for Marine Fouling Assessment during Raft Trials*. DSTO Report MRL-R-808, Defence Science and Technology Organisation, Melbourne.
38. Standards Australia (2004) *Paints and Related Materials – Methods of Test. Method 481.5: Coatings – Durability and Resistance to Fouling – Marine Underwater Paint Systems*. Australian Standard AS 1580.481.5 – 2004. Standards Australia, Sydney.
39. Wahl, M., Molis, M., Davis, A., *et al.* (2004) UV effects that come and go: a global comparison of marine benthic community level impacts. *Global Change Biology*, **10**, 1962–1972.
40. Davis, A.R. & Butler, A.J. (1989) Direct observations of larval dispersal in the colonial ascidian *Podoclavella moluccensis* Sluiter: evidence for closed populations. *Journal of Experimental Marine Biology and Ecology*, **127**, 189–203.
41. Davis, A.R., Butler, A.J. & van Altena, I. (1991) Settlement behaviour of ascidian larvae: preliminary evidence for inhibition by sponge allelochemicals. *Marine Ecology Progress Series*, **72**, 117–123.
42. Rittschof, D., Branscomb, E.S. & Costlow, J.D. (1984) Settlement and behavior in relation to flow and surface in larval barnacles, *Balanus amphitrite* Darwin. *Journal of Experimental Marine Biology and Ecology*, **82**, 131–146.

43. Rittschof, D. & Costlow, J.D. (1989) Surface determination of macroinvertebrate larval settlement. In: *Proceedings of the 21st European Marine Biology Symposium, Gdansk, 14–19 September 1986* (ed. E. Styczynska-Jurewicz), pp. 155–163. Polish Academy of Sciences, Institute of Oceanology, Poland.

44. Rittschof, D. & Costlow, J.D. (1989) Bryozoan and barnacle settlement in relation to initial surface wettability: a comparison of laboratory and field studies. In: *Topics in Marine Biology, Proceedings of the 22nd European Marine Biology Symposium, August 1987* (ed. J.D. Ros), pp. 411–416. Instituto de Ciencias del Mar, Barcelona.

45. Qian, P.Y., Rittschof, D. & Sreedhar, B. (2000) Macrofouling in unidirectional flow: miniature pipes as experimental models for studying the interaction of flow and surface characteristics on the attachment of barnacle, bryozoan and polychaete larvae. *Marine Ecology Progress Series*, **207**, 109–121.

46. Rittschof, D., Clare, A.S., Gerhart, D.J., Avelin, S.M. & Bonaventura, J. (1992) Barnacle *in vitro* assays for biologically active substances: toxicity and settlement assays using mass cultured *Balanus amphitrite amphitrite* Darwin. *Biofouling*, **6**, 115–122.

47. Rittschof, D., Clare, A.S., Gerhart, D.J., Bonaventura, J., Smith, C., & Hadfield, M. (1992) Rapid field assessment of antifouling and foul-release coatings. *Biofouling*, **6**, 181–192.

48. Vasishtha, N., Sundberg, D.C. & Rittschof, D. (1995) Evaluation of release rates and control of biofouling using monolithic coatings containing an isothiazolone. *Biofouling*, **9**, 1–16.

49. Rittschof, D., Orihuela, B., Stafslien, S. *et al.* (2008) Barnacle reattachment: a tool for studying barnacle adhesion. *Biofouling*, **24**, 1–9.

50. Yule, A.B. & Crisp, D.J. (1983) Adhesion of cypris larvae of the barnacle, *Balanus balanoides*, to clean and Arthropodin treated surfaces. *Journal of the Marine Biological Association of the United Kingdom*, **63**, 261–271.

51. Berglin, M., Larsson, A., Jonsson, P.R., & Gatenholm, P. (2001) The adhesion of the barnacle, *Balanus improvisus*, to poly(dimethylsiloxane) fouling-release coatings and poly(methyl methacrylate) panels: the effect of barnacle size on strength and failure mode. *Journal of Adhesion Science and Technology*, **15**(12), 1485–1502.

52. ASTM International (1994) *ASTM D 5618 Standard Test Method for Measurement of Barnacle Adhesion Strength in Shear*. American Society for Testing and Materials, USA.

53. Coutts, A.D.M. & Taylor, M.D. (2004) A preliminary investigation of biosecurity risks associated with biofouling on merchant vessels in New Zealand. *New Zealand Journal of Marine and Freshwater Research*, **38**, 215–229.

A

B

C

Plate I A) Complex life cycle of a broadcast-spawning serpulid polychaete, *Galeolaria caespitose* (Chapter 1), a calcareous tube-building worm that commonly biofouls intertidal marine structures across temperate Australia. Sessile adult males and females release sperm and eggs freely into the sea where, providing gametes drift into each other, fertilisation takes place almost immediately. Alternatively, if spawners/gametes are not in abundance or are released asynchronously, fertilisation may fail (see "sperm limitation" in Chapter 1). An unfertilised egg is shown here (A). Swimming, planktotrophic larvae feed, grow and develop for ~2 weeks in the open water column; (trochophore), larvae shown here two days after spawning/fertilization (B). Larvae then enter a demersal phase where they explore the substratum, settle and start building a tube, eventually growing to a sexually mature adult ~5 cm long (C). Worms settling preferentially near or on other worms can lead to the formation of large colonies of many thousands of worms, extending up 60 cm out from the substratum (D). Such aggregations probably increase the likelihood of high concentrations of released gametes and successful fertilisation taking place during spawning (Copyright: C. A. Styan). B). Use of individual marking in experiments on dispersal and recruitment (Chapter 2). Mussel settlers (*Perna perna*) of approximately 500 μm viewed under UV light. The central individual has been immersed in the fluorochrome calcein stain and shows a greenish growth band where calcein has been incorporated into the shell as a calcium substitute (Copyright: C. von der Meden). C) The intertidal species inhabiting this rocky shore occur in distinct horizontal 'zones' (Chapter 3). The limit of the highest spring tides is clearly marked by a grey band of the barnacle *Chthamalus montagui*, which merges into the zone of *Semibalanus balanoides* beneath it, then a bright green band of *Ulva* sp. is finally followed by the fucoid algal zone, which is exposed only during low tides (Copyright: G. S. Prendergast).

Plate II (A) Biofouling communities on mooring chain, Fara, Scapa Flow, at 5 m depth (Chapter 6) and (B) 10 m depth (Chapter 6; copyright: B. Forbes, Scientific Underwater Logistics and Diving, Stromness, Orkney). (C) Three *Metridium farcimen* positioned directly on a cable in 134 m of water (Chapter 6; reprinted from Reference [41], with permission from Elsevier. Copyright 2007). (D) Subsea cable adjacent to a cluster of porifera, *Florometra serratissima* (crinoids), and *Anthomastus ritteri* (mushroom coral) in 1160 m water depth on Pioneer Seamount (Chapter 6; reprinted from Reference [41], with permission from Elsevier. Copyright 2007). (E) Hydrozoan epibionts (Chapter 7) and (F) Barnacle epibionts on the blue mussel *Mytilus edulis* (Chapter 7; copyright: M. Wahl).

Plate III (A) Benthic sessile organisms need to be epibiont free since they perform their vital functions through their surface. Most of them have developed a variety of antifouling mechanisms (Chapter 8), which range from micro-rugose surface structure, such as the sponge *Corticium candelabrum* (a), continuous cell renewal, such as the calcareous encrusting algae *Lithophyllum incrustans* (b), and release of bioctive chemicals, such as the sponges *Hemimycale colummela* (c) and *Crambe crambe* (d) (copyright: I. Uriz). (B) Barnacle fouling (*Megabalanus coccopoma*, *Sriatobalanus amaryllis*, *Amphibalanus* spp.) on a docking block support strip on the flat bottom of a bulk carrier (Chapter 24; copyright: A. Gillham, Australian Shipowners Association). (C) Barnacles (*Amphibalanus* spp.) and hydroids on the bars of a sea chest intake grate (Chapter 24; copyright: A. Gillham, Australian Shipowners Association). (D) Aggregations of blue mussels (*Mytilus galloprovincialis planulatus*) inside a ship's sea chest (Chapter 24; copyright: J. Lewis, ES Link Services). (E) Shore crab, *Metopograpsus messor*, associated with fouling on the emergency propulsion unit of a ship arriving in south-western Australia after a voyage from the northern Indian Ocean (Chapter 24; copyright: P. Smith, URS Australia).

A

B

C

D

E

F

Plate IV (A–D) Biofouling colonisation residual inside cooling water system (Chapter 12; copyright: A. Terlizzi). (E) Multi-substrata replacement system by plastic net, panels (cembonit and plastic) and rope (natural and artificial) (Chapter 12; copyright: A. Terlizzi). (F) A photograph showing caries of a very advanced stage in a pre-pubertal child (Chapter 11; copyright: D. Spratt, D. Ready and J. Pratten).

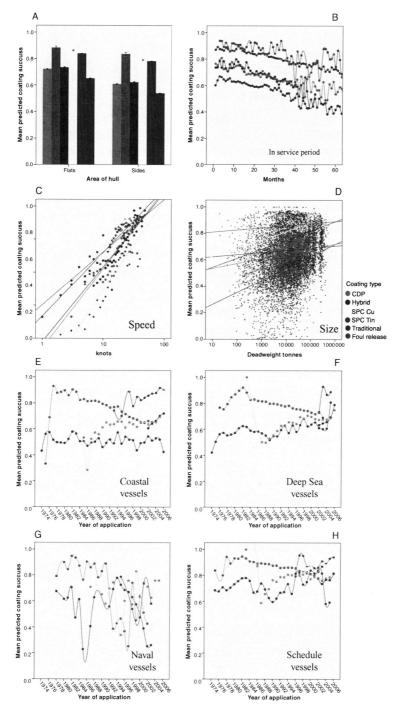

Plate V (A) Coatings performance is better on the shaded flats of vessel compared to the well-lit sides. The odds ratio of flat versus side is 0.576 ($p < 0.001$). Means with error bars = 95 % confidence intervals are shown. Sample size = 152 746 (Chapter 14). (B–D) Coating performance decreases with (B) increasing in-service interval (odds ratio = 0.979 $p < 0.001$), but conversely increases with (C) increasing speed (odds ratio = 1.063, $p < 0.001$) and also increasing (D) vessel size (odds ratio = 1.00000035, $p = 0.004$). Means with fitted regression lines are shown. Sample size = 152 746 (Chapter 14). (E–H) Performance of different coating technologies have changed over the last 30 years with the changes being more or less apparent depending on the type of vessel and type of coating used. Differences between coatings are marked for coastal (E) deep sea (F) and schedule (H) vessels, with the new foul release and MA-SPC coatings performing well when compared with TBT-SPCs. Small sample sizes and stop–start operating schemes obscure the patterns for naval vessels (G). Means with a spline-fit line are shown. Sample size = 152 746 (Chapter 14).

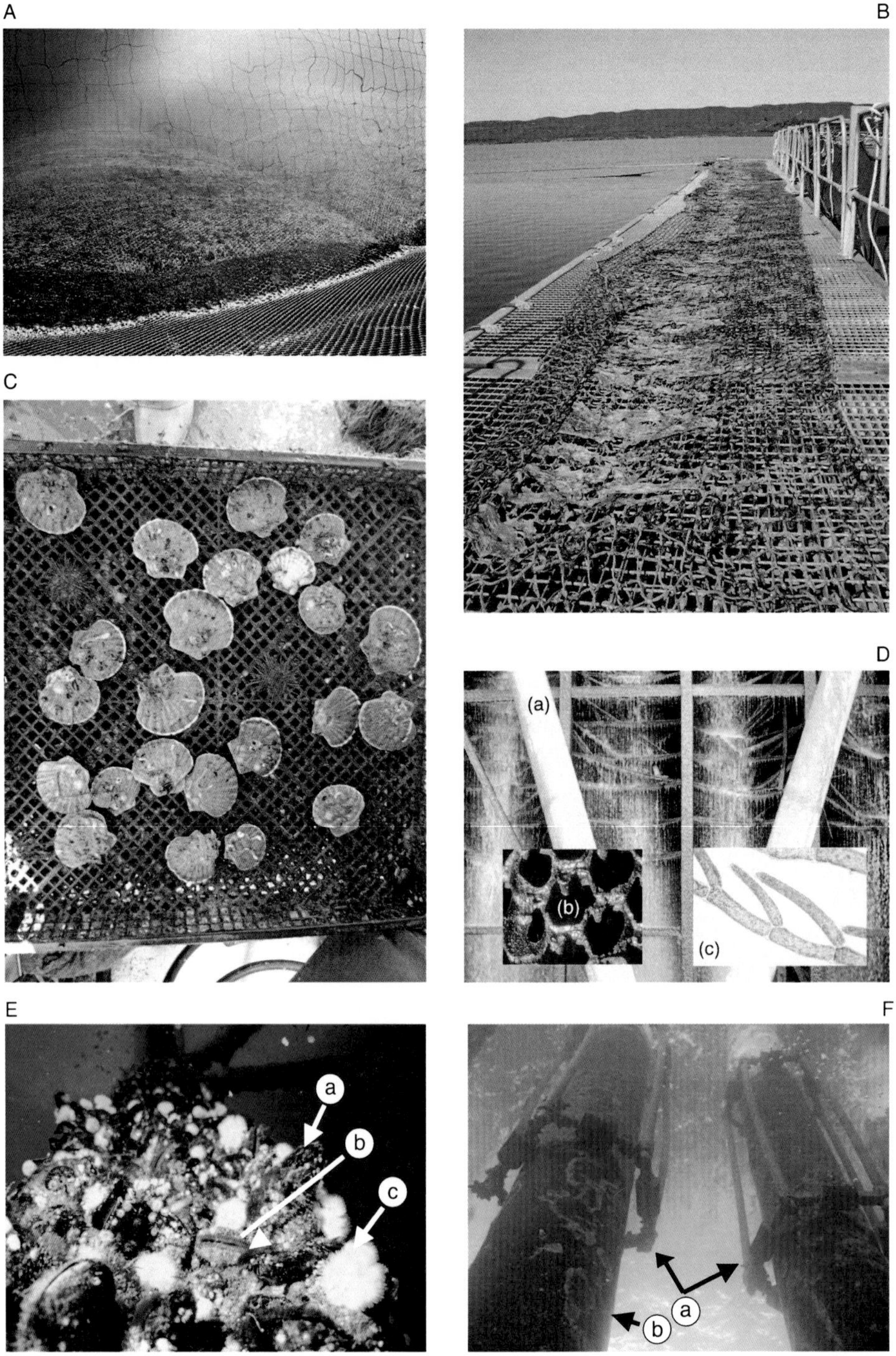

Plate VI (A) Finfish net in the Western Mediterranean densely covered in soft-tube-building amphipods and polychaetes (Chapter 19; copyright: S. Dürr). (B) Air-drying (husbandry) of a part of a fouled finfish net in Western Scotland (Chapter 19; copyright: S. Dürr). (C) Use of sea urchins as biological control in scallop culture in Ireland (Chapter 19; copyright: D. I. Watson). (D) Fouling of a cooling tower at High Marnham Power Station, England. (a) View of the lower section of the tower showing collapsed packing. (b) Close-up of the plastic packing showing *Cladophora* fouling and build-up of debris. (c) View under a microscope of *Cladophora*, the fouling alga (Chapter 20; copyright: P. Henderson). (E) View of the fouling assemblage on a horizontal support member at a depth of 9 m at Platform Holly in the Santa Barbara Channel (Chapter 18). Conspicuous species include the mussel, *Mytilus californianus* (a), rock scallop, *Crassadoma gigantea* (b) and anemone, *Metridium senile* (c). Copyright: D.S. Dugan. (F) Marine growth preventor (MGP) (a) in place on the conductor pipes (b) of Platform Grace in the Santa Barbara Channel (Chapter 18; copyright: G. Sanders, US Minerals Management Service).

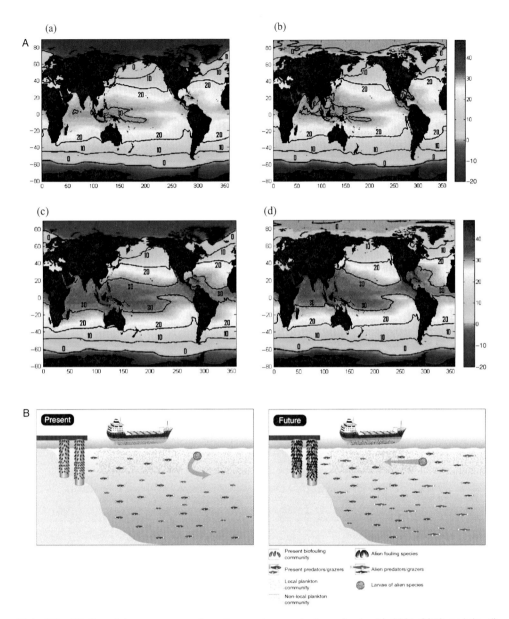

Plate VII (A) Decadal mean sea surface temperature projections for (a, b) 2001–2010 and (c, d) 2091–2100 from CSIRO Mk 3.5 General Circulation Model (GCM) simulation under greenhouse gas emission scenario SRES A2. (a, c) annual and (b, d) June–August. $0°$, $10°$, $20°$ and $30°$ contours shown (Chapter 23). All GCM predictions downloaded from the IPCC data hosted by PCMDI and processed at CSIRO marine research by A. Hobday. (B) Schematic of possible changes in temperate biofouling communities with climate change from present day to future. As climate warms, biofouling growth is more aggressive, alterations of the trophic web change the biofouling communities present, and alteration of ocean currents bring new biofouling species. (Chapter 23).

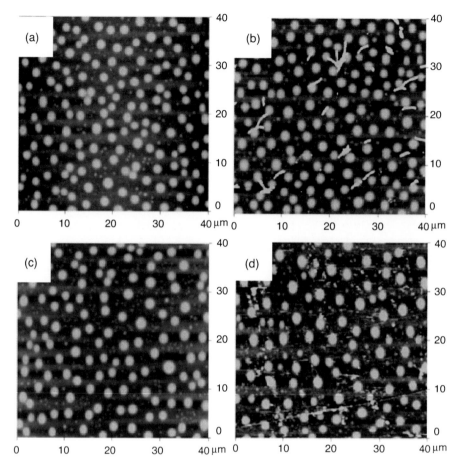

Plate VIII Topographical AFM images of microstructured siloxane–urethane coating surfaces: (a) tapping mode in air before water immersion, (b) tapping mode in air after water immersion, (c) contact mode under water before water immersion and (d) contact mode under water after water immersion (Chapter 25; reprinted with permission from Reference [135]. Copyright 2005 American Chemical Society).

Chapter 23
Biofouling and Climate Change

Elvira S. Poloczanska and Alan J. Butler

The aim of this chapter is to discuss the likely effects of climate change on biofouling. We briefly outline the expected changes in the ocean–atmosphere system over the twenty-first century, and note current understanding of the effects of climate-related variables in regulating biofouling. This enables us to speculate about the likely effects of climate change on biofouling, and to discuss in turn the ecological, economic and environmental impacts of those effects. We conclude that the effects of climate change on biofouling will be highly variable, but that economic and environmental costs may be large and that the control of biofouling will remain a high priority as the climate warms. We comment finally on the need for continuing research, at local scales but coordinated and comparable globally.

23.1 Introduction

Biofouling community structure and biofouling pressure can vary markedly among regions (see Chapter 5). It is generally accepted that marine biofouling is much higher in tropical waters than temperate waters reflecting differences in the nature of the biofouling communities [1]. In tropical marine waters, variations in productivity are related to upwelling events associated with wind conditions, or with heavy precipitation events such as monsoons and tropical storms that wash nutrients off the land. Seasonal changes in biofouling communities are generally not pronounced, reflecting the continual low level of reproduction and settlement throughout the year found in many tropical species [2–4]. Biofouling can thus develop very rapidly on surfaces in tropical regions at any time of the year. In temperate waters, seasonal changes in biofouling communities are more pronounced and biofouling develops most rapidly during the warm spring and summer when growth rates are fastest and larvae and propagules are abundant [5,6]. Exactly how climate change will affect biofoulers in tropical and temperate waters is difficult to predict, especially given that the structure of fouling communities on artificial surfaces can differ greatly among structures and from those growing on nearby natural hard surfaces [7,8]. In this chapter, we discuss potential implications of climate change for biofouling and concentrate on some of the most commonly recorded marine macrofouling organisms.

23.2 Recent and projected changes in the ocean–atmosphere system

According to the Intergovernmental Panel on Climate Change (IPCC) Fourth Assessment Report [9], average global surface temperatures have increased by 0.74°C over the 100 years since 1906 with warming in recent decades being the most rapid. The greatest warming has occurred in the northern hemisphere, particularly at high latitudes. Average Arctic temperatures have risen at almost double the global average rate, while Arctic sea ice extent and snow cover has shrunk. Seasonally, the greatest warming has occurred during winter and spring resulting in extended growing seasons for plants [9, 10]. All of the global climate model simulations under future greenhouse gas emission scenarios used for the IPCC assessment project that warming will continue over the twenty-first century. Under the low greenhouse gas emission scenario family (B1: rapid global economic growth but with introduction of clean and resource efficient technologies), the average global surface warming by the end of this century is likely to be in the range 1.1–2.9°C. Under the highest emission scenarios (A1F1: rapid global economic growth with intensive fossil fuel use), the average global atmospheric surface warming is likely to be in the range 2.4–6.4°C.

Oceans are also warming, with the greatest warming occurring in the surface waters (Plate VII A). The warming trend is widespread over the global ocean, although there has been localised cooling in some regions. Ocean warming is projected to evolve more slowly than surface air temperatures, with the upper ocean warming first then penetration of warming to the deep ocean by the end of the twenty-first century, particularly in mid-latitude regions. Extreme weather is also projected to increase, with heat waves and heavy precipitation events becoming more frequent and tropical cyclones becoming more intense while sea level will continue to rise and spatial patterns of precipitation change. The oceans are projected to become more saline in the tropics while there will be a freshening at higher latitudes.

Presently, the surface ocean is saturated with respect to calcium carbonate. However, the average pH of the oceans has already lowered by about 0.1 since 1750 due to an increased uptake of anthropogenic carbon and currently ranges between 7.9 and 8.3 in the open ocean [9]. The oceans are projected to acidify by a further 0.14–0.35 units over the twenty-first century [9]. Within a few centuries, the projected decrease may lead to an ocean pH estimated to have last occurred a few hundred million years ago. Ocean acidification is not a direct effect of climate change but is a consequence of fossil fuel CO_2 emissions, which are the main driver of projected climate change. Acidification leads to a decrease in the saturation state of calcium carbonate and a shallowing of the depth below which calcium carbonate dissolves. In under-saturated waters marine plants and animals that use calcium carbonate in the form of aragonite and calcite in their shell and body structures, such as corals and pteropods, show reduced calcification [11, 12] while in the longer term, carbonate sediments may dissolve. Ocean carbon cycle models project that Southern Ocean surface waters will become under-saturated in aragonite by 2050, and by 2100 this may extend throughout the entire Southern Ocean and into the sub-arctic Pacific Ocean [12]. In the tropics, the calcium carbonate saturation state of surface waters is projected to decline to 45% of pre-industrial levels by 2100 [12].

23.3   Climate and climate-related factors regulating biofouling communities

Observations of climate change impacts on marine ecosystems are mounting and impacts have been recorded from every ocean [9, 13]. Evidence includes distributional shifts linked to warming temperatures in plankton [14], intertidal fauna [15,16] and temperate fish populations in both the northern [17,18] and southern hemispheres (P. Last, personal communication). The phenology or timing of life cycle events is also responding to warming, with marine turtles and seabirds nesting earlier [19,20], and the peak appearance of phytoplankton changing [21]. Impacts of climate change in the marine environment are likely to act in synergy with other environmental stressors such as pollution, eutrophication and exploitation. The structure of biofouling communities varies considerably with physical and biological factors [22, 23] as well as varying regionally (see Chapter 5), so small alternations in ocean chemistry, temperature and circulation patterns may result in large shifts in biofouling community structure with ecological, economic and environmental consequences for the battle against biofouling (Table 23.1).

23.3.1   Temperature

Temperature is a primary determinant of the distributions of most species, and upper and lower seasonal temperatures can determine the survival of critical life stages and induce spawning or other biological events [24,25]. Temperature regulates biological processes such as growth, metabolism and reproduction [26]. The rate of fouling growth on boats and other artificial surfaces in temperate and polar waters is far greater in the warmer months. In these seas, growth of marine organisms generally slows or halts through the colder winter months. Although this may also be an indirect effect of temperature influencing food availability, temperature influences the length of the breeding season and reproductive output in marine invertebrates such as cirripedes [27, 28] and indirectly the growing season of macroalgae [29,30]. For example, fecundity of the fouling barnacle *Balanus amphitrite* increased fivefold when adults were maintained at $30°C$ compared to $20°C$, although the interbreeding interval also lengthened [4]. The projected warming of the oceans may result in an increase in some fouling groups at higher latitudes particularly in the northern hemisphere where the greatest warming is expected. Note, however, that the impacts of climate change are likely to vary considerably among regions. Canning-Clode and Wahl (Chapter 5) report substantial variation in fouling pressures with latitude with some of the highest and lowest fouling pressures reported at mid-latitude regions. In coastal waters, local warming is likely to be important and may be much greater than the projected average, while seasonal warming will be important at higher latitudes.

23.3.2   Nutrients and salinity

The salinity of sea water tends to be relatively constant in cooler offshore waters, but large fluctuations in salinity can occur in tropical or enclosed seas such as the Mediterranean where evaporation is high. In the monsoonal tropics, the intense bursts of rainfall can drastically reduce salinity of coastal waters and increase sedimentation and organic enrichment through runoff from the land. In the tropics, monsoons regulate productivity and salinity of coastal waters and are thus an important influence on the life cycle of fouling organisms. For example, reproduction in the fouling barnacle *B. amphitrite* in Indian waters is low during the monsoon season, possibly due to large reductions in salinity, and is highest in post-monsoon when productivity of coastal waters is high [4]. Coastal tropical species are often adapted to withstand high salinity and cope with sudden drops in salinity [31], but intense or frequent events can

Table 23.1   Summary table of potential future effects of climate change (including ocean acidification) on biofouling; biophysical effects and impacts on ecology, economy and the environment.

Expected environmental change	Biophysical effects	Ecological implications	Economic implications	Environmental implications
Increased sea surface temperature	Poleward shift in species distribution	Alteration of community structure, shift towards species tolerant of warmer temperatures; increased invasions of exotic species transported by fouling	Rising costs associated with alteration of biofouling community and increased biosecurity threat	Development of new species into marine pests; invasion of new areas by marine pest species
	Shifts in phenology; longer growing seasons, longer reproductive periods	Alteration of trophic food webs; alteration of community structure; longer aggressive fouling season in temperate waters	More frequent cleaning and application of antifouling coatings; increased damage of aquaculture organisms and structures; increased fuel consumption on shipping; increased benefit for aquaculture utilising natural biofouling (e.g. seed mussels)	Increased contamination from chemicals from antifouling coatings; increased greenhouse gas emissions from shipping
Alteration of salinity and nutrients	Alteration of primary productivity; alteration of osmotic stress	Alteration of community structure	Potential impact on efficiency of antifouling coatings; potential of new fouling species	
Wind patterns and ocean currents	Enhanced primary productivity where wind mixing of waters increases, changes in larval advection patterns	Alteration of trophic food webs, changes in species distribution patterns	Potential of new fouling species	
Ocean acidification	Decrease in calcification; dissolution of carbonate structures in undersaturated waters; increased physiological stress; reduction of calcifying plankton	Alteration of trophic food web; reduced growth rates in calcifying species; decline in abundance of calcifying species	Potential impact on efficiency of antifouling coatings	

(Continued)

Table 23.1 (*Continued*)

Expected environmental change	Biophysical effects	Ecological implications	Economic implications	Environmental implications
Increased frequency of extreme weather	Increased frequency of extreme temperature days in coastal systems leading to increased mortality rates	Alteration of communities to species tolerant of high disturbance	Increased risk of mechanical damage to antifouling coatings requiring repairs or resulting in increased fouling; possible benefit of storm removal of fouling from artificial structures	

result in high mortalities of coastal organisms and damage to shell structures. The dissolution of inhabited calcareous tubes of fouling serpulids in the brackish waters of the Brisbane River, Australia, was recorded during a period of increased freshwater input [32].

Nutrient levels are an important factor influencing the development, structure and growth of biofouling communities [33, 34]. Nutrient enrichment may stimulate growth of biofouling organisms particularly in oligotrophic waters, but it may also slow growth. For example, intense terrestrial discharge can lead to anoxia from eutrophication and wide-scale mortalities of benthic organisms. Projected climate change will alter mixed layer depth and wind fields, which drive currents and can influence upwelling and productivity, hence impacting food webs [35], growth rates, reproductive rates and strength of biotic interaction within biofouling communities.

23.3.3   Wind patterns and ocean currents

Wind fields drive major and local currents which can act as facilitators of, or barriers to, pelagic dispersal stages so influencing biogeographical distributions [36, 37]. Projected alterations of wind fields and ocean currents will impact distributions of marine species. For example, the southward penetration of the East Australian Current has increased over the past 60 years [38] and is projected to strengthen further [39]. The effect of this penetration of warm water is evident in the southward extensions into eastern Tasmanian waters of several species including the purple sea urchin *Centrostephanus rodgersii* and the giant rock barnacle *Austromegabalanus nigrescens* [40]. The seasonal appearance of tropical reef fish along the temperate eastern Australian coast has also been linked to larval transport by the East Australian Current [41]. In the Arctic Ocean, the establishment of the blue mussel *Mytilus edulis* in Svalbard, from where it has been absent for at least 1000 years, was probably facilitated by warmer than average water temperatures and increased mass transport of Atlantic water northwards by the West Spitsbergen Current [42].

23.3.4   Extreme weather

As extreme weather events become more frequent, they may play an increasingly important role in regulating population dynamics of coastal fouling organisms. In temperate waters, winter storms can clear areas of benthic organisms, particularly in littoral and sublittoral zones, exposing area for settlement the following spring. Storms have been shown to benefit aquaculture by washing the nets clean of developing biofouling [43], although severe storms may also damage artificial structures. Cyclones can be highly destructive in tropical waters and it is expected that cyclone intensity will increase as global climate warms [9]. The increased frequency of warm days and a reduction of frost and freezing days will influence the survival of organisms in the upper sublittoral zone. Runoff from land after heavy precipitation will also influence salinity, nutrient and suspended sediment loading of coastal waters. Increased turbidity will reduce light intensity leading to a reduction in algal cover if prolonged or severe. Biofouling in shallow waters may thus shift to species adapted for high disturbance regimes.

23.3.5   Ocean chemistry

Increased ocean acidification is likely to negatively affect animals and plants with carbonate (aragonite and calcite) shell and body structures [12,44,45]. Calcifying organisms are sensitive to changes in the saturation state of sea water for calcite and aragonite, which is influenced by temperature and pH. As aragonite is more soluble than calcite, species that use aragonite such as tropical corals, pteropods and coralline red algae will be more vulnerable to increased acidification than species that use calcite [44,45]. Increased acidification will lead to reduced growth rates and a higher susceptibility to erosion and predation as shell matrices weaken. Decreased pH and increased CO_2 levels will also negatively impact the physiology of larger animals leading to slower growth rates and reduced survival of egg and larval stages of marine vertebrates and invertebrates [46].

A study on the long-term effects of a pH reduction to 7.3 (the maximum drop expected within the next 300 years) on the mussel *Mytilus galloprovincialis* indicated that it may be fatal for this species [47]. Growth rate reduced, shell dissolution increased and metabolism slowed. Further studies have also shown reduced calcification in other benthic molluscs with lowered pH [48,49].

Given the high percentage of calcifying organisms in fouling communities and the sensitivity of larval stages to pH, ocean acidification may reduce the biomass of fouling communities. However, there may simply be a shift in communities to non-calcifying species more tolerant of changing pH.

23.4   Ecological impacts of climate change effects on biofouling

Climate impacts on biological systems will be evident as shifts in species distributions, changes in phenology and alteration of trophic structure and community composition and organisation.

23.4.1   Distributions

Species distributions are generally defined by a combination of ranges of environmental variables, often referred to as their 'climate envelope'; they may not survive or reproduce outside these boundaries. As global climate warms, the climate envelope for many species will shift polewards. Poleward range extensions linked to warming temperatures are being observed in an increasing number of terrestrial and marine species [50, 51] including fouling species. For example, the warm-water African barnacle *Solidobalanus fallax* is being recorded with increasing frequency in Western Europe and may become a serious fouling pest of fish cages and aquaculture structures there as it extends its range northwards [52, 53].

The green mussel, *Perna viridis*, is native to tropical and subtropical Asia-Pacific and Indo-Pacific seas where it is a successful fouling species in optimal temperature and salinity conditions, particularly of coastal power plants [54, 55]. It is also a successful invasive species, transported by fouling on ships and other vectors and is causing fouling problems on artificial structures in warmer waters globally [55]. The recent northward spread of this non-indigenous species in the USA has been facilitated by a series of mild winters [56]. This species is of great concern in Australia where there have been outbreaks in northern coastal waters. It may spread southwards, down the Australian western and eastern coastlines, as warmer winter temperatures expand potential habitat and allow successful reproduction [55]. Similar expansions may occur in other invasive fouling species; an increase in the frequency of mild winter temperatures may facilitate further poleward (northwards) expansion of the Pacific oyster *Crassostrea gigas* in the Wadden Sea [57, 58].

Generally, more species have been observed extending their high latitude distributional boundaries polewards than retracting their low-latitude boundaries [51]. However, warmer temperatures will also impact species at their low latitude boundaries and a decline in abundance at low-latitude range edges is expected. For example, a 40-year time series of adult barnacle abundance in south-west England towards the low-latitude boundary of *Semibalanus balanoides* revealed its abundance declined during warm periods, while the warm-water chthamalid species increased [59, 60]. Model projections suggest that *S. balanoides* will disappear from south-east English coasts by the end of the twenty-first century as temperatures continue to rise [61].

23.4.2   Phenology

Changes in phenology are also being recorded from a wide range of taxonomic groups including many marine species [51]. In marine systems, plankton in the north-east Atlantic has advanced the timing of peak appearance by 10–27 days, depending on species group, in only 50 years [21]. Terrestrial vegetative growing seasons are lengthening, particularly in the northern hemisphere where greatest seasonal warming has been recorded [62]. A similar lengthening may occur in marine macroalgae and thus increase the period of aggressive fouling, although growth in marine macroalgae is regulated by the interacting factors of temperature, nutrient availability and light availability [29].

23.4.3   Community structure

Differing responses among species to warming temperatures and other shifts in climate will disrupt biotic interactions resulting in alteration of biological communities [63]. Alteration of the strength of species interactions will mediate both direct and indirect impacts of climate change [61]. Warming temperatures may allow species to establish in areas where they were previously absent, resulting in new biotic interactions.

23.5   Economic impacts of climate change effects on biofouling

Economic impacts of biofouling may increase in many localities, particularly northern hemisphere temperate localities, as global temperatures warm. Warmer temperatures coupled with tightening of legislation controlling the use of biocides are expected to increase fouling problems. Ocean warming will not only potentially increase fouling intensity, but will alter chemical activity rates and dissolution rates of antifouling coatings [64, 65], potentially leading to an increase in the costs of maintaining low levels of fouling on submerged structures. In the aquaculture industry, fouling results in reduction in water exchange, increased incidences of anoxia, increased physical damage to farmed organisms, increased physical damage to nets and equipment and increased threat to workers handling equipment through weight, manipulability and contact with harmful organisms [66] (Chapter 19). Climate change impacts on biofouling may have serious economic consequences in the aquaculture industry, particularly in temperate waters where the aggressive period of fouling (spring and summer) may lengthen or new fouling species establish. For example, in temperate finfish aquaculture, nets require regular cleaning during warmer months [67], while high-performance recreational boat hulls may be cleaned every 3 weeks instead of 4 weeks [68]. Paints used in temperate finfish aquaculture tend to last 6 months, so are usually applied in spring just before the intense fouling season starts [66]. The season may also shift earlier in the future.

Biofouling can be turned to economic gain. For example, fouling by the mussel *M. galloprovincialis* on oil industry offshore structures in the Mediterranean can reach 100 kg m^{-2} after only 1 year if no antifouling treatment is used [69]. These mussels are now harvested by fishermen with considerable economic gain for both the fishermen and the oil company [69]. Climate change may benefit such industries utilising natural fouling if fouling growth is favoured.

The introduction of invasive species through fouling communities can have serious economic consequences [70]. Fouling communities on ships and other submerged structures may harbour pathogens [71–73]. Fouling reduces efficiency and increases fuel consumption in shipping and in industrial plants with sea water intakes [1, 74] and it carries other economic costs [75]. Each of these costs is likely to increase if the rate of growth of biofouling increases with climate change.

23.6   Environmental impacts of climate change effects on biofouling

Biofouling paints containing cuprous oxides have become the predominant form of antifouling used globally to treat artificial surfaces in all marine industries as organotins have been phased out. They are, however, not environmentally benign (see Chapters 16 and 17) and their effects are likely to increase as temperatures increase [68, 76, 77]. The dissolution of copper and other toxic compounds from antifouling coatings is difficult to control and can vary with temperature, salinity, oxygen and pH [76, 78, 79], all of which will be impacted by climate change. The potential for increased fouling of marine artificial surfaces and longer aggressive fouling seasons as marine waters warm adds increasing urgency to the need for environmentally friendly antifouling coatings.

Fouling communities on shipping, recreational boats and towed floating platforms are a major vector for the introduction of exotic species to marine coastal waters [80–84] (see Chapter 24). Successful invasive species generally have a negative impact on local ecosystems through alteration of biotic interactions, trophic disturbance, habitat modification and replacement of native species [82, 85–89].

Global climate change is generally expected to favour invasive species, thus exacerbating their impacts on ecosystems [90–92]. Successful establishment of exotic species is dependent on the physiochemical and biotic properties of the receiving environment [92, 93]. Environmental factors such as cooler temperature thresholds have long been recognised as a barrier to the successful breeding and establishment of species [24]. Climate change may result in greater lengths of temperate coastlines becoming more hospitable to exotic organisms carried by ship fouling as waters warm. This impact may be particularly strong in northern hemisphere mid- to high-latitude waters where the greatest warming is expected [9], and where highest fouling pressures have been recorded (see Chapter 5).

Additionally, the potential opening of new shipping routes through the Arctic circle as summer sea ice diminishes with rising temperatures [94, 95] will allow transfer of cold-water organisms by ship-fouling and other vectors between North Atlantic and North Pacific with major implications for the structure and function of these ecosystems [81, 96]. Such transfer is already likely to have happened with the phytoplankton species *Neodenticula seminae* [97]. The presence of sea ice abrades fouling communities on ship hulls en route so reducing risk to coastlines [96] and conditions in the polar regions are inhospitable to many organisms. Average arctic temperatures have risen at almost double the global average rate in the past 100 years and arctic summer sea ice is projected to disappear entirely by the latter part of the twenty-first century under future climate simulations [9].

23.6.1 *Implications for greenhouse gas emissions*

Fouling of power plants and other industrial installations results in a decrease of operation efficiency and a requirement for extra fuel to maintain operating standards, with a concomitant increase in emissions of greenhouse gases and other pollutants [74]. The rapidly expanding area of ocean-going shipping is the most energy-efficient method to move cargo but is also an often-overlooked emitter of greenhouse gases [98]. Although ocean-going shipping was estimated to account for only 1.5–3% of global emissions of CO_2 in 1996, shipping is a major emitter of nitrogen oxides (14–31% estimated global inventories [98, 99]), which influence tropospheric ozone levels and ultimately contribute to global temperature rise [100]. Fouling can significantly increase ship fuel consumption (estimates range up to 50% [1]), although self-polishing paints and other advancements in antifouling technology have substantially reduced fouling problems on the hulls of these large ocean-going vessels [75] (see Chapters 14 and 15). Recommendations to reduce greenhouse gas emissions from shipping include optimal hull and propeller maintenance to combat fouling and reduce surface roughness, which could reduce fuel usage by 5% [99]. As ship fouling increases with time spent in warmer waters, climate change may exacerbate fouling problems on shipping therefore greenhouse gas emissions, but its impact on emissions is likely to be negligible compared to other measures such as optimising cargo-carrying capacity, more efficient routing and improvements in hull and engine design and technology.

23.7 Conclusions

- Impacts of climate change on biofouling communities are likely to be extremely variable both regionally and latitudinally. In temperate waters, and in higher latitudes, seasonal warming is likely to lengthen the aggressive fouling season and to alter biofouling community structure (Plate VII B). In tropical waters, changes to productivity through alteration of hydrodynamic regimes and precipitation patterns may regulate changes in biofouling.
- Economic and environmental costs of fouling are high and may increase under climate change. However, global costs may not necessarily increase, as technology to reduce biofouling improves.
- Tackling biofouling will remain a priority, especially in regions where climate change will exacerbate fouling aggression. As climate warms and attitudes change, public pressure will increase for alternatives to fossil fuels, such as nuclear power or wave energy, and for desalination plants to combat freshwater shortages – with their associated problems of biofouling.
- Fouling dynamics varies widely, processes are complex, and therefore prediction is highly uncertain.
- To refine the broad predictions we have made here, local research on processes will be needed, and monitoring (to test the predictions, refine process understanding and inform management) will be necessary. Ideally, monitoring programmes would be done in comparable ways at many sites (cf. the meta-analysis of diversity of fouling assemblages outlined in Chapter 5).

References

1. WHOI (1952) *Marine Fouling and its Prevention.* Woods Hole Oceanographic Institute, US Naval Institute, Annapolis.
2. Bauer, R.T. (1992) Testing generalizations about latitudinal variation in reproduction and recruitment patterns with sicyonid and caridean shrimp species. *Invertebrate Reproduction and Development,* **22**, 193–202.
3. Koh, L.L., O'Riordan, R.M. & Lee, W.J. (2005) Sex in the tropics: reproduction of *Chthamalus malayensis* Pilsbury (Class Cirripedia) at the equator. *Marine Biology,* **147**, 121–133.
4. Desai, V., Desai, A.C., Anil, A.C. & Venkat, K. (2006) Reproduction in *Balanus amphitrite* Darwin (Cirripedia: Thoracia): influence of temperature and food concentration. *Marine Biology,* **149**, 1431–1441.
5. Anil, A.C., Chiba, K., Okamoto, K. & Kurokura, H. (1995) Influence of temperature and salinity on larval development of *Balanus amphitrite*: implications in fouling ecology. *Marine Ecology Progress Series,* **118**, 159–166.
6. Greene, J.K. & Grizzle, R.E. (2007) Successional development of fouling communities on open ocean aquaculture fish cages in the western Gulf of Maine, USA. *Aquaculture,* **262**, 289–301.
7. Butler, A.J. & Connolly, R.M. (1996) Development and long term dynamics of a fouling assemblage of sessile marine invertebrates. *Biofouling,* **9**, 187–209.

8. Glasby, T.M. (1999) Differences between subtidal epibiota on pier pilings and rocky reefs at marinas in Sydney, Australia. *Estuarine, Coastal and Shelf Science*, **48**, 281–290.

9. IPCC (2007) *Climate Change 2007: the physical science basis*. Contribution of working group I to the fourth assessment report of the intergovernmental panel on climate change (eds S.D. Solomon, M. Qin, Z. Manning, *et al.*). Cambridge University Press, Cambridge and New York.

10. Sparks, T.H. & Menzel, A. (2002) Observed changes in seasons: an overview. *International Journal of Climatology*, **22**, 1715–1725.

11. Riebesell, U., Zondervan, I., Rost, B., Tortell, P.D., Zeebe, R. & Morel, F.M.M. (2000) Reduced calcification of marine plankton in response to increased atmospheric CO_2. *Nature*, **407**, 364–367.

12. Orr, J.C., Fabry, V.J., Aumont, O., *et al.* (2005) Anthropogenic ocean acidification over the twenty-first century and its impact on calcifying organisms. *Nature*, **437**, 681–686.

13. Parmesan, C. (2006) Ecological and evolutionary responses to recent climate change. *Annual Review of Ecology, Evolution, and Systematics*, **37**, 637–639.

14. Beaugrand, G., Reid, P.C., Ibanez, F., Lindley, J.A. & Edwards, M. (2002) Reorganisation of north Atlantic marine copepod biodiversity and climate. *Science*, **296**, 1692–1694.

15. Barry, J.P., Baxter, C.H., Sagarin, R.D. & Gilman, S.E. (1995) Climate-related, long-term faunal changes in a California rocky intertidal community. *Science*, **267**, 672–675.

16. Mieszkowska, N., Leaper, R., Moore, P., *et al.* (2005) *Assessing and Predicting the Influence of Climatic Change Using Intertidal Rocky Shore Biota*. Marine Biological Association of the UK, Plymouth.

17. Perry, A.L., Low, P.J., Ellis, J.R. & Reynolds, J.D. (2005) Climate change and distribution shifts in marine fishes. *Science*, **308**, 1912–1915.

18. Byrkjedal, I., Godø, O.R. & Heino, M. (2007) Northward range extensions of some mesopelagic fish in the northeastern Atlantic. *Sarsia*, **89**, 484–489.

19. Weishampel, J.F., Bagley, D.A. & Ehrhart, L.M. (2004) Earlier nesting by loggerhead sea turtles following sea surface warming. *Global Change Biology*, **10**, 1424–1427.

20. Moller, A.P., Flensted-Jensen, E. & Mardal, W. (2006) Rapidly advancing laying date in a seabird and the changing advantage of early reproduction. *Journal of Animal Ecology*, **75**, 657–665.

21. Edwards, M. & Richardson, A.J. (2004) Impact of climate change on marine pelagic phenology and trophic mismatch. *Nature*, **430**, 881–884.

22. Benson, P.H., Brining, D.L. & Perrin, D.W. (1973) Marine fouling and its prevention. *Marine Technology*, **1**, 30–37.

23. Hughes, D.J., Cook, E.J. & Sayer, M.D.J. (2005) Biofiltration and biofouling on artificial structures in Europe: the potential for mitigating organic impacts. *Oceanography and Marine Biology: an Annual Review*, **43**, 123–172.

24. Orton, J.H. (1919) Sea-temperature, breeding and distribution in marine animals. *Journal of the Marine Biological Association of the United Kingdom*, **12**, 339–366.

25. Hutchins, L.W. (1947) The bases for temperature zonation in geographical distribution. *Ecological Monographs*, **17**, 325–335.

26. Pörtner, H.O. (2002) Climate variations and the physiological basis of temperature dependant biogeography: systemic to molecular hierarchy of thermal tolerance in animals. *Comparative biochemistry and Physiology A*, **132**, 739–761.

27. Patel, B. & Crisp, D.J. (1960) The influence of temperature on the breeding and the moulting activities of some warm-water species of operculate barnacles. *Journal of the Marine Biological Association of the United Kingdom*, **39**, 667–680.

28. Hines, A.H. (1978) Reproduction in 3 species of intertidal barnacles from central California. *Biological Bulletin*, **154**, 262–281.

29. Kain, J.M. (1989) Seasons in the subtidal. *British Phycological Journal*, **24**, 203–215.

30. Hanisak, M.D. (1979) Growth patterns of *Codium fragile* ssp. *tomentosoides* in response to temperature, irradiance, salinity, and nitrogen source. *Marine Biology*, **50**, 1432–1793.

31. Chung, K. (2001) Ecophysiological adaptability of tropical aquatic organisms to salinity changes. *Revista de Biologia Tropical*, **49**, 9–13.

32. Straughan, D.W. (1972) Ecological studies of *Mercierella engmatica* Fauval (Annelida: Polychaeta) in the Brisbane River. *Journal of Animal Ecology*, **41**, 93–136.

33. Bombace, G. (1989) Artificial reefs in the Mediterranean Sea. *Bulletin of Marine Science*, **44**, 1023–1032.

34. Mayer-Pinto, M. & Junqueira, A.O. (2003) Effects of organic pollution on the initial development of fouling communities in a tropical bay, Brazil. *Marine Pollution Bulletin*, **46**, 1495–1503.

35. Harris, G.P., Griffiths, F.B., Clementson, L.A., Lyne, V. & Vanderhoe, H. (1991) Seasonal and interannual variability in physical processes, nutrient recycling and the structure of the food chain in Tasmanian shelf waters. *Journal of Plankton Research*, **13** (Suppl.), S109–S131.

36. Gaylord, B. & Gaines, S.D. (2000) Temperature or transport? Range limits in marine species mediated solely by flow. *The American Naturalist*, **155**, 769–789.

37. Zacherl, D., Gaines, S.D. & Lonhart, S.I. (2003) The limits to biogeographical distributions: insights from the northward range extension of the marine snail, *Kelletia kelletii* (Forbes 1852). *Journal of Biogeography*, **30**, 913–924.

38. Ridgway, K.R. (2007) Long-term trend and decadal variability of the southward penetration of the East Australian Current. *Geophysical Research Letters*, **34**, L13613.

39. Cai, W., Shi, G., Cowan, T., Bi, D. & Ribbe, J. (2005) The response of the Southern Annular Mode, the East Australian Current, and the southern mid-latitude ocean circulation to global warming. *Geophysical Research Letters*, **32**, L23706.

40. Poloczanska, E.S., Babcock, R.C., Butler, A., *et al.* (2007) Climate change and Australian Marine Life. *Oceanography and Marine Biology: An Annual Review*, **45**, 407–478.

41. Booth, D.J., Figueira, W.F., Gregson, M.A., Brown, L. & Beretta, G. (2007) Occurrence of tropical fishes in temperate southeastern Australia: role of the East Australian Current. *Estuarine, Coastal and Shelf Science*, **72**, 102–114.

42. Berge, J., Johnsen, G., Nilsen, F., Gulliksen, B. & Slagstad, D. (2005) Ocean temperature oscillations enable reappearance of blue mussels *Mytilus edulis* in Svalbard after 1000 year absence. *Marine Ecology Progress Series*, **303**, 167–175.

43. Braithwaite, R.A., Cadavid Carrascosa, M.C. & McEvoy, L.A. (2007) Biofouling of salmon cage netting and the efficiency of a typical copper-based antifoulant. *Aquaculture*, **262**, 219–226.

44. Raven, J., Caldeira, K., Elderfield, H., *et al.* (2005) *Ocean Acidification due to Increasing Atmospheric Carbon Dioxide*. Royal Society Special Report, London.

45. Kleypas, J.A., Feely, R.A., Febry, V.J., Langdon, C., Sabine, C.L. & Robbins, L.L. (2006) *Impacts of Ocean Acidification on Coral Reefs and Other Marine Calcifiers: A Guide for Future Research*. Report of a workshop held 18–20 April 2005, National Science Foundation, National Oceanic and Atmospheric Administration, US Geological Survey, St Petersburg.

46. Pörtner, H.O., Langenbuch, M. & Reipschläger, A. (2004) Biological impact of elevated ocean CO_2 concentrations: lessons from animal physiology and earth history. *Journal of Oceanography*, **60**, 705–718.

47. Michaelidis, B., Ouzounis, C., Paleras, A. & Pörtner, H.O. (2005) Effects of long-term moderate hypercapnia on acid-base balance and growth rate in marine mussels *Mytilus galloprovincialis*. *Marine Ecology Progress Series*, **293**, 109–118.

48. Berge, J.A., Bjerkeng, B., Pettersen, O., Schaanning, M.T. & Oxnevad, S. (2006) Effects of increased sea water concentrations of CO_2 on growth of the bivalve *Mytilus edulis* L. *Chemosphere*, **62**, 681–687.

49. Gazeau, F., Ouiblier, C., Jansen, J.M., Gattuso, J.P., Middelburg, J.J. & Heip, C.H.R. (2007) Impact of elevated CO_2 on shellfish calcification. *Geophysical Research Letters*, **34**, Art. No. L07603.

50. Walther, G-R., Post, E., Convey, P., *et al.* (2002) Ecological responses to recent climate change. *Nature*, **416**, 389–395.

51. Parmesan, C. & Yohe, G. (2003) A globally coherent fingerprint of climate change impacts across natural ecosystems. *Nature*, **421**, 37–42.

52. Southward, A.J. (1995) Occurrence in the English Channel of a warm-water cirripede *Solidobalanus fallax*. *Journal of the Marine Biological association of the United Kingdom*, **75**, 199–210.

53. Southward, A.J., Hiscock, K., Moyse, J. & Elfimov, A.S. (2004) Habitat and distribution of the warm-water barnacle *Solidobalanus fallax* (Crustacea: cirripedia). *Journal of the Marine Biological Association of the UK*, **84**, 1169–1177.
54. Satpathy, K.K. & Rajmohan, R. (2001) Effects of fouling organisms on the water quality of a nuclear power plant cooling system. In: *10th International Congress on Marine Corrosion and Fouling, University of Melbourne February 1999, Additional Papers, DSTO-GD-0287* (ed. J.A. Lewis), pp. 59–71. DSTO Aeronautical and Maritime Research Laboratory, Fishermans Bend, Victoria, Australia.
55. Rajagopal, S., Venugopalan, V.P., Van Der Velde, G. & Jenner, H.A. (2006) Greening of the coasts: a review of the *Perna viridis* success story. *Aquatic Ecology*, **40**, 273–297.
56. Power, A.J., Walker, R.L., Payne, K. & Hurley, D. (2004) First occurrence of the non-indigenous green mussel *Perna viridis* (Linnaeus 1758) in coastal Georgia, United States. *Journal of Shellfish Research*, **23**, 741–744.
57. Nehls, G., Diederich, S. & Thieltges, D.W. (2006) Wadden Sea mussel beds invaded by oysters and slipper limpets: competition or climate control? *Helgoland Marine Research*, **60**, 135–143.
58. Cardoso, J.F.M.F., Langlet, D., Loff, J.F., *et al.* (2007) Spatial variability in growth and reproduction of the Pacific oyster *Crassostrea gigas* (Thunberg 1793) along the west European coast. *Journal of Sea Research*, **57**, 303–315.
59. Southward, A.J., Hawkins, S.J. & Burrows, M.T. (1995) 70 years of observations of changes in distribution and abundance of zooplankton and intertidal organisms in the western English Channel in relation to rising sea temperature. *Journal of Thermal Biology*, **20**, 127–155.
60. Southward, A.J. (1991) 40 years of changes in species composition and population density of barnacles on a rocky shore near Plymouth. *Journal of the Marine Biological Association of the United Kingdom*, **71**, 495–513.
61. Poloczanska, E.S., Hawkins, S.J., Southward, A.J., Burrows, M.T. (2008) Modelling the response of populations of competing species to climate change. *Ecology*, **89**, 3138–3149.
62. Rosenweig, C., Casassa, G., Karoly, D.J., *et al.* (2007) Assessment of observed changes and responses in natural and managed systems. In: *Climate Change 2007: Impacts, Adaptation and Vulnerability. Contribution of Working Group II to the Fourth Assessment Report of the Intergovernmental Panel on Climate Change* (eds M.L. Parry, O.F. Canziani, J.P. Palutikof, P.J. van der Linden & C.E. Hanson), pp. 79–131. Cambridge University Press, Cambridge.
63. Stenseth, N.C. & Mysterud, A. (2002) Climate, changing phenology, and other life history traits: nonlinearity and match-mismatch to the environment. *Proceedings of the National Academy of Sciences*, **99**, 13379–13381.
64. Kiil, S., Dam-Johansen, J., Weinell, C.E. & Pedersen, M.S. (2002) Seawater-soluble pigments and their potential use in self-polishing antifouling paints: simulation-based screening tool. *Progress in Organic Coatings*, **45**, 423–434.
65. Kiil, S., Dam-Johansen, J., Weinell, C.E., Pedersen, M.S. & Codolar, S.A. (2003) Estimation of polishing and leaching behaviour of antifouling paints using mathematical modelling: a literature review. *Biofouling*, **19** (Suppl.), 37–43.
66. Braithwaite, R.A. & McEvoy, L.A. (2005) Marine biofouling on fish farms and its remediation. *Advances in Marine Biology*, **47**, 215–252.
67. Hodson, S.L., Lewis, T.E. & Burke, C.M. (1997) Biofouling of fish-cage netting: efficiency and problems of in situ cleaning. *Aquaculture*, **152**, 77–90.
68. Johnson, L.T. & Miller, J.A. (2002) *What You Need to Know about Non-toxic Antifouling Strategies for Boats*. Report no T-049, California Sea Grant College Program, University of California, La Jolla.
69. Relini, G. & Montanari, M. (2001) Macrofouling role of mussels in Italian seas: a short review. In: *10th International Congress on Marine Corrosion and Fouling, University of Melbourne, February 1999, Additional Papers, DSTO-GD-0287* (ed. J.A. Lewis), pp. 17–32. DSTO Aeronautical and Maritime Research Laboratory, Fishermans Bend, Victoria, Australia.

70. Coutts, A.D.M. & Forrest, B.M. (2007) Development and application of tools for incursion response: lessons learned from the management of the fouling pest *Didemnum vexillum*. *Journal of Experimental Marine Biology and Ecology*, **342**, 154–162.

71. Andrews, J.D. (1988) Epizootiology of the disease caused by the oyster pathogen *Perkinsus marinus* and its effect on the oyster industry. *American fisheries Society Special Publication*, **18**, 257–264.

72. Goggin, C.L. & Lester, R.J.G. (1995) *Perkinsus*, a protistan parasite of abalone in Australia – a review. *Marine and Freshwater Research*, **46**, 639–646.

73. Tan, C.K.F., Nowak, B.F. & Hodson, S.L. (2002) Biofouling as a reservoir of *Neoparamoeba pemaquidensis* (Page 1970), the causative agent of amoebic gill disease in Atlantic salmon. *Aquaculture*, **210**, 49–58.

74. Pritchard, A. (1997) Control of biofouling: its implications for energy utilisation and impacts on the environment within the EEC. *International Biodeterioration and Biodegradation*, **39**, 87.

75. Townsin, R.L. (2003) The ship hull fouling penalty. *Biofouling*, **19** (Suppl.), 9–15.

76. Chambers, L.D., Stokes, K.R., Walsh, F.C. & Wood, R.J.K. (2006) Modern approaches to marine antifouling coatings. *Surface and Coatings Technology*, **201**, 3642–3652.

77. Almeida, E., Diamantino, T.C. & de Sousa, O. (2007) Marine paints: the particular case of antifouling paints. *Progress in Organic Coatings*, **59**, 2–20.

78. Thomas, K.V., Raymond, K., Chadwick, J. & Waldock, M.J. (2001) The effects of changes in environmental parameters on the release of organic booster biocides from antifouling coatings. In: *10th International Congress on Marine Corrosion and Fouling, University of Melbourne, February 1999, Additional Papers, DSTO-GD-0287* (ed. J.A. Lewis), pp. 157–170. DSTO Aeronautical and Maritime Research Laboratory, Fishermans Bend, Victoria, Australia.

79. Yebra, D.M., Kiil, S. & Dam-Johnsen, K. (2004) Antifouling technology – past, present and future steps towards efficient and environmentally friendly antifouling coatings. *Progress in Organic Coatings*, **50**, 75–104.

80. Carlton, J.T. & Cohen, A.N. (2003) Episodic global dispersal in shallow water marine organisms: the case history of the European shore crabs *Carcinus maenas* and *C. aestuarii*. *Journal of Biogeography*, **30**, 1809–1820.

81. Minchin, D. & Gollasch, S. (2003) Fouling and ships' hulls: how changing circumstances and spawning events may result in the spread of exotic species. *Biofouling*, **19** (Suppl.), 111–122.

82. Godwin, L.S., Elredge, L.G. & Gaut, K. (2004) *The Assessment of Hull Fouling as a Mechanism for the Introduction and Dispersal of Marine Alien Species in the Main Hawaiian Islands*. Bishop Museum Technical Report no 28, Bernice Pauahi Bishop Museum, Honolulu, Hawai'i.

83. Nehring, S. (2005) International shipping – a risk for aquatic biodiversity in Germany. *Neobiota*, **6**, 125–143.

84. Mineur, F., Johnson, M.P., Maggs, C.A. & Stegenga, H. (2007) Hull fouling on commercial ships as a vector of macroalgal introduction. *Marine Biology*, **151**, 1299–1307.

85. Ruiz, G.M., Carlton, J.T., Grosholz, E.D. & Hines, A.H. (1997) Global invasions of marine and estuarine habitats by non-indigenous species: mechanisms, extent and consequences. *American Zoologist*, **37**, 621–632.

86. Gordon, D.R. (1998) Effects of invasive, non-indigenous plant species on ecosystem processes: lessons from Florida. *Ecological Applications*, **8**, 975–989.

87. Hooper, D.U., Chapin, F.S., Hector, A., *et al.* (2005) Effects of biodiversity on ecosystem functioning: a consensus of current knowledge. *Ecological Monographs*, **75**, 3–35.

88. Galil, B.S. (2007) Loss or gain? Invasive species and biodiversity in the Mediterranean Sea. *Marine Pollution Bulletin*, **55**, 314–322.

89. Wallentinus, I. & Nyberg, C.D. (2007) Introduced marine species as habitat modifiers. *Marine Pollution Bulletin*, **55**, 323–332.

90. Dukes, J.S. & Mooney, H.A. (1999) Does global change increase the success of biological invaders? *Trends in Ecology and Evolution*, **14**, 135–139.

91. Stachowicz, J.J., Terwin, J.R., Whitlatch, R.B. & Osman, R.W. (2002) Linking climate change and biological invasions: ocean warming facilitates non-indigenous species invasions. *Proceedings of the National Academy of Sciences of the United States of America*, **99**, 15497–15500.

92. Occhipinti-Ambrogi, A. (2007) Global change and marine communities: alien species and climate change. *Marine Pollution Bulletin*, **55**, 342–352.

93. Schaffelke, B., Smith, J.E. & Hewitt, C.L. (2006) Introduced macroalgae – a growing concern. *Journal of Applied Phycology*, **18**, 259–541.

94. MacKenzie, D. (2002) Arctic meltdown, there will be anarchy as northern seas open to shipping. *New Scientist*, **2332** (Mar 07), 5.

95. Howell,.S.E.L. & Yackel, J.J. (2004) A vessel transit assessment of sea ice variability in the Western Arctic 1969–2002: implications for ship navigation. *Canadian Journal of Remote Sensing*, **30**, 205–215.

96. Lewis, P.N., Riddle, M.J. & Hewitt, C.L. (2004) Management of exogenous threats to Antarctica and the sub-Antarctic islands: balancing risks from TBT and non-indigenous marine organisms. *Marine Pollution Bulletin*, **49**, 999–1005.

97. Reid, P.C., Johns, D.G., Edwards, M., Starr, M., Poulin, M. & Snoeijs, P. (2007) A biological consequence of reducing Arctic ice cover: arrival of the Pacific diatom *Neodenticula seminae* in the North Atlantic for the first time in 800 000 years. *Global Change Biology*, **13**, 1910–1921.

98. Friedrich, A., Hienen, F., Kamakaté, F. & Kodjak, D. (2007) *Air Pollution and Greenhouse Gas Emissions from Ocean-going Ships: Impacts, Mitigation Options and Opportunities for Managing Growth*. The International Council on Clean Transportation, Washington.

99. Skjølsvik, K.O., Andersen, A.B., Corbett, J.J. & Skjelvik, J.M. (2000) *Study on Greenhouse Gas Emissions from Ships*, MT Report: MTOO A23-038 to the International Maritime Organization. MARINTEK, Det Norske Veritas (DNV), Centre for Economic Analysis (ECON) and Carnegie Mellon, Trondheim, Norway.

100. Sitch, S., Cox, P.M., Collins, W.J. & Huntingford, C. (2007) Indirect radiative forcing of climate change through ozone effects on the land-carbon sink. *Nature*, **448**, 791–794.

Chapter 24
Biofouling Invasions

John A. Lewis and Ashley D.M. Coutts

This chapter reviews the historical and continuing role of biofouling in marine species invasions through the translocation of marine species between disparate bioregions and/or facilitation of species spread in a recipient region. Factors that influence the process of translocation on a vessel or other biofouled vectors, and how these are influenced by modern antifouling practices and maritime activity, are also reviewed. Most maritime vessels and sectors were found to pose some risk, even on well-maintained ships, often through the establishment of biofouling aggregations in niches such as sea chests, internal sea water systems and areas where antifouling paint is not applied or becomes damaged. Measures that can be used to manage biofouling, and therefore effectively reduce the risk of translocation of invasive marine species, are discussed.

24.1 Invasive species

The introduction of non-indigenous species is considered to be one of the greatest environmental and economic threats and, along with habitat destruction, the leading cause of extinctions and resultant biodiversity decreases worldwide [1–4]. In the marine environment, the introduction of non-indigenous marine species (NIMS) is recognised as one of the top five threats to marine ecosystem function and biodiversity, and of increasing threat to maritime industries [5–7]. Anthropogenic translocation of marine species has presumably followed the movement of humans across oceans since they first embarked on sea voyages of discovery, migration and trade [8]. However, the significance and impact of NIMS was not widely recognised until comparatively recently.

Shipping was considered the source of a number of NIMS introduced to European coastal waters in the mid-twentieth century, such as the Australasian red alga *Asparagopsis armata* and the barnacle *Austrominius* (syn *Elminius*) *modestus* [9]. However, the degree to which NIMS have contributed to regional flora and fauna was first brought to the fore in a survey of San Francisco Bay that recognised almost 100 NIMS [10]. Regional studies elsewhere around the globe have since found similar numbers of exotic species in inshore marine and estuarine communities. For example, 160 introduced and cryptogenic species have been identified in Port Phillip Bay in southern Australia, representing over 13% of the known flora and fauna [11, 12], with 159 introduced species in New Zealand [13], 96 in the Baltic Sea [14], 96 in Pearl Harbour, Hawaii [15], 240 in the Mediterranean Sea [16], and 316 in coastal marine habitats of continental North America [17].

The majority of NIMS around the world are relatively benign, and few have spread widely beyond sheltered ports or harbours [18]. However, a small percentage has caused, and continue to cause, significant economic and environmental impact [3]. Notable examples include the zebra mussel *Dreissena polymorpha* in the Great Lakes [19], the green alga *Caulerpa taxifolia* in the Mediterranean [20], and the Asian clam *Potamocorbula amurensis* in San Francisco Bay [21].

24.2   Vectors

Elton [22] recognised three significant anthropogenic activities that contributed to the redistribution of marine species around the world: (1) the construction of canals (notably the Suez and Panama), (2) accidental transport via ships and (3) deliberate introductions. Many deliberate introductions have been for the purposes of mariculture and this also facilitated the accidental translocation of species associated with brood stock, aquaculture equipment, and living-food consignments, including microbiota, parasites and pests [23]. More recently, the international trade of aquarium species and habitat (e.g. live rock) has been recognised as an additional vector for introduction of non-indigenous species to aquatic environments [24–26]. However, shipping is considered the major vector for the accidental translocation of NIMS [17, 27–29], both for the primary transoceanic movement of NIMS from their native ranges and for the secondary intra-regional spread of species previously introduced by shipping or other vectors such as aquaculture [e.g. 30–32].

The movement of marine species as biofouling on ship hulls, including sessile, sedentary and associated free-living organisms, dates back for centuries [10, 27, 29]. In 1952, close to 2000 biofouling species were listed [33] and many more have been reported since. In more recent times the colonisation, accumulation and consequent transport of fouling organisms on ship hulls were considered to have decreased due to the greatly increased efficiency of antifouling paints, the increased speeds of ocean-going vessels, decreased harbour residence times and increased vessel maintenance [10, 27, 34].

However, ship-related introductions were continuing, and ballast water was proposed as an alternative shipping vector [10, 27]. The discovery of living organisms in ballast water provided support for this hypothesis [e.g. 27, 35, 36] and ballast water was consequently considered the primary vector for the translocation of marine species, with vessel biofouling largely ignored [e.g. 16, 37–41]. It has since been recognised that organisms still attach, survive and are transported between ports and harbours as biofouling on ships [42–46] and recent research suggests that vessel biofouling may be responsible for more NIMS introductions than ballast water [8, 11–13, 47–49]. In Hawaii and Japan, for example, vessel biofouling is considered to be the vector for the continued arrival of NIMS because ships usually arrive with cargo and leave in ballast [15, 48, 49].

24.3   The translocation process for biofouling organisms

For NIMS to be successfully translocated from a donor region and establish in a recipient region, they must pass through and overcome a sequence of events (Figure 24.1). The species must (1) colonise and establish on a vessel's hull in a donor region, (2) survive the translocation

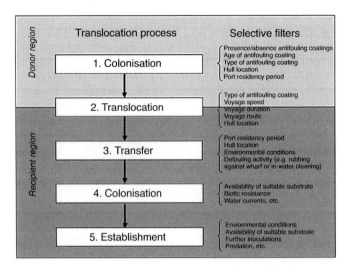

Figure 24.1 Sequence of events in the transfer and establishment of non-indigenous marine species by ship biofouling [50].

on a vessel's hull from the donor to the recipient region, (3) reproduce or be dislodged in the recipient region, (4) colonise available habitat in the recipient region and (5) be able to complete its life cycle in the recipient region (i.e. undergo continued reproduction to become established; Figure 24.1). At each step of the sequence, there is attrition or selective filters (factors) that affect the total number of organisms and species that transition successfully (survive) to the next stage.

The biofouling process on vessels is generally similar to that on other immersed natural and artificial surfaces: initial surface conditioning and microbial biofilm formation, followed by primary surface colonisation by sessile macroalgae and invertebrates, and then secondary colonisation of the biofouling-formed macrohabitat by epibionts, mobile invertebrates and sometimes small fish [51–54]. On a vessel hull, significant environmental influences that differ to other substrata are the presence of antifouling treatments and the movement and activity of the vessel. Antifouling treatments can prevent the recruitment of most microbial, algal and invertebrate species, but favour a few with natural or evolved resistance to antifouling biocides [e.g. 45, 55–58].

Invasion success improves with increasing frequency and density of translocations [59–61], so the greater the quantity, quality and frequency of NIMS translocated to inshore regions with the similar environments, the greater the likelihood of establishment. However, a range of factors are capable of influencing the success of biofouling translocation, including port residency time, and transit speed, duration and route [43, 45, 62, 63].

Higher fouling biomass, diversity and survival occurs on vessels that spend extended periods stationary in ports and harbours, move slowly, travel short distances, and/or ply similar latitudes, than high activity, faster vessels that undertake long voyages through different environmental climes (e.g. trans-equatorial) [42, 43, 45, 46, 64–68]. This is a consequence of the effects of hydrodynamics engendered by the vector movement, environmental fluctuations across water bodies and unfavourable feeding conditions or food supply on organism survival.

Many common biofouling species have life forms that enable their survival on vessels. Barnacles, for example, have a hard shell that protects soft body parts, a conic shape that is hydrodynamically resistant and a strong adhesive and adhesion mechanism to hold on to the surface. Encrusting bryozoans can survive by lying within the boundary layer at the hull surface, protected from turbulent flow that could damage or dislodge the organism. Macroalgal biofouling species commonly have a fast-growing filamentous or sheet-like thallus facilitating rapid maturation, a crustose habit, a heteromorphic life history with a microscopic filamentous or encrusting phase and/or the ability to perennate vegetatively from basal crusts, stolons or thallus fragments, all traits that would facilitate their initial attachment to a vessel hull, survival of transoceanic transport and successful colonisation of a foreign port [69]. Other species are less well adapted to high flow conditions, but can survive because of strong attachment mechanisms, or by hunkering in recesses, niches or within fouling aggregations that provide protection from high flow conditions.

To colonise a recipient location, the biofouling species must release reproductive material in that location. This can be through sporulation or spawning of reproductively mature individuals, dislodgement and fragmentation or, for mobile species, movement off the vector. Many fouling species brood their larvae, retaining fertilised eggs to develop in internal brood structures [18]. Development often takes weeks, so animals do not need to depend on encountering suitable mates when they arrive in a new port. The release of a brood, or brooded larvae, which are thought to have higher survival than larvae that have a long planktonic development, could provide significant initial inoculums to a new port.

Environmental conditions generally influence the triggering of sporulation or spawning and the environment must be suitable to the species to enable larval survival and settlement. Temperature change can induce spawning activity in mussels [70,71] and the water temperature change as a vessel enters a port can trigger spawning of reproductively mature fouling species [72,73]. The invasion of fouling organisms can also be facilitated by the artificial structures in ports, such as pilings and pontoons, by creating habitat unsuitable to native species [74,75]. Such human-produced habitats can be colonised by communities quite different from natural rocky reefs and are often poorly utilised by native species [75,76]. Transient disturbance events and pollution in the recipient environment can also facilitate the establishment, persistence and spread of non-indigenous species [74,77–81].

24.4   Historical translocations

The invertebrate communities typically found on ships, barges, docks, floats and wharf pilings are never found elsewhere and, in the pre-maritime environment, the only similar habitat would have been natural floating materials, mainly drift logs [82]. The evolution and almost cosmopolitan existence of similarly structured fouling communities can therefore be attributed to an anthropogenic selection process and, particularly in and near harbours, the presence of common species within biofouling communities is considered to be a direct consequence of the historical translocation of species by maritime activity.

However, with no records of biofouling or boring communities on ancient ships, apart from general reference to barnacles, shipworms and gribbles (a wood-boring crustacean), discerning historical ship-dispersal events is difficult [34]. Many biofouling related colonisations also predate biological surveys, further confounding historical interpretation of the colonisation events.

For some species, type localities may be from outside of the native range of the species as a consequence of translocation. For example, the type collection of the invasive green alga *Codium fragile* subsp. *tomentosoides* was from the Netherlands in 1900 [83], but the native range is thought to be in the north-west Pacific [84].

For most of history, the driver for the prevention of biofouling on ship hulls has been to minimise the effect of biofouling growth on ship performance. Until the latter part of the nineteenth century, the only effective method of controlling biofouling was to regularly beach vessels to allow the manual removal of unwanted growth [54]. The first authenticated antifouling treatment applied to a vessel was the application of copper sheet to the frigate *HMS Alarm* in 1758 and, by 1780, copper sheathing was in general use by the British Navy [85]. Antifouling paints containing copper compounds first appeared in the 1860s and, over the next century, became the mainstay in antifouling protection.

Today's antifouling coatings allow for docking intervals up to 5 years, and generally prevent the accumulation of extensive levels of biofouling over the hull surface. However, this was not always the case and, in the early part of the twentieth century, it was the practice of shipowners to dry dock and clean the bottoms of their ships every 6–8 months [64]. Even within this time, biofouling growth could reach up to 10 cm in height, and the weight of biofouling on a hull could reach 50–100 ton. Biofouling also caused operational problems in sea water pipework [33] (see Chapter 15).

To obtain some understanding of historical ship-mediated dispersal events, Carlton and Hodder [34] studied fouling on a replica of a sixteenth-century sailing ship, the *Golden Hinde II*, during a transit along 800 km of US coastline from Oregon to California. The vessel travelled at slow speeds (3.5–4 knots), resided in each of four bays for approximately 30 days and spent 1–3 days travelling between bays. Biofouling assemblages were present on the hull, keel and rudder, and 64 taxa were identified within the biofouling community. Of these, 50 (92%) survived and were transported to one or more ports. This was considered to be an effective demonstration of the potential that vessels, travelling along coastlines, and across and between oceans, must have had for the dispersions of marine plants and animals around the world, centuries before the first systematic biological studies.

Charles Darwin, in his monograph on barnacles [86], and Bishop [87] comment on a number of barnacle species that were extremely common on ships' bottoms arriving in England from remote parts of the world, including West Africa, India, China and Australia. Southward and Crisp [88] described 17 barnacle species which were considered to be the more important biofoulers of ships in European waters, 5 of which were considered to have been introduced to Europe by shipping. The apparent ubiquity of numerous biofouling species elsewhere was also speculated to be a consequence of spread by shipping, but the actual vector and time of colonisation of any particular NIMS could rarely be confirmed [66, 89–91]. Recent studies of one ubiquitous biofouling species, the serpulid tubeworm *Hydroides elegans*, found close genetic relatedness between populations scattered around the world, which is best explained by regular and consistent transport of the species on ships [92].

In the subsequent era of more regular study and monitoring of marine and estuarine communities the approximate time of arrival of some non-indigenous biofouling species has been inferred: for example, the arrival of the bryozoan *Watersipora cucullata* in Auckland Harbour, New Zealand, around 1957 [90], the tubeworms *Ficopomatus enigmaticus* in New Zealand in the 1960s, *Hydroides ezoensis* in Southampton, England, in ca. 1980 [93], and *Hydroides sanctaecrucis* in northern Australia in the late 1990s [94], and in Hong Kong the isopod *Sphaeroma walkeri* (ca. 1972), the ascidian *Ciona intestinalis* (ca. 1972), the gastropod *Crepidula onyx* (ca. 1975), the bryozoans *Bugula californica* (ca. 1978), and the bivalves *Mytilopsis sallei* (ca. 1980) and *Mytilus galloprovincialis* (ca. 1981) [95].

Around 50 NIMS are listed to occur in biofouling assemblages in south-eastern Australia [18]. Many of these have a long record of existence in major ports such as Sydney and Adelaide, and are likely to be found in temperate harbours all over the world. Typical of these species are the ascidians *Ciona intestinalis*, *Botryllus schlosseri* and *Botrylloides leachii*, the bryozoans *Bugula neritina*, *Bugula stolonifera* and *Cryptosula pallasiana*, and the barnacle *Amphibalanus amphitrite*, all species described in European catalogues of marine fouling organisms [88, 96, 97]. The listed species were observed to mostly possess short-lived, non-feeding larvae, and could therefore only survive in ballast water for short times, of the order of days. However, by nature, they could travel easily on hulls.

24.5   Modern translocations

As previously mentioned, the role of modern vessels in transporting biofouling organisms was thought to have declined in the late twentieth century as a consequence of increased commercial vessel speeds (and thus decreased retention of biofouling organisms), decreased port residence times (and thus decreased biofouling accumulations), increased use and efficacy of toxic antifouling paints and increased vessel maintenance [98, 99]. The significant advance in coating technology that increased antifouling efficacy was the development of organotin copolymer paints in the mid-1970s [54]. The advantages of this type of coating were an extension of effective life to 5 or more years proportional to coating thickness, controllable biocide release rate, ability to overcoat without loss of activity, no depleted coating to remove before painting, efficient use of biocide, continuous replacement of active surface and a self-smoothing or polishing action that reduces hull roughness and improves hydrodynamic performance in-service [100]. Although organotin antifouling biocides have now been banned [101], modern technology tin-free copolymer antifouling paints do provide equivalent effectiveness [102] (see Chapter 21). However, not all vessels are painted with high-performance coatings. Longer lasting antifouling paints are more expensive, so vessel owners and operators will use a system with a shorter effective life if they anticipate a shorter docking cycle.

Although the outer, immersed hull of a vessel presents the greatest potential surface for colonisation, antifouling paints usually prevent the recruitment and survival of biofouling macroalgae and invertebrate species in this region unless the biocide is depleted, biocide release obstructed by insoluble surface precipitates or deposits, or the paint mechanically damaged or otherwise physically unsound. However, the main purpose of antifouling systems on ships is not to prevent the translocation of exotic species, but to reduce hull roughness caused by biofouling so that vessels travel more efficiently and use less fuel [103]. Biofouling growth is commonly found in so-called niche areas on and around the hull, such as (1) on unpainted surfaces such as cathodic protection anodes, propellers and propeller shafts, rudder stocks and docking block support strips (Plate III B); (2) where antifouling paint is lost or damaged in areas of high turbulence and cavitation, such as propeller stocks, rudders, intake grates (Plate III C), bilge keels and around bow thrusters, can provide colonisation niches; and (3) in areas of low flow in and around complex appendages or equipment, such as in sea chests (Plate III D) and emergency propulsion units [42–46, 104, 105]. The latter are prone to biofouling because effective antifouling paint application is difficult or water flow is insufficient to maintain effective biofouling release from self-polishing antifouling paints [106]. Vessel sea chests have been found to be particularly vulnerable to biofouling [104, 107]. For example, 150 different organisms were identified from sea chests of vessels visiting or operating in New Zealand [106]. Only 40% of these species were confirmed as native species. Significantly, a high proportion of the species collected were mobile adult invertebrates, highlighting that it is not just sessile species that are translocated in biofouling aggregations (Plate III E).

Two other phenomena may have sustained the importance of the ship as an agent for dispersal of biofouling organisms into the twenty-first century [34]. These are the greater ocean-going speeds of vessels, which may effectively reduce the length of time oligohaline–euryhaline species may be subjected to full-strength sea water during ocean transits, and the development of antifouling biocide resistance by some organisms. For example, some biofouling algae, bryozoans and serpulid tubeworms are known to be tolerant of copper antifouling systems [55, 57, 89, 94, 108–112]. The proximity to active antifouling coatings, particularly copper-based systems, can also influence species recruitment and favour NIMS [81, 109].

Faster transit times were considered a factor facilitating trans-equatorial species incursions to New Zealand [44]. James and Hayden [44] found in vessel surveys that many species did survive transit from originating ports, and most of the vessels they surveyed had passed through the tropics. A high diversity of barnacles was also recently collected from the hulls of bulk carriers voyaging to Japan from Australia, with several of the native Australian species not previously reported from Japan [105]. Some of the most environmentally damaging invasive pest species in recent times are species that have successfully crossed significant environmental and biogeographic barriers: for example, the North Pacific sea star (*Asterias amurensis*), the Mediterranean fan worm (*Sabella spallanzanii*) and the Japanese kelp (*Undaria pinnatifida*) that are now established pests in southern Australia [113].

NIMS translocations also continue on slow-moving vessels. Barges, oil exploration rigs, floating dry docks, decommissioned vessels and dredges are renowned for accumulating relatively high levels of biofouling due to low operational activity and long port residency times [47, 72, 114–120], and the international movement of specialised vessels and infrastructure is now a common part of the global operation of many maritime industrial companies.

A number of species provide good examples of significant marine pests that continue to be translocated as biofouling. Here we give the examples of the Asian kelp *U. pinnatifida* and various mussel species.

U. pinnatifida, a native of north-eastern Asia, was first introduced to the French Mediterranean in the 1970s associated with oysters introduced for culture [30, 32, 121, 122]. Within a decade, *U. pinnatifida* had spread to Brittany, southern England and the central Mediterranean, and it has since been found in New Zealand, Australia, Argentina, California, USA and Mexico [30, 112, 123–130].

In New Zealand, *Undaria* was first found in the ports of Wellington and Timaru and was considered to have arrived as biofouling attached to fishing vessels from the north-eastern Pacific [32, 123]. The kelp was subsequently found in 15 other New Zealand ports and harbours. In several ports, *U. pinnatifida* was commonly found growing on floating or suspended objects, including the hulls of resident vessels or vessels laid up pending sale [30]. Hundreds of *U. pinnatifida* sporophytes were found attached to some hulls, and plants were found on the hulls of 25% of moored pleasure craft in one boat harbour. In one instance, plants survived and grew in length during a 4000 km, one-month ship voyage [30].

Recent studies on the genetic diversity and haplotype divergence of *U. pinnatifida* from both its native range and introduced regions worldwide have provided evidence for multiple transoceanic translocations and regional dispersion [32]. In New Zealand, only a single haplotype was found at all locations on the North Island and the northernmost part of the South Island, but a further seven haplotypes were found at different locations elsewhere on the South Island. This suggests that populations introduced to Wellington on the North Island were most likely spread around the coast by commercial and recreational vessels and aquaculture activities, but the South Island has received multiple transoceanic introductions from north-eastern Asia. *Undaria* in Melbourne, Australia, and Argentina had the same haplotype as North Island New Zealand populations, suggesting a latitudinal secondary translocation, but the unique haplotype in Tasmania is likely to be a direct introduction from Japan.

The introduction of *U. pinnatifida* to southern Australia was linked to vessels carrying woodchips from Tasmania to Japan. Ballast water was suggested as the vector, but microscopic brown algal gametophytes are known to live, survive and sporulate on ships' hulls [43], so vessel biofouling offers an alternative hypothesis for this introduction. The upper temperature tolerance of *U. pinnatifida* gametophytes has been determined to be more than 29°C, which would allow present-day passage through the tropics on a ship hull [131]. *U. pinnatifida* spores are also short-lived, which does not favour their transport in ballast [132].

The mussels *Mytilus edulis* and/or *M. galloprovincialis* are native to temperate regions of the northern and southern hemisphere, but their introduction to a number of geographic regions is relatively recent and attributed to maritime activity. *M. edulis* was first reported from northern Japan in 1919, *M. galloprovincialis* from southern Japan in 1935, South Africa in 1972 and Hong Kong in 1983 [95, 133–135]. The translocation of these species as biofouling on ships is supported by their observation in sea chests and clustered around cathodic protection anodes attached to the hulls on ships operating in southern Australia [104, 136]. In 2003, mussels that colonised the sea chests of a Royal Australian Navy frigate in south-western Australia survived a three-month deployment to the Arabian Gulf [137].

Other marine mussels putatively translocated by shipping include the Asian date or bag mussel *Musculista senhousia*, native to the north-west Pacific; the black-striped mussel *Mytilopsis sallei*, native to shores of the Gulf of Mexico and the Caribbean; and the Asian green and brown mussels *Perna viridis* and *Perna perna*. *M. senhousia* was first recorded on the West Coast of the USA in 1941, and is now distributed from Puget Sound to San Diego Bay, in New Zealand in the late 1970s and in Australia in the early 1980s [138, 139]. This mussel commonly grows in byssal nets in mud, but does also attach to boat hulls. Introduction to Australia was considered most likely to be in the sea chests and/or internal sea water piping systems of vessels [139].

The first records of *Mytilopsis* spp. outside their native Atlantic range were of *M. allyneana* from Fiji (ca. 1929) and *M. zeteki* from Panama (ca. 1937), both species now considered likely to be synonymous with *M. sallei* [95]. The species was subsequently introduced to India (ca. 1967), Japan (ca. 1974), Taiwan (ca. 1977) and Hong Kong (ca. 1980). An incursion of *M. sallei* to marinas in Darwin in northern Australia was attributed to biofouling of the hull or internal pipework of an international vessel [140, 141]. The species, along with the *P. viridis*, has since been found attached to the hulls of small foreign fishing vessels and recreational boats entering Australian waters from the north. In 2001, mature aggregations of *P. viridis* were also found in sea water intakes of a ship from south-east Asia moored derelict in Cairns, north-east Australia [142]. *P. viridis* was assessed in Australia as the highest risk exotic marine pest on the basis of potential economic, environmental and human health impacts [143, 144].

24.6 Proposed management measures

Invasive species management has been recognised as a global issue of concern, but the primary focus at an international level has been on the regulation and management of ships' ballast water [e.g.145]. The importance of biofouling as a vector for invasive species translocation is being increasingly recognised.

Preventing the translocation and introduction of NIMS as biofouling can be feasibly effected at any of the stages of the translocation process (Figure 24.1). However, the earlier the intervention, the greater is the effect in managing the invasion process [26]. Eradication of NIMS once they have colonised and established in a new region is rarely successful, because detection of the incursion does not generally occur until the species is well established [7].

The prevention or deterrence of biofouling attachment can be achieved by good antifouling practices. The antifouling systems applied to a ship should prevent biofouling between dockings and, to achieve this, the system specification should take into account the planned docking period, the ship's speed and activity (nautical miles per month) and any projected lay-up periods. To prevent NIMS translocation, attention is also needed to niche areas. For example, the positions of docking blocks and supports should be varied at each docking to ensure that areas under blocks are painted with antifouling, at least at alternate dockings. Consideration should also be given to specifying different antifouling systems to the general hull for areas of low water flow conditions, such as sea chests and rudder stocks, and for areas of high water turbulence and cavitation, such as near propellers, bow thrusters and rudders. Edge retentive coating systems will enhance the durability of antifouling systems on sea intake grates, bilge keel edges and the lips of intake and discharge pipes. Biofouling colonisation and growth in internal sea water cooling and pipework systems, including sea chests, can be prevented by the use of effective marine growth prevention systems, such as electrochlorination or copper-dosing systems.

However, not all niches can be effectively antifouled and, during the period between dockings, antifouling systems can be damaged as a result of grounding, collision or mechanical impact. If the area of damage is relatively minor, in-water repair of the paint system may be possible but, for many niches, regular removal of any biofouling growth represents the most effective strategy to minimise the risk of NIMS translocation. Regular removal of biofouling not only will minimise the risks posed by organisms becoming reproductively mature, but can also reduce the ability of mobile and sedentary species to colonise and survive voyages in the microhabitat formed by sessile colonists.

In-water polishing of propellers to remove biofouling deposits is a common practice on commercial trading vessels to improve ship performance and fuel efficiency by removing hydrodynamic roughness. This activity provides an opportunity for divers to inspect fouling prone niches such as the rudder stocks and hinges, stabiliser fin apertures, rope guards and propeller shafts, cathodic protection anodes, sea chest and bow thrusters tunnel grates, sea chests, echo sounders, velocity probes and overboard discharge outlets and sea inlets. Should significant biofouling growth be detected, it could then be removed, but using appropriate technology to ensure that all material is captured for disposal onshore and no material is allowed to remain in the water column.

In-water cleaning or scrubbing of hulls painted with biocide-containing antifouling paints, for the purpose of delaying dockings or attempting to rejuvenate depleted antifouling coatings, can be counterproductive to the management and prevention of marine pest translocations.

Scrubbing antifouling paints prematurely depletes the antifouling coating and creates a pulse of biocide that can harm the local environment and may impact on future applications by the port authority for the disposal of dredge spoil. Depleted antifouling coatings on hulls will also rapidly re-foul, reducing efficiency and increasing marine pest translocation risks. The cleaning process can also increase the pest incursion risk through the release and dispersal of viable plant and animal fragments, or through stimulation of spawning events.

For these reasons, in-water cleaning has been discouraged by some authorities. For example, in Australia, a 'Code of Practice for In-water Hull Cleaning and Maintenance' was implemented with the intention of minimising the risk of NIMS establishing in Australian waters [146]. The Code essentially bans in-water hull cleaning and only allows the cleaning of sea chests, sea suction grids and other hull apertures under permit and provided that any debris removed (including encrustation, barnacles, weeds) is not allowed to pass into the water column or fall to the seabed. The polishing of ships' propellers may be permitted subject to conditions.

Increasing concerns about the environmental impact of all antifouling biocides have resulted in increasing use of non-toxic, minimally adhesive surfaces [54, 102]. These coatings do not prevent biofouling attachment, but reduce the adhesion strength so that organisms detach under their own weight or are dislodged by water movement. However, biofouling can accumulate on these coatings during periods when the vessel is stationary, and this biofouling may not dislodge during operation. Non-toxic coatings therefore pose a marine pest translocation risk if the biofouling is not effectively managed. This may require removal of biofouling from the hull after periods of low activity. While small craft can be slipped and the biofouling easily removed by high-pressure water washing, this is unlikely to be possible for larger vessels and ships.

The controlled in-water cleaning of vessels that do not or cannot dock regularly is therefore considered necessary to minimise NIMS translocation risks. In addition to the regular inspection and biofouling removal from hull niches, hull cleaning of ships with biocide-free underwater coatings that have been inactive (e.g. polar vessels, ships with fouling release coatings) may be necessary. For the latter, the in-water cleaning would need to be undertaken before the vessel departs port, ensuring that species that settled in that port, stay in that port. For niche areas, which may be fouled by NIMS, measures are needed during the in-water cleaning process to capture all biological debris. New technologies may need to be developed to achieve the latter.

24.7   Conclusions

- Biofouling on ships has provided a vector for the translocation of marine algal and invertebrate species for as long as ships have sailed the seas. Despite the inference that modern antifouling technologies may have reduced this risk and that the uptake and discharge of ballast water may have become the more significant ship-related vector for the movement of marine species, ship biofouling is now widely recognised as continuing as an equal, if not greater, risk.
- A risk is associated with most maritime vessels and sectors, ranging from slow vessels with long port residence times that may not have effective antifouling protection, such as non-trading vessels (e.g. dredges and barges) and international yachts, to high-activity, commercial trading vessels with long docking cycles that, although the outer hull surface

is maintained largely biofouling free, may accumulate significant biofouling aggregations in niche areas such as sea chests, internal sea water systems and in areas of paint damage.

- Measures needed to minimise the risk of species translocation by ship biofouling include closer attention to the effectiveness, appropriateness and maintenance of antifouling systems on vessels, including attention to niche areas and the implementation of regular in-water inspection and removal of biofouling in unprotected hull niches, particularly on ships with long docking cycles.

References

1. Baltz, D.M. (1991) Introduced fishes in marine systems and inland seas. *Biological Conservation*, **56**, 151–177.
2. Vitousek, P.M., D'Antonio, C.M., Loope, L.L. & Westbrooks, R. (1996) Biological invasions as global environmental change. *American Scientist*, **84**, 468–478.
3. Pimental, D., Lach, L., Zuniga, R. & Morrison, D. (2000) Environmental and economic costs of nonindigenous species in the United States. *BioScience*, **50**, 53–64.
4. Nentwig, W. (2007) Biological invasions: why it matters. In: *Biological Invasions* (ed. W. Nentwig), pp. 1–6. Springer, Berlin.
5. Lubchenco, J.A., Olson, M., Brubaker, L.B., *et al.* (1991) The sustainable biosphere initiative: an ecological research agenda. *Ecology*, **72**, 371–412.
6. Carlton, J.T. (2001) *Introduced Species in US Coastal Waters: Environmental Impacts and Management Priorities*. Pew Oceans Commission, Arlington, VA.
7. Hewitt, C.L. & Campbell, M.L. (2007) Mechanisms for the prevention of marine bioinvasions for better biosecurity. *Marine Pollution Bulletin*, **55**, 395–401.
8. Gollasch, S. (2002) The importance of ship hull fouling as a vector of species introductions into the North Sea. *Biofouling*, **18**, 105–121.
9. Bishop, M.W.H. (1947) Establishment of an immigrant barnacle in British coastal waters. *Nature*, **159**, 501.
10. Carlton, J.T. (1979) Introduced invertebrates of San Francisco Bay. In: *San Francisco Bay: The Urbanized Estuary* (ed. T.J. Conomos), pp. 427–444. American Association for the Advancement of Science, Pacific Division, San Francisco, CA.
11. Thresher, R.E., Hewitt, C.L. & Campbell, M.L. (1999) Synthesis: introduced and cryptogenic species in Port Phillip Bay. In: *Marine Biological Invasions of Port Phillip Bay, Victoria*, Technical Report No. 20 (eds C.L. Hewitt, M.L. Campbell, R.E. Thresher & R.B. Martin), pp. 283–295. Centre for Research on Introduced Marine Pests, CSIRO Marine Research, Hobart, Australia.
12. Hewitt, C.L., Campbell, M.L., Thresher, R.E., *et al.* (2004) Introduced and cryptogenic species in Port Phillip Bay, Victoria, Australia. *Marine Biology*, **144**, 183–202.
13. Cranfield, H.J., Gordon, D.J., Willan, R.C., *et al.* (1998) *Adventive Marine Species in New Zealand*. Technical Report No. 34, NIWA, Wellington, New Zealand.
14. Gollasch, S. & Leppäkoski, E. (1999) *Initial Risk Assessment of Alien Species in Nordic Waters*. Nordic Council of Ministers, Copenhagen, Denmark.
15. Coles, S.L., DeFelice, R.C., Eldridge, L.G. & Carlton, J.T. (1999) Historical and recent introductions of nonindigenous species marine species into Pearl Harbour, Oahu, Hawaiian Islands. *Marine Biology*, **135**, 147–158.
16. Ruiz, G.M., Carlton, J.T., Grosholz, E.D. & Hines, A.H. (1997) Global invasions of marine and estuarine environments by non-indigenous species: mechanisms, extent and consequences. *American Zoologist*, **37**, 621–632.
17. Fofonoff, P.W., Ruiz, G.M., Steves, B. & Carlton, J.T. (2003) In ships or on ships? Mechanisms of transfers and invasion for non-native species to the coasts of North America. In: *Invasive Species:*

Vectors and Management Strategies (eds G.M. Ruiz & J.T. Carlton), pp. 152–182. Island Press, Washington, DC.

18. Keough, M.J. & Ross, J. (1999) Introduced fouling species in Port Phillip Bay. In: *Marine Biological Invasions of Port Phillip Bay, Victoria*, Technical Report No. 20 (eds C.L. Hewitt, M.L. Campbell, R.E. Thresher & R.B. Martin), pp. 193–226. Centre for Research on Introduced Marine Pests, CSIRO Marine Research, Hobart, Australia.

19. Pimental, D., Zuniga, R. & Morrison, D. (2005) Update on the environmental and economic costs associated with alien-invasive species in the United States. *Ecological Economics*, **52**, 273–288.

20. Schaffelke, B. & Hewitt, C.L. (2007) Impacts of introduced macroalgae. *Botanica Marina*, **50**, 397–417.

21. Nichols, F.H., Thompson, J.K. & Schemel, L.E. (1990) Remarkable invasion of San Francisco Bay (California, USA) by the Asian clam *Potamocorbula amurensis*. II. Displacement of a former community. *Marine Ecology Progress Series*, **66**, 95–101.

22. Elton, C.S. (1958) *The Ecology of Invasions by Animals and Plants*. Methuen and Co Ltd, London.

23. Minchin, D. (2007) Aquaculture and transport in a changing environment: overlap and links in the spread of alien biota. *Marine Pollution Bulletin*, **55**, 302–313.

24. Bolton, T.F. & Graham, W.M. (2006) Jellyfish on the rocks: bioinvasion threat of the international trade in aquarium live rock. *Biological Invasions*, **8**, 651–653.

25. Calado, R. & Chapman, P.M. (2006) Aquarium species: deadly invaders. *Marine Pollution Bulletin*, **52**, 599–601.

26. Schaffelke, B., Smith, J.E. & Hewitt, C.L. (2006) Introduced macroalgae – a growing concern. *Journal of Applied Phycology*, **18**, 529–541.

27. Carlton, J.T. (1985) Transoceanic and interoceanic dispersal of coastal marine organisms: the biology of ballast water. *Oceanography and Marine Biology: An Annual Review* **23**, 313–371.

28. Grigorovich, I.A., Therriault, T.W. & MacIsaac, H.J. (2003) History of aquatic invertebrate invasions in the Caspian Sea. *Biological Invasions*, **5**, 103–115.

29. Gollasch, S. (2007) Is ballast water a major dispersal mechanism for marine organisms? In: *Biological Invasions* (ed. W. Nentwig), pp. 49–57. Springer, Berlin.

30. Hay, C.H. (1990) The dispersal of sporophytes of *Undaria pinnatifida* by coastal shipping in New Zealand, and implications for further dispersal of *Undaria* in France. *British Journal of Phycology*, **25**, 301–313.

31. Wasson, K., Zabin, C.L., Bedinger, L., Diaz, M.C. & Pearse, J.S. (2001) Biological invasions of estuaries without international shipping: the importance of intraregional transport. *Biological Conservation*, **102**, 143–153.

32. Uwai, S., Nelson, W., Neill, K., *et al.* (2006) Genetic diversity in *Undaria pinnatifida* (Laminariales, Phaeophyceae) deduced from mitochondria genes – origins and succession of introduced populations. *Phycologia*, **45**, 687–695.

33. WHOI (1952) *Marine Fouling and Its Prevention*. Woods Hole Oceanographic Institution, United States Naval Institute, Annapolis, MD.

34. Carlton, J.T. & Hodder, J. (1995) Biogeography and dispersal of coastal marine organisms: experimental studies on a replica of a 16th-century sailing vessel. *Marine Biology*, **121**, 721–730.

35. Middleton, M.J. (1982) The oriental goby, *Acanthogobius flavimanus* (Temminck and Schlegel), an introduced fish in the coastal waters of New South Wales, Australia. *Journal of Fish Biology*, **21**, 513–523.

36. Carlton, J.T. & Geller, J.B. (1993) Ecological roulette: the global transport and invasion of non-indigenous marine organisms. *Science*, **261**, 78–82.

37. Pollard, D.A. & Hutchings, P.A. (1990) A review of exotic marine organisms introduced to the Australian region. I. Fishes. *Asian Fisheries Science*, **3**, 205–221.

38. Pollard, D.A. & Hutchings, P.A. (1990) A review of exotic marine organisms introduced to the Australian region. II. Invertebrates and algae. *Asian Fisheries Science*, **3**, 223–250.

39. Hutchings, P.A. (1992) Ballast water introduction of exotic marine organisms into Australia: current status and management options. *Marine Pollution Bulletin*, **25**, 196–199.

40. Wiley, C.J. & Claudi, R. (2000) The role of ships as a vector of introduction for non-indigenous freshwater organisms, with focus on the Great Lakes. In: *Non-Indigenous Freshwater Organisms* (eds R. Claudi & J.H. Leach), pp. 203–213. Lewis Publishers Inc, Boca Raton, FL.

41. Niimi, A.J. (2004) Role of container vessels in the introduction of exotic species. *Marine Pollution Bulletin*, **49**, 778–782.

42. Rainer, S.F. (1995) Potential for the introduction and translocation of exotic species by hull fouling: a preliminary assessment. Technical Report No. 1, Centre for Research on Introduced Marine Pests, CSIRO, Hobart, Australia.

43. Coutts, A.D.M. (1999) *Hull Fouling as a Modern Vector for Marine Biological Invasions: Investigation of Merchant Vessels Visiting Northern Tasmania.* MSc thesis, Australian Maritime College, Launceston, Australia.

44. James, P. & Hayden, B. (2000) The potential for the introduction of exotic species by vessel fouling: a preliminary study. Client Report No. WLG 00/51, NIWA, Wellington, New Zealand.

45. Lewis, J.A. (2002) Hull fouling as a vector for the translocation of marine organisms: report 1 – hull fouling research. Ballast Water Research Series Report No. 14, Department of Agriculture, Fisheries and Forestry, Canberra, Australia.

46. Coutts, A.D.M. & Taylor, M.D. (2004) A preliminary investigation of biosecurity risks associated with biofouling on merchant vessels in New Zealand. *New Zealand Journal of Marine and Freshwater Research*, **38**, 215–229.

47. Godwin, L.S. (2003) Hull fouling of maritime vessels as a pathway for marine species invasions to the Hawaiian Islands. *Biofouling*, **19** (Suppl.), 123–131.

48. Otani, M. (2004) Introduced marine organisms in Japanese coastal waters, and the processes involved in their entry. *Japanese Journal of Benthology*, **59**, 45–57.

49. Otani, M. (2006) Important vectors for marine organisms unintentionally introduced to Japanese waters. In: *Assessment and Control of Biological Invasion Risks* (eds F. Koike, M.N. Clout, M. Kawamichi, M. De Poorter & K. Iwatsuki), pp. 92–103. Shoukadoh Book Sellers, Kyoto, Japan, and IUCN, Gland, Switzerland.

50. Floerl, O. (2002) *Intracoastal Spread of Fouling Organisms by Recreational Vessels.* PhD thesis, James Cook University, Australia.

51. Wahl, M. (1989) Marine epibiosis. I. Fouling and antifouling: some basic aspects. *Marine Ecology Progress Series*, **58**, 175–189.

52. Wahl, M. (1997) Living attached: aufwuchs, fouling, epibiosis. In: *Fouling Organisms of the Indian Ocean: Biology and Control Technology* (eds R. Nagabhushanam & M.F. Thompson), pp. 31–83. A.A. Balkema, Rotterdam, the Netherlands.

53. Richmond, C.A. & Seed, R. (1991) A review of marine macrofouling communities with special reference to animal fouling. *Biofouling*, **3**, 151–168.

54. Lewis, J.A. (1998) Marine biofouling and its prevention on underwater surfaces. *Materials Forum*, **22**, 41–61.

55. Russell, G. & Morris, O.P. (1973) Ship-fouling as an evolutionary process. In: Proceedings of the Third International Congress on Marine Corrosion and Fouling, Gaithursburg, MD, October 2–6, 1972 (eds R.F. Acker, B. Floyd Brown, J.R. DePalma & W.P. Iverson), pp. 719–730. National Bureau of Standards, Gaithursburg, MD.

56. Evans, L.V. (1981) Marine algae and fouling: a review, with particular reference to ship-fouling. *Botanica Marina*, **24**, 167–171.

57. Reed, R.H. & Moffat, L. (1983) Copper toxicity and copper tolerance in *Enteromorpha compressa* (L.) Grev. *Journal of Experimental Marine Biology and Ecology*, **69**, 85–103.

58. Callow, M.E. (1986) Fouling algae from 'in-service' ships. *Botanica Marina*, **24**, 351–357.

59. Roughgarden, J. (1986) Predicting invasions and rates of spread. In: *Ecology of Biological Invasions of Northern America and Hawaii* (eds H.A. Mooney & J.A. Drake), pp. 179–190. Springer Verlag, New York.

60. Simberloff, D.S. (1989) Which insect introductions succeed and which ones fail? In: *Biological Invasions: A Global Perspective* (eds J.A. Drake, H.A. Mooney, F. di Castri, et al.), pp. 61–67. John Wiley & Sons, New York.

61. Kolar, C.S. & Lodge, D.M. (2001) Progress in invasion biology: predicting invaders. *Trends in Ecology and Evolution*, **16**, 199–204.
62. Floerl, O. (2005) Factors that influence hull fouling on ocean-going vessels. In: *Hull Fouling as a Mechanism for Marine Invasive Species Introductions: Proceedings of a Workshop on Current Issues and Potential Management Strategies*, February 12–13, 2003 (ed. L.S. Godwin), pp. 6–13. Honolulu, Hawaii.
63. Takata, L., Falkner, M. & Gilmore, S. (2006) Commercial vessel fouling in California: analysis, evaluation, and recommendations to reduce non-indigenous species release from the non-ballast water vector. Report to the California State Legislature, Marine Facilities Division, California State Lands Commission, California.
64. Visscher, J.P. (1928) Nature and extent of fouling on ships' bottoms. *Bulletin of the Bureau of Fisheries*, **43**, 193–252.
65. Wood, E.J.F. & Allen, F.E. (1958) *Common Marine Fouling Organisms of Australian Waters*. Department of the Navy, Navy Office, Melbourne, Australia.
66. Skerman, T.M. (1960b) Ship-fouling in New Zealand waters: a survey of marine fouling organisms from vessels of the coastal and overseas trades. *New Zealand Journal of Science*, **3**, 620–648.
67. Floerl, O., Inglis, G.J. & Marsh, H.M. (2005) Selectivity in vector management: an investigation of the effectiveness of measures used to prevent transport of non-indigenous species. *Biological Invasions*, **7**, 459–475.
68. Coutts, A.D.M., Taylor, M.D. & Hewitt, C.L. (2007) Novel method for assessing the *en route* survivorship of biofouling organisms on various vessel types. *Marine Pollution Bulletin*, **54**, 97–116.
69. Lewis, J.A. (1999) A review of the occurrence of exotic macroalgae in southern Australia, with emphasis on Port Phillip Bay. In: *Marine biological Invasions of Port Phillip Bay, Victoria*, Technical Report No. 20 (eds C.L. Hewitt, M.L. Campbell, R.E. Thresher & R.B. Martin), pp. 61–87. Centre for Research on Introduced Marine Pests, CSIRO Marine Research, Hobart, Australia.
70. Buchanan, S. & Babcock, R. (1997) Primary and secondary settlement of the greenshell mussel *Perna canaliculus*. *Journal of Shellfish Research*, **16**, 71–76.
71. Utting, S.D. & Spencer, B.D. (1997) The hatchery culture of bivalve mollusc larvae and juveniles. Laboratory Leaflet No. 68, Ministry of Agriculture, Fisheries and Food, Lowestoft, UK.
72. Apte, S., Holland, B.S., Godwin, L.S. & Gardner, P.A. (2000) Jumping ship: a stepping stone event mediating transfer of a non-indigenous species via a potentially unsuitable environment. *Biological Invasions* **2**, 75–79.
73. Minchin, D. & Gollasch, S. (2003) Fouling and ships' hulls: how changing circumstances and spawning events may result in the spread of exotic species. *Biofouling*, **19**, 111–122.
74. Glasby, T.M. & Creese, R.G. (2007) Invasive marine species management and research. In: *Marine Ecology* (eds S.D. Connell & B.M. Gillanders), pp. 569–594. Oxford University Press, South Melbourne, Australia.
75. Glasby, T.M., Connell, S.D., Holloway, M.G. & Hewitt, C.L. (2007) Non-indigenous biota on artificial structures: could habitat creation facilitate biological invasions? *Marine Biology*, **151**, 887–895.
76. Glasby, T.M. & Connell, S.D. (1999) Urban structures as marine habitats. *Ambio*, **28**, 595–598.
77. Johnston, E.L. & Keough, M.J. (2002) Direct and indirect effects of repeated pollution events on marine hard substrate assemblages. *Ecological Applications*, **12**, 1212–1228.
78. Hutchings, P.A., Hilliard, R.W. & Coles, S.L. (2002) Species introductions and potential for marine pest invasions into tropical marine communities, with special reference to the Indo-Pacific. *Pacific Science*, **56**, 223–233.
79. Clarke, G.F. & Johnston, E.L. (2005) Manipulating larval supply in the field: a controlled study of marine invasibility. *Marine Ecology Progress Series*, **298**, 1–9.
80. Johnston, E.L. (2007) Biological invasions and pollution. In: *Marine Ecology* (eds S.D. Connell & B.M. Gillanders), p. 581. Oxford University Press, South Melbourne, Australia.
81. Dafforn, K.A., Glasby, T.M. & Johnston, E.L. (2007) Differential effects of tributyltin and copper antifoulants on recruitment of non-indigenous species. *Biofouling*, **24**, 23–33.

82. Hadfield, M.G., Carpizo-Ituarte, E., Holm, E.R., Nedved, B. & Unabia, C. (1999) Macrofouling processes: a developmental and evolutionary perspective. Paper presented at *10th International Congress on Marine Corrosion and Fouling*, Melbourne, Australia, February 1999.

83. Silva, P.C. (1955) The dichotomous species of *Codium* in Britain. *Journal of the Marine Biological Association of the United Kingdom*, **34**, 565–577.

84. Trowbridge, C.D. (1998) Ecology of the green macroalga *Codium fragile* (Suringar) Hariot 1889: invasive and noninvasive subspecies. *Oceanography and Marine Biology: An Annual Review*, **36**, 1–64.

85. Laidlow, F.B. (1952) The history of the prevention of fouling. In: *Marine Fouling and Its Prevention*, pp. 211–223. Woods Hole Oceanographic Institution, United States Naval Institute, Annapolis, MD.

86. Darwin, C. (1854) *A Monograph on the Sub-class Cirripedia, with Figures of All the Species: The Balanidae (or Sessile Cirripedes); the Verrucidae, etc., etc., etc.* The Ray Society, London.

87. Bishop, M.W.H. (1951) Distribution of barnacles by ships. *Nature*, **167**, 531.

88. Southward, A.J. & Crisp, D.J. (1963) *Catalogue of Main Marine Fouling Organisms (Found on Ships Coming into European Waters): Barnacles.* Organisation for Economic Co-operation and Development, Paris, France.

89. Allen, F.E. (1953) Distribution of marine invertebrates by ships. *Australian Journal of Marine and Freshwater Research*, **1**, 106–109.

90. Skerman, T.M. (1960a) The recent establishment of the polyzoan *Watersipora cucullata* (Busk) in Auckland Harbour, New Zealand. *New Zealand Journal of Science*, **3**, 615–619.

91. Bagaveeva, E.V., Kubanin, A.A. & Chaplygina, S.F. (1984) Role of ships in settlement of hydroids, polychaetes and bryozoans in the Sea of Japan. *Soviet Journal of Marine Biology*, **10**, 74–79.

92. Pettengill, J.B., Wendt, D.E., Schug, M.D. & Hadfield, M.G. (2007) Biofouling likely serves as a major mode of dispersal for the polychaete tubeworm *Hydroides elegans* as inferred from microsatellite loci. *Biofouling*, **23** (3/4), 161–169.

93. Thorp, C.H., Pyne, S. & West, S.A. (1987) *Hydroides ezoensis* Okuda, a fouling serpulid new to British coastal waters. *Journal of Natural History*, **21**, 863–877.

94. Lewis, J.A., Watson, C. & ten Hove, H.A. (2006) Establishment of the Caribbean serpulid tubeworm *Hydroides sanctaecrucis* Krøyer Mörch, 1863, in northern Australia. *Biological Invasions*, **8**, 665–671.

95. Morton, B. (1987) Recent marine introductions into Hong Kong. *Bulletin of Marine Science*, **41**, 503–513.

96. Ryland, J.S. (1965) *Catalogue of Main Marine Fouling Organisms (Found on Ships Coming into European Waters): Polyzoa.* Organisation for Economic Co-operation and Development, Paris, France.

97. Miller, R.H. (1969) *Catalogue of Main Marine Fouling Organisms (Found on Ships Coming into European Waters): Ascidians of European Waters.* Organisation for Economic Co-operation and Development, Paris, France.

98. Carlton, J.T. & Scanlon, J.A. (1985) Progression and dispersal of an introduced alga *Codium fragile* ssp. *tomentosoides* (Chlorophyta) on the Atlantic coast of North America. *Botanica Marina*, **28**, 155–165.

99. Carlton, J.T. (1992) Introduced marine and estuarine molluscs of North America: an end-of-the-20th-century perspective. *Journal of Shellfish Research*, **11**, 489–505.

100. Evans, C.J. (1987) Organotins combat marine fouling. *Polymers Paint Colour Journal* **177**, 73–76.

101. IMO (2001) *International Convention on the Control of Harmful Anti-Fouling Systems on Ships, 2001.* International Maritime Organization, London.

102. Lewis, J.A. (2002) Hull fouling as a vector for the translocation of marine organisms: report 2 – the significance of the prospective ban on tributyltin antifouling paints on the introduction and translocation of marine pests in Australia. Ballast Water Research Series Report No. 15, Department of Agriculture, Fisheries and Forestry, Canberra, Australia.

103. Minchin, D. (2002) Shipping: global changes and management of bioinvasions. In: *Alien Marine Organisms Introduced by Ships in the Mediterranean and Black Seas*, CIESM Workshop Monographs No. 20 (ed. F. Briand), pp. 99–102. CIESM, Monaco.

104. Australian Shipowners Association (2006) Assessment of introduced marine pest risks associated with niche areas in commercial shipping: final report. Report to Invasive Marine Species Program, Department of Agriculture, Fisheries and Forestry, Canberra, Australia, and Australian Shipowners Association, Port Melbourne, Australia.

105. Otani, M., Oumi, T., Uwai, S., *et al.* (2007) Occurrence and diversity of barnacles on international ships visiting Osaka Bay, Japan, and the risk of their introduction. *Biofouling*, **23**, 277–286.

106. Coutts, A.D.M. & Dodgshun, T.J. (2007) The nature and extent of organisms in vessel seachests: a protected mechanism for marine bioinvasions. *Marine Pollution Bulletin*, **54**, 876–886.

107. Coutts, A.D.M., Moore, K.M. & Hewitt, C.L. (2003) Ships' sea-chests: an overlooked transfer mechanism for non-indigenous marine species. *Marine Pollution Bulletin*, **46**, 1510–1512.

108. Lewis, J.A. & Smith, B.S. (1991) *Hydroides* settlement in Sydney Harbour (Australia) and its control in sea-water cooling systems. *Biodeterioration and Biodegradation*, **8**, 464–466.

109. Johnston, E.L., Keough, M.J. & Qian, P.Y. (2002) Maintenance of species dominance through pulse disturbances to a sessile marine invertebrate assemblage in Port Shelter, Hong Kong. *Marine Ecology Progress Series*, **226**, 103–114.

110. Floerl, O., Pool, T.K. & Inglis, G.J. (2004) Positive interactions between non-indigenous species facilitate transport by human vectors. *Ecological Applications*, **14**, 1724–1736.

111. Piola, R.F. & Johnston, E.L. (2006) Differential resistance to extended copper exposure in four introduced bryozoans. *Marine Ecology Progress Series*, **311**, 103–114.

112. Piola, R.F. & Johnston, E.L. (2006) Differential tolerance to metals among populations of the introduced bryozoan *Bugula neritina*. *Marine Biology*, **148**, 997–1010.

113. Talman, S., Bité, J.S., Campbell, S.J., *et al.* (1999) Impacts of some introduced species in Port Phillip Bay. In: *Marine Biological Invasions of Port Phillip Bay, Victoria*, Technical Report No. 20 (eds C.L. Hewitt, M.L. Campbell, R.E. Thresher & R.B. Martin), pp. 261–274. Centre for Research on Introduced Marine Pests, CSIRO Marine Research, Hobart, Australia.

114. Foster, B.A. & Willan, R.C. (1979) Foreign barnacles transported to New Zealand on an oil platform. *New Zealand Journal of Marine and Freshwater Research*, **13**, 143–149.

115. DeFelice, R.C. (1999) Fouling marine invertebrates on the floating dry dock USS Machinist in Pearl Harbour prior to its move to Apra Harbour, Guam. Report to the U.S. Fish and Wildlife Service, Hawaii Biological Survey Contribution No. 1999–013, Bishop Museum, Hawaii.

116. DeFelice, R.C. & Godwin, L.S. (1999) Records of the marine invertebrates on the hull of USS Missouri on arrival to Pearl Harbor, Oahu, Hawaii. *Bishop Museum Occasional Papers*, **59**, 42–46.

117. Coles, S.L., De Felice, R.C. & Godwin, L.S. (1999) The impact of non-indigenous marine species transported to Pearl Harbor on the hull of the USS Missouri. Hawaii Biological Survey Contribution No. 1999–014, Bishop Museum, Hawaii.

118. Godwin, L.S. & Eldredge, L.G. (2001) The South Oahu marine invasions shipping study. Technical Report No. 20, Bishop Museum, Hawaii.

119. Coutts, A.D.M. (2002) A biosecurity investigation of a barge in the Marlborough Sounds. Report No. 744, Cawthron Institute, Nelson, New Zealand.

120. Lewis, P.N., Bergstrom, D.N. & Whinam, J. (2006) Barging in: a temperate marine community travels to the subantarctic. *Biological Invasions*, **8**, 787–795.

121. Boudouresque, C.F., Gerbal, M. & Knoepffler-Peguy, M. (1985) L'algue japonnaise, *Undaria pinnatifida* (Phaeophyceae, Laminariales) en Mediterranee. *Phycologia*, **24**, 364–366.

122. Floc'h, J.Y., Pajot, R. & Wallentius, I. (1991) The Japanese brown alga *Undaria pinnatifida* on the coast of France and its possible establishment in European waters. *Journal du Conseil – Conseil International pour l'Exploration de la Mer*, **47**, 379–390.

123. Hay, C.H. & Luckens, P.A. (1987) The Asian kelp *Undaria pinnatifida* (Phaeophyta: Laminariales) found in a New Zealand harbour. *New Zealand Journal of Botany*, **25**, 329–332.

124. Sanderson, J.C. (1989) A survey of the distribution of the introduced Japanese macroalga *Undaria pinnatifida* (Harvey) Suringar in Tasmania. Technical Report No. 38, Marine Laboratory, Tasmanian Department of Sea Fisheries, Taroona, Tasmania, Australia.

125. Fletcher, R.L. & Manfredi, C. (1995) The occurrence of *Undaria pinnatifida* (Phaeophyceae, Laminariales) on the south coast of England. *Botanica Marina*, **38**, 355–358.

126. Casas, G.N. & Piriz, M.L. (1996) Surveys of *Undaria pinnatifida* (Laminariales, Pheaophyta) in Golfo Nuevo, Argentina. *Hydrobiologia*, **327**, 213–215.

127. Campbell, S.J. & Burridge, T.R. (1998) Occurrence of *Undaria pinnatifida* (Phaeophyta, Laminariales) in Port Phillip Bay, Victoria, Australia. *Marine and Freshwater Research*, **49**, 379–381.

128. Forrest, B.M., Brown, S.N., Taylor, M.D., Hurd, C.L. & Hay, C.H. (2001) The role of natural dispersal mechanisms in the spread of *Undaria pinnatifida* (Laminariales, Phaeophyceae). *Phycologia*, **39**, 547–553.

129. Silva, P.C., Woodfield, R.A., Cohen, A.N., *et al.* (2002) First report of the Asian kelp *Undaria pinnatifida* in the north-eastern Pacific Ocean. *Biological Invasions*, **4**, 333–338.

130. Aguilar-Rosas, R., Aguilar-Rosas, L.E., Avila-Serrano, G. & Marco-Ramirez, R. (2004) First record of *Undaria pinnatifida* (Harvey) Suringar (Laminariales, Phaeophyta) on the Pacific coast of Mexico. *Botanica Marina*, **47**, 255–258.

131. Peters, A.F. & Breeman, A.M. (1992) Temperature responses of disjunct temperate brown algae indicate long-distance dispersal of microthalli across the tropics. *Journal of Phycology*, **28**, 428–438.

132. Bité, J. (1998) *The Ecology and Reproductive Biology of the Introduced Japanese Macroalga Undaria Pinnatifida (Harvey) Suringar in Port Phillip Bay*. MSc Thesis, Victoria University of Technology, Victoria, Australia.

133. Wilkens, N.P., Fujino, K. & Gosling, E.M. (1983) The Mediterranean mussel *Mytilus galloprovincialis* Lmk. in Japan. *Biological Journal of the Linnaean Society*, **20**, 365–374.

134. Grant, W.S. & Cherry, M.I. (1985) *Mytilus galloprovincialis* in southern Africa. *Journal of Experimental Marine Biology and Ecology*, **90**, 179–191.

135. Lee, S.Y. & Morton, B.S. (1985) The introduction of the Mediterranean mussel *Mytilus galloprovincialis* into Hong Kong. *Malacological Review*, **18**, 107–109.

136. Lewis, J.A. & Gillham, A. (2006) Every nook and cranny: niche biofouling as a potential vector for invasive marine pests. Paper presented at *13th International Congress on Marine Corrosion and Fouling*, Rio de Janeiro, Brazil, July 2006.

137. Polglaze, J.F., Smith, P.R., Hilliard, R.H., *et al.* (2004) Characterisation and management of biofouling in ships of the Royal Australian Navy. Paper presented at 12th International Congress on Marine Corrosion and Fouling, Southampton, UK, July 2004.

138. Willan, R.C. (1987) The mussel *Musculista senhousia* in Australia; another aggressive alien highlights the need for quarantine at ports. *Bulletin of Marine Science*, **41**, 475–489.

139. Slack-Smith, S.M. & Brearley, A. (1987) *Musculista senhousia* (Benson, 1842); a mussel recently introduced into the Swan River estuary, Western Australia (Mollusca; Mytilidae). *Records of the Western Australian Museum*, **13**, 225–230.

140. Willan, R.C., Russell, B.C., Murfet, N.B., *et al.* (2000) Outbreak of *Mytilopsis sallei* (Récluz, 1849) (Bivalvia: Dreissenidae) in Australia. *Molluscan Research*, **20**, 25–30.

141. Willan, R.C. (2007) One in a thousand: case study of a successful eradication. In: *Marine Ecology* (eds S.D. Connell & B.M. Gillanders), pp. 574–575. Oxford University Press, South Melbourne, Australia.

142. Hayes, K.R., Cannon, R., Neil, K. & Inglis, G. (2005) Sensitivity and cost considerations for the detection and eradication of marine pests in ports. *Marine Pollution Bulletin*, **50**, 823–834.

143. Hayes, K.R. & Sliwa, C. (2003) Identifying potential marine pests – a deductive approach applied to Australia. *Marine Pollution Bulletin*, **46**, 91–98.

144. Hayes, K., Sliwa, C., Migus, S., McEnnulty, F. & Dunstan, P. (2004) National priority pests. Part II: Ranking of Australian marine pests. Final Report for the Department of Environment and Heritage, CSIRO Marine Research, Hobart, Australia.

145. International Maritime Organization (2004) *International Convention for the Control and Management of Ships' Ballast Water and Sediments, 2004.* International Maritime Organization, London.

146. Australia and New Zealand Environment Conservation Council (1997) *Code of Practice for Antifouling and In-Water Hull Cleaning and Maintenance.* Australia and New Zealand Environment Conservation Council, Canberra, Australia.

Chapter 25
New Directions in Antifouling Technology

Dean C. Webster and Bret J. Chisholm

The aim of this chapter is to survey approaches being explored to combat or mitigate fouling, with a primary focus on coating systems. Biocidal coating systems (systems which release biocidal chemicals into the environment) are surveyed, beginning with current self-polishing coatings technology, continuing with new approaches being explored for self-polishing and erodible binder systems, and concluding with a discussion on the use of natural antifoulants and their synthetic analogues as a low-toxicity approach to antifouling. Approaches to non-toxic coating systems are then surveyed beginning with a discussion of the fouling-release concept. Non-toxic coating systems containing bound biocides, hybrid-coating systems, coatings having amphiphilic surfaces and other approaches reported in the literature are also discussed. Many of these new approaches are still in the research stage. By reviewing current research efforts in this field, it is hoped that additional innovative approaches will be stimulated, which can lead to environmentally responsible antifouling coating systems.

25.1 Introduction

While antifouling coatings containing tributyltin (TBT) are being phased out due to environmental concerns (see Chapters 16, 17 and 21), biocidal antifouling technologies are likely to be used for the foreseeable future. Most common are coatings containing cuprous oxide as the key antifouling ingredient [1–3]. Ablative and self-polishing copolymer types of coating binder systems are used which degrade slowly in the marine environment in order to release the active ingredient and renew the coating surface over time. A typical self-polishing copolymer contains an alkyl silane functional monomer which hydrolyses slowly over time, converting the insoluble acrylic into a water-soluble polymer (Figure 25.1). The cuprous oxide in the coatings may also be supplemented by organic 'booster' biocides [1]. These organic biocides have a short half-life in the marine environment and thus are not expected to bioaccumulate. Known primarily by their generic or trade names, these are listed in Table 25.1.

Biocidal antifouling coatings are highly effective since the biocides are typically not highly organism specific and can deter a significant fraction of the approximately 4000 marine fouling organisms present in the world's oceans [3]. Concerns over the environmental impact of biocidal antifouling coatings are increasing [4–9], and it is believed that limits will be placed

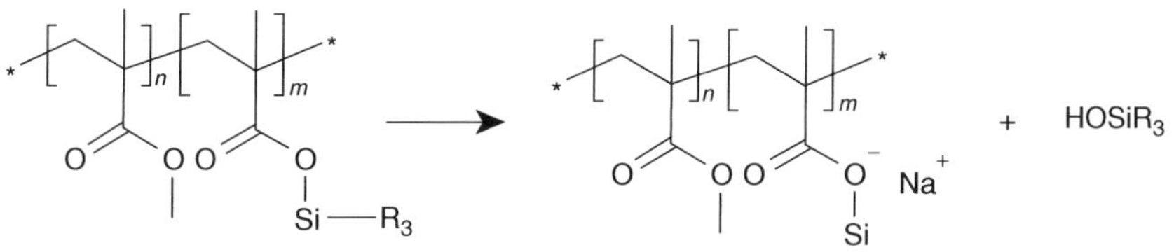

Figure 25.1 Self-polishing copolymer system.

on the amounts of biocide that a coating may emit into the environment, particularly for the case of copper.

Research into new antifouling technologies has proceeded along two primary threads: antifouling coating systems containing less-toxic biocides and totally non-toxic coating systems. The search for natural antifoulants and their incorporation into coating systems has been a highly active research area along with work on the design of new self-polishing binder systems to control the rate of biocide release into the environment.

Significant efforts are also being carried out to identify a completely non-toxic, non-fouling coating system. Much of this effort has been directed toward the fouling-release or easy-release concept, where fouling organisms are not able to make a strong adhesive bond with the coating surface, and thus are easily removed via low levels of shear applied to the coating. Materials which resist fouling through non-toxic means are also being investigated.

25.2   Biocidal antifouling coatings

25.2.1   New biocides

Due to the stringent requirements for the registration of new biocides with government agencies, research efforts in new biocides have been significantly reduced. Indeed, only one new synthetic antifouling biocide, trade name Econea, has been registered recently (see Table 25.1); its use in antifouling coatings has been described in several patents [10–15].

Some research efforts are being carried out to identify low-toxicity antifoulants. Compounds such as benzoic acid and sodium benzoate [16–18], isocyanides [19], isocyanocyclohexane derivatives [20] and quaternary ammonium compounds [21, 22] have been demonstrated to have antifouling properties, but low toxicity.

25.2.2   Erodible and self-polishing polymer systems

The release of biocides into the environment from antifouling coatings is controlled by a number of factors: ingress of water into the polymer binder to extract the biocide, solubility of the biocide in sea water and the hydrolysis/erosion rate of the polymer binder. Thus, the composition of the polymer binder can play a critical role in the biocide release rate, impacting the antifouling properties of the coating as well as its environmental impact. Designing new erodible and self-polishing binder systems can result in minimising the amount of biocide released into the environment. Vallee-Rehel *et al.* [23, 24] have described an approach to a

Table 25.1 Organic 'booster' biocides.

Common/Trade name	Chemical name	CAS number	Structure
Dichlofluanid	1,1-Dichloro-N-[(dimethylamino)sulfonyl]-1-fluoro-N-phenylmethanesulfenamide	1089-98-9	
Sea-Nine 211	4,5-Dichloro-2-n-octyl-4-isothiazolin-3-one	64359-81-5	
Diuron	N'-(3,4-Dichlorophenyl)-N,N-dimethylurea	330-54-1	
Zinc pyrithione	Bis(2-pyridylthio)zinc	13463-41-7	
Irgarol 1051	2-(tert-Butylamino)-4-(cyclopropylamino)-6-(methylthio)-s-triazine	28159-98-0	
Chlorothalonil	2,4,5,6-Tetrachloro-1,3-dicyanobenzene	1897-45-6	
Zineb	Zinc ethylenebisdithiocarbamate	12122-67-7	

Table 25.1   (*Continued*)

Common/Trade name	Chemical name	CAS number	Structure
TCMS pyridine	2,3,5,6-Tetrachloro-4-sulfuronyl pyridine	13108-52-6	
TCMTB	2(Thiocyanomethylthio) benzothiazole	21564-17-0	
Econea	1H-Pyrrole-3-carbonitrile,4-bromo-2-(4-chlorophenyl)-5-(trifluoromethyl)	122454-29-9	

polymer binder system based on three monomers: a hydrophobic monomer, a hydrophilic monomer and a hydrolysable monomer (Figure 25.2). By synthesising copolymers from a combination of the three types of monomers, it is possible to 'tune' the composition so that the erosion rate and biocide release rate of the coating can be precisely controlled.

Several other approaches to erodible or self-polishing polymer binder systems have been reported. A surface-fragmenting self-polishing coating involves a cross-linked polymer matrix linked by hydrolysable moieties in the backbone of the polymer network [26]. A copper-containing coating based on this technology had good antifouling performance. In a coating based on a copper complex of an acrylic copolymer, it was found that the release of copper from the coating was lower and more uniform compared to a conventional coating containing

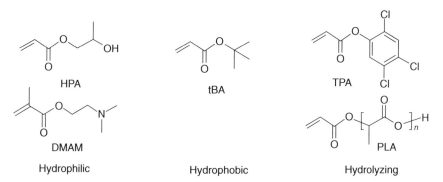

Figure 25.2   A tunable self-polishing binder system can be prepared from a polymer composed of three monomers [25] DMAM, 2-(dimethylamino)ethyl methacrylate; HPA, 2-hydroxypropyl acrylate; PLA, poly-lactic acid acrylate; tBA, *tert*-butyl acrylate; TPA, 2,4,5-trichlorophenyl acrylate.

cuprous oxide [27]. Graft copolymers containing hydrolysable oligoester side chains [28], copolymers of ε-caprolactone with α-hydroxy acids [29] and poly(ester anhydride)s [30] have also been reported.

Another potential route to controlling biocide emission into the environment is through encapsulation. Zhang *et al.* [31] encapsulated Sea-Nine 211 and incorporated the encapsulated biocide into a coating. The release rate of the biocide was sufficient to deter fouling.

25.3   Natural antifoulants

For most organisms that live in the marine environment, their survival depends on their ability to deter settlement of other marine life on them [32]. Thus, many marine plants and animals do not become fouled themselves (Chapters 7 and 8). The mechanism of deterrence is through chemical means: the organism synthesises and exudes a metabolite that acts as a deterrent to the settlement of other fouling bacteria or marine organisms. Thus, a significant effort has been directed toward extracting compounds from various marine sources and screening these for their antifouling activity. It is believed that the fact that these compounds occur naturally will not lead to the environmental problems that have been associated with synthetic biocides.

Marine algae are ubiquitous in the environment and produce metabolites that deter fouling. Thus, extracts from marine algae are being explored as naturally occurring antifouling compounds by a number of research groups [33–35]. Much of the work that has been reported on antifouling extracts from marine microalgae and macroalgae has been described in a review [36]. Marine sponges can also yield antifouling extracts [37, 38]. Boshale *et al.* [39] studied extracts from a variety of marine plant and animal life and found a number of compounds that were found to have antifouling activity. Antifouling compounds have also been extracted from the surface of a blue mussel [40].

Bacterial colonisation of surfaces may be an important mechanism of fouling deterrence for many types of marine life [41]. Bacterial biofilms that form on the surfaces of many marine organisms may release metabolites that deter settlement of other fouling organisms. Burgess *et al.* [42] isolated 650 bacteria from the surfaces of various marine plants and animals and found four species which had antifouling properties. Five metabolites were extracted from the most active bacterium and one compound was found to have a high minimum inhibitory concentration against a number of fouling bacteria. Paint formulations were made from extracts of metabolites and were found to have good antifouling properties based on laboratory assays with algae and barnacle cyprids.

Diketopiperzines isolated from a deep-sea bacterium were found to have antifouling properties [43]. Reviews of additional studies of the antifouling potential of metabolites isolated from bacteria have also appeared recently [44, 45]. Rather than isolating bacterial metabolites and using these as antifouling compounds in coatings, Yee *et al.* [46] prepared surface coatings containing immobilised antifouling marine bacteria. Other naturally occurring compounds have been explored as antifoulants. These include capsaicin and zosteric acid [47–49], tannins [18], tannin–aluminium complexes [50] and cupric tannate [51]. Synthetic analogues of naturally occurring antifoulants have also been found to deter fouling [52, 53].

While there have been many natural antifouling compounds isolated from various non-fouling organisms, there are a number of barriers to the development of antifouling technology using this approach [54]. Often the composition of the extracts is not well-characterised; thus, the nature of the active ingredient is not known. In addition, in most of these studies reported, only limited antifouling experiments have been carried out. Large-scale field experiments at multiple test sites around the world are needed to demonstrate broad-based antifouling efficacy. An additional significant barrier is that even though these antifoulants may be naturally occurring compounds, for use in a commercial antifouling application the compounds will still have to go through a time-consuming and expensive registration process with government agencies.

25.4   Non-toxic non-fouling approaches

A long-term goal of antifouling research is to design non-toxic approaches to mitigate fouling. One area that has received the most attention has been the so-called fouling-release coatings – coatings which do not deter fouling but only allow a weak bond to form between the coating surface and the fouling organism. Fouling-release coating systems based on silicone elastomers, fluoropolymers, fluorosilicones and hybrids of siloxanes with other materials have been explored. In addition, some early research is being carried out on approaches to deter fouling through non-toxic means.

25.4.1   Fouling-release coatings – overview

Biofouling on ships occurs primarily when a vessel is in port. Ideally, a fouling-release coating would allow for fouling to be dislodged as the vessel is taken to relatively high speed. Fundamental research has shown that several material factors are critical to fouling-release performance. These factors include surface energy, elastic modulus and coating thickness.

According to conventional theory, adhesion of a biomass to a substrate should be proportional to the surface energy of the substrate [55]. However, it was found experimentally by Baier and DePalma [56] that a minimum in biomass adhesion occurs at a surface energy in the range of 22–24 mN m^{-1}, as shown in Figure 25.3. This surface energy is lower than most surfaces derived from hydrocarbon polymers or organic polymers containing heteroatoms and higher than most surfaces generated from fluorinated polymers [57]. The class of polymers that allow for surface energies in the range of 22–24 mN m^{-1} is the polysiloxanes. For example, polydimethylsiloxane (PDMS) and polyphenylmethylsiloxane have surface energies of 20 and 26 mN m^{-1}, respectively.

Fundamental studies conducted by Newby et al. [58] comparing the adhesion strength of a viscoelastic adhesive to a fluorocarbon surface, a hydrocarbon surface and PDMS surface showed results consistent with the Baier curve. Adhesion strength to the PDMS surface was much lower than the fluorocarbon surface in spite of the fluorocarbon surface having a significantly lower surface energy. Visualisation of the peeling event showed that the mechanism of adhesion loss was quite different for the polysiloxane surface. Adhesive failure involving the PDMS surface occurred by a fingering process that involved interfacial slippage between the two materials. These results indicated that contributions from variations in mechanical properties must be considered in addition to variations in surface energy when explaining the better fouling-release properties of polysiloxanes as opposed to lower surface energy perfluorinated polymers.

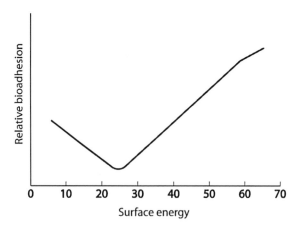

Figure 25.3 The Baier curve relating relative bioadhesion to coating surface energy [56].

Compared to hydrocarbon based polymers, polysiloxanes possess a more flexible polymer backbone, which can be attributed to the longer bond length of the Si—O bond (0.164 nm) as compared to the C—C bond (0.153 nm) and larger bond angle associated with the Si—O—Si linkage (143°) as compared to the C—C—C linkage (112°) [59]. The energy required for rotation around siloxane bonds is almost zero compared to 14 kJ mol^{-1} required for rotation about C—C bonds in polyethylene. The high polymer backbone mobility of polysiloxanes results in the production of cross-linked networks that exhibit very low glass transition temperatures and very low elastic moduli.

The contribution of elastic modulus to fouling-release performance has been extensively investigated and shown to be a key factor contributing to the exceptionally good fouling-release performance of siloxane coatings. For example, Stein et al. [60], by varying filler content, prepared an array of siloxane coatings that varied with respect to elastic modulus but had essentially the same surface energy. Results obtained for pseudobarnacle [61] adhesion showed that increasing the elastic modulus resulted in an increase in the adhesion strength of the pseudobarnacle to the coating surface. By plotting relative adhesion data as a function of the square root of the product of critical surface free energy and elastic modulus for a variety of polymers, Brady [62] and Singer [63] found a linear relationship illustrating the interaction between both surface energy and elastic modulus in predicting fouling-release performance (Figure 25.4).

In addition to elastic modulus, coating thickness has been shown to be critical to fouling-release performance. The influence of coating thickness on fouling-release can best be understood by considering the Kendall equation, which models the force required to pull off a rigid disc adhered to an elastomeric surface [64]:

$$P_c = \pi a^2 \left(\frac{2 I_c K}{t} \right)^{1/2}$$

where P_c is the pull-off force, a is the contact radius, I_c is the interfacial fracture energy, K is the elastomeric coating bulk modulus and t is the coating thickness. This model was tested by Granghoffer and Gent [65] as well as by Kohl and Singer [66], and experimental results were found to be in good agreement with Kendall's model. Thus, maximum fouling-release

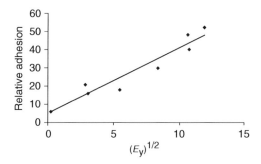

Figure 25.4 The relationship of relative bioadhesion to the square root of the product of critical surface energy and modulus. (Reprinted with permission from Reference [63]. Copyright 2000 Federation of Societies for Coatings Technology.)

performance requires the production of relatively thick coating films, with thicknesses, in some cases, in excess of 1 mm [67].

25.4.2 Fouling-release coatings based on silicone elastomers

The first reported use of silicones as fouling-release coatings was by Muller and Noacki [68] in a US patent issued in 1972 to Battelle Institute. Serious commercial development of silicone fouling-release coatings, however, did not begin until the 1980s [69]. Much of the early coating development work, which resulted in various US patents, was conducted by Milne and co-workers [70–72]. A key discovery described by Milne in his 1977 US patent was the enhanced fouling-release performance obtained by adding non-reactive silicone oil to a silicone elastomer coating composition. The concept of using a non-reactive silicone oil to enhance fouling release was later investigated in detail by investigators at the General Electric Company [73, 74]. A recent report by Meyer *et al.* [75] concluded that the improved fouling-release performance of oil-containing coatings may result from an inhibition of bioadhesive cross-linking by the oil.

25.4.3 Bound biocides

Current antifouling coatings for marine applications function by a leaching mechanism in which biocides are slowly released from the coating into the aquatic environment [1]. In an attempt to produce environmentally friendly antifouling coatings, an approach based on the concept of 'tethering' biocide moieties to a coating surface has been investigated by several researchers. As described in a recent review by Kenawy *et al.* [76], a number of non-leaching, antimicrobial coatings have been developed for biomedical applications. Most of these contact active antimicrobial coatings inhibit biofilm growth by disrupting the cytoplasmic membrane of cells, causing leakage of the cytoplasmic constituents and subsequent death of the cell. Quaternary ammonium salts (QASs) are an example of effective contact active biocide moieties. The high charge density associated with QAS moieties exerts a strong electrostatic interaction with the negatively charged cell walls, resulting in rupture of the cell [76].

As discussed by Callow and co-workers [77, 78], biofouling in the marine environment occurs in stages with the formation of a microbial biofilm occurring at an early stage. The presence of a microbial biofilm induces the settlement of macrofoulers such as *Ulva* (formerly *Entermorpha*), a grass-like alga commonly seen on rocks present along shorelines [78]. Since the formation of a microbial biofilm precedes the settlement of macrofoulers, contact active biocide moieties chemically bound or tethered to the coating surface may provide antifouling character without leaching toxic components into the aquatic environment.

Antimicrobial activity toward marine microfoulers such as bacteria and diatoms has been demonstrated for coatings containing tethered biocide moieties. For example, as early as 1973, Walters *et al.* [79] showed that functionalisation of non-woven fibers with QAS moieties resulted in biocidal activity towards six representative species of algae. Mellouki *et al.* [80] prepared coatings based on a vinyl copolymer containing QAS moieties and found improved resistance to microfouling as compared to a control after 4 months of sea water immersion. Thomas *et al.* [81] tethered the common biocide triclosan (5-chloro-2-(2,4-dichlorophenoxy)phenol) to polysiloxane-based coatings and found a significant reduction in macrofouling after 1 month of sea water immersion.

A primary limitation of using a tethered biocide approach is that the microorganism must come into physical contact with the coating surface. As a result, the surface must be relatively free of debris that can mask the surface bound biocide moieties. It is for this reason that the coatings must also possess good fouling-release characteristics so that the surface can be easily cleaned periodically to regenerate antifouling efficacy. The creation of hybrid antifouling/fouling-release coatings based on tethered biocide moieties is currently being investigated [82]. Polysiloxane-based matrix-forming materials are the focus of the investigations since previous work has clearly shown that this class of materials provides the best fouling-release character.

25.4.4   *Fluoropolymers*

Fluoropolymers have very low surface energies in the range of 10–20 m Nm^{-1}, and it was expected that fluoropolymers would make suitable non-adherent coatings for marine fouling. Indeed, poly(tetrafluorethylene), Teflon®, is used in a large number of applications for non-stick surfaces. However, Teflon® is a highly crystalline, insoluble polymer that is difficult to process and, thus, cannot be used in a field-applied coating such as a ship hull. In addition, the Baier curve (Figure 25.3) indicates that relative bioadhesion increases as the surface energy is reduced below 22 m Nm^{-1}, which has tended to deter researchers from exploring fluoropolymers for fouling-release coatings applications.

However, about 30 years ago, Griffith and co-workers [83,84] prepared coatings based on fluorinated epoxy resins and some formulations also included particulate poly(tetrafluorethylene). Due to the extreme hydrophobicity of these coatings, they had excellent anticorrosion properties and some formulations were promising as antifouling coatings. Bonafede and Brady [85] prepared a series of fluorinated polyurethane coatings and determined their biofouling resistance, surface and mechanical properties. While the surface energies of the coatings varied from 12 to 33 m Nm^{-1}, fouling performance of all of the coatings was similar. Coatings that had lower modulus, however, were easier to clean, supporting the observation discussed earlier that modulus plays an important role in fouling-release coatings in addition to surface energy. Elastomeric fluorinated polyurethane coatings were prepared by Brady and Aronson [86] from several different fluorinated polyols. The best performing coating had the lowest modulus and a moderate surface energy. A series of copolymers were prepared from fluorinated acrylate and methacrylate monomers and evaluated as non-fouling coatings by Tsibouklis *et al.* [87]. In laboratory assays with algae, barnacle cyprids, yeasts and bacteria, most of the fluorinated coatings deterred settlement compared to glass and acrylic controls.

Yarbrough *et al.* [88] prepared and characterised a series of coatings based on a cross-linkable terpolymer containing perfluoropolyether side chains. When evaluated in a laboratory assay with *Ulva* spores, spore settlement was lower on the fluorinated coatings and spore removal was higher, in comparison to glass and an elastomeric PDMS.

Youngblood *et al.* [89] described the preparation of styrene–isoprene block copolymers, which were then modified with either poly(ethylene glycol) (PEG) or semi-fluorinated side chains. In experiments with *Ulva* spores, spore settlement was lower compared to a glass surface for both the fluorinated polymer and the polymers with PEG side chains; however, spore removal was higher for the fluorinated polymer.

25.4.5 *Amphiphilic/hybrid systems*

To determine the interaction of two marine algae, the diatom *Navicula* and the green alga *Ulva*, with hydrophilic and hydrophobic surfaces, Krishnan *et al.* [90] synthesised block copolymers of polystyrene with monomers having either fluorinated side chains or PEG side chains. Interestingly, it was found that the two organisms had opposite adhesion behaviour on these surfaces: the diatom *Navicula* was readily removed from the hydrophilic surface but adhered well to the hydrophobic surface, while *Ulva* had low adhesion on the hydrophobic fluoropolymer coating but adhered strongly to the polar surface. This result illustrates the challenge of designing an easy-release coating for the marine environment: since different organisms have different mechanisms of adhesion, designing a surface that resists *all* possible marine organisms is challenging. Thus, interest in the design of materials having amphiphilic surfaces has arisen as a potential method for discouraging the settlement of a wide variety of fouling organisms.

Krishnan *et al.* [91] have reported the synthesis and antifouling properties of block copolymers having amphiphilic side chains composed of PEG and fluoroalkyl groups. Due to the mobility and amphiphilic nature of the side chains, surface reconstruction can occur when the material is exposed to water (Figure 25.5). Evaluation of the coatings with respect to *Navicula* (diatom) and *Ulva* (alga) indicated that the adhesion of both organisms to the surface was weak, demonstrating a potential approach to a non-toxic coating surface that does not allow a range of organisms to adhere strongly. Amphiphilic coatings based on fluoropolymers cross-linked with either PEG or PDMS cross-linkers are being explored by the Wooley group [92–97].

25.4.6 *Siloxane–urethane hybrid systems*

While silicones appear to be the leading fouling-release coating system to date, silicone elastomer coating systems suffer from several drawbacks. First, silicones are mechanically weak and thus easily damaged by cleaning or docking procedures. Silicone elastomers also have poor adhesion to most substrates and a tie-coat primer is needed to improve the adhesion of the fouling-release coating layer to the anticorrosion primer. This complicates the application of these systems. To overcome these limitations, several approaches to the design of hybrid coating systems are being explored.

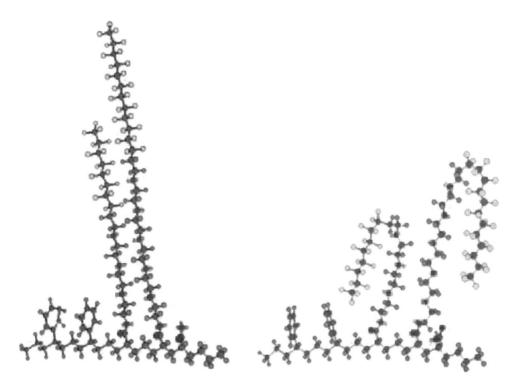

Figure 25.5 Model of surface reconstruction of polymer having amphiphilic side chains. (Reprinted with permission from Reference [91]. Copyright 2006 American Chemical Society.)

Due to the low surface energy of PDMS, it has long been recognised that copolymer systems containing PDMS segments will be surface-enriched with siloxane [98]. Thus, any siloxane-containing system will have the surface properties of PDMS, but tend to retain the bulk properties of the other material used. This is an ideal situation for designing coatings that have minimally adhesive surfaces, but also have the tough bulk mechanical properties required for an underwater marine application.

A number of approaches to low-surface-energy coatings containing siloxane segments have been reported in the literature, but not all of these have been evaluated as fouling-release coatings in the marine environment. Siloxane–polyurethane systems have been the subject of several investigations. Polyurethanes are high-performance coating systems and have excellent mechanical properties and good adhesion to most substrates. In addition, polyurethane coatings can undergo rapid cure under ambient conditions.

Brooks [99] described an interpenetrating polymer network made from co-curing a siloxane elastomer and a polyurethane that inhibits fouling of underwater surfaces. Adkins *et al.* [100] described the preparation of several different types of siloxane–urethane coating systems and many were reported to have easy-clean properties in field immersion studies. Siloxane–polyurethane block copolymer systems have been studied by Wynne and collaborators [101–105], and the effect of composition and other variables on the bulk and surface properties determined. However, a particular challenge with using the siloxane–urethane system for marine applications is that upon water immersion, the more hydrophilic urethane groups tend to migrate to the surface, rendering the coating hydrophilic [106–108].

To overcome this limitation, Webster and co-workers [109] are exploring the preparation of cross-linked siloxane–polyurethane coating systems. Using a systematic, combinatorial and high-throughput approach [110], the effect of key compositional variables such as siloxane composition and polyol composition on the surface and bulk properties of the coating are being explored [111–114]. Rapid laboratory bioassays have also been developed to permit the screening of large numbers of coating samples in an efficient time frame [115, 116]. Using this approach, coating compositions have been identified that have a combination of excellent release properties, stability in water, good adhesion and high bulk modulus, and are candidates for field testing.

25.4.7 *Surface microtopography*

Biofouling involves the interaction of bioadhesives with a surface. The adhesive must be able to wet the surface in order for a good adhesive bond to be formed. In recent years, the discovery of superhydrophobic surfaces has highlighted the important role that surface topography plays in surface wetting. Superhydrophobic surfaces typically involve surface roughness on the nanometer scale and are responsible for the so-called Lotus-leaf effect. It also appears that many non-fouling marine organisms use surface roughness as at least a component of their fouling deterrent mechanism [117–119]. Genzer and Efimenko [120] have recently reviewed the subject of superhydrophobic surfaces and their implication for marine fouling. Marmur [121, 122] has also discussed superhydrophobicity and biofouling from a theoretical perspective.

Surface topography on a larger scale may also limit the surface area available for the attachment of fouling organisms, reducing the strength of the adhesive bond formed [123, 124]. Thus, the design and exploration of coating systems that have micro- and nanotopography has become a focus of research into non-toxic, non-fouling coating surfaces.

Berntsson *et al.* [125] compared the settlement of the barnacle *Balanus improvisus* on several smooth and microtextured polymers and found that recruitment was lower on the microtextured surfaces. PDMS elastomers having microtextured surfaces were prepared by Andersson *et al.* [126] by molding the PDMS against various mesh materials. Short-term field tests showed that the microtextured surfaces were less fouled by barnacles than the smooth control surfaces. Behavioural experiments with barnacle cyprids showed a preference for smooth surfaces over the microtextured surfaces [127]. Petronis *et al.* [128] prepared a series of PDMS elastomers having various defined microtopographies – riblets and pyramids – using a molding process. Field testing of these coatings showed that the surface with the largest riblet pattern (69-µm profile height) had the lowest barnacle settlement compared to a smooth PDMS surface.

Bers and Wahl [129] prepared replicated surfaces of a variety of naturally occurring marine organisms and found that a number of the topographies resulted in reduced settlement compared to smooth controls, but that the reduction in fouling was not general, but specific microtopographies rejected specific classes of organisms (see also Chapter 8).

The Brennan group has prepared silicone elastomer surfaces with well-defined surface microtopography. In studies involving the settlement of *Ulva*, it was found that some microtopographies resulted in higher levels of spore settlement while other topographies, in particular the Sharklet™ pattern, resulted in reduced settlement of spores on the surface [130–133]. Figure 25.6 shows an image of a smooth surface microchannel and the Sharklet™ pattern with settled *Ulva* spores.

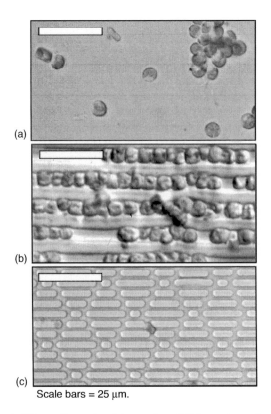

(a)

(b)

(c)

Scale bars = 25 μm.

Figure 25.6 Settlement of *Ulva* settlement on (a) smooth and (b) 5-μm channel and (c) Sharklet patterned silicone elastomer surfaces. (Reprinted with permission from Reference [130]. Copyright 2006 Taylor & Francis Ltd.)

A particular challenge in designing microtopographical surfaces to resist fouling is that the settling form of marine organisms can occur in a wide range of sizes. Thus, a topographical surface having a particular length scale may resist fouling from only one size of fouling organism. A recent report by Schumacher *et al.* [134] showed that engineered microtopographical surfaces resistant to barnacle settlement could be designed, but that the feature spacing required was an order or magnitude larger than that for *Ulva* spores. However, the pattern effective for reducing barnacle settlement enhances the settlement of *Ulva* spores [131, 132]. Thus, it appears that no single microtopography can be employed to deter fouling by the range of organisms found in nature; a surface containing a hierarchical set of topographies on multiple length scales may be required [134].

Majumdar and Webster [135] have reported the discovery of a unique siloxane–urethane coating that spontaneously develops a regular surface microstructure consisting of siloxane domains surrounded by polyurethane, which is stable upon water immersion because of the constraints imposed due to chemical cross-linking (Plate VIII A). The size and area distribution of the microdomains can be controlled by varying the solvent used in preparing the coatings and the time of mixing in solution prior to deposition also affects the formation of microdomains as well as their size [136]. The presence of the surface microdomains was shown to decrease the adhesive strength of barnacles [137]. This represents a feasible system for commercial application since the formation of the microtopographical surface occurs spontaneously on film formation.

25.4.8   *Other approaches*

A number of alternative approaches to controlling fouling have been reported recently. Many of these concepts are just now emerging and additional work will be required to determine the broad feasibility of these approaches. One new concept that is beginning to emerge is to use a self-polishing, or degradable coating by itself (without biocides) as a self-renewing non-fouling coating. As the coating material hydrolyses or degrades, it will carry away any fouling that may have become built-up on the coating. For example, a biodegradable polymer has been explored as a method for providing a surface that is constantly renewed in the marine environment [138]. Verborgt and Webb [139] have recently disclosed a self-polishing polyurethane coating system which uses novel hydrolysable polyols.

Tang *et al.* [140] prepared a series of hybrid xerogel coatings having a variation in chain length of the alkyl silane. The xerogel based on the longest alkyl chain (C8) had the lowest surface energy, showed low settlement of both *Ulva* spores and barnacle cyprids, and had good release properties. In a series of patents, Simendinger [141–143] has disclosed antifouling coating compositions composed of a glassy matrix formed from a silicone and silicate along with another component that can phase separate, typically another silicone, a hydrocarbon or a fatty acid. Zhang *et al.* [144] have demonstrated that sulfobetaine and carboxybetaine polymers can resist fouling by proteins. Interpenetrating polymer networks consisting of a polyurethane and a sulfobetaine polymer were also resistant to protein fouling and may have potential as fouling-resistant surfaces in the marine environment [145].

Since many fouling species produce adhesive proteins, it has been proposed that certain enzymes might be capable of degrading these adhesives. Pettit *et al.* [146] screened a number of commercially available enzymes for their effect on the adhesion of several marine organisms and found several that appeared to be effective. Dobretsov *et al.* [147] screened several proteases for their effectiveness at inhibiting the settlement of a bryozoan larva and found several that inhibited settlement significantly.

A rather unique approach for deterring fouling of surfaces is the use of pulsed electrical fields [148]. Pulsed low-voltage fields supplied through interdigitated electrodes could significantly reduce the formation of a bacterial biofilm.

25.5   Conclusions

- Designing new antifouling technologies is a challenging task and a number of approaches are currently being pursued.
- In the near term, it appears that coatings containing biocides will continue to be used; however, an emerging emphasis appears to be on optimising the coating system to moderate or control the rate of biocide release to environmentally acceptable levels.
- Alternative low-toxicity biocides are being explored as antifouling chemicals extracted from marine organisms, but studies demonstrating the broad efficacy of these approaches have not yet been reported.

- Significant efforts are being directed toward non-toxic antifouling approaches with a focus on fouling-release coatings.
- While silicone elastomer coatings have been introduced commercially, they are challenging to apply and have a relatively short useful service life. Therefore, there are significant opportunities in improving the application properties and lifetime of fouling-release coatings. Research is being conducted along several lines to generate tough and durable fouling-release coatings.
- A number of promising emerging technologies are being explored including materials having inherently deterrent surface properties, coatings with surface micro- and nanotopography, amphiphilic surfaces and novel fluoropolymers. While all of these technologies appear to have some promise, many of these need to be demonstrated in a practical coating system.

References

1. Omae, I. (2003) General aspects of tin-free antifouling paints. *Chemical Reviews*, **103** (9), 3431–3448.
2. Almeida, E., Diamantino, T.C. & De Sousa, O. (2007) Marine paints: the particular case of antifouling paints. *Progress in Organic Coatings*, **59** (1), 2–20.
3. Yebra, D.M., Kiil, S. & Dam-Johansen, K. (2004) Antifouling technology – past, present and future steps towards efficient and environmentally friendly antifouling coatings. *Progress in Organic Coatings*, **50** (2), 75–104.
4. Konstantinou, I.K. & Albanis, T.A. (2004) Worldwide occurrence and effects of antifouling paint booster biocides in the aquatic environment: a review. *Environment International*, **30** (2), 235–248.
5. Voulvoulis, N., Scrimshaw, M.D. & Lester, J.N. (2002) Comparative environmental assessment of biocides used in antifouling paints. *Chemosphere*, **47** (7), 789–795.
6. Harino, H., Mori, Y., Yamaguchi, Y., Shibata, K. & Senda, T. (2005) Monitoring of antifouling booster biocides in water and sediment from the port of Osaka, Japan. *Archives of Environmental Contamination and Toxicology*, **48** (3), 303–310.
7. Srinivasan, M. & Swain, G.W. (2007) Managing the use of copper-based antifouling paints. *Environmental Management*, **39** (3), 423–441.
8. Steen, R.J.C.A., Ariese, F., van Hattum, B., Jacobsen, J. & Jacobson, A. (2004) Monitoring and evaluation of the environmental dissipation of the marine antifoulant 4,5-dichloro-2-*N*-octyl-4-isothiazolin-3-one (DCOIT) in a Danish Harbor. *Chemosphere*, **57** (6), 513–521.
9. Smith, L.D., Negri, A.P., Philipp, E., Webster, N.S. & Heyward, A.J. (2003) The effects of antifoulant-paint-contaminated sediments on coral recruits and branchlets. *Marine Biology*, **143** (4), 651–657.
10. Kempen, T.M.J. (2007) Synergistic antifouling combinations of 4-bromo-2-(4-chlorophenyl)-5-(trifluoromethyl)-1H-pyrrole-3-carbonitrile and metal compounds. WO Patent No. 2007088172.
11. Kramer, J.P. & Vos, M. (1998) Light- and bright-colored antifouling paints. EP Patent No. 831134.
12. Maartensson, L. (2006) Use of a combination of substances to prevent marine biofouling. WO Patent No. 2006080890.
13. Quaiser, S., Lynam, I., Wind, M., Celik, A. & Kampf, G. (2006) Fouling-resistant polyurethanes containing biocide for maritime uses. WO Patent No. 2006097425.
14. Van Der Flaas, M.A.J. & Crawley, L.S. (1996) Use of pyrrole compounds as antifouling agents. EP Patent No. 746979.
15. Van Der Flaas, M.A.J. & Nys, J.R. (2003) Synergistic antifouling compositions comprising 4-bromo-2-(4-chlorophenyl)-5-(trifluoromethyl)-1H-pyrrole-3-carbonitrile. WO Patent No. 2003039256.

16. Al-Juhni, A.A. & Zhang, N.B.-M. (2006) Incorporation of benzoic acid and sodium benzoate into silicone coatings and subsequent leaching of the compound from the incorporated coatings. *Progress in Organic Coatings*, **56** (2–3), 135–145.

17. Haque, H., Cutright, T. & Zhang Newby, B.-M. (2005) Effectiveness of sodium benzoate as a freshwater low toxicity antifoulant when dispersed in solution and entrapped in silicone coatings. *Biofouling*, **21** (2), 109–119.

18. Stupak, M.E., Garcia, M.T. & Perez, M.C. (2003) Non-toxic alternative compounds for marine antifouling paints. *International Biodeterioration and Biodegradation*, **52** (1), 49–52.

19. Nogata, Y., Kitano, Y., Yoshimura, E., Shinshima, K. & Sakaguchi, I. (2004) Antifouling activity of simple synthetic isocyanides against larvae of the barnacle *Balanus amphitrite*. *Biofouling*, **20** (2), 87–91.

20. Kitano, Y., Nogata, Y., Shinshima, K., *et al.* (2004) Synthesis and anti-barnacle activities of novel isocyanocyclohexane compounds containing an ester or an ether functional group. *Biofouling*, **20** (2), 93–100.

21. Cowling, M.J., Hodgkiess, T., Parr, A.C.S., Smith, M.J. & Marrs, S.J. (2000) An alternative approach to antifouling based on analogues of natural processes. *Science of the Total Environment*, **258** (1–2), 129–137.

22. Cowie, P.R., Smith, M.J., Hannah, F., Cowling, M.J. & Hodgkeiss, T. (2006) The prevention of microfouling and macrofouling on hydrogels impregnated with either Arquad 2C-75 or benzalkonium chloride. *Biofouling*, **22** (3–4), 173–185.

23. Vallee-Rehel, K., Langlois, V. & Guerin, P. (1998) Contribution of pendant ester group hydrolysis to the erosion of acrylic polymers in binders aimed at organotin-free antifouling paints. *Journal of Environmental Polymer Degradation*, **6** (4), 175–186.

24. Vallee-Rehel, K., Mariette, B., Hoarau, P.A., *et al.* (1998) A new approach in the development and testing of antifouling paints without organotin derivatives. *Journal of Coatings Technology*, **70** (880), 55–63.

25. Thouvenin, M., Langlois, V., Briandet, R., *et al.* (2003) Study of erodible paint properties involved in antifouling activity. *Biofouling*, **19** (3), 177–186.

26. Kuo, P.-L., Chuang, T.-F. & Wang, H.-L. (1999) Surface-fragmenting, self-polishing, tin-free antifouling coatings. *Journal of Coatings Technology*, **71** (893), 77–83.

27. Samui, A.B., Hande, V.R. & Deb, P.C. (1997) Synthesis and characterization of copoly(MMA-MA)-Cu complex and study on its leaching behavior. *Journal of Coatings Technology*, **69** (867), 67–72.

28. Vallee-Rehel, K., Langlois, V., Guerin, P. & Le Borgne, A. (1999) Graft copolymers, for erodible resins, from α-hydroxyacids oligomers macromonomers and acrylic monomers. *Journal of Environmental Polymer Degradation*, **7** (1), 27–34.

29. Faye, F., Linossier, I., Langlois, V., Renard, E. & Vallee-Rehel, K. (2006) Degradation and controlled release behavior of ε-caprolactone copolymers in biodegradable antifouling coatings. *Biomacromolecules*, **7** (3), 851–857.

30. Faye, F., Linossier, I., Langlois, V. & Vallee-Rehel, K. (2007) Biodegradable poly(ester-anhydride) for new antifouling coating. *Biomacromolecules*, **8** (5), 1751–1758.

31. Zhang, M., Cabane, E. & Claverie, J. (2007) Transparent antifouling coatings via nanoencapsulation of a biocide. *Journal of Applied Polymer Science*, **105** (6), 3824–3833.

32. Fusetani, N. (2004) Biofouling and antifouling. *Natural Product Reports*, **21**, 94–104.

33. Hellio, C., Berge, J.P., Beaupoil, C., Le Gal, Y. & Bourgougnon, N. (2002) Screening of marine algal extracts for anti-settlement activities against microalgae and macroalgae. *Biofouling*, **18** (3), 205–215.

34. Cho, J.Y., Kowon, E.-H., Choi, J.-S., *et al.* (2001) Antifouling activity of seaweed extracts on the green alga *Enteromorpha prolifera* and the mussel *Mytilus edulis*. *Journal of Applied Phycology*, **13**, 117–125.

35. Da Gama, B.A.P., Pereira, R.C., Carvalho, A.G.V., Coutinho, R. & Yoneshigue-Valentin, Y. (2002) The effects of seaweed secondary metabolites on biofouling. *Biofouling*, **18** (1), 13–20.

36. Bhadury, P. & Wright, P.C. (2004) Exploitation of marine algae: biogenic compounds for potential antifouling applications. *Planta*, **219** (4), 561–578.

37. Yang, L.H., Lee, O.O., Jin, T., Li, X.C. & Qian, P.Y. (2006) Antifouling properties of 10β-formamidokalihinol-A and kalihinol A isolated from the marine sponge *Acanthella cavernosa*. *Biofouling*, **22** (1–2), 23–32.

38. Faimali, M., Sepcic, K., Turk, T. & Geraci, S. (2003) Non-toxic antifouling activity of polymeric 3-alkylpyridinium salts from the mediterranean sponge *Reniera sarai* (Pulitzer-Finali). *Biofouling*, **19** (1), 47–56.

39. Bhosale, S.H., Nagle, V.L. & Jagtap, T.G. (2002) Antifouling potential of some marine organisms from India against species of *Bacillus* and *Pseudomonas*. *Marine Biotechnology*, **4** (2), 111–118.

40. Bers, A.V., D'Souza, F., Klijnstra, J.W., Willemsen, P.R. & Wahl, M. (2006) Chemical defence in mussels: antifouling effect of crude extracts of the periostracum of the blue mussel *Mytilus edulis*. *Biofouling*, **22** (4), 251–259.

41. Armstrong, E., Boyd, K.G. & Burgess, J.G. (2000) Prevention of marine biofouling using natural compounds from marine organisms. *Biotechnology Annual Review*, **6**, 221–241.

42. Burgess, J.G., Boyd, K.G., Armstrong, E., *et al.* (2003) The development of a marine natural product-based antifouling paint. *Biofouling*, **19** (Suppl.), 197–205.

43. Li, X., Dobretsov, S., Xu, Y., *et al.* (2006) Antifouling diketopiperazines produced by a deep-sea bacterium, *Streptomyces fungicidicus*. *Biofouling*, **22** (3), 201–208.

44. Dahms, H.-U., Ying, X. & Pfeiffer, C. (2006) Antifouling potential of cyanobacteria: a mini-review. *Biofouling*, **22** (5), 317–327.

45. Dobretsov, S., Dahms, H.-U. & Qian, P.-Y. (2006) Inhibition of biofouling by marine microorganisms and their metabolites. *Biofouling*, **22** (1–2), 43–54.

46. Yee, L.H., Holmstroem, C., Fuary, E.T., *et al.* (2007) Inhibition of fouling by marine bacteria immobilised in k-carrageenan beads. *Biofouling*, **23** (4), 287–294.

47. Xu, Q., Barrios, C.A., Cutright, T. & Zhang Newby, B.-M. (2005) Evaluation of toxicity of capsaicin and zosteric acid and their potential application as antifoulants. *Environmental Toxicology*, **20** (5), 467–474.

48. Xu, Q., Barrios, C.A., Cutright, T. & Zhang Newby, B.-M. (2005) Assessment of antifouling effectiveness of two natural product antifoulants by attachment study with freshwater bacteria. *Environmental Science and Pollution Research International*, **12** (5), 278–284.

49. Angarano, M.-B., McMahon, R.F., Hawkins, D.L. & Schetz, J.A. (2007) Exploration of structure–antifouling relationships of capsaicin-like compounds that inhibit zebra mussel (*Dreissena polymorpha*) macrofouling. *Biofouling*, **23** (5), 295–305.

50. Perez, M., Garcia, M., Blustein, G. & Stupak, M. (2007) Tannin and tannate from the Quebracho tree: an eco-friendly alternative for controlling marine biofouling. *Biofouling*, **23** (3), 151–159.

51. Perez, M., Blustein, G., Garcia, M., del Amo, B. & Stupak, M. (2006) Cupric tannate: a low copper content antifouling pigment. *Progress in Organic Coatings*, **55** (4), 311–315.

52. Dahlstroem, M., Lindgren, F., Berntsson, K., *et al.* (2005) Evidence for different pharmacological targets for imidazoline compounds inhibiting settlement of the barnacle *Balanus improvisus*. *Journal of Experimental Zoology, Part A: Comparative Experimental Biology*, **303A** (7), 551–562.

53. Sjogren, M., Johnson, A.-L., Hedner, E., *et al.* (2006) Antifouling activity of synthesized peptide analogues of the sponge metabolite barettin. *Peptides*, **27** (9), 2058–2064.

54. Rittschof, D. (2000) Natural product antifoulants: one perspective on the challenges related to coatings development. *Biofouling*, **15** (1–3), 119–127.

55. Gent, A.N. & Schultz, J. (1972) Effect of wetting liquids on the strength of adhesion of viscoelastic materials. *Journal of Adhesion*, **3** (4), 281–294.

56. Baier, R.E. & DePalma, V.A. (1971) The relation of the internal surface of grafts to thrombosis. In: *Management of Occlusive Arterial Disease* (ed. W.A. Dale), pp. 147–163. Yearbook Medical Publishers, Chicago.

57. Brandrup, J. & Immergut, E.H. (1989) *Polymer Handbook*, 3rd edn. John Wiley & Sons, New York.

58. Zhang Newby, B.-M., Chaudhury, M.K. & Brown, H.R. (1995) Macroscopic evidence of the effect of interfacial slippage on adhesion. *Science*, **269** (5229), 1407–1409.

59. Owen, M.J. (2000) Surface properties and applications. In: *Silicon-Containing Polymers* (eds R.G. Jones, W. Ando & J. Chojnowski), pp. 213–231. Kluwer, The Netherlands.

60. Stein, J., Truby, K., Darkangelo Wood, C., *et al.* (2003) Structure–property relationships of silicone biofouling-release coatings: effect of silicone network architecture on pseudobarnacle attachment strengths. *Biofouling*, **19** (2), 87–94.

61. Swain, G.W.J., Schultz, M.P., Griffith, J.R. & Snyder, S. (1997) The relationship between barnacle and pseudo-barnacle adhesion measurements: a method to predict the foul release properties of silicones? Paper presented at *US Pacific Rim Workshop on Emerging Nonmetallic Materials for the Marine Environment*, Honolulu, Hawaii, 18–20 March 1997, Office of Naval Research, USA.

62. Brady, R.F., Jr. & Singer, I.L. (2000) Mechanical factors favoring release from fouling release coatings. *Biofouling*, **15** (1–3), 73–81.

63. Brady, R.F. (2000) Clean hulls without poisons: devising and testing nontoxic marine coatings. *Journal of Coatings Technology*, **72** (900), 45–56.

64. Kendall, K. (1971) The adhesion and surface energy of elastic solids. *Journal of Physics D: Applied Physics*, **4**, 1186–1195.

65. Ganghoffer, J.F. & Gent, A.N. (1995) Adhesion of a rigid punch to a thin elastic layer. *Journal of Adhesion*, **48** (1–4), 75–84.

66. Kohl, J.G. & Singer, I.L. (1999) Pull-off behavior of epoxy bonded to silicone duplex coatings. *Progress in Organic Coatings*, **36** (1–2), 15–20.

67. Wendt, D.E., Kowalke, G.L., Kim, J. & Singer, I.L. (2006) Factors that influence elastomeric coating performance: the effect of coating thickness on basal plate morphology, growth and critical removal stress of the barnacle *Balanus amphitrite*. *Biofouling*, **22** (1–2), 1–9.

68. Muller, M. & Noacki, L. (1972) Ship's hull coated with antifouling silicone rubber. US Patent No. 3,702,778.

69. Candries, M., Altar, M. & Anderson, M.D. (2001) Foul-release systems and drag. In: *Consolidation of Technical Advances in the Protective and Marine Coatings Industry: Proceedings of the PCE 2001 Conference*, Antwerp.

70. Milne, A. & Callow, M.E. (1985) Non-biocidal antifouling processes. In: *Polymers in a Marine Environment* (ed. R. Smith), pp. 229–233. The Institute of Marine Engineers, London.

71. Milne, A. (1977) Anti-fouling marine compositions. US Patent No. 4,025,693.

72. McLearie, J., Finnie, A.A., Andrews, A.F., Millichamp, I.S. & Milne, A. (1994) Anti-fouling coating composition. US Patent No. 5,302,192.

73. Stein, J., Truby, K., Wood, C.D., *et al.* (2003) Silicone foul release coatings: effect of the interaction of oil and coating functionalities on the magnitude of macrofouling attachment strengths. *Biofouling*, **19** (Suppl.), 71–82.

74. Truby, K., Wood, C., Stein, J., *et al.* (2000) Evaluation of the performance enhancement of silicone biofouling-release coatings by oil incorporation. *Biofouling*, **15** (1–3), 141–150.

75. Meyer, A., Baier, R., Wood, C., *et al.* (2006) Contact angle anomalies indicate that surface-active eluates from silicone coatings inhibit the adhesive mechanisms of fouling organisms. *Biofouling*, **22** (6), 411–423.

76. Kenawy, E.-R., Worley, S.D. & Broughton, R. (2007) The chemistry and applications of antimicrobial polymers: a state-of-the-art review. *Biomacromolecules*, **8** (5), 1359–1384.

77. Callow, M.E. & Callow, J.E. (2002) Marine biofouling: a sticky problem. *Biologist*, **49** (1), 10–14.

78. Joint, I., Tait, K., Callow, M.E., *et al.* (2002) Cell-to-cell communication across the prokaryote–eukaryote boundary. *Science*, **298** (5596), 1207.

79. Walters, P.A., Abbott, E.A. & Isquith, A.J. (1973) Algicidal activity of a surface-bonded organosilicon quaternary ammonium chloride. *Applied Microbiology*, **25** (2), 253–256.

80. Mellouki, A., Bianchi, A., Perichaud, A. & Sauvet, G. (1989) Evaluation of antifouling properties of non-toxic marine paints. *Marine Pollution Bulletin*, **20** (12), 612–615.

81. Thomas, J., Choi, S.-B., Fjeldheim, R. & Boudjouk, P. (2004) Silicones containing pendant biocides for antifouling coatings. *Biofouling*, **20** (4–5), 227–236.

82. Majumdar, P., Lee, E., Ward, K. & Chisholm, B. (2007) Incorporation of quaternary ammonium salts in silanol terminated polydimethylsiloxane using a high-throughput combinatorial approach. *Polymer Preprints*, **48** (1), 165–166.

83. Bultman, J.D. & Griffith, J.R. (1994) Fluoropolymer and silicone fouling-release coatings. In: *Recent Developments in Biofouling Control* (eds M.-F. Thompson, R. Nagabhushanam, R. Sarojini & M. Fingerman), pp. 383–389. A.A. Balkema, Rotterdam.

84. Griffith, J.R. & Bultman, J.D. (1978) Fluorinated naval coatings. *Industrial and Engineering Chemistry Product Research and Development*, **17** (1), 8–9.

85. Bonafede, S.J. & Brady, R.F., Jr. (1998) Compositional effects on the fouling resistance of fluorourethane coatings. *Surface Coatings International*, **81** (4), 181–185.

86. Brady, R.F., Jr. & Aronson, C.L. (2003) Elastomeric fluorinated polyurethane coatings for nontoxic fouling control. *Biofouling*, **19** (Suppl.), 59–62.

87. Tsibouklis, J., Stone, M., Thorpe, A.A., *et al.* (2002) Fluoropolymer coatings with inherent resistance to biofouling. *Surface Coatings International, Part B: Coatings Transactions*, **85** (4), 301–308.

88. Yarbrough, J.C., Rolland, J.P., DeSimone, J.M., *et al.* (2006) Contact angle analysis, surface dynamics, and biofouling characteristics of cross-linkable, random perfluoropolyether-based graft terpolymers. *Macromolecules*, **39** (7), 2521–2528.

89. Youngblood, J.P., Andruzzi, L., Ober, C.K., *et al.* (2003) Coatings based on side-chain ether-linked poly(ethylene glycol) and fluorocarbon polymers for the control of marine biofouling. *Biofouling*, **19** (Suppl.), 91–98.

90. Krishnan, S., Wang, N., Ober, C.K., *et al.* (2006) Comparison of the fouling release properties of hydrophobic fluorinated and hydrophilic pegylated block copolymer surfaces: attachment strength of the diatom *Navicula* and the green alga *Ulva*. *Biomacromolecules*, **7** (5), 1449–1462.

91. Krishnan, S., Ayothi, R., Hexemer, A., *et al.* (2006) Anti-biofouling properties of comblike block copolymers with amphiphilic side chains. *Langmuir*, **22** (11), 5075–5086.

92. Gan, D., Mueller, A. & Wooley, K.L. (2003) Amphiphilic and hydrophobic surface patterns generated from hyperbranched fluoropolymer/linear polymer networks: minimally adhesive coatings via the crosslinking of hyperbranched fluoropolymers. *Journal of Polymer Science, Part A: Polymer Chemistry*, **41** (22), 3531–3540.

93. Gudipati, C.S., Greenlief, C.M., Johnson, J.A., Prayongpan, P. & Wooley, K.L. (2004) Hyperbranched fluoropolymer and linear poly(ethylene glycol) based amphiphilic crosslinked networks as efficient antifouling coatings: an insight into the surface compositions, topographies, and morphologies. *Journal of Polymer Science, Part A: Polymer Chemistry*, **42** (24), 6193–6208.

94. Gudipati, C.S., Finlay, J.A., Callow, J.A., Callow, M.E. & Wooley, K.L. (2005) The antifouling and fouling-release performance of hyperbranched fluoropolymer (HBFP)-poly(ethylene glycol) (PEG) composite coatings evaluated by adsorption of biomacromolecules and the green fouling alga *Ulva*. *Langmuir*, **21** (7), 3044–3053.

95. Xu, J., Bohnsack, D.A., Mackay, M.E. & Wooley, K.L. (2007) Unusual mechanical performance of amphiphilic crosslinked polymer networks. *Journal of the American Chemical Society*, **129** (3), 506–507.

96. Powell, K.T., Cheng, C., Wooley, K.L., Singh, A. & Urban, M.W. (2006) Complex amphiphilic networks derived from diamine-terminated poly(ethylene glycol) and benzylic chloride-functionalized hyperbranched fluoropolymers. *Journal of Polymer Science, Part A: Polymer Chemistry*, **44** (16), 4782–4794.

97. Powell, K.T., Cheng, C. & Wooley, K.L. (2007) Complex amphiphilic hyperbranched fluoropolymers by atom transfer radical self-condensing vinyl (co)polymerization. *Macromolecules*, **40** (13), 4509–4515.

98. Ha, C.-S. & Gardella J.A., Jr. (2005) X-ray photoelectron spectroscopy studies on the surface segregation in poly(dimethylsiloxane) containing block copolymers. *Journal of Macromolecular Science, Part C: Polymer Reviews*, **45** (1), 1–18.

99. Brooks, R.R. (1991) Process for inhibiting fouling of an underwater surface. US Patent No. 5,017,322.

100. Adkins, J.D., Mera, A.E., Roe-Short, M.A., Pawlikowski, G.T. & Brady, R.F., Jr. (1996) Novel non-toxic coatings designed to resist marine fouling. *Progress in Organic Coatings*, **29** (1–4), 1–5.

101. Chen, X., Gardella, J.A., Jr., Ho, T. & Wynne, K.J. (1995) Surface composition of a series of dimethylsiloxane urea urethane segmented copolymers studied by electron spectroscopy for chemical analysis. *Macromolecules*, **28** (5), 1635–1642.

102. Ho, T. & Wynne, K.J. (1996) A method to assess the average molecular weight for surface soft segments in poly[(dimethylsiloxane)–urea]s. *Macromolecules*, **29** (11), 3991–3995.

103. Ho, T., Wynne, K.J. & Nissan, R.A. (1993) Polydimethylsiloxane–urea–urethane copolymers with 1,4-benzenedimethanol as chain extender. *Macromolecules*, **26** (25), 7029–7036.

104. Wynne, K.J., Ho, T., Nissan, R.A., Chen, X. & Gardella, J.A., Jr. (1994) Poly(dimethylsiloxane)–urea–urethane copolymers: synthesis and surface properties. In: *Inorganic and Organometallic Polymers II*. (ACS Symposium Series 572), pp. 64–80. American Chemical Society, Washington, DC.

105. Gardella, J.A., Jr., Ho, T., Wynne, K.J. & Zhuang, H.-Z. (1995) Using solubility difference to achieve surface phase separation in dimethylsiloxane–urea–urethane copolymers. *Journal of Colloid and Interface Science*, **176** (1), 277–279.

106. Pike, J.K., Ho, T. & Wynne, K.J. (1996) Water-induced surface rearrangements of poly(dimethylsiloxane–urea–urethane) segmented block copolymers. *Chemistry of Materials*, **8** (4), 856–860.

107. Tezuka, Y., Kazama, H. & Imai, K. (1991) Environmentally induced macromolecular rearrangement on the surface of polyurethane–polysiloxane block copolymers. *Journal of the Chemical Society, Faraday Transactions*, **87** (1), 147–152.

108. Tezuka, Y., Ono, T. & Imai, K. (1990) Environmentally induced macromolecular rearrangement on the surface of polyurethane–polysiloxane graft copolymers. *Journal of Colloid and Interface Science*, **136** (2), 408–414.

109. Majumdar, P., Ekin, A. & Webster, D.C. (2007) Thermoset siloxane–urethane fouling release coatings. *ACS Symposium Series*, **957**, 61–75.

110. Webster, D.C., Chisholm, B.J. & Stafslien, S.J. (2007) Mini-review: combinatorial approaches for the design of novel coating systems. *Biofouling*, **23** (3–4), 179–192.

111. Ekin, A. & Webster, D.C. (2006) Library synthesis and characterization of 3-aminopropyl-terminated poly(dimethylsiloxane)s and poly(ε-caprolactone)-*b*-poly(dimethylsiloxane)s. *Journal of Polymer Science, Part A: Polymer Chemistry*, **44** (16), 4880–4894.

112. Ekin, A. & Webster, D.C. (2007) Combinatorial and high-throughput screening of the effect of siloxane composition on the surface properties of crosslinked siloxane–polyurethane coatings. *Journal of Combinatorial Chemistry*, **9** (1), 178–188.

113. Ekin, A. & Webster, D.C. (2006) Synthesis and characterization of novel hydroxyalkyl carbamate and dihydroxyalkyl carbamate terminated poly(dimethylsiloxane) oligomers and their block copolymers with poly(ε-caprolactone). *Macromolecules*, **39** (25), 8659–8668.

114. Pieper, R., Ekin, A., Webster, D.C., *et al.* (2007) A combinatorial approach to study the effect of acrylic polyol composition on the properties of crosslinked siloxane–polyurethane fouling-release coatings. *Journal of Coatings Technology and Research*, **4** (4), 453–461.

115. Casse, F., Ribeiro, E., Ekin, A., *et al.* (2007) Laboratory screening of coating libraries for algal adhesion. *Biofouling*, **23** (3–4), 267–276.

116. Stafslien, S., Daniels, J., Mayo, B., *et al.* (2007) Combinatorial materials research applied to the development of new surface coatings. IV: A high-throughput bacterial retention and retraction assay for screening fouling-release performance of coatings. *Biofouling*, **23** (1), 45–54.

117. Baum, C., Meyer, W., Stelzer, R., Fleischer, L.-G. & Siebers, D. (2002) Average nanorough skin surface of the pilot whale (*Globicephala melas*, Delphinidae): considerations on the self-cleaning abilities based on nanoroughness. *Marine Biology*, **140**, 653–657.

118. Scardino, A., De Nys, R., Ison, O., O'Connor, W. & Steinberg, P. (2003) Microtopography and antifouling properties of the shell surface of the bivalve molluscs *Mytilus galloprovincialis* and *Pinctada imbricata*. *Biofouling*, **19** (Suppl.), 221–230.

119. Scardino, A.J. & de Nys, R. (2004) Fouling deterrence on the bivalve shell *Mytilus galloprovincialis*: a physical phenomenon? *Biofouling*, **20** (4–5), 249–257.

120. Genzer, J. & Efimenko, K. (2006) Recent developments in superhydrophobic surfaces and their relevance to marine fouling: a review. *Biofouling*, **22** (5), 339–360.

121. Marmur, A. (2006) Underwater superhydrophobicity: theoretical feasibility. *Langmuir*, **22** (4), 1400–1402.

122. Marmur, A. (2006) Super-hydrophobicity fundamentals: implications to biofouling prevention. *Biofouling*, **22** (1–2), 107–115.

123. Scardino, A.J., Harvey, E. & De Nys, R. (2006) Testing attachment point theory: diatom attachment on microtextured polyimide biomimics. *Biofouling*, **22** (1–2), 55–60.

124. Howell, D. & Behrends, B. (2006) A review of surface roughness in antifouling coatings illustrating the importance of cutoff length. *Biofouling*, **22** (6), 401–410.

125. Berntsson, K.M., Andreasson, H., Jonsson, P.R., *et al.* (2000) Reduction of barnacle recruitment on micro-textured surfaces: analysis of effective topographic characteristics and evaluation of skin friction. *Biofouling*, **16** (2–4), 245–261.

126. Andersson, M., Berntsson, K., Jonsson, P. & Gatenholm, P. (1999) Microtextured surfaces: towards macrofouling resistant coatings. *Biofouling*, **14** (2), 167–178.

127. Berntsson, K.M., Jonsson, P.R., Lejhall, M. & Gatenholm, P. (2000) Analysis of behavioural rejection of micro-textured surfaces and implications for recruitment by the barnacle *Balanus improvisus*. *Journal of Experimental Marine Biology and Ecology*, **251**, 59–83.

128. Petronis, S., Berntsson, K., Gold, J. & Gatenholm, P. (2000) Design and microstructuring of PDMS surfaces for improved marine biofouling resistance. *Journal of Biomaterials Science, Polymer Edition*, **11** (10), 1051–1072.

129. Bers, A.V. & Wahl, M. (2004) The influence of natural surface microtopographies on fouling. *Biofouling*, **20** (1), 43–51.

130. Carman, M.L., Estes, T.G., Feinberg, A.W., *et al.* (2006) Engineered antifouling microtopographies – correlating wettability with cell attachment. *Biofouling*, **22** (1–2), 11–21.

131. Callow, M.E., Jennings, A.R., Brennan, A.B., *et al.* (2002) Microtopographic cues for settlement of zoospores of the green fouling alga *Enteromorpha*. *Biofouling*, **18** (3), 237–245.

132. Hoipkemeier-Wilson, L., Schumacher, J.F., Carman, M.L., *et al.* (2004) Antifouling potential of lubricious, micro-engineered, PDMS elastomers against zoospores of the green fouling alga *Ulva* (*Enteromorpha*). *Biofouling*, **20** (1), 53–63.

133. Schumacher, J.F., Carman, M.L., Estes, T.G., *et al.* (2007) Engineered antifouling microtopographies – effect of feature size, geometry, and roughness on settlement of zoospores of the green alga *Ulva*. *Biofouling*, **23** (1–2), 55–62.

134. Schumacher, J.F., Aldred, N., Callow, M.E., *et al.* (2007) Species-specific engineered antifouling topographies: correlations between the settlement of algal zoospores and barnacle cyprids. *Biofouling*, **23** (5–6), 307–317.

135. Majumdar, P. & Webster, D.C. (2005) Preparation of siloxane–urethane coatings having spontaneously formed stable biphasic microtopograpical surfaces. *Macromolecules*, **38** (14), 5857–5859.

136. Majumdar, P. & Webster, D.C. (2006) Influence of solvent composition and degree of reaction on the formation of surface microtopography in a thermoset siloxane–urethane system. *Polymer*, **47** (11), 4172–4181.

137. Majumdar, P., Stafslien, S., Daniels, J. & Webster, D.C. (2007) High throughput combinatorial characterization of thermosetting siloxane–urethane coatings having spontaneously formed microtopographical surfaces. *Journal of Coatings Technology and Research*, **4**, 131–138.

138. Yu, J. (2003) Biodegradation-based polymer surface erosion and surface renewal for foul-release at low ship speeds. *Biofouling*, **19** (Suppl.), 83–90.

139. Verborgt, J. & Webb, A. (2007) Solvent-free, self-polishing polyurethane matrix for use in solvent-free antifoulings. US Application Patent No. 2007014753.

140. Tang, Y., Finlay, J.A., Kowalke, G.L., *et al.* (2005) Hybrid xerogel films as novel coatings for antifouling and fouling release. *Biofouling*, **21** (1), 59–71.

141. Simendinger, W.H. (2003) Antifouling coating composition. US Patent No. 6,559,201.

142. Simendinger, W.H. (2001) Antifouling coating composition. US Patent No. 6,313,193.

143. Simendinger, W.H. (2002) Antifouling coating composition. US Patent No. 6,476,095.

144. Zhang, Z., Chao, T., Chen, S. & Jiang, S. (2006) Superlow fouling sulfobetaine and carboxybetaine polymers on glass slides. *Langmuir*, **22** (24), 10072–10077.

145. Chang, Y., Chen, S., Yu, Q., *et al.* (2007) Development of biocompatible interpenetrating polymer networks containing a sulfobetaine-based polymer and a segmented polyurethane for protein resistance. *Biomacromolecules*, **8** (1), 122–127.

146. Pettitt, M.E., Henry, S.L., Callow, M.E., Callow, J.A. & Clare, A.S. (2004) Activity of commercial enzymes on settlement and adhesion of cypris larvae of the barnacle *Balanus amphitrite*, spores of the green alga *Ulva linza*, and the diatom *Navicula perminuta*. *Biofouling*, **20** (6), 299–311.

147. Dobretsov, S., Xiong, H., Xu, Y., Levin, L.A. & Qian, P.-Y. (2007) Novel antifoulants: inhibition of larval attachment by proteases. *Marine Biotechnology*, **9** (3), 388–397.

148. Perez-Roa, R.E., Tompkins, D.T., Paulose, M., *et al.* (2006) Effects of localised, low-voltage pulsed electric fields on the development and inhibition of *Pseudomonas aeruginosa* biofilms. *Biofouling*, **22** (5–6), 383–390.

Chapter 26
Implications of International and European Regulatory Developments for Marine Antifouling

Ilona Cheyne

Future legal regulation may affect marine antifouling systems directly or indirectly, in both the international and national legislative spheres. This chapter examines how changes may be introduced through the International Convention on the Control of Harmful Antifouling Systems on Ships, 2001 (ICAFS), and the possible effect of new rules on the management of ballast water. It also considers how indirect effects may come about through other policy areas and discusses some of them in the context of EC law and policy.

26.1   The introduction of new antifouling restrictions

There are two main ways in which new restrictions may be imposed on antifouling systems, through amendments to the ICAFS and the coming into force of the International Maritime Organisation (IMO) International Convention for the Control and Management of Ships' Ballast Water and Sediments, 2004.

26.2   Amendments to the ICAFS

New controls on antifouling substances may be introduced by amending Annex 1 of the ICAFS[1]. Any party to the Convention may propose an amendment containing specified information, including identification of active ingredients or components, information on how the system or its transformation products pose a risk to human health or non-target organisms, evidence of the potential to cause adverse effects in the environment and a recommendation about what restrictions might be effective in reducing that risk[2]. The proposal is considered by the Marine Environment Protection Committee (MEPC) of IMO, which may ask for a more detailed review based on the proposal. If so, the proposing party must submit a comprehensive proposal with more detailed information, including developments in data since the initial proposal, findings on environmental fate and effect, data on any unintended effects on non-target organisms and data on the potential for human health effects[3]. It must also include a summary of the results of studies conducted on the adverse effects of the antifouling system, evidence from any monitoring studies, a qualitative statement of the level of uncertainty in the evaluation, recommendations for specific control measures and a summary of any studies on the potential effects of those control measures on air quality, shipyard conditions, international shipping and so on. Finally, it should contain information on various physical and chemical properties of the component[4].

The MEPC will establish a technical group to assist it with the evaluation of a proposal[5]. The technical group comprises representatives of the parties, IMO members, the UN and specialised agencies, intergovernmental organisations which have agreements with IMO, and NGOs which have consultative status with IMO. The members of the technical group must have sufficient relevant scientific expertise to be able to objectively review the technical merits of the proposal[6]. The technical group has responsibility for evaluating the comprehensive proposal and any other information submitted by interested entities. The purpose of the evaluation is to determine whether the proposal has demonstrated 'a potential for unreasonable risk of adverse effects on non-target organisms or human health such that the amendment of Annex 1 is warranted'[7]. Although the evaluation is concerned with a risk assessment, the factors to be considered by the technical group include the technical feasibility of control measures and the cost-effectiveness of the proposal, and the effects of the proposed control measures on the environment, shipyard health and safety concerns, and the cost to international shipping and other relevant sectors[8]. Thus the technical group not only carries out the scientific risk assessment, but also forms a view about the acceptability of the risk and the proposed measures. It submits a written report to the MEPC in which it recommends whether the proposed control measures are warranted and suitable and whether alternative control measures would be more suitable[9]. However, only representatives of the parties may be involved in formulating a recommendation. If unanimity cannot be achieved, the minority opinion must be included in the report[10]. The decision whether to approve the proposed amendment is taken by the MEPC, taking into account the technical group's report. Thus the report is not decisive, and the power of the MEPC to make the final decision means that considerations other than risk assessment or cost–benefit analysis may affect the final outcome. If the proposed amendment is not approved, another proposal on the same antifouling system may be brought if new information becomes available[11]. Antifouling systems which are placed in Annex 1 by an amendment after the Convention comes into force can be retained on ships until they are scheduled to be renewed, provided that the period is no more than 60 months or the MEPC decides that exceptional circumstances warrant earlier implementation of the control[12].

One interesting point is that the MEPC must apply the precautionary principle if the technical group's report has found that there exists a threat of serious or irreversible damage[13]. This means that lack of full scientific certainty must not be used as a reason for not including an antifouling system in Annex 1[14]. The prohibition on the use of organotins was based on very clear evidence and therefore did not depend on the precautionary principle as its justification[15]. However, the precautionary principle means that other substances may be considered for inclusion in Annex 1 even if the evidence against them is weaker. The precautionary principle requires some evidence before it is triggered, but not full scientific certainty about the risk. In the EC, the threshold has been described as more than a 'purely hypothetical approach to the risk, founded on mere conjecture which has not been scientifically verified'[16]. However, beyond that or an equivalent threshold, the precautionary principle may be used to justify preventive measures in the face of a plausible though uncertain threat. The EC is likely to argue for a precautionary approach towards other antifouling substances[17].

Even if the case is made on the basis of the precautionary principle, however, considerations of cost and feasibility will be powerful counterarguments in practice. In addition, competing environmental considerations will need to be taken into account, such as efficient fuel consumption and effective protection against the carriage and release of alien invasive species. The time it has taken for the Convention to come into force also suggests that it is likely to be a difficult process before Annex 1 will be amended to include restrictions on new antifouling substances.

26.3   Ballast water management

Another relevant policy area is the need to control pollution such as the introduction of alien invasive species and pathogens through the release of ballast water. The International Convention for the Control and Management of Ships' Ballast Water and Sediments, 2004 (BWC) is an IMO convention which will introduce a ballast water management regime when it comes into force[18]. The main obligation is for each party to implement the requirements of the Convention in order to prevent, minimise and ultimately eliminate the transfer of harmful aquatic organisms and pathogens through the control and management of ships' ballast water and sediments[19]. It is also possible to introduce more stringent measures[20]. Parties should ensure that ballast water management practices do not cause greater harm than they prevent[21]. Neither should their actions under the Convention cause damage to the environment, human health, property or resources of other States[22]. The main provisions of the Convention relate to the provision of appropriate facilities for the reception of sediments, scientific and technical research and monitoring, survey certification and inspection, and technical assistance and cooperation[23]. The Annex contains a number of regulations which cover, inter alia, management and control requirements for ships, special requirements in certain areas and standards for ballast water management. The BWC therefore regulates a wide range of management techniques and monitoring arrangements, and the development and specific implementation of the Convention is still ongoing. Of particular interest is the need for IMO to approve ballast water management systems which employ active substances or preparations containing one or more active substances[24]. Systems which require approval include those that make use of chemicals or biocides, make use of organisms or biological mechanisms or which alter the chemical or physical characteristics of the ballast water[25]. Guidelines on the procedure for approval of these systems have been adopted by the MEPC governing, inter alia, risk assessment focusing on persistency, bioaccumulation and toxicity[26].

It is not entirely clear how the restrictions under the BWC will relate to the ICAFS with regard to the use of biocides. Both ballast water management systems and antifouling systems will be evaluated in a case-by-case manner, on the basis of considerations such as persistency, toxicity and bioaccumulation. At the same time, however, rising concern about the introduction of alien invasive species in ballast water must have implications for the use of biocides in ballast water management systems. There must be a balance between the damage caused to human health and non-target organisms by biocides and the damage caused to local biodiversity by alien invasive species or to human health by pathogens. It is worth noting again in this regard that parties to the BWC are required to ensure that they do not cause more harm than they prevent[27]. As substances are reviewed for acceptability in ballast water management systems, consideration will have to be given to potential overlaps and conflicts.

26.4   Other relevant policy areas in EC Law

Although there is a specific antifouling regime in EC law, other regulatory regimes may also have an effect on the use of biocides. Developments in those regimes may therefore have implications on the use of marine antifouling systems. A broad policy context is given by the Commission's Green Paper on Maritime Policy and the Marine Strategy Framework Directive[28]. More specifically, the discussion here focuses on three main areas: aquaculture, water quality regulation and waste management[29].

26.4.1   Aquaculture

The use of antifouling substances in aquaculture is already covered by EC legislation. Although the ban on organotin compounds in the ICAFS applies to shipping hulls or external parts or surfaces, EC legislation goes further. Annex V of the Biocides Directive[30] covers antifouling products 'used to control the growth and settlement of fouling organisms (microbes and higher forms of plant or animal species) on vessels, aquaculture equipment or other structures used in water'. Thus the ban on organotins applies to equipment used in aquaculture. Although there is no specific mention of antifouling or the use of biocides in the Community's strategy for sustainable development of aquaculture, the Commission has recognised the need to take steps to mitigate pollution from intensive fish farming[31]. In particular, it is possible that aquaculture will be included in the remit of the Integrated Pollution Prevention and Control (IPPC) Directive[32]. The consequence of this would be that aquaculture would be brought under a centralised system of regulation which takes into account a wide range of environmental impacts from a single activity and imposes conditions designed to ensure a high level of protection of the environment through best available techniques. The integrated approach includes monitoring and contamination clean-up. The IPPC Directive relies on a dynamic set of standards which are produced through a dialogue between the Commission and the relevant industry[33]. If necessary, standard emission limits can be introduced across the Community[34].

26.4.2   Water quality regulation

Biocidal substances in ballast water management systems and used as antifouling systems for other purposes will also potentially fall under the Water Framework Directive[35]. This directive regulates the management of surface waters, groundwater, and sea areas within the Member States' land and territorial sea. The purpose of the Directive is to establish a framework for the prevention and reduction of pollution, the promotion of sustainable water use, the protection of the aquatic environment and the improvement of the status of aquatic ecosystems[36]. This specifically includes the protection of territorial and marine waters. The Directive is also intended to help achieve the objectives of relevant international agreements, including those aiming to prevent and eliminate pollution of the marine environment, by providing for the cessation or phasing out of discharges, emissions and losses of 'priority hazardous substances'[37]. The ultimate aim is to achieve concentrations 'near background values for naturally occurring substances and close to zero for man-made synthetic substances' in the marine environment[38].

Much of the Directive is concerned with the establishment of water management plans for each river basin district, but the strategy which has the most immediate implications for marine antifouling systems is the identification and control of pollutants. Thirty-three priority hazardous substances have so far been listed in Annex X of the Directive, including tributyltin compounds[39]. The identification of these substances should be linked with hazardous substances that are controlled under international agreements, including IMO conventions such as the ICAFS and BWC[40]. Quality standards for concentrations of the substances have now been proposed[41].

26.4.3   *Waste management*

Waste management is also relevant to the control of marine antifouling systems. There has been uncertainty about whether the definition of waste should include by-products of normal industrial processes, but the European Court of Justice has tended towards the more inclusive definition[42]. Leaching of biocidal substances into the water might therefore be held to constitute waste under the EC's Waste Framework Directive[43], and the disposal or recovery of antifouling systems that are removed from ships' hulls and surfaces or released with ballast water would certainly be covered. The Waste Framework Directive imposes a duty on Member States to ensure that waste is recovered or disposed of without endangering human health and without using processes or methods which could harm the environment. In particular, recovery or disposal should be carried out without risk, inter alia, to water, plants or animals[44]. Member States should also take necessary measures to prohibit the abandonment, dumping or uncontrolled disposal of waste[45]. It is likely that specific legislation will eventually be introduced to cover the waste management of antifouling substances, ballast water and sediment because of the advantages of developing specific rules for the waste management of specialised streams[46].

26.5   Conclusions

- The field of marine antifouling regulation is relatively new, but the ICAFS, BWC and certain EC regulatory regimes will have a significant impact on the practices of the shipping industry and recreational sailing.
- The impact on activities other than shipping, such as aquaculture and certain industrial plants, is also likely to be significant.
- It is unlikely that a consistent or coherent legal picture will emerge for some time.
- It is likely that a gradual scheme of prohibition and restriction through international instruments will be the best and only way forwards.
- Marine antifouling systems must also be viewed in the wider context of other types of environmental and safety standards and procedures, as illustrated in EC law and policy in aquaculture, water quality regulation and waste management.

Notes

1. Article 6.
2. See Annex 2.
3. See Annex 3(1) and (3).

4. Annex 3(2) lists melting point, boiling point, density (relative density), vapour pressure, water solubility/pH/dissociation constant (pK_a), oxidation/reduction potential, molecular mass, molecular structure and other physical and chemical properties identified in the initial proposal.
5. Article 7.
6. Article 2(10).
7. Article 6(4).
8. Article 6(4)(a).
9. Article 6(4)(c).
10. Article 7(4).
11. Ibid.
12. Article 4(2).
13. The ICAFS as a whole is based on the precautionary principle as expressed in Principle 15 of the Rio Declaration on Environment and Development. The Preamble cites the Rio Declaration and a reference to the principle in resolution MEPC.67(37) adopted by MEPC on 15 September 1995. Principle 15 states, 'Where there are threats of serious or irreversible damage, lack of full scientific certainty shall not be used as a reason for postponing cost-effective measures to prevent environmental degradation'.
14. Article 6(5).
15. Santillo D., Johnston, P. & Langston W.J. (2002) Tributyltin (TBT) antifoulants: a tale of ships, snails and imposex. In: *Late Lessons from Early Warnings: The Precautionary Principle 1896–2000* (eds P. Harremoës, D. Gee, M. MacGarvin, A. Stirling, J. Keys, B. Wynne & S. Guedes Vaz). European Environment Agency. Available from http://reports.eea.europa.eu/environmental _issue_report_2001_22/en. Accessed 3 July 2009.
16. Case T-13/99, *Pfizer v Council* [2002] ECR II-03305, para 143; Case C-236/01, *Monsanto* [2003] ECR I-8105, para 106; Case E-3/00, *EFTA Surveillance Authority v Norway* EFTA Court Reports 2000–2001, paras 36–38.
17. The Commission's legislative proposal explicitly referred to the use of the precautionary principle when considering potential restrictions of other antifouling substances. See Commission Proposal for a Regulation of the European Parliament and of the Council on the Prohibition of Organotin Compounds on Ships, COM(2002) 396 final, p. 4.
18. The Convention will come into force 12 months after ratification by 30 States, representing 35% of world merchant shipping tonnage. As of 30 June 2008, 14 States had ratified, representing 3.55% of world merchant shipping tonnage. In EC law, biocidal substances used to control harmful organisms in ballast water will fall under Directive 98/8/EC of the European Parliament and of the Council of 16 February 1998 concerning the placing of biocidal products on the market (Biocides Directive), OJ L 123/1, and therefore under the normal authorisation procedures including the need for the active substance to be listed in Annex 1.
19. Article 2(1). Article 1(8) defines 'Harmful Aquatic Organisms and Pathogens' as 'aquatic organisms or pathogens which, if introduced into the sea including estuaries, or into freshwater courses, may create hazards to the environment, human health, property or resources, impair biological diversity or interfere with other legitimate uses of such areas'.
20. Article 2(3).
21. Article 2(7).
22. Article 2(6).
23. Articles 5–12.
24. Annex, Regulation D-3.2.
25. IMO Summary of the BWC, available from http://www.imo.org. Regulation D-2 provides for a Ballast Water Performance Standard which requires ships to discharge 'less than 10 viable organisms per cubic metre greater than or equal to 50 micrometres in minimum dimension and less than 10 viable organisms per milliliter less than 50 micrometres in minimum dimension and greater than or equal to 10 micrometres in minimum dimension; and discharge of the indicator microbes shall not exceed the specified concentrations described in paragraph 2'. The indicator microbes include toxicogenic

Vibrio cholerae (O1 and O139) with less than 1 colony forming unit (cfu) per 100 mL or less than 1 cfu per 1 g (wet weight) zooplankton samples; *Escherichia coli* less than 250 cfu per 100 mL; and intestinal enterococci less than 100 cfu per 100 mL.

26. MEPC.126(53).
27. See footnote 21.
28. See Communication from the Commission to the Council and the European Parliament – Thematic Strategy on the Protection and Conservation of the Marine Environment, COM(2005) 504 final, 24.10.2005; Green Paper – Towards a Future Maritime Policy for the Union: A European Vision for the Oceans and Seas, COM(2006) 275 final, 7 June 2006; Proposal for a Directive of the European Parliament and of the Council establishing a Framework for Community Action in the field of Marine Environmental Policy (Marine Strategy Directive), COM(2005) 505 final, 24 October 2005. And see also Directive 2008/56/EC of the European Parliament and of the Council of 17 June 2008 establishing a framework for community action in the field of marine environmental policy (Marine Strategy Framework Directive), OJ L 164/19.
29. Other potentially relevant areas of regulation include REACH, a programme of registration, evaluation and authorisation of chemicals which extends to existing and new substances including substances already covered by Directive 76/769/EEC of 27 July 1976 on the approximation of the laws, regulations and administrative provisions of the Member States relating to restrictions on the marketing and use of certain dangerous substances and preparations, OJ L 262/201(under which organostannic compounds on vessels have been banned). Directive 76/769 was repealed with effect from 1 June 2009. See Regulation 1907/2006 concerning the Registration, Evaluation, Authorisation and Restriction of Chemicals (REACH), establishing a European Chemicals Agency, OJ L 396/1.
30. See footnote 18. Substances included in the Biocides Directive will be regarded as already registered for the purposes of REACH, see Regulation 1907/2006, footnote 29, Article 15(2).
31. Communication from the Commission to the Council and the European Parliament – A Strategy for the Sustainable Development of European Aquaculture (Communication on aquaculture), COM(2002) 511 final, 19 September 2002, pp. 18–20.
32. Council Directive 96/61/EC of 24 September 1996 concerning integrated pollution prevention and control, OJ L 257/26. See Communication on aquaculture, footnote 31, p. 19.
33. Each industry is to have a Best Available Technique Reference (BREF) document, which will be taken into account by Member States when they set permit conditions. They can be found at http://eippcb.jrc.es/. Accessed 10 July 2009.
34. Article 18.
35. Directive 2000/60/EC establishing a framework for Community action in the field of water policy, OJ L 327/1.
36. Article 1.
37. Article 16.
38. Ibid.
39. Decision No 2455/2001/EC establishing the list of priority substances in the field of water policy and amending Directive 2000/60/EC, OJ L 331/1. These substances were identified using both monitoring of surface waters in the Member States and associated modelling, a method known as Combined Monitoring-based and Modelling-based Priority Setting (the COMMPS procedure). And see Proposal for a Directive of the European Parliament and of the Council on environmental quality standards in the field of water policy and amending Directive 2000/60/EC, COM(2006) 398 final, 17 July 2006.
40. Ibid., Article 16(3).
41. Directive 2008/105/EC of the European Parliament and of the Council of 16 December 2008 on environmental quality standards in the field of water policy, amending and subsequently repealing Council Directives 82/176/EEC, 83/513/EEC, 84/156/EEC 84/491/EEC, 86/280/EEC and amending Directive 2000/60/EC of the European Parliament and of the Council, OJ L 348/84.

42. Cheyne, I. (2002) The definition of waste in EC Law, *Journal of Environmental Law*, **14**, 61; Scotford, E. (2007) Trash or treasure: policy tensions in EC waste regulation, *Journal of Environmental Law*, **19**, 367.
43. Directive 2006/12/EC on waste, OJ L 114/9. But see also Directive 2008/98/EC of the European Parliament and of the Council of 19 November 2008 on waste and repealing certain Directives, OJ L 312/3.
44. Directive 2006/12/EC, Article 4.
45. Ibid.
46. A more specialised approach has already been adopted in a number of fields. Regulated waste streams include waste oils, titanium dioxide, sewage sludge used in agriculture, batteries and accumulators, packaging and packaging waste, PCB/PCTs, end-of-life vehicles, urban waste water, waste electrical and electronic equipment, and extractive industries.

Chapter 27
Research on Practical Environmentally Benign Antifouling Coatings

Dan Rittschof

The aim of this chapter is to provide a stationary target to help focus questions and hypotheses for those with practical interests in fouling management. For academics the hope is to provide insight into business models, markets, regulatory considerations and the importance of pre-emptive rather than post hoc attention to potential environmental impacts. For business people and regulators the goal is to provide perspective on the necessity for considering new business models and regulatory approaches. For all, the goal is to support the concept of cooperative multidisciplinary approaches for a complex global problem whose solutions must be considered in the context of having global environmental consequences.

27.1   Environmentally benign antifouling coatings

The idea of generating environmentally benign antifouling coatings has engrossed me for 25 years. During my active research in antifouling, I have watched several generations of new researchers reinvent many aspects of the antifouling research wheel. Some of this reinvention is a necessary part of learning. Other reinventions are a genuine waste of time. My hope is to reduce some of the unnecessary reinvention by sharing thoughts, a skeleton amount of key references and sharing results from my quest for environmentally benign solutions to manage fouling. I am not alone in this quest; every contributor to this book as well as the authors cited in the reference lists are making substantive contributions. From my perspective there is the need and plenty of room for more research and researchers to attack especially the practical problems of fouling and antifouling into the foreseeable future.

Predicting the future directions of research in the management of fouling is art not science. New research directions depend upon the teams of researchers with the passion and resources to go forward, upon the selected knowledge base, on the funding climate, upon limiting assumptions and ultimately upon societal, governmental and business objectives and priorities. As issues with the environment come to the fore, there is hope that rational approaches and practical solutions will prevail. One very attractive option is the over 30-year-old idea of copying natural solutions to fouling management aptly described in Chapter 8.

In the discourse that follows, I will provide a personal view of the processes and assumptions of fouling and its management that dictate how research might go forward. Although I will make statements as if they are true, my intent is to provide researchers in antifouling, those interested in working in antifouling, managers and businessmen a stationary target from which to formulate their own assumptions and apply their own creative energies. I am most comfortable with the idea that the reader may agree with issues I raise and find solutions, disagree with issues I raise and find solutions, and identify areas that have not occurred to me and find solutions.

27.2   Fouling and fouling control

Biofouling, colonisation of fluids and surfaces such as sensors, toilet seats, ship hulls, water intake pipes, heat exchangers, reverse osmosis membranes, foodstuffs and food preparation surfaces, teeth, arteries and implants in humans is a multifaceted, global-scale problem (see Chapters 5, 11, 12, 15 and 20). Fouling directly impacts the surfaces being fouled and the transport of foulants from one location to the other. If one would like to gain insight into the development of these still incomplete ideas, one can visit on the topic from my research group and collaborators over the last 20 years [1–9].

Solutions to hull fouling have been attempted since the time of the Phoenicians and have long entertained the global scientific and business communities. All submerged surfaces foul [10–12]. Living organisms attached to surfaces take advantage of flow by the surface to gain nutrients and food and expel wastes. In most aquatic environments, space on submerged surfaces is a limiting resource. The specifics of fouling community development are complex and highly variable [2].

Biofouling involves molecular, microbial and macroorganisms. In water, there is a boundary layer between a surface and water with no net movement; molecules, microbes and organisms that are sufficiently small, less than 500 µm, in at least one dimension can be passively trapped or actively enter the boundary [12]. Even very poor glues enable organisms to attach within the boundary layer.

Sometimes there are relationships between these different levels of fouling and sometimes there is no relation [2]. The spectrum of mechanisms of fouling of surfaces range from purely probabilistic, what is in the water column attaches or settles, to succession, requiring a series of molecules and then organisms attach before the next more complex level of fouling can occur (Figure 27.1).

Although there are surfaces, such as silicones [13], that bioadhesives attach to less strongly, bacteria and some kinds of macroorganisms settle on all surface chemistries and surface energies [14–16]. Mechanisms that result in attachment of fouling molecules and organisms are similar and are based on molecular bonding (Figure 27.2) – some combination of ionic, hydrogen, van der Waals and covalent bonding. Because all surfaces are composed of molecules and all molecules undergo at least one kind of bonding, all surfaces foul. Some specialised bioadhesives are mainly protein [17] while others are mainly carbohydrate [18]. Adhesives usually exclude water, bond with the surface and cross-link into refractory adhesive plaques [19].

Many of the fouling organisms in ports and on ship hulls are weed species [6,20]. These weed species are rapidly maturing, highly fecund organisms that do not require fouling succession.

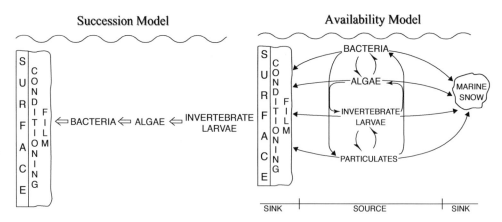

Figure 27.1 The spectrum of biological fouling. Classically, biofouling was thought to be mainly a successional process. This may have been the case prior to shipping activity. The availability model is appropriate for cosmopolitan fouling organisms now found in the world's ports. These foulers were introduced through shipping and dominate because they mature rapidly and are highly fecund and analogous to terrestrial weeds. (Modified from [2, 6].)

For example, a barnacle like *Amphibalanus* (=*Balanus* [21]) *amphitrite* completes its larval development in 3–4 days, settles on a surface and is sexually mature, releasing its own larvae in about 3 weeks [8, 22].

Fouling mechanisms are highly variable at the physical and biological levels due to behaviour of particulates in flow and due to the sensory capabilities, surface selection behaviour and

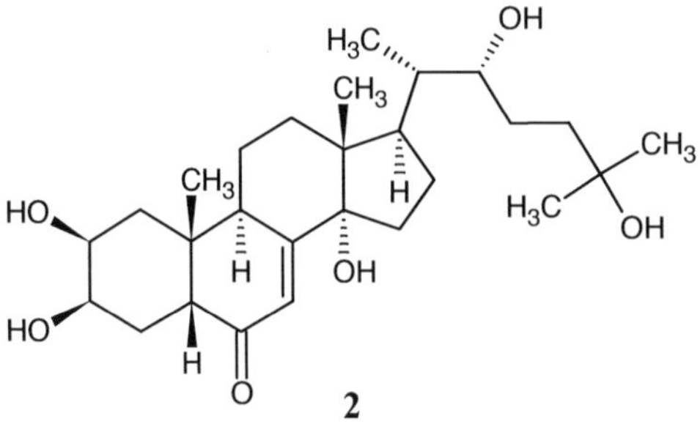

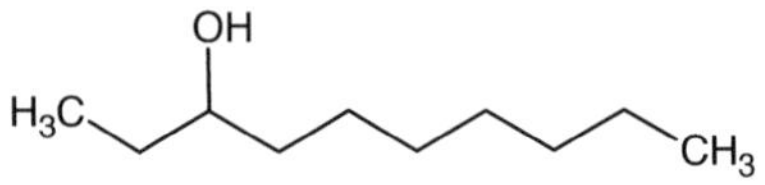

Figure 27.2 Biologically active molecules that would bind using bonding forces other than covalent bonding. (1) 3-Decanol, a molecule from hermit crab blood that signals shell availability to other hermit crabs; it would bind readily to a low-surface-energy surface like Teflon. (2) Ecdysone, a steroid that would bind readily to a low-surface-energy surface like Teflon. (3) Gly-Met-Arg, a tripeptide that mimics pheromones and signal molecules in many phyla of aquatic invertebrates. This tripeptide can rearrange itself to bind to any-surface-energy surface.

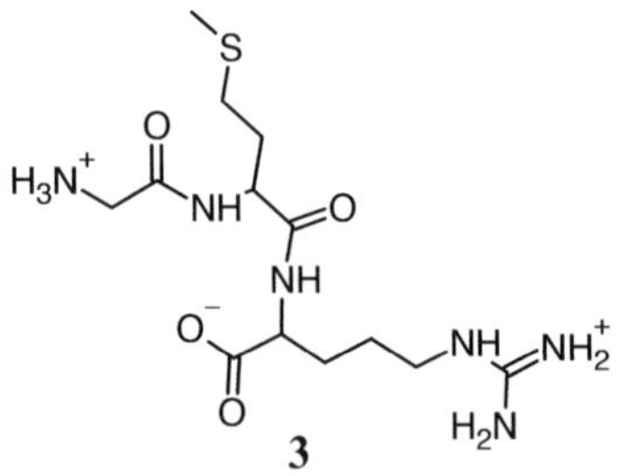

swimming abilities of microbes, invertebrate larvae and plant propagules. Molecular fouling is abiotic, while microfouling (cyanobacteria, bacteria, diatoms and fungi) is a combination of living organisms and exopolymers. Marine snow is microfoulers that aggregate in the water column and then often attach to surfaces. Marine snow attaches to all surfaces when delivered in flow. Living organisms trapped in the snow grow after attachment. Macrofouling is accomplished by individual propagules which have variable motility. Many fouling organisms respond to shear by settling on surfaces [22–27].

27.3   Biofouling management

I hope one take-home lesson from the incomplete review above is that biofouling is a 'catch-all' term that encompasses multiple phenomena and multiple mechanisms. The only unifying factors are all biofouling is biologically based and all biofouling involves basic chemical interactions. As a result, there is room for a large number of commercial management options tailored to the specific fouling problem to be addressed.

It is no surprise that commercial fouling control methods either oxidise organic molecules and kill propagules (chlorine, ferrate, etc.) or use broad-spectrum biocides. Historically, organotins gave the desired control but had unacceptable environmental impacts [28]. The present solution is the use of copper ion alone or in combination with broad-spectrum organic biocides to kill foulers [6, 7]. These broad-spectrum biocides are released from the polymer coating, enter organisms and disrupt essential life processes [29, 30] (see also Chapters 16 and 17).

Some toxin-release polymer systems are spongy resin/rosin coatings that act as reservoirs to release trapped toxins over time. These coatings tolerate molecular and a small amount of microfouling that are not impacted by the toxins. Other polymer systems are called ablative copolymer systems [31]. The polymer matrix of ablative copolymers hydrolyses and polishes releasing trapped toxins and minimising build-up of resistant forms. Resistant organisms eventually prevail and provide a less toxic platform for colonisation by less-resistant species.

Although the biocides used in addition to copper are referred to as 'booster biocides' by the coatings industry, in reality the additional biocides are approximately as toxic as the primary biocides. Four common organic biocides are Sea 9211N®, zinc pyrithione (Omadine®), Diuron® and Irgarol 1051® (Figure 27.3). Sea 9211N® [32] and zinc (copper in sea water) pyrithione [32, 33] are short-lived. Diuron® [34] and Irgarol 1051® [34–36] are much more stable and build up to measurable levels in the environment [30, 37, 38].

Antifouling is just one of several very important functions of hull coatings. Hull coatings are complex multicomponent systems which include anticorrosive and antifouling components [31] (see Chapter 13). Coatings have important physical and anticorrosive properties. Hull coatings have necessary physical properties that include maintaining coating integrity and that have physical properties which maximise hull performance.

Existing commercial solutions to fouling are an uncomfortable and increasingly unacceptable compromise between fouling management, corrosion and environmental degradation. Oxidation control measures cause corrosion and have unacceptable environmental impacts. Similarly, broad-spectrum biocides that must be released and diffuse into organisms to kill them have extensive impacts on non-target species and ecosystems. Pressure to find alternative fouling control measures increases as governments become aware of unacceptable environmental impacts [39, 40].

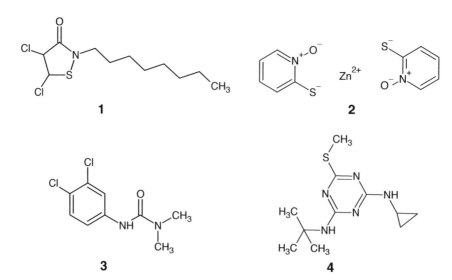

Figure 27.3 Commercial organic biocides. (**1**) Sea 9211N® is an isothiazalone that is purported to break down quickly once released from a coating. (**2**) Zinc Omadine® (zinc pyrithione) is rapidly converted to copper Omadine® in sea water and then dissociates. This compound is also used in a variety of personal hygiene products. (**3**) Diuorne® and (**4**) Irgarol 1051 are photosynthesis inhibitors that are rapidly building up in the environment. One must ask if inhibiting or shutting off photosynthesis in regions of the ocean is prudent.

27.4 The antifouling problem in a nutshell

As governments recognise and legislate against uses of broad-spectrum biocide antifouling coatings, the industry and antifouling researchers are actively looking for alternatives. Because antifouling is just one of several very important functions of hull coatings, existing coating formulations and business models tolerate only small changes. In fact, to date the only solutions that fit formulation and business models are those that substitute one broad-spectrum biocide for another. The new coatings contain several broad-spectrum biocides, usually some form of copper that releases copper ion and one or more organic biocides. Generally the new co-biocide coatings are a response to requirements to reduce the release of copper to conform to some kind of environmental or clean water standard. However, as coatings containing organic biocides gain popularity and market share, unacceptable environmental impacts due to the build-up of the organic biocides and their metabolites will become evident, the environment will be adversely affected, legislation will be generated and the quest for alternative biocides will begin anew.

27.5 Control of biofouling and environmental impacts

One way to think of biofouling of ships is to consider them as habitats that move from one port or country to another. There are two major habitats on a ship, the hull and the ballast tanks (Chapter 24). Hull fouling has been carefully managed because fouling on hulls results in expensive decreases in performance and increases in maintenance costs [41]. Thus, hull fouling management results in build-up of toxic compounds in the environment [30, 34, 36, 38, 42]. Although science tracks known bioactive compounds, all polymers leach a large number of compounds whose fates and effects are unknown. For example, leachates from a commercial silicone, a model foul-release coating we use to rear genetic lines of barnacles in the laboratory [43], are not chronically toxic to barnacles but contain a suite of molecules whose fates and effects are entirely unstudied (Figure 27.4).

Thus, biofouling and its management represent a fine line between decreased performance, the risk of environmental introductions of invasive species due to inadequate fouling management and the biological and societal consequences of environmental degradation due to excessive use of toxic control measures. A ship that is a good citizen would minimally impact the environments that it visits. It should be no surprise that existing technologies were developed to maximise efficiency and result in short-term economic gain and to meet the letter rather than the spirit of government regulations. Since biologists have not been involved in the business processes, one cannot fault the businessmen and chemists for areas beyond their expertise.

Ballast water has not been managed effectively. Ballast water is implicated globally as a source of invasive species [44]. Ballast water is not managed effectively because (a) effective management has not been developed and (b) managing such large volumes of water, roughly a little less than the volume of the Atlantic Ocean each year, would have a large impact on shipping costs. The use of less toxic hull fouling management may increase the risk of invasive species introductions from hulls.

27.6   Practical solutions

The ideal antifouling solutions would be ones that fit within existing business models and polymer systems that managed fouling on hull with minimal impact on other organisms or the environment [4, 6, 9, 45]. Such a solution is possible but would require major changes in research, business and government regulations. This solution requires infrastructure and expertise lacking in industry, government and academia and would require an approach which recognises the needs of business and the environment [6, 7]. At present, there is no appropriate infrastructure anywhere in the world that would facilitate generations of ideal antifouling technology. Research teams would be composed of businessmen, chemists, biologists and government officials with the charge of ensuring products minimally impacted environments.

27.7   Considerations for a novel antifouling coating

Developing a novel antifouling coating is not practical from a commercial point of view because a novel coating system does not fit within existing business plans [7, 8]. This lack of practicality, combined with the lack of an appropriate infrastructure, is a major reason that novel hull coating systems are slow to develop. To understand the complexity of the issue for coatings manufacturers and suppliers, one must consider how coatings are formulated, manufactured, applied and repaired. Then, with this information as background and in the context of globalisation, one can consider how a government or organisation might expedite the development of new and novel formulations.

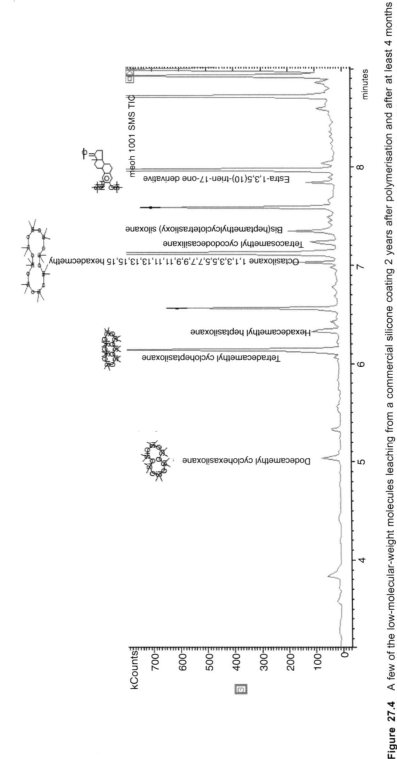

Figure 27.4 A few of the low-molecular-weight molecules leaching from a commercial silicone coating 2 years after polymerisation and after at least 4 months of exposure to sea water. Gas chromatography trace with mass spectroscopy (MS) trace of a commercially available silicone coating. Fifteen-minute methanol extraction was performed on coated glass panels that had been used for barnacle settlement for 2 years. Each peak on the trace represents a molecule released from the silicone surface. Peaks with matches in the NIST MS library are labelled with molecular names and structure. This trace shows that cyclic silicone monomers are being released from the surfaces of the coating. Unidentified peaks did not show up in the NIST database.

In the existing business plan, marine paint companies buy ingredients from specialty chemical companies, and then formulate, mix and distribute coatings systems. The coatings systems are multilayered (primers, anticorrosive layers, tie coats, antifouling layers) and, when applied correctly, result in coatings (usually with a warranty) to perform as promised. Polymer chemists, chemical engineers and chemists experienced with broad-spectrum biocides do the formulations and are continually making small adjustments that improve general performance or performance in a specific geographic area or due to a specific governmental restriction. Because the chemistry of toxin release is temperature dependent, formulations are designed for specific uses and regions. Formulations are routinely tested at sites located in key market regions [11].

Historically there has been no need for biologists in the formulation and antifouling evaluation process because successful coatings do not foul. Anyone can be trained to do industry assessments of coatings because their criteria are straightforward, non-scientific and do not require statistics. Assessment involves measures of coating dissolution (how the coating rubs off on your thumb) and visible slime, grass and hard fouling. Most assessments are done photographically with digital cameras, and sent back to the people doing the formulations for evaluation. In industry, coatings fail if a certain relatively low percentage of fouling occurs.

Even within a very limited spectrum of changes and with regional testing, changes that theoretically should not impact product performance can and do result in changes in properties and performance. This is a testament to the complexity of the chemistry and properties of antifouling coatings. An example of an unexpected impact occurred when companies changed the antifouling coating from an organic solvent to a water-based application to comply with volatile organic compound regulations in California [46]. Although no change in performance was expected, the new coatings failed because the standard cathodic protection for corrosion on ships interfered with the release of toxic metals such as copper. Such mistakes are costly to coatings suppliers and are part of the reason for the high level of conservatism by companies in introducing new coatings.

Given the costly surprises that occur, even with well-understood formulations, it should be no surprise that development of a novel coating might take a long time to reach market. An optimistic estimate for the length of time it might take to test and develop a new coating for market is at least a decade once the basic formulation is in place. It is more likely that commercialisation would take at least two decades.

27.8   Coating application and removal

Coatings removers and appliers are yet another stumbling block to developing novel coatings. These business people have to keep their workers safe, comply with environmental regulations and support a functional product at a profit. For a coatings applier, changing processes is risky. Switching, for example, from spraying a solvent-based ablative copolymer to a silicone foul-release coating is a nightmare. The foul-release coatings are so slippery that they present hazards to workers that include falling as well as boats and ships falling of their stanchions. Most application facilities specialise in one kind of coating. It is a given that application errors will result as part of the process involved in learning to apply a new coating. Business pressure discourages new coatings technologies.

27.9   Governmental oversight

Although government regulations are generated in good faith, regulations are flawed and miss the mark for protecting society and the environment. Registration procedures are by in large, far removed from actual environmental impacts and routinely socio-political attempts to manage complex problems with measurable means. The result is, with the exception of the progressive environmentally aware European governments (Sweden, for example), regulations usually protect the health of workers but have no relation to environmental degradation [30] (Chapters 21 and 26).

Even so, a final and major hurdle for development of novel coatings is biocide registration. Various parts of the world have specific definitions of a biocide. In the European Union the definition is so broad that any compound that leaches from a coating is considered a biocide. Registration in the European Union is relatively quick and inexpensive. For example, registration of Sea Nine 211® took several years and cost several hundreds of thousands of euros. Registration of Sea Nine 211® in the USA was slow and expensive. Registration of Sea Nine 211® took over a decade and cost approximately US$11 000 000 [4]. If one combines this estimate of time and cost to registration with the time to development and testing, unless development and commercialisation are carefully synchronised, it could easily take two decades to bring a product to market.

27.10   Perspective

From an ethical perspective, registration and compliance with the laws are actually meaningless if the laws do not take into account the actual impact if a coating garners a significant market share increasing the release of the registered compounds. An ethical way to proceed with respect to the environment would be to generate coatings that, at least theoretically, would have minimal impact on the environment.

27.11   Theoretical considerations for a novel antifouling coating

So, in this imperfect world, what would it take to actively pursue and commercialise environmentally benign novel antifouling coatings? A partial list includes topics that involve close working relations between governments, businesses and academia [2,4,6,47]. At the fine scale for producing products one needs materials engineers, chemical engineers, polymer chemists, synthetic chemists, biologists, government regulators and funding agencies [9]. In short, the world needs a new infrastructure that attends to global problems and enables their solution.

27.12   Environmentally benign coatings

The future novel coating will be applicable on top of existing anticorrosive coatings systems. From a fantasy/fashion of today perspective, it will be based on biomimetic concepts. It will be an easy clean coating that has a micro-texture (mimicking non-wettable plant leaves or other surfaces [48–50]), bioadhesive-resistant, low-surface-energy surface that contains environmentally benign compounds that target problem foulers and their adhesives and that is periodically mechanically cleaned and rejuvenated rather than replaced. The reasons for each component are justified below.

Micro-textured surfaces were first patented in the 1950s because they inhibited certain kinds of macrofoulers (and may also stimulate settlement of others). These kinds of surfaces have taken on a new life in academic materials research with modern nanotechnology [51]. By using micro-textured surfaces one can reduce the target population of potential foulers. Bioadhesive-resistant, low-surface-energy surfaces, such as silicones, further reduce the fouling target organisms and because adhesives adhere relatively weakly, they are easy to clean. Thus, the majority of the antifouling properties of the coating are based upon mechanical and physiochemical properties.

Coatings might contain three types of environmentally benign compounds that degraded in 5 minutes to 2 hours after release from the coating. The ideal compounds would be added to coatings at catalytic levels ($\mu g\ g^{-1}$) and would be released at $\leq ng\ cm^{-1}\ day^{-1}$; they would breakdown quickly and not add extensively to the nutrient load in the water column. One type of compound would be a biocide, used with the principal purpose of controlling microfouling and enabling release of the other two types of compounds. The other compounds would disrupt bioadhesive polymerisation [52] and molecules that disrupt the metamorphic cascade of macrofouling organisms [53]. These would be delivered to the surface of the coating, diffuse into adhesives and organisms and be degraded entirely within 2 hours (a biological half-life of about 5 minutes).

Chemical compounds can degrade because they are chemically unstable or because they are biologically consumed. The target compounds would be chemically simple and engineered to be biologically consumed. They would be protected from biological degradation until release by the coating polymer matrix. These compounds might be pendant, chemically linked to the polymer matrix, or trapped and diffusionally released. In Singapore, Tropical Marine Science Institute at National University Singapore and International Council for the Exploration of the Sea have a joint project for developing and designing these kinds of molecules. The project started in 1999 with a workshop at National University Singapore and is now in the intellectual property development and protection phase.

The reason for cleaning and renewal of the coatings is simple. We have no idea of the impacts of hydrolysis products of ablative copolymers on the environment. The easiest and most practical solution is low leach products having components that have known environmental fates and effects. Thus, a stable matrix that can be periodically reloaded with control agents is an admirable goal.

27.13 Commercialisation

The bottom line for developing environmentally benign antifouling coatings is developing the ethics; correct business models; environmental regulations; government support systems and research infrastructure to support the concept. This will probably be the case for all problems that require global-scale solutions because the solutions are contrary to best business practices. This is because initially they require cooperation (business + government + academia) rather than the usual competition, adversarial, non-interactive or uncooperative positions that are standard between these three branches of society. One possibility might be a global solution developed by something like a UN program that is then commercialised at national or regional levels. Such an approach would require that countries recognise that it is in their self-interest to participate cooperatively in attempts to solve global-scale problems. The Kyoto Protocol [54] is the first partial attempt at such an approach. One can expect the process to be incomplete because of the lack of support for Kyoto Protocol by the US government. However, this is acceptable as long as the products are generated because when money can be made, recalcitrant countries will come to the table after the fact. One can only hope that globalisation will result in global awareness and a new cooperative infrastructure that recognises respects and is compatible with the diversity of existing societal and cultural paradigms.

27.14   Conclusions

- Environmentally benign antifouling coatings are theoretically possible but unlikely because they are not compatible with cultural, business and regulatory assumptions and philosophies.
- Fouling is a plethora of interrelated phenomena from the molecular to the macro level.
- Fouling management reflects the complexity of fouling.
- Classic antifouling management is based upon broad-spectrum biocides and a limited number of release technologies.
- Broad-spectrum organic biocides employed to replace environmentally damaging heavy metal biocides may be as or more destructive to ecosystems.
- Booster biocides are in reality potent broad-spectrum organic biocides.
- Antifouling coatings are complex and encompass essential physical and mechanical properties beyond just the release of toxins to manage fouling.
- Fouling management business models tolerate only small changes.
- Fouling management that requires new or dedicated infrastructure is not economically viable in the existing business environment.
- The existing infrastructure for coatings application and removal must be considered in the development of new coatings.
- Governmental oversight and regulations are slow in implementation and do not necessarily protect society or the environment because they do not take actual rates of breakdown or impact of market share on the level of antifoulants in the environment.
- Ideally, novel antifouling coating development would require government, business and academia to work together.
- Environmentally benign coatings will probably contain at least three different mechanisms of action.
- Commercialisation requires society find a way to pay for costs of development and of required new infrastructure.

References

1. Rittschof, D. & Costlow, J.D. (1987) Macrofouling and its management by nontoxic means. In: *Advances in Aquatic Biology and Fisheries: Prof. Balakrishnan Nair Felicitation Volume* (ed. Prof.N. Balakrishnan Nair Felicitation Committee), pp. 1–11. Department of Aquatic Biology and Fisheries, University of Kerala, India.

2. Clare, A.S., Rittschof, D., Gerhart, D.J. & Maki, J.S. (1992) Molecular approaches to non-toxic antifouling. *Journal of Invertebrate Reproduction and Development*, **22**, 67–76.

3. Rittschof, D. (1999) Fouling and natural product antifoulants. In: *Recent Advances in Marine Biotechnology*, Vol. **3** (eds M. Fingerman, R. Nagabhushanam & M F. Thompson), pp. 245–257. Oxford & IBH Publishing Company, New Delhi.

4. Rittschof, D. (2000) Natural product antifoulants: one perspective on the challenges related to coatings development. *Biofouling*, **15**, 199–207.

5. Rittschof, D. (2001) Natural product antifoulants and coatings development. In: *Marine Chemical Ecology* (eds J. McClintock & P. Baker), pp. 543–557. CRC Press, New York.

6. Rittschof, D. (2008) Novel antifouling coatings: a multiconceptual approach. In: *Marine and Industrial Biofouling* (eds H.-C. Flemming, R. Venkatesan, P.S. Murthy & K. Cooksey). Springer, New York.

7. Rittschof, D. (2008) Ships as habitats: biofouling, a problem that requires global solutions. *Cosmos*, **4** (1), 71–81.

8. Rittschof, D. & Holm, E.R. (1997) Antifouling and foul-release: a primer. In: *Recent Advances in Marine Biotechnology, Vol. 1: Endocrinology and Reproduction* (eds M. Fingerman, R. Nagabhushanam & M.F. Thompson), pp. 497–512. Oxford & IBH Publishing Company, New Delhi.

9. Rittschof, D. & Parker, K.K. (2001) Cooperative antifoulant testing: a novel multisector approach. In: *Recent Advances in Marine Biotechnology*, Vol. **6** (eds M. Fingerman & R. Nagabhushanam), pp. 239–253. Science Publishers, Enfield, CT.

10. Costlow, J.D. & Tipper, R.C. (1984) *Marine Biodeterioration: An Interdisciplinary Study*. Naval Institute Press, Annapolis, MD.

11. Price, R.R., Patchan, M., Rittschof, D., *et al.* (1992) Special issue on the US Navy Biofouling Research Program. *Biofouling*, **6**.

12. Crisp, D.J. (1984) Overview on research on marine invertebrate larvae, 1940–1980. In: *Marine Biodeterioration: An Interdisciplinary Study* (eds J.D. Costlow & R.C. Tipper), pp. 103–126. Naval Institute Press, Annapolis , MD.

13. Stein, J., Truby, K., Darkangelo-Wood, C., *et al.* (2003) Silicone foul release coatings: effect of the interaction of oil and coating functionalities on the magnitude of macrofouling attachment strengths. *Biofouling*, **19** (Suppl.), S71–S82.

14. Baier, R.E. (1970) Surface properties influencing biological adhesion. In: *Adhesion in Biological Systems* (ed. R.S. Manley), pp. 15–48. Academic Press, New York.

15. Baier, R.E. (1981) Early events of microbiofouling of all heat transfer equipment. In: *Fouling of Heat Transfer Equipment* (eds E.F.C. Somerscales & J.G. Knudsen), pp. 293–304. Hemisphere, New York.

16. Roberts, D., Rittschof, D., Holm, E. & Schmidt, A.R. (1991) Factors influencing initial larval settlement: temporal, spatial and molecular components. *Journal of Experimental Marine Biology and Ecology*, **150** (2), 203–211.

17. Nakano, M., Shen, J.-R. & Kamino, K. (2007) Self-assembling peptide inspired by a barnacle underwater adhesive protein. *Biomacromolecules*, **8**, 1830–1835.

18. Callow, M.E. & Callow, J.A. (2002) Marine biofouling: a sticky problem. *Biologist*, **49**, 10–14.

19. Waite, J.H. & Qin, X.-X. (2001) Polyphenolic phosphoprotein from the adhesive pads of the common mussel. *Biochemistry*, **40**, 2887–2893.

20. Carlton, J.T. (2001) *Introduced Species in U.S. Coastal Waters: Environmental Impacts and Management Priorities*. Pew Oceans Commission, Arlington, TX. Available from http://www.pewtrusts.org/uploadedFiles/wwwpewtrustsorg/Reports/Protecting_ocean_life/env_oceans_species.pdf. Cited 23 January 2008.

21. Pitombo, F.B. (2004) Phylogenetic analysis of the balinidae (Cirripedia, Balanomorpha). *Zoologica Scripta*, **33**, 261–276.

22. Rittschof, D., Branscomb, E.S. & Costlow, J.D. (1984) Settlement and behavior in relation to flow and surface in larval barnacles, *Balanus amphitrite* Darwin. *Journal of Experimental Marine Biology and Ecology*, **82**, 131–146.

23. Crisp, D.J. (1955) The behavior of barnacle larvae in relation to water movement over a surface. *Journal of Experimental Biology*, **32**, 569–590.

24. Mullineaux, L.S. & Butman, C.A. (1990) Recruitment of encrusting benthic invertebrates in boundary layer flows: a deep water experiment on Cross Seamount. *Limnology and Oceanography*, **35**, 409–423.

25. Walters, L.J., Hadfield, M.G. & del Carmen, K.A. (1997) The importance of larval choice and hydrodynamics in creating aggregations of *Hydroides elegans* (Polychaeta: Serpulidae). *Invertebrate Biology*, **116**, 102–114.

26. Qian, P.Y., Rittschof, D. & Streeder, B. (2000) Macrofouling in unidirectional flows: miniature pipes as experimental models for studying the effects of hydrodynamics on invertebrate larval settlement. *Marine Ecology Progress Series*, **207**, 191–121.

27. Rittschof, D., Sin, T.-M., Teo, S.L.-M. & Coutinho, R. (2007) Fouling in natural flows: cylinders and panels as collectors of particles and barnacle larvae. *Journal of Experimental Marine Biology and Ecology*, **348** (1–2), 85–96.

28. Champ, M. (2000) A review of organotin regulatory strategies, pending actions, related costs and benefits. *Science of the Total Environment*, **258**, 21–71.

29. Fisher, E.C., Castelli, V.J., Rodgers, S.D. & Bleile, H.R. (1984) Technology for control of marine biofouling – a review. In: *Marine Biodeterioration: An Interdisciplinary Study* (eds J.D. Costlow & R.C. Tipper), pp. 261–299. Naval Institute Press, Annapolis, MD.

30. Readman, J.W. (1996) Antifouling herbicides – a threat to the marine environment? *Marine Pollution Bulletin*, **32**, 320–321.

31. Preiser, H.S., Ticker, A. & Bohlander, G.S. (1984) Coating selection or optimum ship performance. In: *Marine Biodeterioration: An Interdisciplinary Study* (eds J.D. Costlow & R.C. Tipper), pp. 223–228. Naval Institute Press, Annapolis, MD.

32. Galvin, R.M., Mellado, J.M.R. & Neihof, R.A. (1998) A contribution to the study of the natural dynamics of pyrithione (ii): deactivation by direct chemical and adsorptive oxidation. *European Water Management*, **4**, 61–64.

33. Dahllöf, I., Grunnet, K., Haller, R., Hjorth, M., Maraldo, K. & Petersen, D.G. (2005) Analysis, fate and toxicity of zinc and copper pyrithione in the marine environment. *Tema Nord*, **5**, 550–583.

34. Gough, M.A., Fothergill, J. & Hendrie, J.D. (1994) A survey of southern England coastal waters for the s-triazine antifouling compound Irgarol 1051. *Marine Pollution Bulletin*, **28**, 613–620.

35. Callow, M.E. & Willingham, G.L. (1996) Degradation of antifouling biocides. *Biofouling*, **10** (1/3), 239–249.

36. Liu, D., Pacepavicius, G.J., Maguire, R.J., Lau, Y.L., Okamura, H. & Aoyama, I. (1999) Survey for the occurrence of the new antifouling compound Irgarol 1051 in the aquatic environment. *Water Research*, **33**, 2833–2843.

37. Tolosa, I. & Readman, J.W. (1996) Simultaneous analysis of the antifouling agents: tributyltin, triphenyltin and Irgarol 1051 used in antifouling paints. *Marine Pollution Bulletin*, **335**, 267–274.

38. Tolosa, I., Readman, J.W., Blaevoet, A., Ghilini, S., Bartocci, J. & Horvat, M. (1996) Contamination of Mediterranean (Cote d'Azur) coastal waters by organotins and Irgarol 1051 used in antifouling paints. *Marine Pollution Bulletin*, **32**, 335–341.

39. Kegley, S., Hill, B. & Orme, S. (2007) Chemical identification and use for Diuron. *PAN Pesticides Database* [database on the Internet]. Pesticide Action Network, North America, San Francisco, USA. Available from http://www.pesticideinfo.org/Detail_Chemical.jsp?Rec_Id =PC33293. Cited 23 January 2008.

40. IMO (2002) *International Convention for the Prevention of Pollution from Ships, 1973, as Modified by the Protocol of 1978 Relating Thereto* (MARPOL 73/78) International Maritime Organisation, London. Available from http://www.imo.org/Conventions/ contents.asp?doc_id=678&topic_id=258. Cited 6 April 2003.

41. Alberte, R.S., Snyder, S. & Zahuranec, B. (1992) Biofouling research needs for the United States Navy: program history and goals. *Biofouling*, **6**, 91–95.

42. Brancato, M.S., Toll, J., DeForest, D. & Tear, L. (1999) Aquatic ecological risks posed by tributyltin in United States surface waters: pre-1989 to 1996 data. *Environmental Toxicology and Chemistry*, **18**, 567–577.

43. Holm, E.R., Orihuela, B., Kavanagh, C.J. & Rittschof, D. (2005) Variation among families for characteristics of the adhesive plaque in the barnacle *Balanus amphitrite*. *Biofouling*, **21**, 121–126.

44. Willingham, G.L. & Jacobson, A.H. (1996) Designing an environmentally safe marine antifoulant. In: *Designing Safer Chemicals, American Chemical Society Symposium Series 640* (eds S.C. DeVito & R.L. Garrett), pp. 224–233. American Chemical Society, Washington, DC.

45. US Geological Survey Toxic Substances Hydrology Program (2007) Volatile organic compounds (VOCs). A Database on the Internet. Available from http://toxics.usgs.gov/definitions/vocs.html. Cited 23 January 2008.

46. Clare, A.S. (1997) Towards nontoxic antifouling (mini-review). *Journal of Marine Biotechnology*, **6**, 3–6.

47. de Nys, R. & Steinberg, P.D. (2002) Linking marine biology and biotechnology. *Current Opinion in Biotechnology*, **13**, 244–248.

48. Baum, C., Meyer, W., Stelzer, R., Fleischer, L.G. & Siebers, D. (2002) Average nanorough skin surface of the pilot whale (*Globicephala melas*, Delphinidae): considerations on the self-cleaning abilities based on nanoroughness. *Marine Biology*, **140**, 653–657.

49. Scardino, A.J. & de Nys, R. (2004) Fouling deterrence on the bivalve shell *Mytilus galloprovincialis*: a physical phenomenon? *Biofouling*, **20**, 249–257.

50. Schumacher, J.F., Aldred, N., Callow, M.E., *et al.* (2007) Species-specific engineered antifouling topographies: correlations between the settlement of algal zoospores and barnacle cyprids. *Biofouling*, **23**, 307–317.

51. AMBIO (2007) *Advanced Nanostructured Surfaces for the Control of Biofouling*. Centrum Techniki Okrętowej S.A. (Ship Design and Research Centre), Poland. Available from http://cto.gda.pl/index.php?id=116&L=1. Cited 23 January 2008.

52. Kavanagh, C.J., Swain, G.W., Kovach, B.S., *et al.* (2003) The effects of silicone fluid additives and silicone elastomer matrices on barnacle adhesion strength. *Biofouling*, **19** (6), 381–390.

53. Rittschof, D., Lai, C.H., Kok, L.M. & Teo, S.L.M. (2003) Pharmaceuticals as antifoulants: concept and principle. *Biofouling*, **19** (Suppl.), S207–S212.

54. UNFCCC (1998) *Kyoto Protocol to the United Nations Framework Convention on Climate Change*, United Nations. Available from http://unfccc.int/resource/docs/convkp/kpeng.pdf. Cited 23 January 2008.

Index

Printed and bound in the UK by
CPI Antony Rowe, Eastbourne